CUNNINGHAM'S MANUAL OF PRACTICAL ANATOMY

Volume 1

Cunningham's Manual of Practical Anatomy

CUNNINGHAM'S MANUAL OF PRACTICAL ANATOMY

Seventeenth edition

Volume 1 General Anatomy, Upper and Lower Limbs

Dr Rachel Koshi MBBS, MS, PhD

OXFORD
UNIVERSITY PRESS

Great Clarendon Street, Oxford, OX2 6DP,
United Kingdom

Oxford University Press is a department of the University of Oxford.
It furthers the University's objective of excellence in research, scholarship,
and education by publishing worldwide. Oxford is a registered trade mark of
Oxford University Press in the UK and in certain other countries

Published in the United States of America by Oxford University Press
198 Madison Avenue, New York, NY 10016, United States of America

British Library Cataloguing in Publication Data

Data available

Library of Congress Control Number is on file at the Library of Congress

ISBN 978–0–19–892334–3

DOI: 10.1093/med/9780198923343.001.0001

Printed in India by Thomson Press

I fondly dedicate this book to late Dr KG Koshi for his encouragement and support when I chose a career in anatomy, and to Dr Mary Jacob, under whose guidance I learnt the subject and developed a love for teaching.

Oxford University Press would like to dedicate this book to the memory of the late George John Romanes, Professor of Anatomy at Edinburgh University from 1954 to 1984, who brought his wisdom to previous editions of *Cunningham's*.

Foreword to the 16th edition

It gives me great pleasure to pen down the Foreword to the 16th edition of *Cunningham's Manual of Practical Anatomy*. Just as the curriculum of anatomy is incomplete without dissection, so also learning by dissection is incomplete without a manual.

Cunningham's Manual of Practical Anatomy is one of the oldest dissectors, the first edition of which was published as early as 1893. Since then, the manual has been an inseparable companion to students during dissection.

I remember my days as a first MBBS student, the only dissector known in those days was *Cunningham's* manual. The manual helped me to dissect scientifically, step by step, explore the body, see all structures as mentioned, and admire God's highest creation—the human body— so perfectly. As a postgraduate student I marvelled at the manual and learnt details of structures, in a way as if I had my teacher with me telling me what to do next. The clearly defined steps of dissection, and the comprehensive revision tables at the end, helped me personally to develop a liking for dissection and the subject of anatomy.

Today, as a Professor and Head of Anatomy, teaching anatomy for more than 30 years, I find *Cunningham's* manual extremely useful to all the students dissecting and learning anatomy.

With the explosion of knowledge and ongoing curricular changes, the manual has been revised at frequent intervals. The 16th edition is more student friendly. The language is simplified, so that the book can be comprehended by one and all. The objectives are well defined. The clinical application notes at the end of each chapter are an academic feast to the learners. The lucidly enumerated steps of dissection make a student explore various structures, the layout, and relations and compare them with the simplified labelled illustrations in the manual. This helps in sequential dissection in a scientific way and for knowledge retention. The text also includes multiple-choice questions for self-assessment and holistic comprehension.

Keeping the concept of 'Adult Learning Principles' in mind, i.e. adults learn when they 'DO', and with a global movement towards 'Competency - based Curriculum', students learn anatomy when they dissect; *Cunningham's* manual will help students to dissect on their own, at their own speed and time, and become competent doctors, who can cater to the needs of the society in a much better way.

I recommend this invaluable manual to all the learners who want to master the subject of anatomy.

Dr Pritha S Bhuiyan
Professor and Head, Department of Anatomy
Professor and Coordinator, Department of Medical Education
Seth GS Medical College and KEM Hospital, Parel, Mumbai

Preface to the 16th edition

Cunningham's Manual of Practical Anatomy has been the most widely used dissection manual in India for many decades. This edition is extensively revised to meet the needs of the present-day medical student.

Firstly, at the start of each chapter and at the beginning of the description of a region, introductory remarks have been added in order to provide context to the whole human body and to the practice of medicine. In order to appreciate the 'big picture', Chapter 1 (General introduction) has been expanded and supplemented by new artwork. Throughout all three volumes, all anatomical terms are updated and explained using the latest terminology, and the language has been modernized.

Dissection forms an integral part of learning anatomy, and the practice of dissection enables students to retain and recall anatomical details learnt in the first year of medical school during their clinical practice. To make the dissection process easier and more meaningful, in this edition, each dissection is presented with a heading, and a list of objectives to be accomplished. The details of dissections have been retained from the earlier edition but are presented as numbered, stepwise easy-to-follow instructions that help students navigate their way through the tissues of the body, and to isolate, define, and study important anatomical structures.

This manual contains a number of old and new features that enable students to integrate the anatomy learnt in the dissection hall with clinical practice. Each region has images of living anatomy to help students identify on the skin surface bony or soft tissue landmarks that lie beneath. Numerous X-rays and magnetic resonance imaging further enable the student to visualize internal structures in the living. Matters of clinical importance, when mentioned in the text, are highlighted.

A brand new feature of this edition is the presentation of one or more clinical application notes at the end of each chapter. Some of these notes focus attention on the anatomical basis of commonly used physical diagnostic tests such as palpation of the arterial pulse or measurement of blood pressure. Others deal with the underlying anatomy of clinical findings in diseases such as breast cancer or the cervical rib syndrome. Common joint injuries to the knee and other limb joints are discussed with reference to the intra- and periarticular structures described and dissected. Effects of some common nerve injuries along the course of the nerve are described in a clinical context. Many clinical application notes are in a Q&A format that challenges the student to brainstorm the material covered in the chapter. Multiple-choice questions on each section are included at the end to help students assess their preparedness for the university examination.

It is hoped that this new edition respects the legacy of Cunningham in producing a text and manual that is accurate, student friendly, comprehensive, and interesting, and that it will serve the community of students who are beginning their career in medicine to gain knowledge and appreciation of the anatomy of the human body.

Dr Rachel Koshi

Contributors and reviewers

Contributors to the 17th edition

Dr J Suganthy, Professor of Anatomy, Christian Medical College, Vellore, India.
Dr Suganthy wrote the MCQs for the online material and Chapter 2 General Anatomy MCQs.

Dr Ritwik Baidya, MD, Assistant Professor of Anatomy in Radiology, Weill Cornell Medicine, Radiology/Program of Gross Anatomy, New York, NY, USA.
Dr Baidya wrote the online image-based questions.

Reviewers to the 17th edition

Oxford University Press would like to thank all of those who reviewed new material for the 17th edition and provided valuable feedback:

CS Ramesh Babu, Associate Professor of Anatomy, Muzafffarnagar Medical College, Muzaffarnagar, India.

Dr Aby S Charles, Senior Resident, Department of Anatomy, Christian Medical College, Vellore, India.

Professor Sunil Jonathan Holla, MBBS, MD (Anatomy), Professor of Anatomical Sciences, St Matthew's University School of Medicine, Grand Cayman, Cayman Islands (formerly of Christian Medical College, Vellore, India).

Contributors to the 16th edition

Dr J Suganthy, Professor of Anatomy, Christian Medical College, Vellore, India.
Dr Suganthy wrote the MCQs, reviewed manuscripts, and provided help and advice with the artwork, and most importantly gave much moral support.

Dr Aparna Irodi, Professor, Department of Radiology, Christian Medical College and Hospital, Vellore, India.
Dr Irodi kindly researched, identified, and contributed the radiology images.

Contents

Additional online access and MCQs

Purchasers of the 17th edition have access to the online version of *Cunningham's*, including bonus multiple-choice questions (MCQs) for 12 months. Enhance your understanding and be ready for anatomy examinations by visiting the online appendix website where you will find MCQs.

Search for *Cunningham's Manual of Practical Anatomy Volume 1 General Anatomy, Upper and Lower Limbs* 17th edition at https://academic.oup.com/ and go to the online appendix at the end of the book. Use your scratch-off code on the inside cover to access the material.

Table of competencies required by the National Medical Commission of India gross anatomy curricula

See Tables 0.1, 0.2, 0.3, and 0.4 for competencies required by the National Medical Commission of India gross anatomy curricula.

General anatomy

Table 0.1 Competencies in general anatomy

Competency no.	Competency	Chapter in 17th edition
AN2.1	Describe parts, and blood and nerve supply of a long bone	1. General anatomy
AN2.3	Enumerate special features of a sesamoid bone	1. General anatomy
AN2.4	Describe various types of cartilage with its structure and distribution in the body	1. General anatomy
AN2.6	Explain the concept of nerve supply of joints and Hilton's law	1. General anatomy
AN3.1	Classify muscle tissue according to structure and action	1. General anatomy
AN3.3	Explain shunt and spurt muscles	1. General anatomy
AN4.1	Describe different types of skin and dermatomes in body	1. General anatomy
AN4.5	Explain the principles of skin incisions	1. General anatomy
AN5.1	Differentiate between blood, vascular, and lymphatic systems	1. General anatomy
AN5.2	Differentiate between pulmonary and systemic circulation	1. General anatomy
AN5.5	Describe the portal system, giving examples	1. General anatomy
AN5.7	Explain the function of meta-arterioles, precapillary sphincters, and arteriovenous anastomoses	1. General anatomy
AN5.8	Define thrombosis, infarction, and aneurysm	1. General anatomy
AN6.3	Explain the concept of lymphoedema and spread of tumours via the lymphatic and venous systems	1. General anatomy
AN7.1	Describe the general plan of the nervous system with components of the central, peripheral, and autonomic nervous systems	1. General anatomy
AN7.2	List the components of nervous tissue and their functions	1. General anatomy
AN7.3	Describe the parts of a neuron and classify them based on the number of neurites, size, and function	1. General anatomy
AN7.7	Describe various types of synapses	1. General anatomy
AN7.8	Describe differences between sympathetic and spinal ganglia	1. General anatomy

The upper limb

Table 0.2 *Continued*

The lower limb

Table 0.3 Competencies for the lower limb

Table 0.3 *Continued*

Competency no.	Competency	Chapter in 17th edition
AN18.5	Explain the anatomical basis of locking and unlocking of the knee joint	23. The joints of the lower limb
AN18.6	Describe knee joint injuries with applied anatomy of the knee	23. The joints of the lower limb
AN18.7	Explain the anatomical basis of osteoarthritis	23. The joints of the lower limb
AN19.1	Describe and demonstrate the major muscles of the back of the leg with their attachment, nerve supply, and actions	22. The leg and foot
AN19.2	Describe and demonstrate the origin, course, relations, branches (or tributaries), and termination of important nerves and vessels of the back of the leg	22. The leg and foot
AN19.3	Explain the concept of 'peripheral heart'	22. The leg and foot
AN19.4	Explain the anatomical basis of rupture of the calcaneal tendon	22. The leg and foot
AN19.5	Describe factors maintaining the important arches of the foot, with their importance	22. The leg and foot
AN19.6	Explain the anatomical basis of flat foot and club foot	22. The leg and foot
AN19.7	Explain the anatomical basis of metatarsalgia and plantar fasciitis	22. The leg and foot
AN20.1	Describe and demonstrate the type, articular surfaces, capsule, synovial membrane, ligaments, relations, movements and muscles involved, and blood and nerve supply of the tibiofibular and ankle joints	23. The joints of the lower limb
AN20.2	Describe the subtalar and transverse tarsal joints	23. The joints of the lower limb
AN20.3	Describe and demonstrate the fascia lata, venous drainage, lymphatic drainage, retinacula, and dermatomes of the lower limb	17. The front and medial side of the thigh 23. The joints of the lower limb 22. The leg and foot 24. The nerves and nerve injuries of the lower limb
AN20.4	Explain the anatomical basis of enlarged inguinal lymph nodes	17. The front and medial side of the thigh
AN20.5	Explain the anatomical basis of varicose veins and deep vein thrombosis	17. The front and medial side of the thigh 22. The leg and foot
AN20.6	Identify the bones and joints of the lower limb seen in anteroposterior and lateral view radiographs of various regions of the lower limb	21. The hip joint 23. The joints of the lower limb
AN20.7	Identify and demonstrate important bony landmarks of the lower limb: vertebral levels of the highest point of the iliac crest, posterior superior iliac spine, iliac tubercle, pubic tubercle, ischial tuberosity, and adductor tubercle; tibial tuberosity, head of the fibula; medial and lateral malleoli, condyles of the femur and tibia; sustentaculum tali, tuberosity of the fifth metatarsal, and tuberosity of the navicular	17. The front and medial side of the thigh 18. The gluteal region 22. The leg and foot
AN20.8	Identify and demonstrate palpation of the femoral, popliteal, post-tibial, anti-tibial, and dorsalis pedis blood vessels in a simulated environment	17. The front and medial side of the thigh 19. The popliteal fossa 22. The leg and foot
AN20.9	Identify and demonstrate palpation of vessels (femoral, popliteal, dorsalis pedis, post-tibial), mid-inguinal point, surface projection of: femoral nerve, saphenous opening, sciatic, tibial, and common peroneal and deep peroneal nerves, great and small saphenous veins	25. Surface marking of the lower limb

Competencies in Medical Ethics

Table 0.4 Ethics in Anatomy

Competency no.	Competency	Chapter in 17th edition
AN82.1	Demonstrate respect and follow the correct procedure when handling cadavers and other biological tissues	1. General anatomy

PART 1

General anatomy

CHAPTER 1
General anatomy

Human anatomy is the study of the structure of the human body. For descriptive purposes, the human body is divided into regions: head, neck, trunk, and limbs. General anatomy is common to all the regions and introduces a vocabulary to define the positions of each anatomical structure, and an elementary knowledge of the kinds of structures you will encounter.

Terms of position

The body usually lies horizontally on a table during dissection, but the dissector must remember that terms describing positions are always used as though the body is in the **anatomical position**. In this position, the person is standing upright, with the upper limbs by the sides and the palms directed forwards, and the eyes looking straight ahead.

Descriptive terms are used to indicate the position of structures as if the body were in the anatomical position [Fig. 1.1]. **Superior**, or **cephalic**, refers to the position of a part that is nearer the head, whereas **inferior** means nearer the feet. **Caudal** (towards the tail) can replace inferior in the trunk. **Anterior** means nearer the front of the body, and **posterior** means nearer the back. **Ventral** and **dorsal** may be used instead of anterior and posterior in the trunk and have the advantage of also being appropriate for four-legged animals (*venter*=belly; *dorsum*=back). In the hand, **dorsal** commonly replaces posterior, and **palmar** replaces anterior. In the foot, the corresponding surfaces are superior and inferior in the anatomical position, but these terms are usually replaced by **dorsal** (**dorsum** of the foot) and **plantar** (*planta*=the sole), respectively.

Median means in the middle. Thus, the **median plane** is an imaginary plane that divides the body into two equal halves—right and left. The **anterior** and **posterior median lines** are where the median plane meets the anterior and posterior surfaces of the body, respectively. A structure is said to be median when it is bisected by the median plane. **Medial** means nearer the median plane, and **lateral** means further away from that plane. The presence of two bones, one lateral and the other medial, in the forearm (radius and ulna) and leg (fibula and tibia) have resulted in the terms **ulnar** or **radial** side of the forearm, and **fibular** or **tibial** side of the leg. The words outer and inner, or their equivalents **external** and **internal**, are used only in the sense of nearer the surface or further away from it in any direction; they are not synonymous with medial and lateral. **Superficial**, meaning nearer the skin, and **deep**, meaning further from it, are the terms most usually used when the direction is of no importance. When describing the surfaces of a hollow organ, **external** refers to the outer surface, and **internal** to the inner surface.

A **sagittal plane** may pass through any part of the body, parallel to the median plane. A **coronal plane** is a vertical plane at right angle to the median plane. A transverse plane is a horizontal plane (perpendicular to both the above). All other planes are oblique planes.

Proximal (nearer to) and **distal** (further from) indicate the relative distances of structures from the root of that structure (e.g. the elbow is proximal to the wrist, but distal to the shoulder). **Middle**, or its Latin equivalent *medius*, is used to indicate a position between superior and inferior

Terms of position

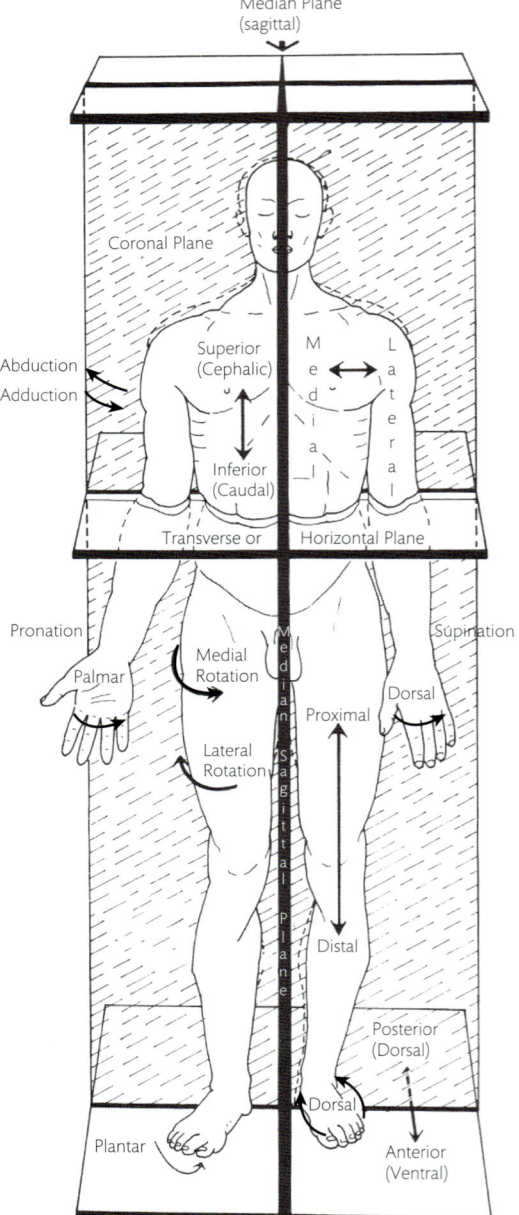

Fig. 1.1 Diagram illustrating some anatomical terms of position and movement.

or between anterior and posterior. **Intermediate** is used to indicate a position between lateral and medial.

The terms **superolateral** and **inferomedial**, or **anteroinferior** and **posterosuperior**, or any other combination of the standard terms, may be used to show intermediate positions.

Terms of movement

Movements take place at joints and may occur in any plane, but are usually described in the sagittal and coronal planes [Fig. 1.1]. Movements of the trunk in the sagittal plane are **flexion** (bending anteriorly) and **extension** (straightening or bending posteriorly). In the limbs, flexion is the movement which carries the limb anteriorly and folds it; extension is the movement which carries it posteriorly and straightens it. (Note flexion and extension for the knee joint do not follow this rule. Flexion of the knee folds the limb but results in the leg being carried posteriorly.) At the ankle, the terms used are **plantar flexion** (movement towards the sole) and **dorsiflexion** (movement towards the dorsum). Movements of the trunk in the coronal plane (i.e. side-to-side movement) are known as **lateral flexion**. Movement of the limb away from the median plane is **abduction**, and movement towards the median plane is **adduction**. In keeping with this definition, at the wrist, abduction refers to movement of the hand away from the median plane towards the radial (thumb) side. Abduction of the wrist is also referred to as **radial deviation**. Similarly, adduction of the wrist is also referred to as **ulnar deviation**. In the fingers and toes, abduction means the spreading apart of, and adduction the drawing together of, the digits. In the hand, this movement is in reference to the line of the middle finger. In the foot, it is in reference to the line of the second toe. The thumb lies at right angles to the fingers. Hence, abduction and adduction carry the thumb anteriorly and posteriorly, respectively.

Rotation is the term applied to the movement in which a part of the body is turned around its own longitudinal axis. In the limbs, lateral and medial rotation refers to the direction of movement of the anterior surface. (When the front of the arm or thigh is turned laterally, it is lateral rotation, and when turned medially, it is medial rotation.) A special movement in the forearm is the rotation of the radius on the stationary ulna. This movement is **pronation**. The hand moves with the radius and is turned so that the palm faces posteriorly. The opposite movement is **supination**, and it turns the hand back to the anatomical position. Inversion and eversion are movements specific to the foot. Inversion turns the sole of the foot medially, and eversion turns the sole laterally.

Skeletal system

The skeletal system includes bones, cartilages, and joints. It can be divided into two functional parts: (1) the **axial skeleton**, consisting of the skull bones (including the ear ossicles), hyoid, vertebrae, sacrum, ribs, and sternum; and (2) the **appendicular skeleton**, consisting of the bones of the pectoral and pelvic girdles and the free upper and lower limbs [Fig. 1.2].

Bone

Bone is a form of connective tissue in which the intercellular substance consists of dense, white fibres embedded in a hard calcium phosphate matrix. The fibres impart resilience to the bone, whereas the calcium salts resist compression forces.

The **periosteum** is a dense layer of fibrous tissue which covers the surfaces of bones, except where they articulate with other bones. The **articular surfaces** are covered by **articular cartilage**. The periosteum is continuous with muscles, tendons,

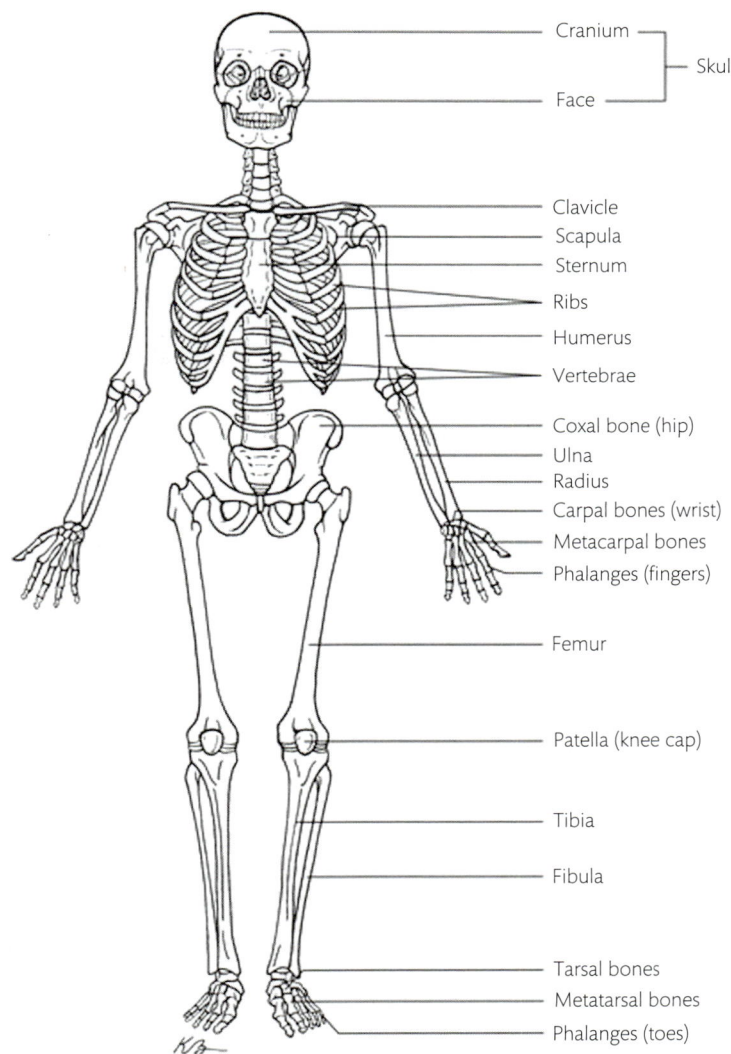

Cranium
Skull
Face

Clavicle
Scapula
Sternum
Ribs
Humerus
Vertebrae

Coxal bone (hip)
Ulna
Radius
Carpal bones (wrist)
Metacarpal bones
Phalanges (fingers)

Femur

Patella (knee cap)

Tibia

Fibula

Tarsal bones
Metatarsal bones
Phalanges (toes)

Fig. 1.2 The human skeleton.

ligaments, fibrous capsules of joints, intermuscular septa, and the deep fascia where a bone is subcutaneous. The periosteum has an inner cellular **osteogenic layer** and an outer **fibrous layer**. The osteogenic layer is the source of bone deposition in growing bone. Osteogenesis from the periosteum can begin again in adult life when increased strength of a bone is required (e.g. when the weight or muscularity of the individual increases, or when new bone is formed at the site of a healing fracture). Remodelling of newly laid bone is accomplished by removal of excess bone by osteoclasts.

Classification of bones according to type

Bone is found in two forms: (1) **compact bone**, which is dense and forms the tubular shafts of long bones; and (2) **cancellous bone**, which is a lattice of bone spicules that occurs at the ends of long bones and fills flat and irregular bones. Spaces between the spicules are filled with bone marrow.

Classification of bones according to shape

Bones can be classified according to their shape: (1) **long bones** of the limbs, which have a narrow, tubular **body** (shaft) made up of compact bone, and enlarged articular ends composed largely of cancellous bone (e.g. humerus, femur); (2) **short bones**, which are roughly cuboidal in shape (e.g. bones of the wrist and foot, carpal and tarsal bones); (3) **flat bones**, which are composed of a layer of spongy bone, sandwiched between two layers of compact bone (e.g. sternum, scapula, vault of the skull); (4) **irregular bones**, which do not conform to any set shape (e.g. vertebrae); and (5) **pneumatic bones** of the skull, which contain air spaces.

Classification of bones according to mode of ossification

Bones are formed in two ways, by: (1) **endochondral ossification**; or (2) **intramembranous ossification**.

Endochondral ossification

In **endochondral ossification**, bones are pre-formed in cartilage. The earliest precursor of bone is a cartilaginous model that increases in size by proliferation of its cells and production of matrix [Fig. 1.3A]. This model is then replaced by bone by a sequence of changes.

Sequential changes in the cartilaginous model:

1. A supporting shell of bone is laid down by the periosteum on the external surface of the body of the model [Fig. 1.3B].
2. The matrix of the cartilage deep to this shell becomes calcified and the cells die, leaving empty spaces in the calcified cartilage.

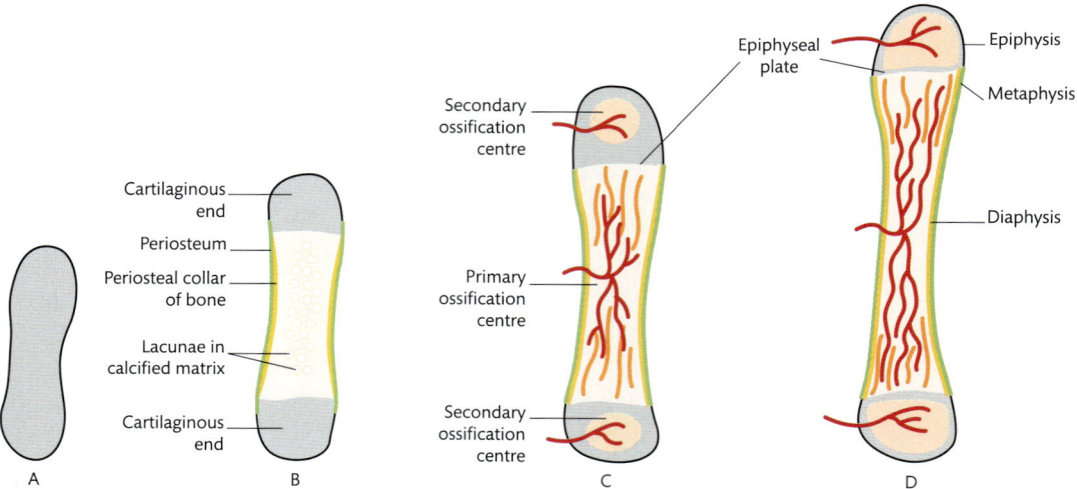

Fig. 1.3 Endochondral ossification. Diagram of the four stages in the development of a long bone. (A) Cartilaginous model of the long bone before ossification begins; 8 weeks of intrauterine life (IUL). (B) Ossification has begun in the centre where empty spaces and spicules of calcified cartilage are seen. Compact bone is laid down by the periosteum; 2–3 months of IUL. (C) Blood vessels invade the centre of the bone, and ossification of calcified cartilage begins. Later, blood vessels invade the ends of the bone which begin to ossify; childhood. (D) Most of the cartilage is replaced by bone. Growth plate is seen between the bones ossified from the primary and secondary ossification centres; adolescence–early adulthood.

3. These spaces coalesce (join), leaving longitudinal spicules of calcified cartilage between them.
4. This calcified cartilage is invaded by blood vessels from the surrounding shell, and bone is laid down on the spicules by the action of bone-forming cells—**osteoblasts brought in by invading blood vessels**. This process begins at the centre of the body of the cartilaginous model, in the part which is destined to become the centre of the shaft of the long bone. This centre (of ossification) is called the **primary ossification centre**. From the primary ossification centre, ossification spreads towards the ends, which remain cartilaginous for a while after the centre has been ossified (Fig. 1.3C).
5. In time, **secondary ossification centres** develop at each end of the cartilaginous model of the long bone, and ossification in them proceeds in all directions (Fig. 1.3D). For a period of time, the bone formed at each end is separated from the ossifying body by a zone of growing cartilage called the **epiphyseal plate** or **growth plate**. The growth plate serves an important function of adding new cartilage to the shaft, thus providing material for growth. As growth proceeds, the ends of the growing bone move away from the centre of the body. The external shell of bone increases in length at the same rate and results in growth in length of the long bone.
6. When growth in length of bone is complete (in early adulthood), the growth cartilages in the long bones become quiescent and stop growing. Ossification from the body spreads into the growth cartilage, resulting in fusion of the bone in the body (formed from the primary ossification centre) and the bone at the ends (formed from secondary ossification centres). The arteries supplying the shaft and the ends unite. This brings growth in length to a halt. After this has occurred in all bones, there is no further increase in height of the individual.

Growth in diameter of the long bone is brought about by the addition of bone to the external surface of the enclosing shell. This bone is produced by a highly cellular osteogenic layer of periosteum. As the shell increases in thickness, the bone on the inner surface is removed by osteoclasts and the marrow cavity increases in diameter.

In short and irregular bones, ossification starts in the middle of the cartilaginous model and proceeds outwards. No external shell of bone is formed. The bone continues to grow until the adult size is reached, at which time the bone has replaced all of the cartilage, except that which persists on the articular surfaces.

Parts of a growing long bone

Specific terms are used to designate the parts of a growing long bone. The bone developed from the primary ossification centre is the **diaphysis**. The **epiphysis** is the bone developed from the secondary ossification centre and lies at the end of the bone. (A bone can have more than one epiphysis at each end.) The **metaphysis** is the epiphyseal end of the diaphysis. It is the zone of active ossification [Fig. 1.3D]. The nutrient artery enters the diaphysis. Branches of the nutrient artery to the metaphysis are the metaphyseal arteries. Arteries to the epiphysis are the epiphyseal arteries.

Laws of ossification

Endochondral ossification of long bones is governed by certain principles known as the laws of ossification. They are:

1. **Primary ossification centres** (in the shaft of the long bones) appear at approximately 8 weeks of intrauterine life. **Secondary ossification centres**, (at the ends of the long bones) appear much later, at or after birth. In short and irregular bones, the single ossification centre (primary centre) appears after birth [Fig. 1.4].
2. Although long bones have epiphyses at both ends, growth in length occurs mainly at one end, known as the '**growing end**'. The epiphysis of the **growing end usually appears earlier and fuses with the body later** than that at the other end. The growing ends in the upper limb bones are at the shoulders and wrists, and in the lower limbs at the knees.

➲ In a growing child, injury to the growing end of a long bone is more serious than injury to the other end.
➲ Since epiphyses are visible on radiographs and are separated from the body of the bone by a clear region of growth cartilage, they have to be differentiated from fractures. To do so, the physician reading the radiographs of a child needs to know where epiphyses normally appear and till when they are normally present.

Intramembranous ossification

Intramembranous (membranous) ossification happens when bone is formed in connective tissue. In this type of ossification, osteoblasts invade the fibrous membrane to form many separate spicules of bone. These spicules fuse with each other to form a lattice around the capillaries of connective tissue. This lattice-work may persist as can-

(A)

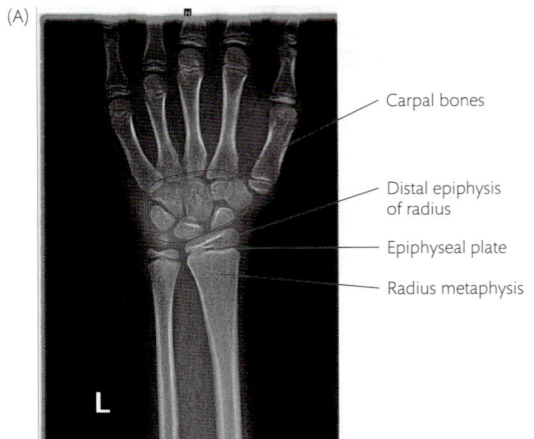

Carpal bones

Distal epiphysis of radius

Epiphyseal plate

Radius metaphysis

L

(B)

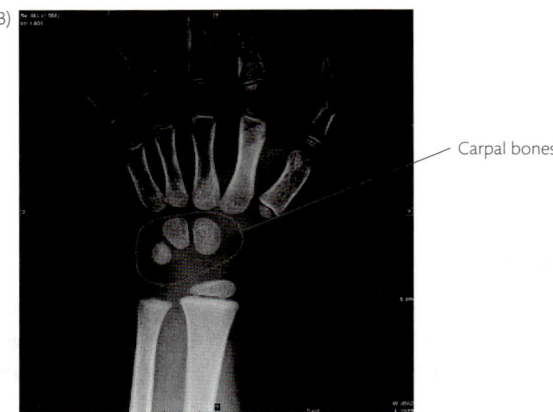

Carpal bones

Fig. 1.4 (A) X-ray of the hand of a 12-year-old child. The epiphyses at the lower end of the radius and ulna are still separated from the diaphysis by the growth cartilage. Ossification centres for all eight carpal bones have appeared. (B) X-ray of the hand of a 2-year-old child. Only three carpal bones have begun to ossify. The ossification centre for the lower end of the ulna has not yet appeared.

cellous bone, or continued deposition of bone in the cavities of the lattice can turn it into compact bone. Flat bones of the skull (e.g. parietal, frontal), are ossified in membrane.

Sesamoid bones

Sesamoid bones are bones that develop within a tendon. This happens in situations where the tendon is compressed against a bony prominence. The patella is the largest and most consistently found sesamoid bone and develops in the tendon of the quadriceps femoris. Other examples of sesamoid bones are the pisiform formed in the tendon of the flexor carpi ulnaris, and medial and lateral hallux sesamoid bones found in the two heads of

the flexor hallucis brevis. Sesamoid bones have an articular surface which slides on (articulates) the surface under pressure. Sesamoid bones do not have periosteum.

Blood and nerve supply of long bones

Bones have a rich blood supply. The main artery to the bone is the nutrient artery which enters the shaft of the bone through the nutrient foramen. It divides in the medullary cavity into ascending and descending branches that run towards the ends of the bone. Numerous small periosteal arteries supply the peripheral parts of compact bone. The ends of long bones are supplied by metaphyseal and epiphyseal arteries.

The periosteum and bone have sensory and autonomic (vasomotor) nerves which accompany the arteries.

Markings on dried bones

Dried bones used in the study of anatomy have a number of important surface markings. A dried bone: (1) is smooth where it is covered, in life, by articular cartilage; (2) gives a fleshy attachment to muscles; and (3) is subcutaneous. It is often roughened where ligaments, aponeuroses, and tendons are attached. It has grooves lodging blood vessels, and holes (foramina) where blood vessels enter and leave the bone. Many of these features are more easily felt than seen. Table 1.1 lists common terms used in describing bony landmarks.

You will note that some terms have similar meanings (e.g. tubercle and tuberosity). A tuberosity is usually more prominent than a tubercle.

Cartilage

Cartilage is another form of connective tissue. Compared to bone, it is less rigid and more flexible. The intercellular substance consists of a semi-rigid matrix in which the chondrocytes are embedded. In later years, cartilage in some situations becomes ossified to form bone. Cartilage is classified as **hyaline**, **elastic**, or **fibrocartilage**, depending on the composition and type of connective tissue fibres in the matrix. Hyaline cartilage is found in the nose and larynx, and covering the articulating surface of synovial joints [Fig. 1.5B]. Elastic cartilage is found in the pinna of the ear. Fibrocartilage is found in interver-

Table 1.1 Terms describing bony landmarks

Term	Definition	Example
Condyle	Rounded elevation, usually articular	Femoral condyles
Crest or ridge	Linear elevation	Crest of greater tubercle
Epicondyle	Non-articular bony elevation usually above a condyle	Lateral epicondyle of humerus
Facet	Smooth, flat area	Costal facet
Foramen	Opening or passage through the bony	Nutrient foramen
Fossa	Shallow depression	Glenoid fossa
Groove or sulcus	Linear depression	Intertubercular sulcus
Head or capitulum	Rounded elevation, usually articular	Head of humerus
Line	Linear elevation, less defined than crest	Soleal line
Notch	Indentation of edge or border	Scapular notch
Protuberance	Projection	External occipital protuberance
Spine or spinous process	Pointed elevation	Spine of vertebra
Trochanter	Large, blunt elevation	Greater trochanter of femur
Trochlea	Pulley-shaped articular prominence	Trochlea of humerus
Tubercle	Small, rounded elevation	Greater tubercle of humerus
Tuberosity	Large, rounded elevation	Ischial tuberosity

tebral discs and the pubic symphysis [Fig. 1.5B]. The structural differences among the three types of cartilage are beyond the scope of this book.

Joint

A **joint** is where two or more bones make contact with each other to allow movement. The bones taking part in a joint are held against each other by an intervening tissue or substance. Joints are classified according to the: (1) substance between the articulating surfaces; and (2) degree of movement they permit.

Classification of joints based on morphology

Joints are classified according to the type of substance between the bones: (1) fibrous joints; (2) cartilaginous joints; and (3) synovial joints [Fig. 1.5]. Joints where the adjacent bones are united by a thin layer of dense fibrous tissue are **fibrous joints**. Others where the adjacent bones are united by fibrocartilage or hyaline cartilage are **cartilaginous joints**. In **synovial joints**, the articulating surfaces of the bones are covered with firm, slippery **articular cartilage**, and they slide on each other within a narrow joint cavity containing a lubricant—the **synovial fluid**. The articular surfaces of these bones are held in apposition by a tubular sheath of fibrous tissue—the **fibrous capsule**.

Fibrous joints

Fibrous joints are those in which two or more articulating bones are held together by fibrous tissue. Fibrous joints are further classified as: (1) sutures; (2) syndesmosis; and (2) gomphosis.

1. Sutures are restricted to the skull. The fibrous tissue between the bones is continuous with the periosteum on the inner and outer surfaces of the bone [Fig. 1.5]. In later life, some sutures are obliterated by fusion of the two bones across a suture. This process is known as **synostosis**. A **schindylesis** is a special type of suture where a ridge on one bone fits into a groove on the other bone (e.g. the suture between the vomer and the sphenoid). Sutures allow very little movement between the bones.

2. In the syndesmosis type of fibrous joint, the bones are united by a sheet of fibrous tissue. This type of joint is partially movable (e.g. the interosseous membrane between the radius and the ulna, the inferior tibiofibular joint).

3. A gomphosis is a peg-and-socket joint between the tooth and its bony socket. The two elements are in intimate contact with each other and are held together by collagen fibres. There is limited movement between the two elements.

Cartilaginous joints

Cartilaginous joints are those in which two articulating bones are held together by cartilage [Fig.

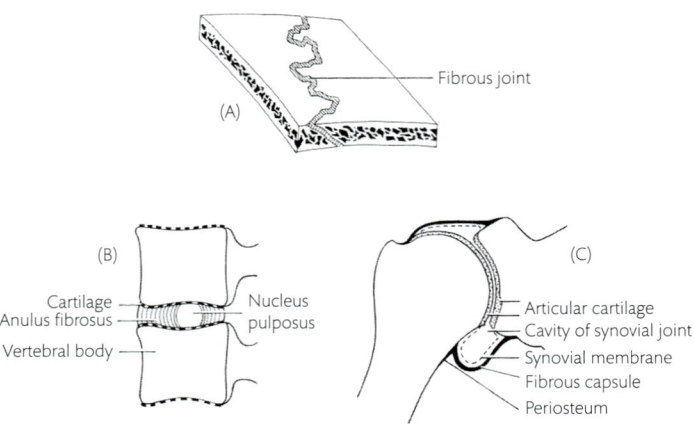

(A) ——— Fibrous joint

(B)

Cartilage ———
Anulus fibrosus ———
Vertebral body ———
Nucleus
pulposus

(C)

Articular cartilage
Cavity of synovial joint
Synovial membrane
Fibrous capsule
Periosteum

Fig. 1.5 Diagrams showing the three types of joints. (A) Fibrous joint between two skull bones. (B) Cartilaginous joint between two adjacent vertebrae (the annulus fibrosus and nucleus pulposus are parts of the intervertebral disc). (C) Synovial joint between the scapula and the humerus—shoulder joint.

1.5B]. In general, they allow more movement than fibrous joints. Cartilaginous joints are further classified as: (1) **primary cartilaginous joints** or **synchondroses**; and (2) **secondary cartilaginous joints** or **symphysis**.

Primary cartilaginous joints are seen in situations where a bone ossifies in cartilage from more than one ossification centre. They are the joints between the epiphysis and diaphysis, or between two epiphyses of a growing bone. When growth of a long bone is complete, this primary cartilaginous joint is obliterated by a process known as synostosis.

Secondary cartilaginous joints, or symphyses, are strong, slightly movable joints where the bones involved are united by fibrocartilage. Examples include the pubic symphysis between the two pubic bones and intervertebral discs between adjacent vertebrae [Fig. 1.5B].

Synovial joints

Synovial joints are those in which two or more articulating bones are held together by an articular capsule and are separated from each other by synovial fluid. They are joints with the maximum amount of movement between the bones [Figs. 1.5C, 1.6].

All synovial joints have some common features: fibrous capsule, synovial membrane, synovial fluid, articular cartilage, and ligaments. In addition, some synovial joints have an articular disc. An understanding of these is essential. The joint cavity is surrounded by an **articular capsule** or a **fibrous capsule**. The articular capsule is a tubular sheath of fibrous tissue that is firmly adherent to the ends of the articulating bones, beyond the articulating surfaces. It is lined on the inner surface by the **synovial membrane**. At the junction of the fibrous capsule and bone, the synovial membrane lining the fibrous capsule is reflected onto the bone and covers the intracapsular non-articulating parts of the bone. The synovial membrane also covers intracapsular tendons and ligaments when present. It secretes synovial fluid. The articulating surfaces of the bones lie within the fibrous capsule and are covered by hyaline cartilage known as **articular cartilage**. The articular cartilage is firm, and its surface is slippery and bathed in the synovial fluid. It has a smooth surface for the bones to slide over in movements of the joint. A narrow joint cavity filled with synovial fluid separates the bones [Fig. 1.6].

The fibrous capsule may be strengthened by **ligaments**, which are strong bands of inelastic fibrous tissue connecting bones at joints [Fig. 1.7]. Ligaments may be inside the fibrous capsule (intracapsular) or outside the fibrous capsule (extracapsular). They are often found in situations where they will not interfere with movement. For example, at the elbow joint, strong collateral ligaments are found on the medial and lateral sides. They lie approximately as radii of the arc of movement and thus remain tight in all positions, effectively holding the bones together. In contrast, the anterior and posterior parts of the capsule of the elbow joint are thin and loose to allow easy movement. Some ligaments, like the iliofemoral ligament of the hip joint, act to limit excessive movement. Ligaments are often named for their position. For

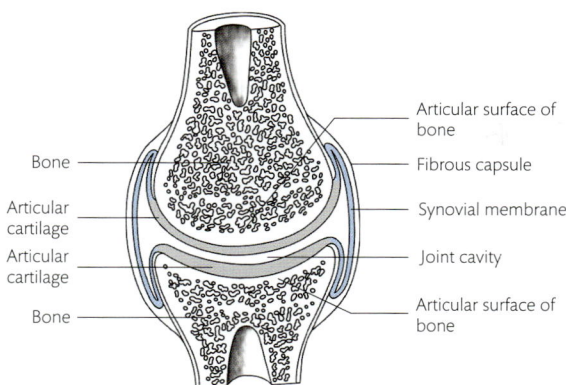

Fig. 1.6 Schematic section through a synovial joint.

example, the ligaments on the side of the elbow joint are called lateral and medial collateral ligaments, or radial and ulnar collateral ligaments, as they lie on the radial and ulnar sides of the elbow, respectively.

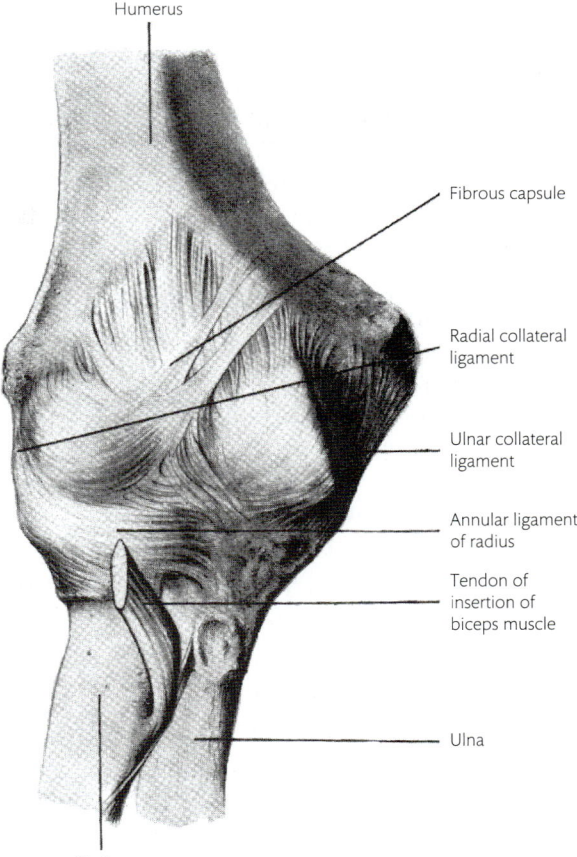

Fig. 1.7 The fibrous or articular capsule of the elbow joint, strengthened by ligaments.

In joints where considerable movement is required in many different directions (e.g. the shoulder joint), the fibrous capsule is thin and lax throughout. The joint is supported by muscles which closely surround the joint and are able to stretch or tighten in any position. Where extreme mobility in only one direction is required (e.g. at the knuckles or knee), the appropriate part of the fibrous capsule is entirely replaced by the tendon of a muscle.

An **articular disc** is a disc composed of fibrous tissue or cartilage and is interposed between opposing surfaces of synovial joints. This disc may have different curvatures on its two surfaces and thus convert a single joint cavity into two, with each cavity allowing for a different type of movement. The articular disc may also act as a shock absorber within the joint or assist with spreading of the synovial fluid between the joint surfaces. Stability and complexity of movement at a joint may be increased by having an articular disc between the bones (e.g. temporomandibular joint [Fig. 1.8D], sternoclavicular joint).

Types of synovial joints

Synovial joints can be subclassified based on the shape of the articulating surfaces and the number of axes around which movements can take place. The joint surfaces of the bones in synovial joints are of many different shapes to allow particular movements while preventing other movements. In some situations, as described below, certain mechanical factors may limit free movement around all axes, even if the bony configuration is conducive to free movement.

Based on the shape of the articulating surface, synovial joints are further subclassified as follows:

1. In **plane synovial joints**, the surfaces of the bones are flat, permitting only slight gliding movements (e.g. joints between the carpal bones of the hand and tarsal bones of the foot) [Fig. 1.8A]. The function of these joints is to provide some resilience to an otherwise rigid structure. More usually, the surfaces of the articulating bones are curved.

2. The **ball-and-socket** type of joint (e.g. shoulder and hip joints) allows the greatest range of movement. In this type of joint, the spherical end of one bone fits into a cup-shaped recess in the other. In the shoulder, the hemispherical head of the humerus fits into the shallow glenoid fossa of the

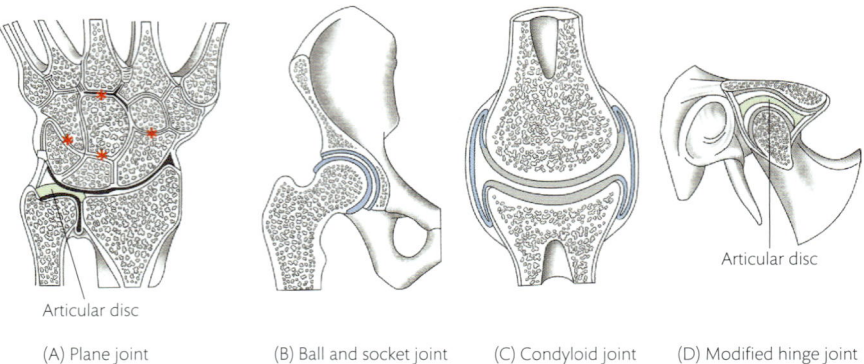

Articular disc

(A) Plane joint   (B) Ball and socket joint   (C) Condyloid joint   (D) Modified hinge joint

Articular disc

Fig. 1.8 Schematic section to show the different types of articulating surfaces in synovial joints. Asterisks indicate the articular surfaces of plane joints of the hand.

scapula. In the hip, the nearly spherical head of the femur fits deep into the cup-shaped acetabulum of the hip bone [Fig. 1.8B]. In ball-and-socket joints where the socket is shallow (e.g. shoulder joint), the range of movements is great, but with less stability (e.g. shoulder joint). In ball-and-socket joints where the socket is deep, movements are restricted, but the joint is stable (e.g. hip joint). The bony contours of a ball-and-socket joint allow for movement in three perpendicular axes, and these joints are classified as **multiaxial joints** [Table 1.2].

3. **Condyloid joints** (e.g. metacarpophalangeal joints of the hand where the fingers meet the palm) have a bony configuration similar to the ball-and-socket type of joint, but rotation is limited by ligaments.

4. **Ellipsoid joints** (e.g. wrist joint) also have a ball-and-socket type of bony configuration, but the radius of curvature of the bony surfaces is long in one direction and short in the other. This disparity makes rotation impossible.

5. In **saddle joints** (e.g. carpometacarpal joint of the thumb), the articular surface is concave in one direction and convex at right angles to this—that is, the convex surface of one bone fits into the concave surface of the other. Condyloid, ellipsoid, and saddle joints allow movements in only two directions, at right angles to each other—flexion

and extension, abduction and adduction (but no rotation). These are **bi-axial joints**.

6. **Hinge joints** (e.g. interphalangeal joints of digits, ankle joint) have a bony configuration and arrangement of ligaments, permitting flexion and extension while preventing all other movements.

7. In **pivot joints** (e.g. proximal radioulnar joint), the cylindrical head of the radius rotates within a ring formed by the ulna and the annular ligament [see Fig. 11.7]. At such a joint, only rotation is possible. Hinge and pivot joints are **uniaxial joints**.

Classification of synovial joints based on the number of axes around which movement takes place is summarized in Table 1.2.

Blood and nerve supply of joints

Joints have a rich arterial supply from surrounding blood vessels. Joints also have a rich supply of sensory nerves that supply the joint capsule and periarticular structures. The nerve supply to a joint is from nerves that supply the muscles crossing the joint. Hilton's law states that the nerves supplying a joint also supply the muscles moving the joint and the skin covering the joint.

Table 1.2 Type of synovial joint and number of axes of movement

Number of axes	Type	Example
Uniaxial	Hinge	Elbow joint
	Pivot	Atlanto-axial joint
Bi-axial	Condyloid	Metacarpophalangeal joint
	Ellipsoid	Wrist joint
	Saddle	First carpometacarpal joint
Multiaxial	Ball and socket	Shoulder joint, hip joint

Muscle

The muscular system consists of all muscles in the body. Muscle fibres, or cells, have contractile elements which bring about movement or alter the shape of a part or an organ. Muscle fibres are closely associated with connective tissue, which forms a covering for them. The connective tissue around the individual muscle fibre is the endomysium. A group of muscle fibres together form a fasciculus, and the fasciculus is covered by perimysium. The entire muscle is covered by connective tissue called epimysium. Muscles are classified on the basis of location and function into smooth, cardiac, and skeletal muscles. The structural differences between the three types of muscle are beyond the scope of this book. However, in brief, cardiac and skeletal muscles have cross-striations, when seen under a microscope, and are called striated muscles. Smooth muscle is non-striated. Cardiac and smooth muscles are innervated by nerves from the autonomic nervous system and act involuntarily. Skeletal muscle is voluntary (i.e. of an individual's own volition). Cardiac muscle is found only in the heart. Smooth muscle is found in many viscera, blood vessels, and skin. Skeletal muscle is found in all parts of the body—head and neck, trunk, and limbs. Further discussion in this section will be devoted to skeletal muscle, as skeletal muscle forms an integral part of the study of gross anatomy.

Skeletal muscle

Some of the skeletal muscles of the body are shown in Fig. 1.9. Skeletal muscles produce movements at

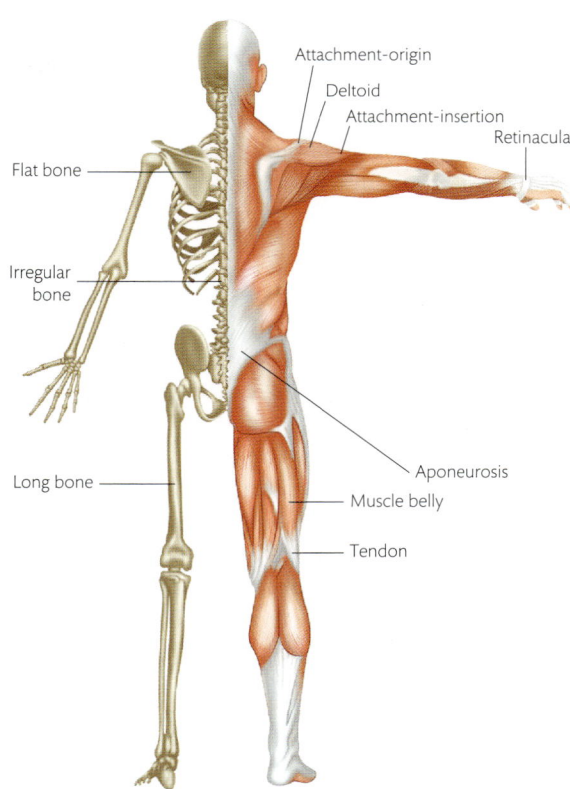

Fig. 1.9 General features of skeletal muscles and bones.

joints by approximating bones (or other structures) to which they are attached, when they contract. Each muscle has at least two attachments, one at each end, and in general cross at least one joint. The action of a muscle on the joint can be worked out from its attachments and its relation to the joint. For example, a muscle which crosses the front of the elbow will produce flexion of the elbow (e.g. biceps brachii). A muscle which crosses behind the elbow will produce extension (e.g. triceps brachii). Skeletal muscles are innervated by motor nerves.

Parts of skeletal muscle

Skeletal muscles are made up of two elements. The fleshy part of a muscle is composed of bundles of red **muscle fibres**, or **muscle cells**. Muscle fibres are red, fleshy, and contractile. The other element in muscles is **fibrous tissue** which surrounds muscle fibres and hold them together and as they slide during contraction. In contrast to muscle fibres, these fibres are white and non-contractile [Fig. 1.10A]. Fibrous tissue attaches muscle fibres to bone. Fibrous tissue may be so short that the belly appears to be attached directly to bone (e.g. attachment of the brachialis to the humerus). In other situations, fibrous tissue forms long, inelastic cords known as **tendons** (e.g. biceps brachii). **Tendons** usually extend over the surface or into the substance of the muscle and thus increase the surface area for attachment of muscle fibres. Tendons also enable muscles to: (1) act at a considerable distance from the muscle belly (e.g. muscles of the forearm that act on the fingers); and (2) change the direction of its pull by passing round a fibrous or bony pulley. **Aponeuroses** are flattened tendons [Fig. 1.10]. Where two aponeuroses meet each other edge-to-edge, their fibres interlock (interdigitate) to form a linear tendinous **raphe**. Such raphes can be stretched along their length by separation of their interdigitating fibres, even though the muscles forming them cannot be pulled apart. The flat muscles on the two sides of the abdominal wall meet in the anterior median plane, forming the largest raphe in the body—the linea alba. The linea alba stretches freely in extension of the trunk but still holds the muscles together.

Skeletal muscles are most commonly attached to bone, ligament, or fascia. They can also be attached to the skin (e.g. face); mucous membrane (e.g. tongue); and eyeball (e.g. muscles that move the eyeball). Muscles have at least two distinct attachments: the **origin** and **insertion**. When a

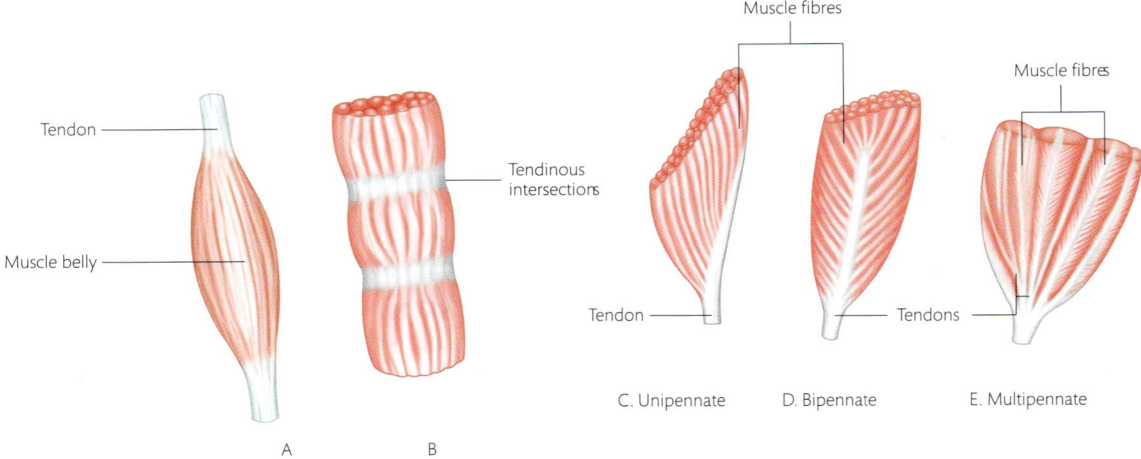

Fig. 1.10 Schematic diagram showing various arrangements of muscle fibres and tendons.

muscle contracts, it is usual to consider one end as relatively fixed—the **origin**, and the other as relatively mobile—the **insertion**. Note that the origin and insertion are not intrinsic properties of an individual muscle. The attachment which moves is determined by other forces in action at the time. For example, a muscle passing from the leg into the foot will move the foot (keeping the leg steady) when the foot is off the ground. The same muscles will move the leg on the foot when the foot is placed steadily on the ground. Similarly, the upper limb muscle latissimus dorsi can be used to pull a rope downwards, as when using a pulley to draw water from a well, or it can also be used to pull oneself up when climbing on a suspended rope.

Classification of muscles according to shape and architecture

Skeletal muscles can be classified according to their shape and orientation of the fibres to the direction of pull. The orientation of muscle fibres within a muscle can be **parallel, oblique,** or **spiral**.

Muscles with fibres arranged in parallel vary in shape and can be: (1) **flat** and **short** (e.g. thyrohyoid); (2) **strap-like** (e.g. sternohyoid, sartorius); or (3) **fusiform** (e.g. biceps brachii). The fibres in the belly of a fusiform muscle are parallel, but at the ends, they converge onto a tendon [Fig. 1.10A and B]. Some fusiform muscles have multiple heads (e.g. biceps brachii, triceps brachii). Others have more than one belly (e.g. omohyoid, digastric).

In muscles with an oblique orientation of fibres, the individual fibre is oblique to the line of pull.

Such muscles may be: (1) **triangular** (e.g. temporalis); or (2) **pennate**, where fibres are short and run obliquely to insert into a tendon—like the barbs of a feather. Pennate muscles are further subclassified as: (1) unipennate, when all fibres converge from one direction (e.g. extensor digitorum longus); (2) bipennate, when fibres converge onto the tendon from two directions (e.g. dorsal interossei); and (3) multipennate, when the muscle has a series of intramuscular tendons (e.g. deltoid).

Muscles may exhibit a **spiral** or **twisted** appearance (e.g. pectoralis major, sternocleidomastoid). Some skeletal muscles are **circular** in shape (e.g. the orbicularis oculi which closes the eyes).

Naming of muscles

Names of muscles are based on their: (1) **shape** (e.g. *deltoid* or triangular; *teres* or round); (2) **size** (e.g. *major* or big; *minor* or small; *longus* or long; *brevis* or short); (3) **number of heads or bellies** (e.g. *biceps* or two heads; *quadriceps* or four heads; *digastric* or two bellies); (4) **depth from the surface** of the body (e.g. *superficialis* or superficial; *profundus* or deep; *external* or on the outside; *internal* or on the inside); (5) **attachment** (e.g. brachioradialis, from brachium (arm) to radius); (6) **position** (e.g. *pectoralis* on the chest; *femoris* on the femur); and (7) **action** (e.g. *adductor* and *abductor*; *flexor* and *extensor*). Most muscle names have more than one descriptive feature in them (e.g. *extensor carpi radialis longus*—long extensor of the carpal bones on the radial side; *quadriceps femoris*—muscle with four heads related to the femur; *pronator teres*—the rounded pronator).

Different types of muscle contraction

Muscles contract in two different ways to meet the demands placed on them: (1) **isometric contraction**—when the length of the muscle remains the same, but the muscle undergoes a change in tension; and (2) **isotonic contraction**—when the tension of the muscle remains the same, but the muscle undergoes a change in length. Isometric contraction without movements occurs in all anti-gravity muscles when a person is standing still. The tension developed in an anti-gravity muscle, like the quadriceps femoris, needs to be equal to the pull of gravity to keep the body straight. This happens without any change in length of muscle. There are two types of isotonic contraction: **concentric contraction** and **eccentric contraction**. **Concentric contraction** is when a muscle shortens to produce a movement. In this situation, the tension developed in the muscle is greater than the load on it and movement is brought about. On the other hand, **eccentric contraction** is when the tension developed in a muscle is less than the load acting against it and the muscle lengthens to allow movement to occur. For example, the deltoid muscle which passes over the shoulder contracts concentrically to abduct the arm from the side of the body. When the outstretched arm is lowered to the side, the deltoid *lengthens under constant tension*—eccentric contraction—so as to control the descent of the arm.

Exercise 1.1: to test these actions of the deltoid, place your left hand over your right deltoid muscle (i.e. on the lateral surface of the shoulder below its tip) [Fig. 1.9]. Now abduct one arm till it is horizontal, and feel the deltoid muscle hardening as it contracts (concentric contraction). Hold the arm in this position and note that the deltoid remains contracted and hard (isometric contraction). Now slowly lower the arm to the side, and note that the deltoid remains contracted for most of this action (eccentric contraction).

Function of muscles

Movement across a joint is brought about by a group of muscles, rather than by a single one. Muscles in the group may be classified as prime movers, antagonists, synergists, and fixators. **Prime movers**, or **agonists**, initiate and carry out the movement. Muscles that oppose the movement, or initiate and carry out the opposite movement are antagonists. In flexion of the elbow joint, the brachialis is the prime mover and the triceps brachii (which extends the elbow joint) is the antagonist.

Fixators work together to steady a proximal joint, so that movement can take place effectively at a more distal joint. For example, for the digits of the hand (fingers) to be able to grasp an object tightly, the extensors and flexors of the wrist contract together to fix and keep the wrist joint steady.

When a muscle crosses multiple joints, it may not be able to shorten sufficiently to produce the full range of movement at all joints. This is because muscle fibres can only contract to 40% of their fully stretched length. This limitation is **active insufficiency** of a muscle. For example, the digits cannot be fully flexed when the wrist is flexed, due to active insufficiency. Ascertain this on your own wrist and fingers. At the same time, the opposing muscles—the extensors of the wrist and fingers—may not be able to stretch sufficiently to allow full flexion of the wrist and fingers. This is known as **passive insufficiency**. In such a situation, a third set of muscles maybe used to fix a proximal joint (to keep it steady), so that muscles producing movement distally can act effectively. These muscles are called **synergists**.

Classification of muscles according to force of action

Muscles are classified as spurt and shunt muscles, based on the predominant effect caused on the joint they cross. A **spurt** muscle has its proximal attachment far from the joint on which it acts, and its distal attachment close to the joint. It does little to maintain the integrity of the joint but has great advantage in speed and range of movement of the bones. Spurt muscles tend to be prime movers (e.g. biceps brachii) [Fig. 1.11].

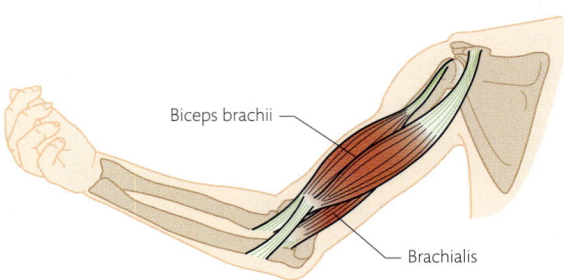

Fig. 1.11 Insertion of the spurt muscle biceps brachii, close to the elbow joint.

By contrast, a **shunt** muscle has its proximal attachment near the joint on which it acts, and its distal attachment at a greater distance away from the joint. When a shunt muscle contracts, it applies force along the length of the bones and pulls the joint surfaces together to stabilize the joint. Shunt muscles are less capable of bringing about movement but help in maintaining stability in all positions. They act as ligaments of variable length and tension, in the place of the usual ligaments which would restrict movement. The rotator cuff muscles of the shoulder joint are a good example of shunt muscles. The brachioradialis is also a shunt muscle.

Artery and nerve supply to muscles

Muscles are supplied by numerous arteries and veins. The main artery and motor nerve enter the muscle at a distinct **neurovascular hilum**. Smaller arteries enter at other points as well. Motor nerves entering the muscles carry nerve impulses which cause the muscle to contract. They also carry sensory information from the muscle and tendon on the amount of tension in the muscle, the degree of contraction of the muscle, and position of the joints. Sympathetic nerve fibres reach the muscle around the blood vessels. It is possible to stimulate contraction in individual muscles by applying an electrical impulse to the skin overlying the neurovascular hilum—the **motor point**.

Naming and innervation of muscle groups

Limb muscles are often classified in groups by the principal action they have on a particular joint (e.g. flexors, extensors, abductors, adductors). Although this classification is commonly used, it is unsatisfactory because a single muscle may be a flexor of one joint and an extensor of another (e.g. rectus femoris).

The terms flexors and extensors for a group of muscles are also used to designate groups of limb muscles which develop, respectively, from the ventral and dorsal sheets of primitive muscles (irrespective of the actual functions of the individual muscles). The anterior divisions of the ventral rami of the spinal nerves (see below) supply flexor muscles. The posterior divisions of the ventral rami supply the extensors. Hybrid muscles are limb muscles that are supplied by two separate nerves, one from the anterior and the other from the posterior division of the ventral rami; for example, the brachialis is supplied by the musculocutaneous nerve (ventral division) and radial nerve (dorsal division). It should be noted that all muscles with dual nerve supply are not hybrid muscles.

Bursae and synovial sheath

Where two adjacent structures, like muscle and bone, tendon and bone, or skin and bone, slide over each other, a synovial sac, called a **bursa**, is often developed between them to reduce friction. The **bursa** is a closed sac lined with a synovial membrane, which secretes synovial fluid into the sac. Tendons which rub against bone or firm tissue, like retinacula, are enclosed in **synovial sheaths** containing synovial fluid. Like bursae, synovial sheaths are closed sacs filled with synovial fluid.

❯ When there is irritation or infection of the bursa or synovial sheath, secretion is increased and the cavity becomes swollen and tender—a condition known as **bursitis** or **tenosynovitis**.

Skin

The skin is the largest organ of the body and covers its entire surface. It provides protection from deleterious agents in the environment, and information of touch, pain, pressure, heat, and cold through the sensory nerves. It helps to regulate heat, and synthesizes and stores vitamin D.

The **skin** consists of a superficial layer of avascular, stratified squamous epithelium—the **epidermis**—and a deeper vascular, dense fibrous tissue layer—the **dermis**. The dermis sends small peg-like protrusions into the epidermis. These protrusions help to bind the epidermis to the dermis by increasing the area of contact between them. The dermis is rich in collagen and elastic fibres. In any specified area of the body, the collagen fibres run in the same direction to form **tension lines**. These tension lines, or lines of Langer, run transversely in the neck and trunk, and in longitudinal spirals in the limbs. ❯ Surgical skin incisions made along the tension lines heal faster and produce less scar tissue. By contrast, skin incisions made across tension lines tend to gape at the wound, heal slowly, and produce more scar tissue.

Skin ligaments are connective tissue fibres that run from the dermis to the underlying deep fascia. In your dissection, you will see that these ligaments are short and firmly attached to the deep fascia in some areas, like in the palms of the hand. They are thin and loosely attached in areas like the dorsum of the hand.

Examine the skin of the palm of your hand. The skin on the palm of the hand and sole of the feet is different from the skin covering the rest of the body. It is thick and hairless, whereas the skin over the rest of the body is thin and contains hair follicles. In general, the skin on the dorsum—back of the trunk, neck, and scalp—is thicker and has less fine sensation than the skin on the ventral aspect of the body. Two types of glands are found in the skin. Sweat glands are found all over the body in hairy and hairless skin. Sebaceous glands are found only in hairy skin and in relation to the hair follicles [Fig. 1.12].

Skin appendages

The skin appendages include sweat glands, nails, and the pilosebaceous unit. The **pilosebaceous unit** comprises the hair shaft, hair follicle, sebaceous gland, and arrector pili muscle. The hair shaft is the part of hair which projects outside the surface of the skin. The hair follicle is a sac-like structure that contains the root of the hair. It extends from the epidermis into the dermis. Each hair follicle is associated with one or more **sebaceous glands** that secrete sebum to moisturize the skin. A band of smooth muscle—the **arrector pili**—extends deep to the sebaceous gland, from the hair follicle to the

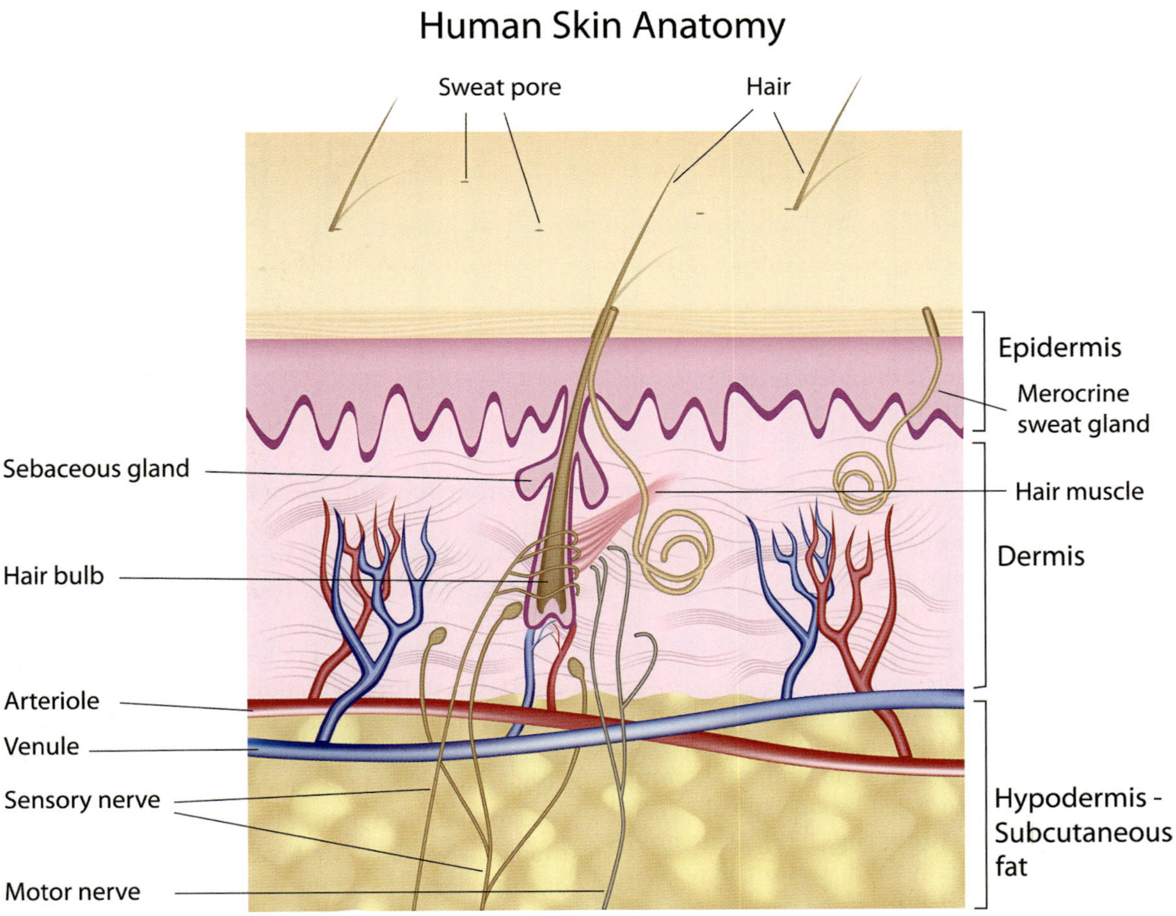

Human Skin Anatomy

Fig. 1.12 Schematic section through thin, hairy skin.

Alila Medical Media/Shutterstock.com

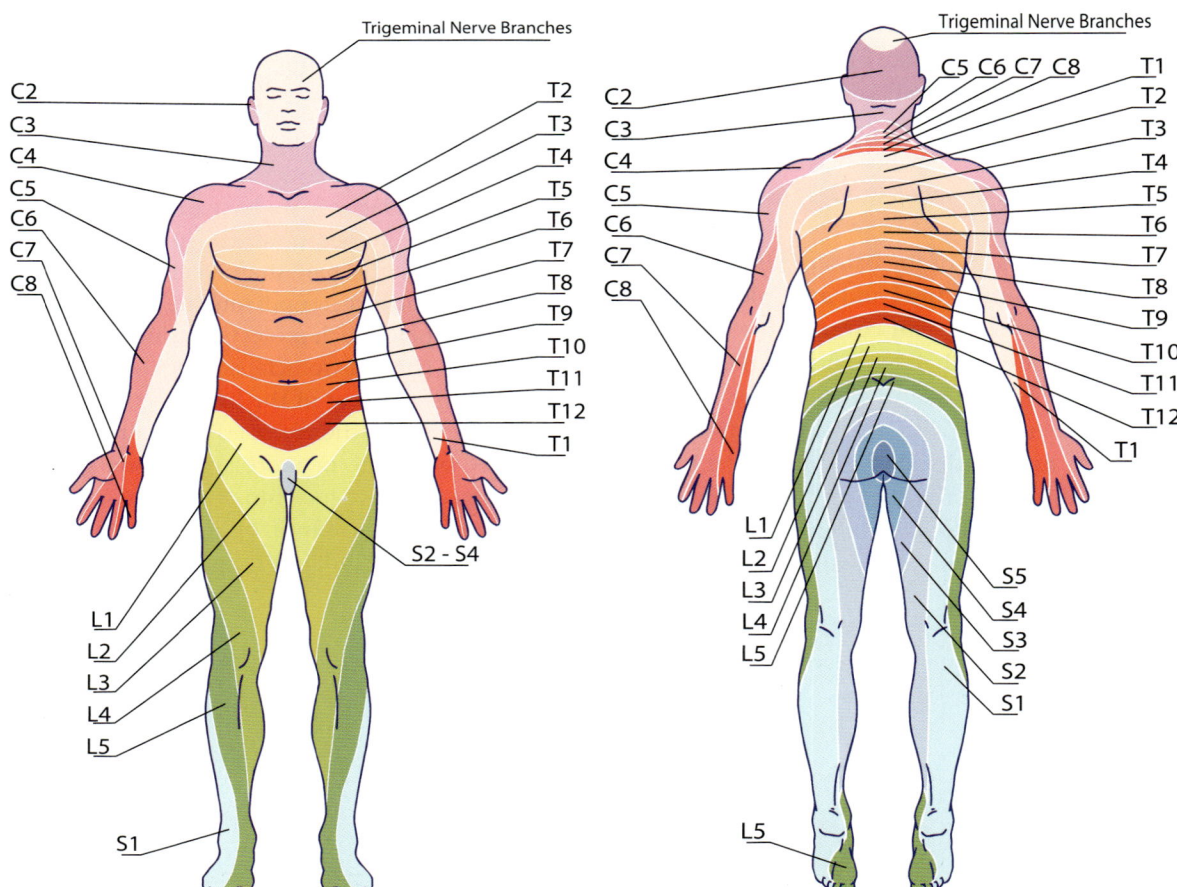

Fig. 1.13 Dermatomes of the body. (A) Ventral aspect. (B) Dorsal aspect. C, cervical; T, thoracic; L, lumbar; S, sacral.
Stihii/Shutterstock.com

dermis. The arrector pili is under control of sympathetic nerves and functions to raise the hairs on the skin, under sympathetic stimulation, in response to threat or cold [Fig. 1.13].

Nails are plates of keratinized tissue found at the tip of the digits. They are important elements in the protection of fingertips and contribute to tactile sensation by acting as a counterforce to the fingertip pad.

Nerve supply of skin

The skin has an abundance of sensory nerve endings, and can appreciate touch, pain, pressure, temperature, and movement of air and water around it. A strip of skin supplied by one spinal nerve is known as a **dermatome** [Fig. 1.13]. Dermatomes are described more fully in the section on nerves.

The skin is separated from the deeper structures (muscles and bones) by two layers of connective tissue—the superficial and deep fasciae [Fig. 1.14].

Superficial fascia

The **superficial fascia** is a fibrous mesh consisting of loose areolar tissue, fat, cutaneous blood vessels, and nerves. It connects the dermis to the underlying deep fascia. It is particularly dense in the scalp, back of the neck, palms of the hands, and soles of the feet, and binds the skin firmly to the deep fascia in these situations. In other parts of the body, it is loose and elastic, and allows the skin to move freely.

The thickness of the superficial fascia varies with the amount of fat in it. It is thinnest in the eyelids, nipples, areolae of the breasts, and in some parts of the external genitalia where there is no fat. In a well-nourished body, the **fat** in the superficial fascia rounds off the contours. Fat distribution and amount vary between the sexes. The smooth-

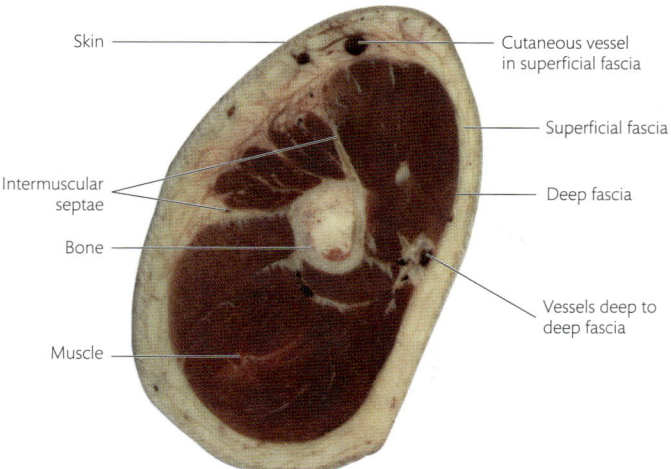

Skin — Cutaneous vessel in superficial fascia

Superficial fascia

Intermuscular septae — Deep fascia

Bone

Muscle — Vessels deep to deep fascia

Fig. 1.14 Cross-section through the arm showing the relationship of the skin, superficial fascia, cutaneous vessels and nerves, deep fascia, muscles, intermuscular septa, and bone to each other.

Image courtesy of Visible Human Project of the US National Library of Medicine.

er outline of a woman's figure is due to the greater amount of subcutaneous fat.

Deep fascia

The **deep fascia** is the dense, inelastic membrane which separates the superficial fascia from the underlying structures. The deep fascia also surrounds muscles, and blood vessels and nerves which lie between them. It sends fibrous partitions, or **septa**, known as **intermuscular septa**, between the muscles to the periosteum of the bones. It forms a major source of attachment for many muscles. The deep fascia also forms tunnels within which the muscles of a group can slide independent of each other. **Intermuscular septa** are better developed between adjacent muscles having different actions [Fig. 1.14]. In the wrists and ankles, the deep fascia is thickened to form **retinacula** [Fig. 1.9] (e.g. flexor and extensor retinacula of the wrist, flexor and extensor retinacula of the ankles). These retinacula hold the tendons against the bones during movements of the wrist and ankle (and prevent them from bowstringing). The **iliotibial tract** is another thickening of the deep fascia on the lateral side of the thigh. It helps to stabilize the knee when a person is standing. The **plantar aponeurosis** on the sole of the foot is thickened deep fascia which supports the foot and protects

the vital structures in it from the weight of the body on the foot.

Circulatory system

The **circulatory system** is made up of the **cardiovascular system** (heart and blood vessels) and the **lymphatic system**. Blood circulates continuously through the cardiovascular system. The lymph, a clear fluid formed in the interstitial tissue spaces, is transported centrally to the large veins in the neck by the lymphatic system.

Blood vessels

This section covers the description of blood vessels in the systemic and pulmonary circulation. Systemic circulation carries oxygenated blood from the left ventricle to all parts of the body, and returns deoxygenated blood to the right atrium. Pulmonary circulation carries deoxygenated blood from the right ventricle to the lungs, and returns oxygenated blood to the left atrium [Fig. 1.15].

The **blood vessels** you will see and identify in dissection are the **arteries** and **veins**. For the sake of completion, a note is added about the capillaries which connect the smallest of the arteries—the **arterioles**—to the smallest of the veins—the **venules**. The wall of arteries and veins is made up of three layers of tissue. From inside to out, these

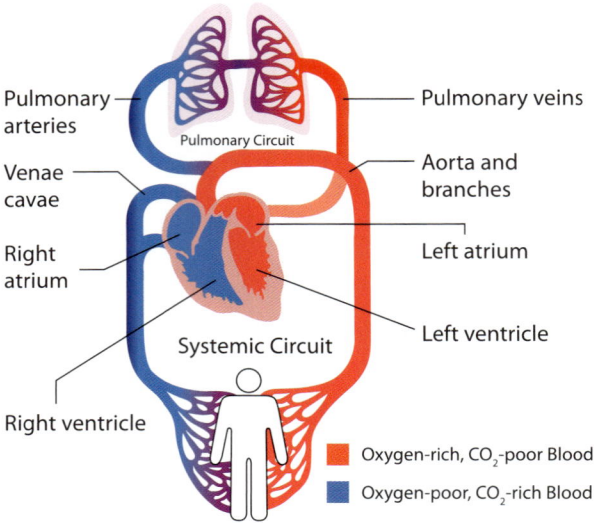

Pulmonary arteries

Pulmonary veins

Pulmonary Circuit

Venae cavae

Aorta and branches

Right atrium

Left atrium

Left ventricle

Systemic Circuit

Right ventricle

Oxygen-rich, CO_2-poor Blood

Oxygen-poor, CO_2-rich Blood

Fig. 1.15 Schematic diagram showing the heart and the pulmonary and systemic circulation.

yaruna/Shutterstock.com

layers are the **tunica intima**, **tunica media**, and **tunica externa**. The tunica intima is lined by a single layer of flattened epithelial cells—the **endothelium**.

Arteries

Arteries are blood vessels which carry blood away from the heart [Fig. 1.16]. The large arteries in the body are the aorta and pulmonary trunk. They begin at the heart and are approximately 2.5 cm in diameter. They give rise to a series of branches which vary in size with the volume of tissue each has to supply. These branch and rebranch repeatedly and become successively smaller in calibre. Larger arteries have many elastic fibres in the tunica media and are known as **elastic arteries**. They expand during ventricular contraction and recoil or return to their normal size during ventricular relaxation. They are also called **conducting arteries**. Medium-sized arteries have more smooth muscle fibres in the tunica media and are known as **muscular** or **distributing arteries**. The smooth muscle in their walls contracts rhythmically to create a pulsatile flow and distribute blood to all parts of the body. The smallest arterial vessels (<0.1 mm in diameter) are known as **arterioles**. They transmit blood into the capillaries [Fig. 1.17].

Collateral circulation

Collateral circulation is nature's way of ensuring that a second source of blood flow is present to take over blood supply to a region or organ if the primary artery is compromised.

In many tissues, small arteries unite with one another to form tubular loops called **arterial anastomoses**. Such anastomoses occur especially around the joints of the limbs, in the gastrointestinal tract, and at the base of the brain. When one of the arteries taking part in the anastomosis is blocked, the remaining arteries enlarge gradually to produce a **collateral circulation** and maintain blood flow to the tissue otherwise supplied by the blocked artery.

In some tissues, the degree of anastomosis between adjacent arteries may be so minimal that blockage of one vessel cannot be compensated for by the opening up of the others. Arteries which are solely responsible for perfusion of a segment of tissue are called **end arteries**. When an end artery is blocked, the tissue supplied by it dies (for lack of collateral circulation). End arteries are found in the eye, brain, lungs, kidneys, and spleen.

Capillaries

Blood capillaries are microscopic tubes that form a network of channels connecting arterioles and venules. The capillary wall consists of a single layer of flattened **endothelial cells**, through which substances are exchanged between blood and tissues.

Human circulatory system

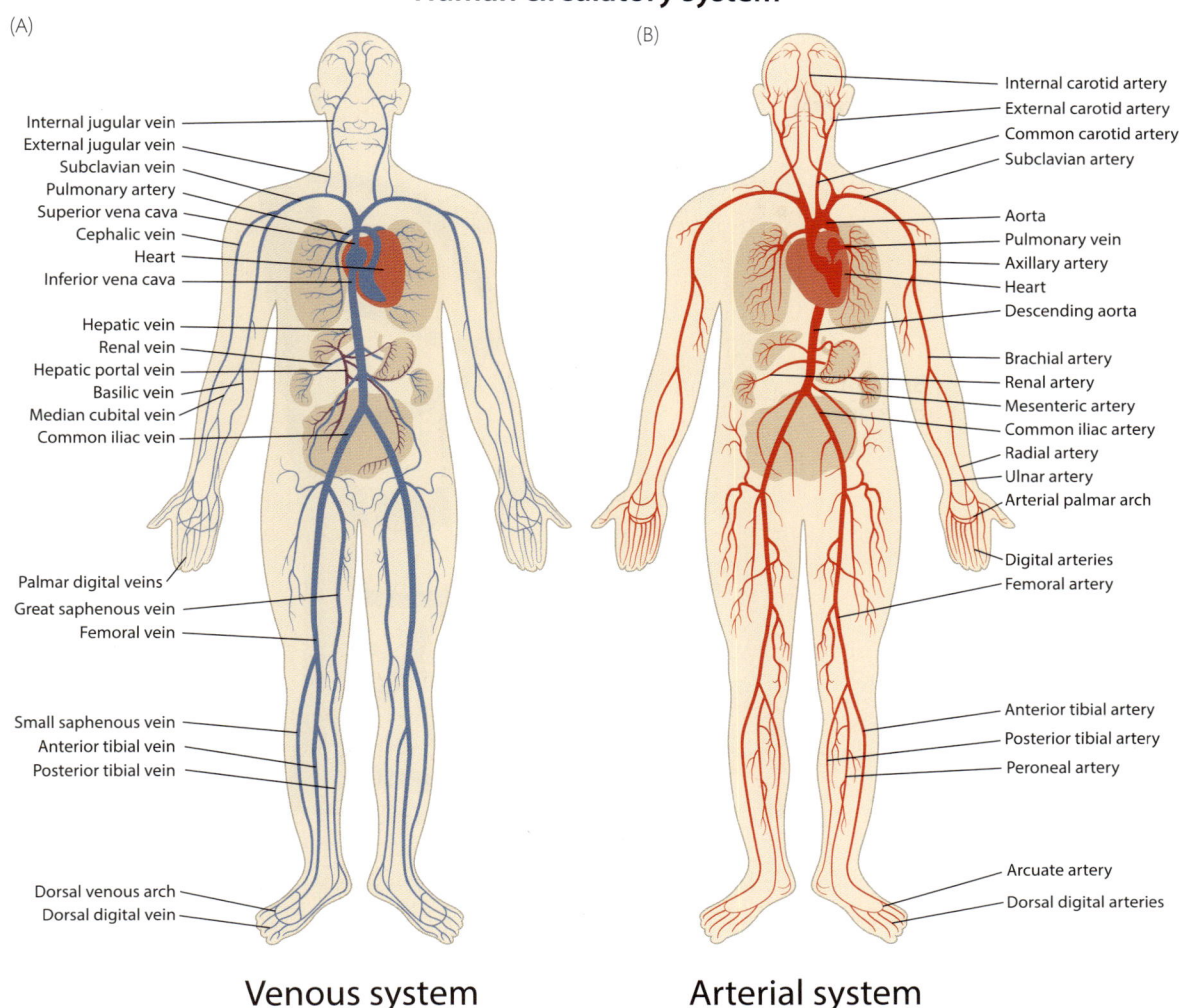

(A)

Internal jugular vein
External jugular vein
Subclavian vein
Pulmonary artery
Superior vena cava
Cephalic vein
Heart
Inferior vena cava

Hepatic vein
Renal vein
Hepatic portal vein
Basilic vein
Median cubital vein
Common iliac vein

Palmar digital veins
Great saphenous vein
Femoral vein

Small saphenous vein
Anterior tibial vein
Posterior tibial vein

Dorsal venous arch
Dorsal digital vein

(B)

Internal carotid artery
External carotid artery
Common carotid artery
Subclavian artery

Aorta
Pulmonary vein
Axillary artery
Heart
Descending aorta

Brachial artery
Renal artery
Mesenteric artery
Common iliac artery
Radial artery
Ulnar artery
Arterial palmar arch

Digital arteries
Femoral artery

Anterior tibial artery
Posterior tibial artery
Peroneal artery

Arcuate artery
Dorsal digital arteries

Venous system

Arterial system

Fig. 1.16 Circulation of blood through the arterial and venous systems. (A) Arteries. (B) Veins.

Olga Bolbot/Shutterstock.com

Veins

Veins are blood vessels that carry blood to the heart. The smallest vein is the **venule**. Venules and small veins unite with each other successively to form larger veins. A smaller vein that opens into a larger vein is known as a **tributary** of the larger vein. The largest veins—the pulmonary veins and the superior and inferior venae cavae—ultimately open into the heart [Fig. 1.16]. Blood flow in the systemic veins is sluggish, and venous return to the heart is aided by: (1) the pressure applied on veins by contracting leg muscles; and (2) the suction force generated by the fall in intrathoracic pressure during inspiration. **Valves** in the veins of the lower limb prevent backflow of blood due to gravity. The positions of valves in the superficial veins can be seen as localized swellings along their course when the veins are distended with blood. Communications between superficial and deep veins permit the superficial veins to drain into deep veins. Whenever possible, you should slit open the veins in the different parts of the body to see the position and structure of the valves. In the distal parts of the limbs, two or more deep veins accompany the named arteries. These are known as **venae comitantes**. Differences between arteries and veins are shown in Table 1.3.

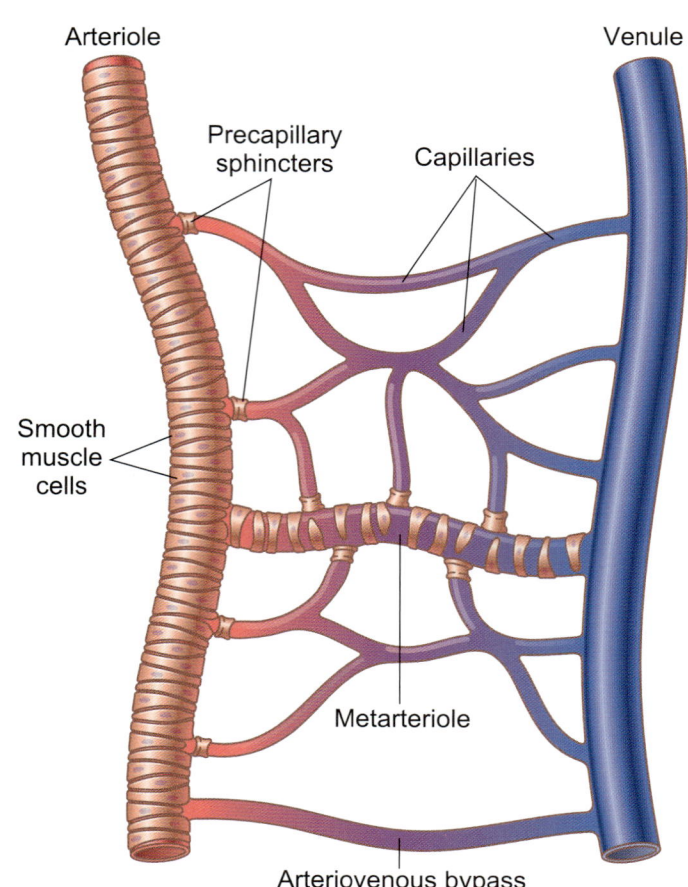

Arteriole

Venule

Precapillary
sphincters

Capillaries

Smooth
muscle
cells

Metarteriole

Arteriovenous bypass

Fig. 1.17 Schematic diagram of a microcirculatory unit.

periyanayagam/Shutterstock.com

Microcirculation

Arterioles, capillaries, and venules constitute the **microcirculatory units** and are not seen by the naked eye. Fig. 1.17 illustrates the arterioles which feed the capillaries, capillary network, and venules that drain the microcirculatory bed. In the capillary network, exchange of gases, nutrients, and waste products takes place between the intravascular and extravascular spaces. (The extravasated fluid will be collected as lymph.) The short segment of vasculature between the arteriole and the capillaries is the **metarteriole**. A circular band of smooth muscle fibres, called the **precapillary sphincter**, surrounds the metarteriole and regulates the amount of blood entering the capillary network. The capillaries may be bypassed in some situations when a small artery connects directly with a small vein—an **arteriovenous anastomosis**.

Portal system

Most veins end in bigger veins or in the heart. In a few specific instances, veins end by breaking down again into capillaries. A portal system is one in which a vein begins and ends in capillaries. The most common example of a portal venous system is

Table 1.3 Differences between arteries and veins

Characteristics	Arteries	Veins
Direction of blood flow	Away from the heart	Towards the heart
Pressure	High pressure	Low pressure
Valves	Valves are absent	Valves are present
Thickness of wall	Walls are thicker	Walls are thinner
Shape of lumen	Lumen is round	Lumen is collapsed

seen in the digestive system. The portal vein begins by the union of veins that drain the stomach and intestines, and ends in the capillary network within the liver. (After traversing the liver, the blood from the portal vein is collected into the hepatic veins which open into the inferior vena cava.) Similarly, the hypothalamus—a part of the brain—is connected to the pituitary gland by the veins of the hypothalamohypophyseal portal system.

Applied anatomy

⮑ Proper functioning of blood vessels is essential to ensure that organs and tissues get the oxygen and nutrients they need. Blood vessels can develop problems, such as blockages or enlargement which can compromise blood flow, and can even be life-threatening.

A thrombus is a blood clot attached to the intima of the blood vessel. It can develop in an artery or a vein. It remains at the site of formation and hinders blood flow. An embolus is a detached mass that travels through the bloodstream, and is capable of clogging arterial capillary beds at a site distant from its point of origin.

Infarction is tissue death or necrosis due to inadequate blood supply to the affected area. It may be caused by blockage, rupture, mechanical compression, or vasoconstriction of an artery. For example, myocardial infarction is death of cardiac muscle due to loss of blood supply.

An aneurysm is the enlargement of an artery caused by weakness in the arterial wall. An aneurysm may occur in any blood vessel but is most often seen in an artery. A ruptured aneurysm can be fatal.

Lymphatic system

The **lymphatic system** is complementary to the circulatory system. Lymph is a clear fluid formed in the interstitial tissue spaces. It contains waste products and cellular debris, bacteria, and proteins. It is transported centrally to the large veins in the neck by the lymphatic system [Fig. 1.18]. The lymphatic system plays an important part in the immune defence of the body. The cells of the lymph are mostly lymphocytes.

The lymphatic system consists of a large network of lymph vessels, lymph nodes (where the highest lymphocyte concentration is found), lymphoid organs, lymphoid tissues, and lymph. Lymph vessels are not demonstrated by dissection but are described because of the importance of this system in clinical practice.

Lymph vessels

Lymph capillaries, or **lymphatics**, have a structure similar to that of blood capillaries but are wider and less regular in shape. They are more permeable to particulate matter and cells than blood capillaries.

Lymphoid organs

Associated lymphoid organs are composed of lymphoid tissue and are the sites of lymphocyte production or activation. These include the **lymph nodes** (where the highest lymphocyte concentration is found), **spleen**, **thymus**, and **tonsils**. Lymphoid tissue is also seen in the mucosa of the gut, bronchi, and pharynx (mucosa-associated lymphoid tissue, or MALT).

Lymph nodes are firm, gland-like structures which filter the lymph. They vary in size, from a pinhead to a large bean, and lie along the path of lymph vessels. Small lymph vessels unite to form larger lymph vessels, many of which converge on a lymph node. The lymph passes through the node and leaves it in a vessel which usually drains into a secondary, and through it into a tertiary lymph node. Thus, the lymph drains through a series of lymph nodes and is gathered into larger lymph vessels which enter a great vein at the root of the neck. The vessels that carry the lymph to a node are called **afferent vessels**; those that carry it away from a node are **efferent lymph vessels** (*ad* = to; *ex* = from; *fero* = carry).

In the limbs, the lymph nodes are largest and most numerous in the armpit or axilla and groin. They are usually found in groups which are linked to each other by lymph vessels.

Lymph vessels in the superficial fascia drain the lymph capillary plexuses of the skin. They converge directly on the important groups of lymph nodes situated mainly in the axilla, the groin, and the neck. In deeper tissue, most lymph vessels and nodes are situated along the deep veins.

Lymph nodes filter lymph, trapping viruses and bacteria and enabling the body to mount an immune response against invading organisms. Lymph nodes also trap cancer cells.

LYMPHATIC SYSTEM

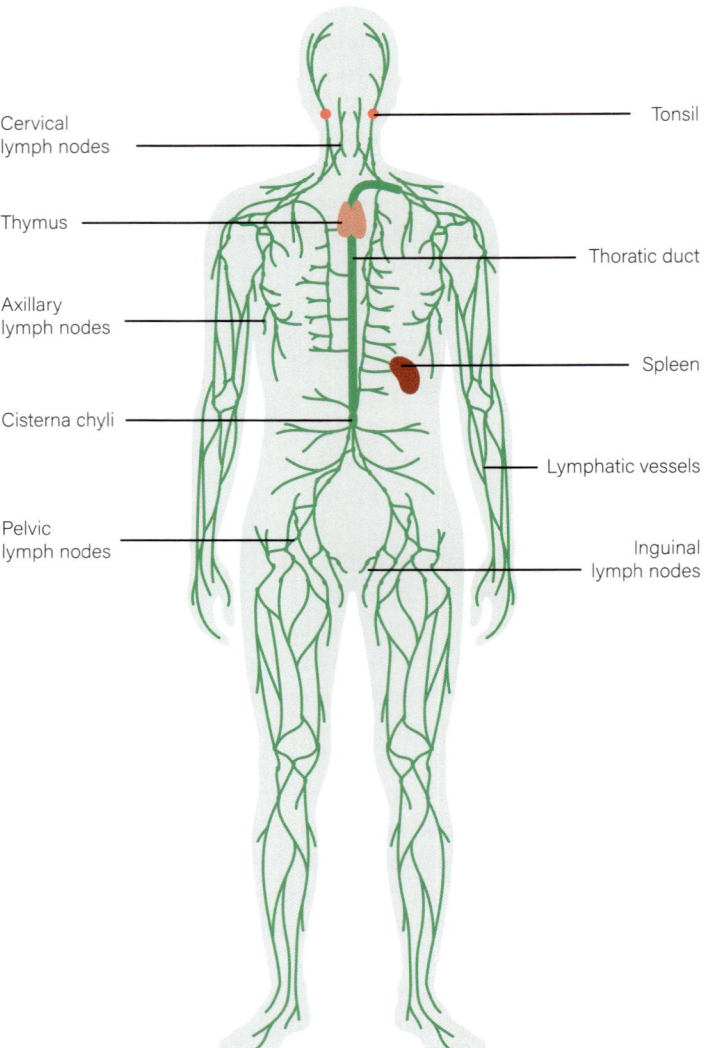

Cervical
lymph nodes

Thymus

Axillary
lymph nodes

Cisterna chyli

Pelvic
lymph nodes

Tonsil

Thoratic duct

Spleen

Lymphatic vessels

Inguinal
lymph nodes

Fig. 1.18 Schematic diagram showing the lymphatic system.

Pikovit/Shutterstock.com

➔ When lymph vessels are blocked or damaged, there is a build-up of fluid in soft tissues in the region, resulting in swelling known as **lymphoedema**.

➔ Tumour cells may also enter the lymphatic system and spread to distant regions in the lymph. For example, cancer cells from the breast can spread to the axillary, and even the cervical, lymph nodes.

Nervous system

The nervous system consists of two main parts—the **central nervous system** (**CNS**) made up of

the brain and spinal cord, and the **peripheral nervous system** (**PNS**) made up of spinal nerves, cranial nerves, and the peripheral part of the autonomic nervous system [Fig. 1.19]. The **autonomic nervous system** controls visceral activities, such as contraction of involuntary muscles in the skin, eyeball, blood vessels, heart, and other viscera, and secretion of glands. It has two components—the **sympathetic nervous system** and the **parasympathetic nervous system**.

The nervous system is concerned with the receipt and integration of sensory information (sensory system), production of movement (motor system), memory, emotions, and other complex higher

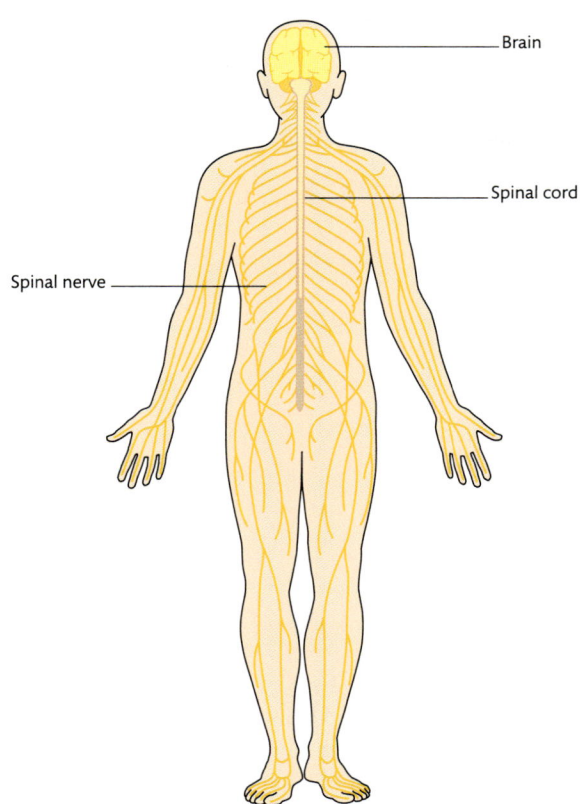

Fig. 1.19 Schematic diagram showing spinal nerves.

Reproduced with permission from Drake, R. L. and Vogl, W., Mitchell, A. W. M. *Gray's Anatomy for Students*. Copyright © 2005 Elsevier.

functions. These complex activities are achieved by passage of impulses through the interconnected networks of cells in the nervous system.

The nervous system contains: (1) **nerve cells** or **neurons**; and (2) connective tissue cells—the **neuroglia**. **Neurons** are variable in size and shape. They have a **cell body** and **cytoplasmic processes** called **neurites**, which extend from the cell body [Fig. 1.20]. These processes are of two types—axons and dendrites. **Dendrites** receive stimuli. They vary greatly from cell to cell but are commonly branched in a complicated fashion and restricted to the vicinity of the cell body. **Axons** are usually thinner than dendrites and transmit impulses away from the cell body, either to other nerve cells or to peripheral tissues through the peripheral nerves. Most of the axons in the nervous system are covered with a fatty **myelin sheath**. Each neuron is linked into the system by cell junctions, or **synapses**, where nerve cells come together and neuronal impulse from one cell is transferred to another.

Within the CNS, nerve cell bodies are grouped together and make up the **grey matter**. Axons are grouped together to form the **white matter**. Dendrites and nerve synapses are usually confined to the grey matter.

Neuroglia, or glial cells in the nervous system, support, insulate, and nourish the neurons. In the CNS, there are four types of glial cells—**astrocytes**, **oligodendrocytes**, **ependyma**, and **microglia**. Astrocytes, oligodendroglia, and ependymal cells are derived from the ectoderm, in common with most of the nervous system. Microglia are derived from the mesoderm, as are blood vessels.

Astrocytes are star-shaped cells with cell processes radiating from them. One or more of the processes extend to cover a segment of an adjacent capillary, so that processes of multiple astrocytes together cover the capillaries of the CNS. There are two main types of astrocytes: (1) **protoplasmic astrocytes**, which have processes branching repeatedly to form a dense bush and are found predominately in the grey matter; and (2) **fibrillary astrocytes**, which have long, thin, sparsely branching processes and are found in the white matter. **Oligodendroglia** are small cells with few short processes. They form myelin sheaths, a lipid-protein covering for axons, within the CNS. **Microglia** are small rod-shaped cells. In nerve fibre injury, they ingest particles of degenerating myelin and develop foam-like cytoplasm. The **ependyma** forms the epithelial lining for the cavities of the brain and spinal cord. It also covers the vascular pia mater which invaginates the ventricles to form the choroid plexuses.

In the PNS, there are two types of glial cells—**satellite cells** covering the neurons of ganglia, and **Schwann cells** that form myelin around axons.

Classification of neurons

1. Based on the length of processes, neurons are classified as: (1) **Golgi type I**, which are neurons with long processes; and (2) **Golgi type II**, which are neurons with short processes.
2. Based on the number of neurites, the cell body of a neuron can appear to have 'poles' where the neurite arises from [Fig. 1.21]. Neurons are classified as: (1) **multipolar cells** having a single long axon and multiple dendrites (e.g. anterior horn

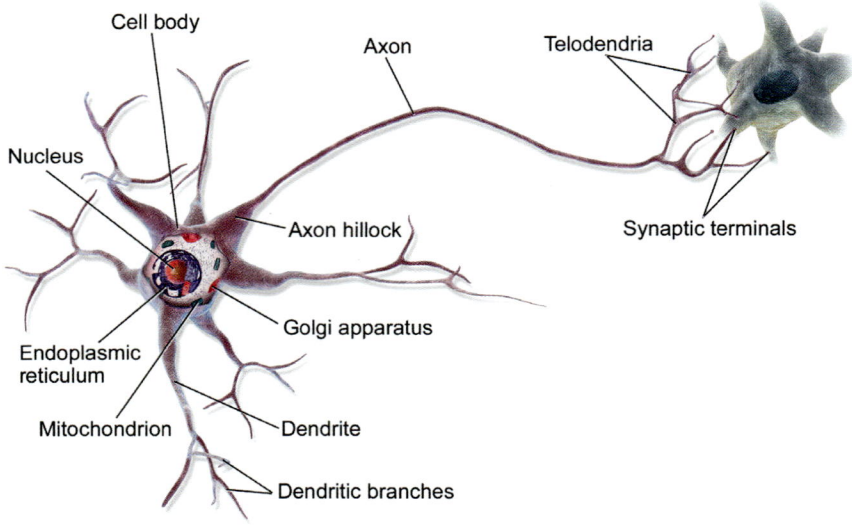

Fig. 1.20 Schematic diagram of a neuron. Note the axon synapsing with another neuron.

Reproduced from Bruce Blaus Blausen.com staff (2014). 'Medical gallery of Blausen Medical 2014'. *WikiJournal of Medicine* 1 (2). DOI:10.15347/wjm/2014.010. ISSN 2002-4436.

cells of the spinal cord); (2) **bipolar cells** having two neurites—a dendrite and an axon arising from opposite sides of the cell body (e.g. bipolar cells of the retina and olfactory epithelium); and (3) **pseudounipolar neurons** also having an axon and a dendrite, but with the two arising from the cell body as a single unit and then split into two (e.g. dorsal root ganglion).

3. Based on function, neurons are classified as: (1) sensory or **afferent nerves** which carry sensory information from the skin and deeper tissue to the CNS (e.g. neurons in the dorsal root ganglion); (2) motor or **efferent nerves** which carry impulses from the CNS—they supply muscles and glands (e.g. neurons in the anterior horn cells of the spinal cord); and (3) **integrative** (e.g. small neurons

Types of neurons

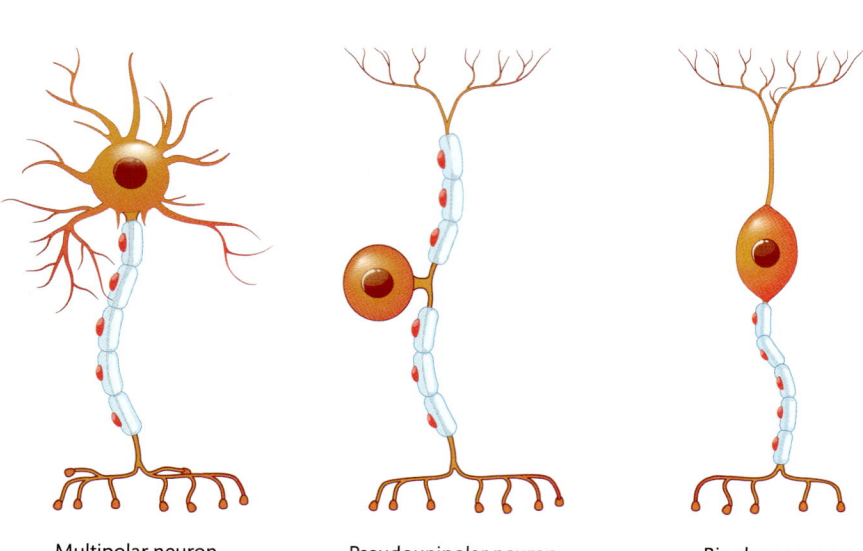

Multipolar neuron   Pseudounipolar neuron   Bipolar neuron

Fig. 1.21 Schematic diagram of multipolar, pseudounipolar, and bipolar neurons.

Designua/Shutterstock.com

that connect the afferent and efferent nerves of a reflex arc).

Nerves

In the course of your dissection, you will encounter a lot of nerves, and your study of the human body includes learning about important nerves in the body like the phrenic nerve, median nerve, vagus, and others. Nerves are made up of large numbers of **nerve fibres** or **axons** bound together in bundles by fibrous tissue. The fibrous tissue forms a delicate sheath—the **endoneurium**—around each nerve fibre. Bundles of nerve fibres are enclosed in a cellular and fibrous sheath—the **perineurium**. And a collection of nerve bundles is enclosed in a dense, fibrous layer—the **epineurium**.

Most nerves of the PNS are myelinated nerves. Each nerve is enclosed in a series of cells—the **Schwann cells**—which are arranged end-to-end on the nerve. In a large-diameter nerve fibre, each Schwann cell forms one segment of a discontinuous, laminated fatty sheath—the **myelin sheath**. Such nerves are referred to as **myelinated nerves** and are white in colour. The gaps between the segments of myelin are known as **nodes**. Thinner nerves are simply embedded in Schwann cells. They are grey in colour and are called **non-myelinated nerves**.

Nerves are described as branching and uniting with one another. It is important to note that this description is misleading, and in reality, there is usually no division of the nerve fibre at the point of branching, and never any fusion of individual nerve fibres at the points of uniting. At points described as branching of a nerve, nerve fibres from the parent stem pass into two or more separate bundles. At points where two nerves seemingly unite, two or more bundles merge into a single one.

Nerves may be classified as: (1) **cranial nerves** when they are attached to the brain (cranial nerves emerge from the skull or cranium); and (2) **spinal nerves** when they arise from the spinal cord. Spinal nerves emerge from the vertebral column through the intervertebral foramina [Fig. 1.19].

Spinal nerves

There are 31 pairs of spinal nerves, named after the groups of vertebrae between which they emerge. There are eight pairs of cervical, twelve pairs of thoracic, five pairs of lumbar, five pairs of sacral, and one pair of coccygeal nerves. All, except the cervical nerves, emerge caudal to the corresponding vertebrae. The first seven cervical nerves emerge cranial to the corresponding vertebrae. The eighth cervical nerve emerges between the seventh cervical and first thoracic vertebrae.

Spinal nerves are attached to the spinal cord by ventral and dorsal roots [Fig. 1.22]. The **ventral root** consists of bundles of efferent fibres which arise from nerve cells in the spinal medulla. The **dorsal root** consists of bundles of afferent fibres and a swelling formed by nerve cells—the **dorsal root** or **spinal ganglion**. The fibres in the dorsal root are processes of the cells in the spinal ganglion.

The ventral and dorsal roots unite in the intervertebral foramen and form the **trunk** of the spinal nerve. The trunk is short and consists of a mixture of efferent and afferent fibres. It divides into a **ventral** and a **dorsal ramus** as it emerges from the intervertebral foramen. (Do not confuse the rami with the roots which form the spinal nerve. Both ventral and dorsal rami contain both efferent and afferent fibres.)

The small **dorsal ramus** passes backwards into the muscle on either side of the vertebral column (erector spinae). Here it divides into lateral and medial branches which supply the muscle, and one of them sends a branch to the overlying skin. These cutaneous branches of the dorsal rami form a row of nerves on each side of the midline of the back [Fig. 6.2].

The large **ventral rami** run laterally from the spinal trunk. In the thoracic region, the thoracic ventral rami run along the lower border of the corresponding ribs. They form the intercostal (between ribs) and subcostal (*below* = rib) nerves (*costa* = rib). Each ventral ramus supplies the muscle of the intercostal space in which it lies, and gives off lateral and anterior cutaneous branches. The lateral and anterior cutaneous branches, together with the cutaneous branch of the dorsal ramus, supply a strip of skin from the posterior median line to the anterior median line. The strip of skin supplied by a single spinal nerve is known as a **dermatome** [Fig. 1.13]. In practice, no area of skin is supplied solely by a single spinal nerve, because adjacent dermatomes overlap. The total mass of muscle supplied by a single spinal nerve is a **myotome**.

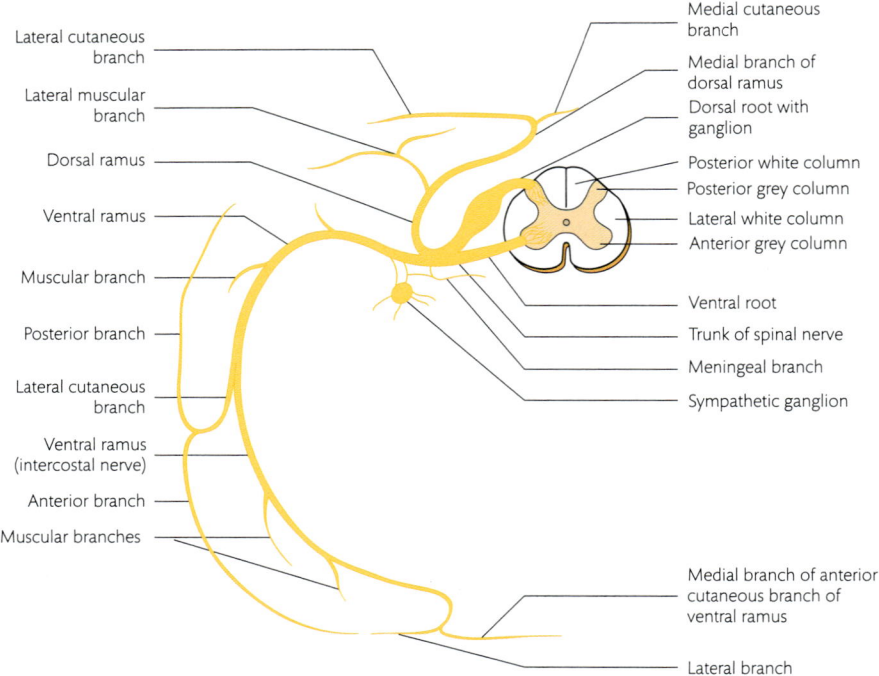

Lateral cutaneous branch

Lateral muscular branch

Dorsal ramus

Ventral ramus

Muscular branch

Posterior branch

Lateral cutaneous branch

Ventral ramus (intercostal nerve)

Anterior branch

Muscular branches

Medial cutaneous branch

Medial branch of dorsal ramus

Dorsal root with ganglion

Posterior white column

Posterior grey column

Lateral white column

Anterior grey column

Ventral root

Trunk of spinal nerve

Meningeal branch

Sympathetic ganglion

Medial branch of anterior cutaneous branch of ventral ramus

Lateral branch

Fig. 1.22 Diagram of a typical spinal nerve.

The ventral rami of the cervical, first thoracic, lumbar, sacral, and coccygeal nerves unite and divide repeatedly to form **nerve plexuses**. The ventral rami of the upper cervical nerves form the cervical plexus. The ventral rami of the lower cervical and first thoracic nerves form the brachial plexus which supplies the upper limb. The lumbar, sacral, and coccygeal ventral rami form plexuses with names which correspond to the names of the ventral rami. The lumbar and sacral plexuses are mainly concerned with the nerve supply of the lower limb.

Loss of motor innervation to skeletal muscle

Motor and sensory innervation to skeletal muscle has been described in the section on muscles. Motor nerves supplying skeletal muscles cause the muscle to contract, and damage to these nerves results in loss of function or weakness of muscles (i.e. paralysis).

➲ In a patient with muscle weakness, it is important to determine whether the primary problem is in the muscle or the nerve supplying it. Electromyography is a diagnostic procedure where the non-functioning muscle is stimulated to contract by applying an electrical impulse to the skin overlying the neurovascular hilum—the motor point. A denervated, but other-wise healthy, muscle will contract in response to the electrical stimulus. A dystrophic muscle will not.

➲ In a patient with suspected nerve injury, a neurological examination requires testing of individual spinal nerves. By testing the strength and contraction of selected muscles, the neurologist can ascertain the integrity of the spinal nerve supplying it. For example, if the strength and tone of the biceps brachii are good, it indicates that the C5 nerve is intact [see Tables 12.11 and 24.11].

Synapses

Synapses are junctions between two or more neurons where they communicate with each other. At least two neurons are involved in a synapse—the presynaptic neuron and the post-synaptic neuron. The nerve impulse from the presynaptic neuron is transmitted to the post-synaptic neuron across the synaptic cleft. When the impulse reaches the synapse through the presynaptic neuron, neuro-chemicals are released into the synaptic cleft which stimulate post-synaptic receptors.

Based on the part of the neurons involved, synapses can be classified into axosomatic, axodendritic, and axo-axonal synapses. Axodendritic synapses are most common. They are also classified according to function into excitatory or inhibitory synapses. Synapses can be chemical or electrical.

Differences between sympathetic and spinal ganglia

A collection of nerve cell bodies outside the CNS constitutes a ganglion. Sympathetic ganglia are seen on the sympathetic chain, and in relation to the front of the aorta. They are motor in function. Under the microscope, these neurons are seen to be multipolar and scattered among the nerve fibres. They receive impulses through the preganglionic nerve that arises within the CNS and transmit to the post-ganglionic nerve which supplies the smooth muscle or gland under its control.

Spinal or sensory ganglia are located on sensory nerves. They are made up of pseudounipolar neurons, the peripheral processes of which pick up sensations from the periphery and transmit them through the central process to the CNS. (No synapse occurs in the spinal ganglia. When viewed under the microscope, the neurons are seen to be in groups separated by bundles of nerve fibres.)

The cadaver as our first teacher

For the medical student, dissection is an important way of getting a fuller understanding of the structure and function of the human body. It aids in learning simple structures like the valves in the veins, and more complex ones like the heart. Without a sound knowledge of these, normal and abnormal circulation of the blood through the body cannot be properly understood. Similarly, knowledge of movements occurring at joints, the muscles which cause them, and the nerves innervating these muscles is essential to understand and address the effects of injury or disease in any of these elements of the musculoskeletal system.

As you enter the dissection hall to start your study of gross anatomy, you will have your first and lasting encounter with the cadaver. The cadaver is your first patient, as well as your first teacher. Human dissection will help you observe the body as a whole being and not just as the bones and organs. You will have the opportunity to make careful observations without medical equipment, and be able to formulate your observations into plausible hypotheses. You may also be able to identify how lifestyle might have contributed to chronic disease, as evidenced by the condition of the body. At all times, be appreciative of the contribution to your education made by a once living person and treat the cadaver with respect.

General instructions for dissection

The cadaver for dissection is embalmed with fixatives to preserve it. The whole body has been kept moist by storing it with moist wrappings or immersing it in fluid. Be careful to ensure that the body is kept moist during the entire time you will be working on it.

Dissection instruments

You will require a scalpel, two pairs of forceps, preferably with rounded tips, and a strong blunt hook or seeker. In addition, you may find it useful to have a hand lens to examine small structures and a torchlight.

Organization of dissections and commonly used terms

Under the skin, the human body consists of a number of structures embedded in fibrous connective tissue which varies in density, from a loose mesh to tough sheets of fibres. Dissection is the process of freeing the structures from this tissue and demonstrating them clearly.

In this manual, dissections are organized in a stepwise manner to enable you to systematically explore the region under study and learn important anatomical details. The objectives for each dissection are stated at the start to enable you to focus your attention on the particular area. Before you start dissecting, learn the meaning of the following terms:

1. Dissect—to cut or tear apart. In the laboratory, dissecting an area requires you to separate the tissue in such a way as to expose the structure under study—muscle, vessel, nerve, etc. This can be done best by blunt dissection with a hook or a pair of forceps by isolating and pulling the structure through loose layers of connective tissue. In this way, it is possible to free organs without damaging blood vessels or nerves. Use of sharp instruments, such as scalpels or scissors, should be reserved for cutting the skin and dense layers of the deep fascia which enclose many organs and partially conceal them.

2. Cut or transect—to divide by using a sharp instrument, usually to expose deeper lying structures.

3. Clean—to remove fat and fascia from the surface of a muscle, or to define the edge of a muscle, or to remove the connective tissue covering of a nerve or vessel.

4. Define—to remove the connective tissue masking the border of a structure, so that the extent of the structure is more clearly seen.

5. Retract—to pull aside or separate one structure from an adjacent structure. It is a temporary displacement done to visualize an underlying structure.

6. Reflect—to fold back a cut structure, usually skin or transected muscle.

Removal of the skin

Remove the skin from the superficial fascia in a series of flaps which can be replaced to prevent drying of the part. It is probably better to cut through both the skin and superficial fascia and remove both of them in one layer from the underlying deep fascia by blunt dissection. The blood vessels and nerves entering the superficial fascia through the deep fascia are easily found in this way and can be traced for some distance. The alternative of searching for their minute branches in the superficial fascia is a tedious, and often unrewarding, process. The student should be aware that the distribution of cutaneous nerves is of considerable clinical importance, but this is best learnt by reference to diagrams, except in the case of the larger branches which are easily followed. In the superficial fascia, the nerves are almost always accompanied by a small artery and one or more minute veins. Larger veins are also found in the superficial fascia. They run a solitary course to pierce the deep fascia and drain into the deep veins. At such junctions, these superficial veins contain valves which prevent the reflux of blood from the deep veins.

Deep dissection

When the deep fascia has been uncovered and examined, proceed to remove it. This is made difficult because it sends fibrous sheets between the muscles, enclosing each in a separate compartment. Where a number of muscles arise together, the walls of these compartments also give origin to the muscle fibres. Thus, they form a tendinous sheet which appears to bind together adjacent muscles. In other places, it is relatively easy to strip the deep fascia from muscles, because only delicate strands pass between the individual bundles of muscle fibres. It is important to follow each muscle to its attachments and to define these accurately, for it is only in this way that the action of a muscle can be determined.

As each muscle is exposed and lifted from its bed, look for the neurovascular bundle entering it. Follow the structures in the neurovascular bundle back to the main nerve trunk and vessel from which they arise. In many situations, the arteries are accompanied by tributaries of the main vein which often obscure the artery and nerve. In these cases, it is advisable to remove the vein, so that a clearer view of the artery and nerve can be obtained. In any case, it will be found that there are usually multiple venous channels and that their arrangement is much less standard than that of the arteries. The arteries are less constant in their arrangement than the nerves.

Variations

We all know that the external appearance of individuals varies greatly. The same type of variation exists in the size, position, and shape of the internal organs among different individuals. Therefore, no single account of the structure of the body exactly fits every individual, and students must expect to find variations from the descriptions given in this book. For this reason, students should take every opportunity to look at the other bodies being dissected at the same time. Some of the variations are of considerable clinical importance (e.g. differences in the anastomotic arrangement between the arteries at the base of the brain), whereas others have little significance (e.g. an extra belly of a particular muscle or a marked difference in the arrangement of superficial veins, even on the two sides of the same body). Congenital anomalies, variations that arise from defects in intrauterine development, are not commonly seen in the dissection hall. Many of these are so severe that they lead to early death. Other congenital anomalies may be present throughout life without any overt sign. The student should understand the main processes of development and the effects of its abnormalities on the structure and function of the various systems.

Anatomy of the living body

In the dead preserved body, the texture and appearance of the organs have been altered. The student should remember that the purpose of studying formalin-fixed cadavers is just a tool to help visualize the living body in action, so that the effects of injury or disease can be appreciated and abnormalities can be recognized. Dissection is only a means to the end of a fuller understanding of function. In addition to studying the body by dissection, the living body should be observed and palpated.

Special radiological techniques

An increasing number of techniques are being established to visualize the internal structure of the body without surgical intervention. Of these, the oldest is the use of X-rays.

X-rays

This technique depends on the differential absorption of X-rays by the various tissues of the body on their way through it to a sensitive film (or other recording apparatus) which is blackened by development in direct proportion to the amount of X-rays reaching it. Thus, if the exposure is correct, the film is deeply blackened outside the area shaded by the body and is completely clear where a tissue intervenes which absorbs all the penetrating X-rays. Compact bone and teeth are the most absorbent tissues, whereas the lungs, windpipe, and intestine which contain air are the least. Most other tissues, including those containing fluid, have intermediate absorption per unit of thickness, except fat which has a lower level of absorption, though not as low as air.

Each point on the film (radiograph) has a density (blackness) which is directly proportional to the amount of X-rays reaching it, and hence inversely proportional to the sum of absorptions of all tissues which lie between the source of X-rays and that point on the radiograph. Hence, on a posteroanterior radiograph of the chest, the lung fields are dark because the air they contain does not absorb X-rays as much as the vertebral column, the breast bone (sternum), and the tissues lying between the two lungs. Also, absorption by the ribs, added to absorption by tissue in the lungs, etc., makes the ribs appear as lighter strips in the lung fields, and the fuzzy, whiter areas in the medial parts of each lung are due to absorption by the larger blood vessels (fluid-filled) in these parts of the lungs.

On examining a plain X-ray of the chest, the following points should be noted.

1. The outlines of the heart and great vessels which arise from these are obvious on the right (left side of the patient) of the median white area, because they project beyond the vertebral column and sternum into the lung fields and make a sharp contrast in absorption by comparison with that on the left where the vertebral column and sternum overlap the heart and vessels.
2. On each side, the lower part of the lung field becomes lighter.
3. Air introduced into the abdominal (peritoneal) cavity makes the lower surface of the diaphragm (a thin partition between the thorax and the abdomen) and the upper surfaces of the organs immediately below it obvious, though, without that air, only the upper surface of the diaphragm would have been seen because of its contrast with the lungs. The presence of air in the upper part of the peritoneal cavity shows that this radiograph was taken with the patient in the erect position.

If the intensity of the X-rays or the length of exposure had been increased, the lung fields would have become darker and the breast and rib 'shadows' much less obvious, though some detail of the vertebral column would have been visible. Conversely, it is possible to show minor differences in tissue absorption (e.g. fat vs tumour), most easily with low-intensity X-rays.

Thus, radiographs show the outline of structures where there is a change in X-ray absorption but cannot show the outline of two adjacent structures which have the same X-ray density. Clearly, any hollow organ (e.g. blood vessel, gut tube, etc.) which can be filled with a substance that absorbs X-rays more effectively than bone (e.g. heavy metals such as barium) or less effectively than the surrounding tissue (e.g. air) can make its outline obvious. The combination of both (double contrast), where a small amount of X-ray-opaque material is introduced into a cavity followed by air, allows the first to outline the internal surface, so that irregularities are made visible by the contrast with air. (Using the X-ray-opaque material alone merely produces a silhouette.)

Three more recent techniques all produce pictures representing slices taken through the body at any desired level. All of these depend on special features of the tissues and can be recorded in digital form and reproduced in any desired manner that the information permits.

Computerized tomography

Computerized tomography (CT) is a technique which uses X-rays and depends on the differences in absorption by different tissues—a feature which can be enhanced by the introduction of special materials such as iodinated contrast media. Once a series of transverse sections have been made and the results recorded in digital form, it is possible to combine these in the computer and construct images in a different plane (such as sagittal or coronal, or even three-dimensional (3D), reconstructions), if required. By varying the timing of image acquisition following intravenous administration of iodinated contrast media, CT images can be obtained in arterial, venous, and delayed phases. CT angiograms of different arterial systems can be easily obtained.

Ultrasound

Ultrasound (sonar) can also be used to produce electronic pictures which represent the reflection of ultrasound from surfaces where one tissue meets another. Unfortunately, ultrasound cannot be used where there is air or bone, because it is not transmitted through these, but it has the advantage that there is no evidence, unlike X-rays, of it having any damaging effects on even the most sensitive tissues. It is used therefore as a method of choice to scan the pelvis where there is the possibility of pregnancy and to determine gross abnormalities at an early stage. The method consists of passing an ultrasound transmitter and recorder over the skin and showing the computerized reflection on a cathode ray tube. The pictures produced are more difficult to read than CT images but represent sections taken through the body under the path of travel of the instrument. They are most useful when shown as a continuous recording (real-time) which can take account of movements of tissues such as heart valves.

Magnetic resonance imaging

In magnetic resonance imaging (MRI), protons in the tissue are made to resonate in a strong magnetic field by subjecting them to an appropriate radio wavelength. Such resonations are recorded and computed, and converted electronically to pictures. Different areas in these pictures show different intensities, depending on the amount of protons or water in the tissues—their T1 and T2 relaxation times. Imaging parameters can be adjusted to obtain different pulse sequences (such as T1-weighted, T2-weighted, fluid-attenuated inversion recovery (FLAIR), etc.), with varying image contrast. MRI has better soft tissue contrast than CT scans and does not use harmful ionizing radiation like CT. By using magnetic resonance (MR) spectroscopy, it is possible to assess the biochemical composition and metabolism of the tissue. Specific parts of the brain that are activated when performing certain tasks can be mapped by using functional MRI. MR angiography images can be obtained with or without the use of contrast media (gadolinium-based agents). The main disadvantages of MRI are limited availability, the expense involved, and the longer time taken in image acquisition.

CHAPTER 2
MCQs on general anatomy

The following questions have four options. You are required to choose the most correct answer.

1. **The mid-sagittal plane is an imaginary plane which divides the body into:**
    A. anterior and posterior parts
    B. medial and lateral parts
    C. right and left parts
    D. superior and inferior parts

2. **Which of the following is true of blood capillaries?**
    A. they are absent from the dermis
    B. they are approximately 8–10 μm in diameter
    C. they are lined by mesothelial cells
    D. they do not allow exchange of gases

3. **Which cell type produces myelin in the central nervous system?**
    A. astrocytes
    B. microglia
    C. oligodendrocytes
    D. Schwann cells

4. **Bones are classified according to their shape. Which of the following are short bones?**
    A. carpals
    B. metacarpals
    C. metatarsals
    D. phalanges

5. **Which of the following is a uniaxial joint?**
    A. condyloid joint
    B. ellipsoid joint
    C. pivot joint
    D. saddle joint

6. **Which one of the following is true regarding endochondral ossification?**

   A. at the growing end of the long bone, the epiphysis usually appears later and fuses with the body later than at the other end

   B. the epiphysis is the bone developed from the primary centre of ossification

   C. the epiphysis is the zone of active ossification

   D. the primary centre of ossification appears after birth in short and irregular bones

7. **The extensors of the limb are innervated by which part of the spinal nerve?**

   A. the anterior division of the ventral ramus

   B. the dorsal ramus

   C. the dorsal root

   D. the posterior division of the ventral ramus

8. **An example of a multipennate muscle is the:**

   A. deltoid

   B. dorsal interossei

   C. flexor pollicis longus

   D. palmar interossei

9. **Which of the following muscles act as an antagonist during flexion of the elbow?**

   A. biceps brachii

   B. brachialis

   C. pronator teres

   D. triceps brachii

10. **Which of the following is true of the articular cartilage in synovial joints?**

    A. does not ossify

    B. has a rich blood supply

    C. lined by synovial membrane

    D. made of fibrocartilage

Please see Chapter 27 for the answers.

Find additional MCQs online by searching for *Cunningham's Manual of Practical Anatomy Volume 1 General Anatomy, Upper and Lower Limbs*, 17th edition at https://academic.oup.com/ and go to the online appendix at the end of the book. Use your scratch-off code on the inside cover to access the material. The code will work for 12 months.

PART 2

The upper limb

CHAPTER 3
Introduction to the upper limb

Parts of the Upper Limb

There are four parts to the upper limb: the shoulder, the arm or brachium, the forearm or antebrachium, and the hand.

The term **shoulder** includes a number of smaller regions: the shoulder joint, the axilla or armpit, the scapular region around the shoulder blade, and the pectoral or breast region on the front of the chest. The scapula (or shoulder blade) and the clavicle (or collarbone) are the bones of the shoulder girdle. The scapula and clavicle articulate with each other at the acromioclavicular joint, but the only articulation between the shoulder girdle with the rest of the skeleton is the articulation of the clavicle with the upper end of the sternum at the sternoclavicular joint. The mobile scapula is otherwise held in position entirely by muscles.

The **arm** is the part of the upper limb between the shoulder and the elbow. The arm bone is the humerus, and it articulates with the scapula at the shoulder joint, and with the radius and ulna at the elbow joint.

The **forearm** extends from the elbow to the wrist. Its bones—the radius and ulna—articulate with the humerus at the elbow joint, as mentioned above, and with each other at the radioulnar joints, and distally the radius (but not the ulna) articulates with the carpal bones at the radiocarpal joint. In the anatomical position (supine position of the forearm), the bones are parallel, and the radius is lateral to the ulna. When the palm of the hand faces posteriorly (prone position of the forearm), the distal end of the radius has rotated around the distal end of the ulna, so that the radius lies obliquely across the ulna. These movements are called **pronation** and **supination**, respectively.

The **hand** consists of the wrist or carpus, the hand proper or metacarpus, and the digits (thumb and fingers). The eight small wrist, or carpal, bones are arranged in two rows (proximal and distal), each consisting of four bones. The carpal bones articulate: (1) with one another at the intercarpal joints; (2) proximally, with the radius at the radiocarpal joint; and (3) distally, with the metacarpal bones at the carpometacarpal joints. The articulation of the carpal bones with the radius accounts for the movement of the hand with the radius in pronation and supination. The small movements that occur at each of these joints add up to allow a considerable range of movement. Posteriorly, the carpal bones are close to the skin, but anteriorly they are covered by muscles of the ball of the thumb (**thenar eminence**) and of the little finger (**hypothenar eminence**), and, between these, by the long tendons entering the hand from the forearm.

The hand proper has five metacarpal bones numbered 1 to 5, beginning from the thumb side. Proximally, the base of the metacarpal bones articulates with the distal row of carpal bones (carpometacarpal joints), and the second to fifth metacarpal bones also articulate with each other (intermetacarpal joints). Distally, each metacarpal bone articulates with the proximal phalanx of the corresponding digit.

The **digits** are: the thumb or pollex, the forefinger or index, the middle finger or digitus medius, the ring finger or annularis, and the little finger or minimus. Each finger has three phalanges—the thumb has only two. The proximal phalanx of each finger articulates with the corresponding metacarpal head at the metacarpophalangeal joint. The phalanges articulate with one another at the proximal and distal interphalangeal joints.

Adjacency of the thorax, neck, and upper limb

The walls of the thorax form a conical structure, which is flattened anteroposteriorly. It has an apex superiorly that is cut obliquely to form the **superior aperture of the thorax**. This is continuous above with the root of the neck and has, as its margins, the first thoracic vertebra, the first ribs, and the upper part of the sternum (manubrium). The upper limb is attached to the trunk by muscles and bones, which spread out from the proximal part of the limb to the anterior and posterior surfaces of the thorax.

CHAPTER 4
Osteology of the upper limb

Introduction

The study of the upper limb begins with the study of the bones: the clavicle, scapula, humerus, radius, ulna, and hand bones. By the end of the study, the student should be able to name each bone and determine its position in the body, identify the side to which it belongs, hold it in the anatomical position, describe the important bony features, and demonstrate important muscle attachments on the bone.

Clavicle

General features of the clavicle

The clavicle is a long bone that extends laterally and slightly upwards from its articulation with the sternum (sternoclavicular joint) to its articulation with the scapula (acromioclavicular joint) [Fig. 5.6]. It has three parts: the medial or sternal end, the shaft, and the lateral or acromial end. The **medial end** is rounded and articulates with the manubrium sternum and first costal cartilage. The **shaft** has a double curve and is convex forwards in its medial two-thirds and concave forwards in its lateral one-third [Fig. 4.1]. The medial part of the shaft has four surfaces: anterior, posterior, superior, and inferior surfaces. In contrast to the medial part, the lateral part of the shaft is flat and has superior and inferior surfaces The **lateral end** is flat and articulates with the medial surface of the acromion of the scapula at the acromioclavicular joint. It has well-defined anterior and posterior borders.

Bony landmarks and muscle attachments

By using Figs. 4.1, 4.2, and 4.3, and dry bones, identify the following bony landmarks and muscle attachments on the clavicle.

Medial end

The rounded medial end has an articular facet for articulation with the manubrium sternum on the medial surface. Inferiorly, the medial end of the clavicle articulates with the first costal cartilage and has the markings for the costoclavicular ligament [Fig. 4.2].

Shaft

On the shaft, the medial half of the anterior surface gives origin to the clavicular head of the pectoralis major; and the lateral one-third gives origin to the deltoid. The medial part of the superior surface gives origin to the sternocleidomastoid [Fig 4.3]. Most of the posterior surface is smooth and lies in front of the important nerves and vessels entering the axilla. The lateral one-third of this surface gives insertion to the trapezius. In the middle of the inferior surface is a well-defined groove for insertion of the subclavius, and the foramen for the nutrient artery [Fig. 4.2]. The lateral end of the inferior surface overlies the coracoid process of the scapula. It bears the conoid tubercle—near the posterior border—and the trapezoid line, directed laterally and forwards from the conoid tubercle. The tubercle and line give attachment to the conoid and trapezoid parts of the powerful coracoclavicular ligament which attaches the clavicle to the coracoid process.

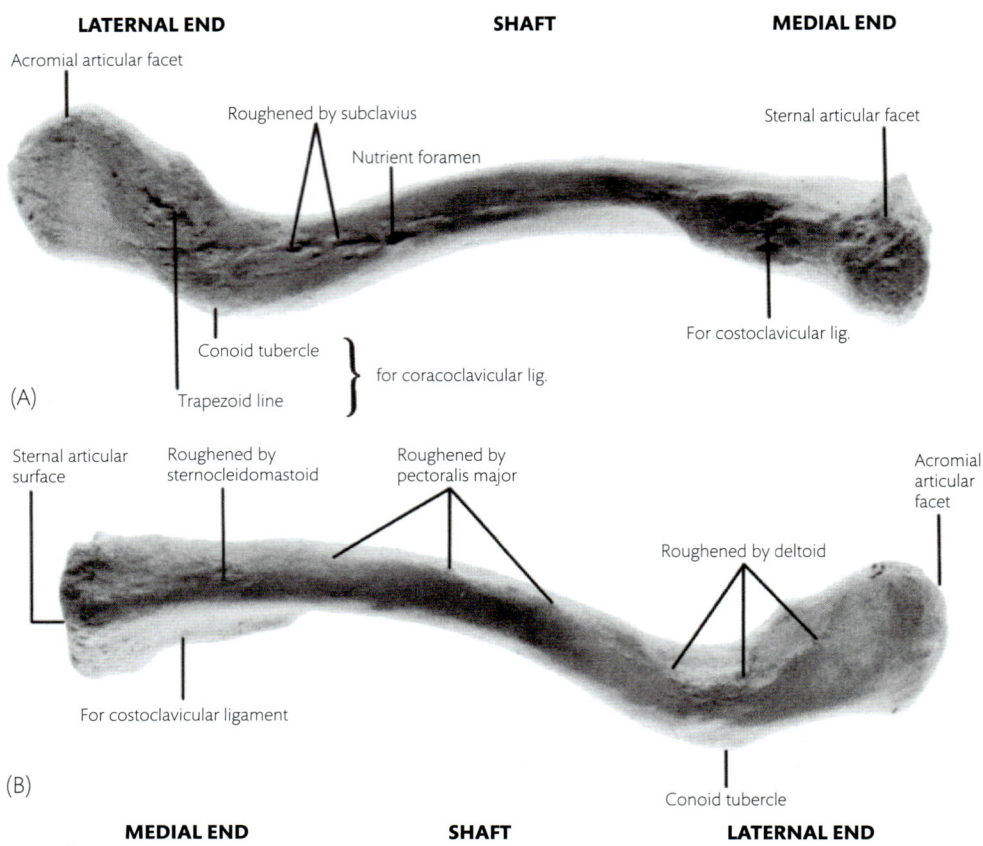

Fig. 4.1 (A) Right clavicle, inferior surface. (B) Right clavicle, superior surface.

Lateral end

The lateral end of the clavicle has a small articular facet for articulation with the acromion of the scapula.

Unique features of the clavicle

The clavicle has some unique features. It is:

- Subcutaneous throughout its length
- The first bone to start ossifying
- The only long bone that ossifies in membrane

- The only long bone to have two primary centres of ossification
- The only long bone placed horizontally
- The most commonly fractured long bone, and
- It can be pierced by cutaneous nerves.

Applied anatomy

The clavicle acts as a strut/support which transmits forces from the upper limb to the trunk and prevents the scapula, and hence the shoulder, from

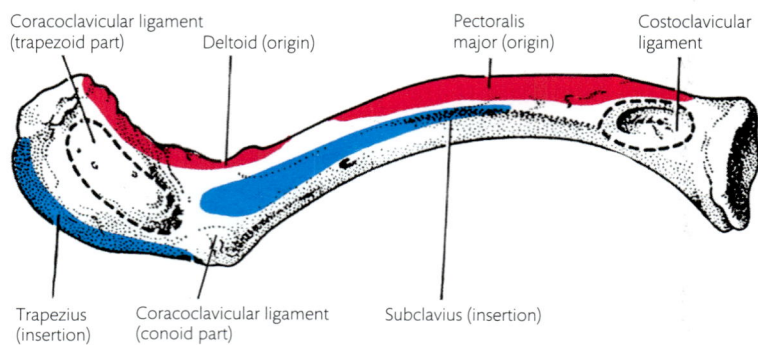

Fig. 4.2 Muscle attachments of the inferior surface of the right clavicle.

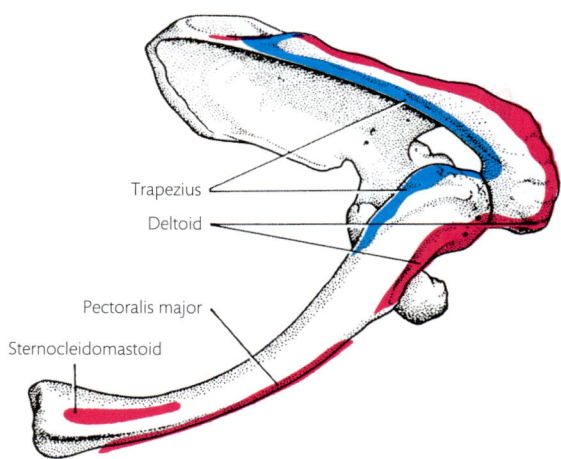

Fig. 4.3 Shoulder girdle from above, showing muscle attachments.

Trapezius
Deltoid
Pectoralis major
Sternocleidomastoid

2. The medial two-thirds of the shaft is convex forwards
3. The subclavius groove and conoid tubercle are directed inferiorly.

Exercise 4.1B: to hold the clavicle in the anatomical position

Hold the clavicle in the anatomical position and in the hand of the side to which it belongs.

Scapula

General features of the scapula

The **scapula** is a thin, triangular plate of bone which lies at a tangent to the posterolateral surface of the thorax over the second to seventh ribs [Fig. 4.4]. It is roughly triangular in shape and has two surfaces—ventral or costal and dorsal surfaces; three borders—superior, medial, and lateral borders; three angles—superior, lateral, and inferior angles; and three projections or processes—spinous, acromion, and coracoid processes. The most striking feature on the dorsal surface is a shelf-like projection—the **spine of the scapula** [Figs. 4.4, 4.5].

sagging downwards and medially under the weight of the limb. Sagging down of the upper limb is seen when the clavicle is fractured.

Exercise 4.1A: to determine the side of the clavicle

Identify the side of the clavicle by using the following instructions. Hold the clavicle such that:

1. The rounded sternal end is directed medially

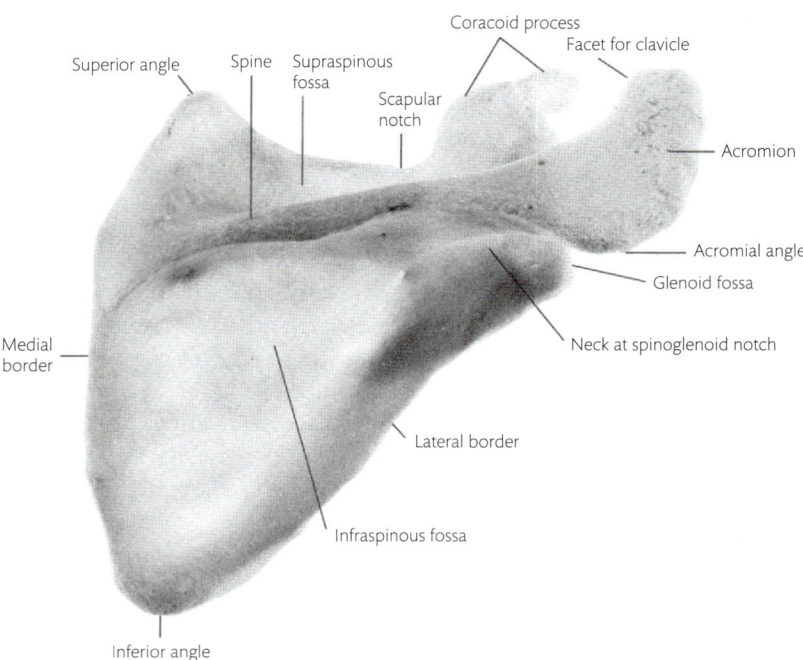

Superior angle
Spine
Supraspinous fossa
Scapular notch
Coracoid process
Facet for clavicle
Acromion
Acromial angle
Glenoid fossa
Neck at spinoglenoid notch
Medial border
Lateral border
Infraspinous fossa
Inferior angle

Fig. 4.4 Dorsal surface of the right scapula.

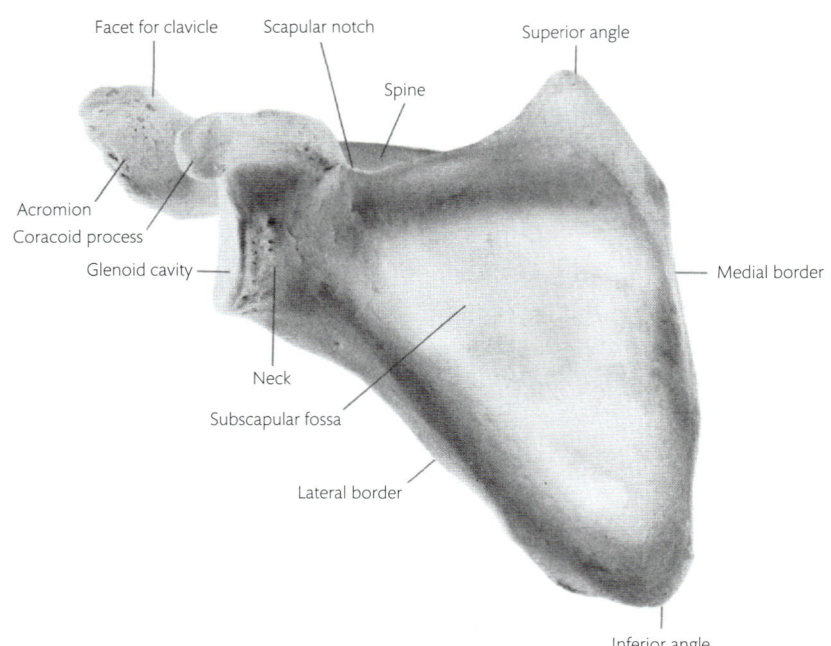

Fig. 4.5 Ventral surface of the right scapula.

The lateral angle is the thickest part of the bone. It is truncated (the apex of the angle is cut), so that the end forms the shallow, pear-shaped **glenoid cavity** for articulation with the head of the humerus at the glenohumeral or shoulder joint. The glenoid fossa has an oval articular surface which is slightly indented at the upper anterior margin. The glenoid cavity is continuous with the rest of the scapula through the **neck of the scapula** and faces anterolaterally to articulate with the hemispherical head of the humerus.

The thin, concave **superior border** of the scapula extends from the glenoid fossa to the superior angle. The root of the coracoid process lies at the lateral end. The long, slightly thickened **medial border** extends from the superior to the inferior angle. The **lateral border**, between the inferior angle and the glenoid fossa, is thick and rounded. The **costal** or **ventral surface** is concave and the **dorsal surface** is convex. The **spine of the scapula** is attached to the dorsal surface. The **acromion process** is continuous with the spine of the scapula and overhangs the glenoid fossa. It articulates with the lateral end of the clavicle. The **coracoid process** arises from the lateral end of the superior border. The tip of the coracoid is directed anteriorly.

Bony landmarks and muscle attachments

By using Figs. 4.3, 4.4, 4.5, 4.6, and 4.7, and dry bones, identify the following bony landmarks and muscle attachments on the scapula:

Ventral surface

Much of the ventral or costal surface of the scapula is concave and has rough markings for the attachment of the subscapularis. Most of it (excluding a small area at the root of the coracoid process) gives origin to the subscapularis. The serratus anterior is attached all along the medial border [Fig. 4.6].

Dorsal surface

The dorsal surface of the scapula is divided into the **supraspinous** and **infraspinous fossae** by the **spine of the scapula**. The spine runs from the medial border of the scapula to its neck and increases in height laterally. The posterior surface of the spine (**crest of the spine**) widens laterally to become the flattened **acromion** which projects forwards at the palpable **acromial angle** [Fig. 4.2]. The supraspinous and infraspinous fossae are continuous with each other laterally between the lateral border of the spine and the glenoid fossa through the **spinoglenoid notch**. The supraspinatus and infraspinatus are attached to the medial

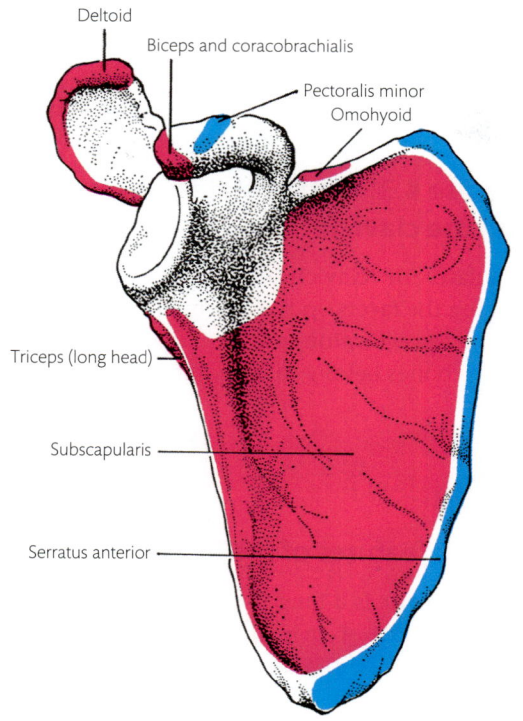

Fig. 4.6 Muscle attachments to the costal surface of the right scapula.

Superior border

The thin **superior border** extends between the lateral and superior angles. The **root of the coracoid process** lies at its lateral end, just medial to the glenoid fossa. Medial to the root of the coracoid is the **scapular notch**. The inferior belly of the omohyoid takes origin from the ligament that spans the scapular notch and the adjoining bone.

Lateral border

The thick **lateral border** extends from the glenoid cavity to the inferior angle. A small elevation—the **infraglenoid tubercle**—lies adjacent to the inferior border of the glenoid cavity and gives origin to the long head of triceps.

The **medial margin** is thin and extends from the superior to the inferior angle. It gives attachment on its dorsal aspect to the levator scapulae in the area between the superior angle and the root of the spinous process, the rhomboid minor at the root of the spinous process, and the rhomboid major between the root of the spinous process and the inferior angle [Fig. 4.7].

Lateral angle

The **lateral angle** is broad and truncated, and includes the **glenoid fossa**, neck, and coracoid process of the scapula. In the anatomical position, the glenoid fossa is directed laterally and forwards. A small elevation at the upper margin of the glenoid fossa is the **supraglenoid tubercle** and gives origin to the long head of the biceps brachii. The slightly constricted part of the scapula medial to the glenoid fossa is the **neck of the scapula**.

Coracoid process

The **coracoid** is a thick buttress which extends anterosuperiorly from the neck of the scapula and the upper part of the glenoid cavity to end in the **coracoid process**. The tip of the coracoid process gives attachment to the short head of the biceps brachii and the coracobrachialis. The pectoralis minor is inserted onto the superior surface of the coracoid. On the superior aspect of the coracoid process, note the roughened area for attachment of the coracoclavicular ligaments [Fig. 4.6]. The coracoid process arches forwards, leaving a space beneath. The subscapularis passes from the subscapular fossa of the scapula beneath the coracoid process.

Spinous process

The anterior border of the spine is attached to the dorsal surface of the scapula. The posterior border is free

---

parts of the supraspinous and infraspinous fossae. The supraspinatus is also attached to the superior surface of the spine of the scapula. Close to the lateral border, the dorsal surface gives attachment to the teres minor in the upper part, and to the teres major in the lower part. Some fibres of the latissimus dorsi take origin from the dorsal surface of the inferior angle [Fig. 4.7].

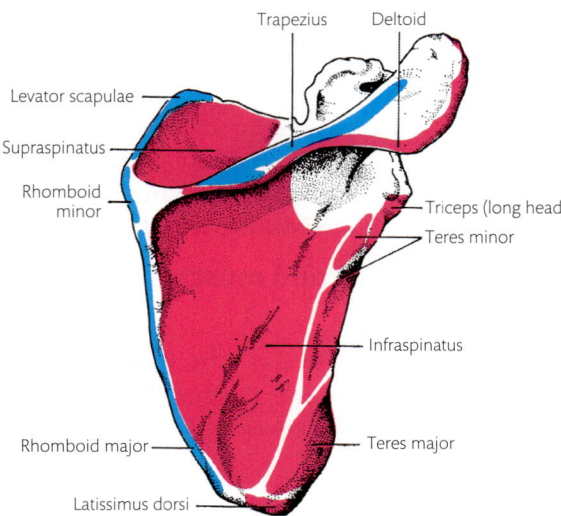

Fig. 4.7 Muscle attachments to the dorsal surface of the right scapula.

and forms the **crest of the spine of the scapula**. The crest of the spine of the scapula gives insertion to the trapezius at its upper border, and origin to the deltoid at its lower border. The lateral border of the spine stops short of the glenoid fossa, leaving a **spinoglenoid notch** between the spine and the glenoid fossa. At the medial end, the spine meets the medial border of the scapula in a small, smooth triangular area.

Acromion process

The **acromion process** is continuous with the spine of the scapula and projects forwards, overhanging the glenoid cavity. The superior surface is subcutaneous. The lateral margin gives origin to the deltoid and the medial margin gives insertion to the trapezius. By using Fig. 4.3, appreciate how the trapezius and deltoid have a continuous attachment to the spine of the scapula, acromion, and lateral part of the clavicle. The small articular facet on the medial surface of the acromion articulates with the acromial end of the clavicle. The acromion process arches forwards, leaving a space beneath, and the supraspinatus passes from the supraspinous fossa through this space.

The acromion and coracoid form two parts of an incomplete bony arch above the glenoid cavity and the head of the humerus which articulates with it. This **coracoacromial arch** is completed by the **coracoacromial ligament**. Above the coracoacromial ligament, the lateral end of the clavicle articulates with the acromion.

Applied anatomy

The scapula lies posterior to the axilla and is almost entirely covered by muscles. Movements of the scapula are limited only by its articulation with the clavicle and, through it, with the sternoclavicular joint around which these movements take place. The scapula slides freely on the thoracic wall in the absence of bony articulations between it and the wall. The muscles of the scapula either attach the scapula to the humerus or hold it against the thorax.

Exercise 4.2A: to determine the side of the scapula

Identify the side of the scapula by holding the scapula such that:

1. The glenoid cavity is directed laterally
2. The smallest (most acute) angle is directed inferiorly
3. The spine of the scapula is directed dorsally.

Exercise 4.2B: to hold the scapula in the anatomical position

Hold the scapula in the anatomical position and in the hand of the side to which it belongs.

Exercise 4.3: to articulate the scapula with the clavicle

Articulate the clavicle and scapula together by bringing the facets on the acromial end of the clavicle and the acromion together [Fig. 4.3] (make sure the two bones belong to the same side).

Humerus

General features of the humerus

The **humerus** is a long bone, and the bone of the arm. It consists of an **upper end**, a **shaft**, and a **lower end** [Fig. 4.8]. The expanded upper end of the **humerus** consists of the hemispherical articular **head**, which is directed medially, posteriorly, and upwards to articulate with the glenoid fossa of the scapula at the shoulder joint. The **greater tubercle** is directed laterally and the **lesser tubercle** is directed anteriorly. The head is separated from the two tubercles by a shallow groove—the **anatomical neck of the humerus**. The upper end is separated from the shaft of the humerus by the narrow **surgical neck of the humerus**. The **shaft** has three surfaces—**anteromedial**, **anterolateral**, and **posterior surfaces**; and three borders—**anterior**, **medial**, and **lateral borders**. In the lower third of the shaft, the medial and lateral borders become more clearly defined and form the **medial** and **lateral supracondylar ridges**. The lower end of the humerus articulates with the upper ends of the radius and ulna to form the elbow joint. There are four bony prominences on the lower end. From lateral to medial, they are the **lateral epicondyle**, **capitulum**, **trochlea**, and **medial epicondyle**.

Bony landmarks and muscle attachments

By using Figs. 4.8, 4.9, and 4.10, and dry bones, identify the following bony landmarks and muscle attachments on the humerus.

Upper end

The head is convex and rounded, and forms less than half of a sphere. The large articular surface of

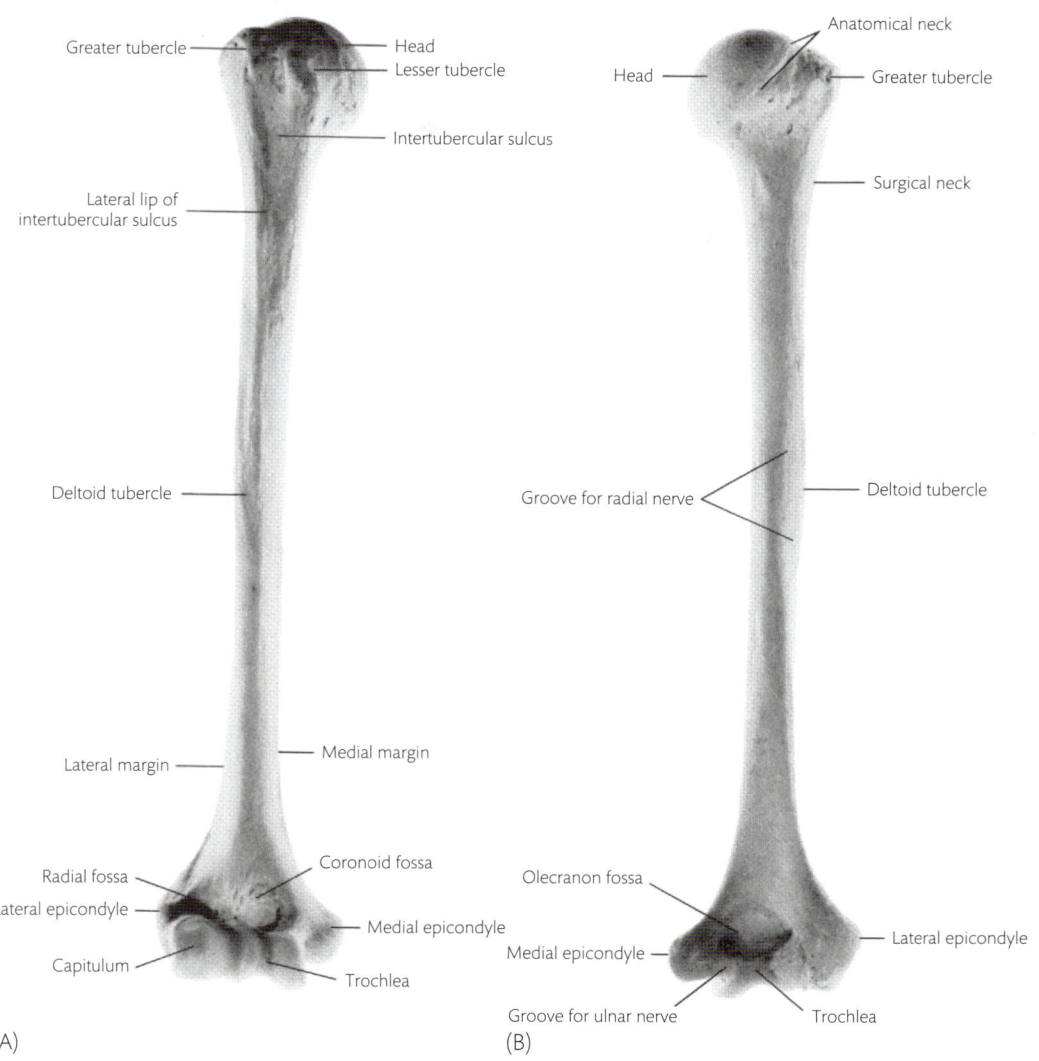

Greater tubercle — Head
— Lesser tubercle
— Intertubercular sulcus

Lateral lip of
intertubercular sulcus

Deltoid tubercle

Lateral margin — — Medial margin

Radial fossa — — Coronoid fossa
Lateral epicondyle
Capitulum — — Medial epicondyle
— Trochlea

(A)

Anatomical neck
Head — — Greater tubercle

— Surgical neck

Groove for radial nerve — — Deltoid tubercle

Olecranon fossa
Medial epicondyle — — Lateral epicondyle
Groove for ulnar nerve — Trochlea

(B)

Fig. 4.8 Right humerus. (A) Anterior aspect. (B) Posterior aspect.

the head articulates with the glenoid fossa of the scapula. The **greater tubercle** is the most lateral bony part of the shoulder and is overhung by the acromion process of the scapula. It has three smooth facets on the posterior–superior surface. The supraspinatus is attached on the **superior facet**, the infraspinatus on the **posterosuperior** facet, and the teres minor on the **posterior** facet. The smaller, **lesser tubercle** of the humerus is directed anteriorly. It gives insertion to the subscapularis. The tubercles are separated from each other by a shallow groove, the **intertubercular sulcus** or **bicipital groove**. The sulcus is limited on either side by the **medial** and **lateral lips of the intertubercular sulcus**. Traced superiorly, the medial lip of the intertubercular sulcus continues with the

lesser tubercle and is also called the **crest of the lesser tubercle**. The lateral lip of the intertubercular sulcus continues with the greater tubercle and is also called the **crest of the greater tubercle**. The teres major is attached to the medial lip of the intertubercular sulcus, and the pectoral major to the lateral lip. The latissimus dorsi is attached to the floor of the intertubercular sulcus. The tendon of the long head of the biceps brachii lies in the intertubercular sulcus and in the groove between the crests of the greater and lesser tubercles.

Shaft

The smooth **posterior surface** of the shaft lies between the medial and lateral borders. At the upper end, the axillary nerve and posterior circum-

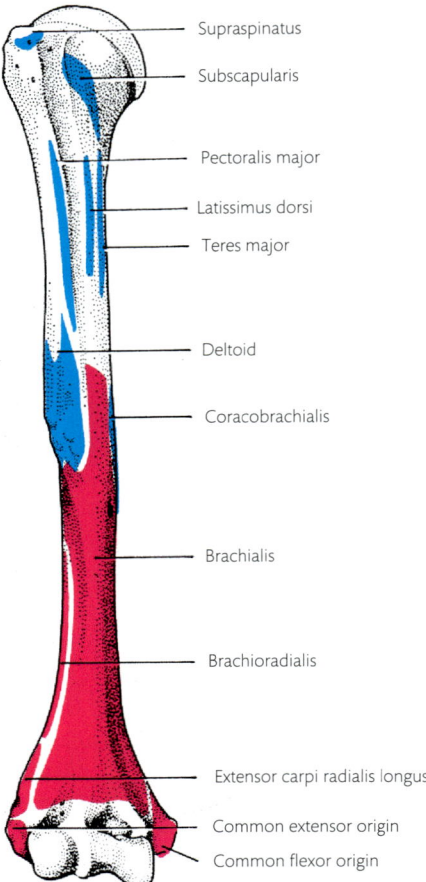

Fig. 4.9 Anterior aspect of the humerus to show muscle attachments.

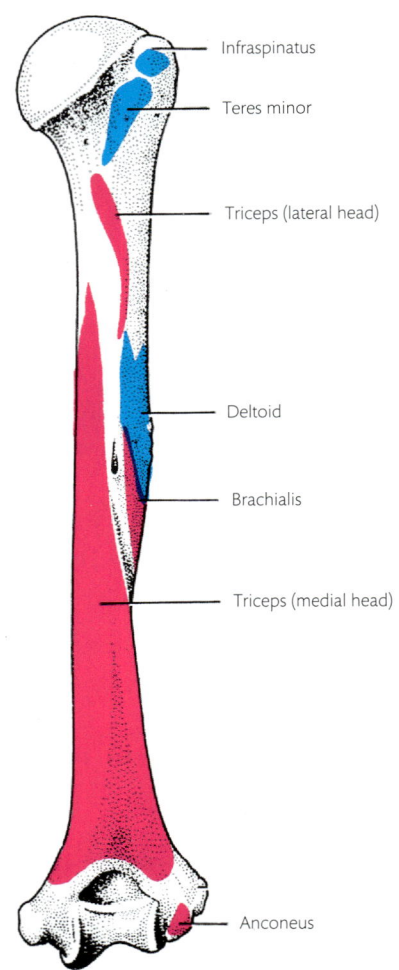

Fig. 4.10 Posterior aspect of the humerus to show muscle attachments.

flex humeral artery run transversely across the posterior surface. A shallow groove, the **spiral groove** or **radial groove**, runs downwards from the medial to the lateral aspect on the posterior surface of the shaft. The radial nerve and profunda brachii artery lie in this groove. The lateral head of the triceps has a long, linear attachment on the posterior surface, above the radial groove. The entire area below the radial groove gives origin to the medial head of the triceps.

The **anterolateral surface** lies between the lateral and anterior borders. The small elevation midway down the lateral aspect of the shaft is the **deltoid tubercle**. It gives insertion to the deltoid. The lower part of the anterolateral surface gives origin to the brachialis. The lower part of the lateral margin is continuous with the lateral supracondylar ridge and gives origin to the brachioradialis. The lower part of the **anteromedial surface** gives

attachment to the brachialis. The coracobrachialis is attached to the middle of the medial border. The medial border is continuous with the medial supracondylar lines inferiorly.

Lower end

The distal end of the humerus is wide transversely and has two articular and two non-articular parts. The **lateral epicondyle** is a rounded, non-articular projection which continues upwards with the lateral supracondylar line. The anterior and lateral surfaces are roughened for attachment of the superficial muscles of the extensor compartment of the forearm. The posterior surface gives origin to the anconeus. The part of the lateral supracondylar line just above the lateral epicondyle gives origin to the extensor carpi radialis longus.

The capitulum is rounded and has an articular facet on its inferior and anterior aspects for articulation with the head of the radius. On the anterior surface of the lower end of the humerus, the bone immediately above the capitulum is hollowed out to form the radial fossa. The head of the radius fits into the radial fossa during flexion.

The trochlea is pulley-shaped and articulates with the trochlear notch of the ulna. The medial flange of the pulley projects further down than the lateral flange. It has a continuous curved articular facet on the anterior, inferior, and posterior aspects, for articulation with the trochlear notch of the ulna. The bone above the trochlea is hollowed out anteriorly to form the coronoid fossa. The coronoid process of the ulna fits into this during full flexion. On the posterior aspect of the humerus, the bone above the trochlea is hollowed out to form the olecranon fossa and the olecranon process of the ulna fits into this in full extension.

The medial epicondyle is a blunt, subcutaneous medial projection at the lower end of the medial supracondylar ridge. Its posterior surface has a shallow groove that houses the ulnar nerve. The medial epicondyle and adjoining part of the medial supracondylar ridge give origin to the superficial flexors of the front of forearm.

Exercise 4.4A: to determine the side of the humerus

Identify the side of the humerus by using the following instructions: hold the humerus such that:

1. The head of the humerus is directed superiorly
2. The head of the humerus is directed medially
3. The lesser tubercle is directed anteriorly.

Exercise 4.4B: to hold the humerus in the anatomical position

Hold the humerus in the anatomical position and in the hand of the side to which it belongs.

Exercise 4.5: to articulate the humerus with the scapula

Articulate the scapula and humerus together by bringing the facets on the glenoid fossa of the scapula and the head of humerus together. (Make sure the two bones you are using belong to the same side.)

Radius

General features of the radius

The **radius** is a long bone. It is one of the two bones of the forearm and lies lateral to the ulna. Like all long bones, it consists of an **upper end**, a **shaft**, and a **lower end** [Figs. 4.11, 4.12].

The upper end has a **head**, **neck**, and **radial tuberosity**. The **head** is disc-shaped and is separated from the radial tuberosity by the narrow neck. The **shaft** is convex laterally. It has three surfaces—**anterior**, **posterior**, and **lateral**; and three borders—**anterior**, **posterior**, and **medial**. The medial border is sharp and is the **interosseous border**. The lower end of the radius is expanded and has five surfaces. The **anterior surface** is smooth and concave. The **lateral surface** has a downward prolongation—the **styloid process of the radius**. The **medial surface** is concave and has an articular facet for the head of the ulna. The **dorsal surface** has a prominent **dorsal tubercle**. The **inferior surface** articulates with the scaphoid and lunate to form the wrist joint.

Bony landmarks and muscle attachments

By using Figs. 4.11, 4.12, 4.13, and 4.14, and dry bones, identify the following bony landmarks and muscle attachments on the radius.

Upper end

The **head** of the radius is disc-shaped. The proximal surface is covered by articular cartilage for articulation with the capitulum of the humerus at the elbow joint. The circumference of the head of the radius is also smooth and covered by articular cartilage. It articulates with the radial notch on the ulnar and the annular ligament to form the superior radioulnar joint.

The **neck** is narrow and short. The **radial tuberosity** projects anteriorly and medially. The anterior aspect is smooth and the posterior aspect is roughened for attachment of the biceps brachii tendon.

Shaft

The **medial border** or **interosseous border** of the radius is sharp. Superiorly, it extends to just below the radial tuberosity. Inferiorly, it is contin-

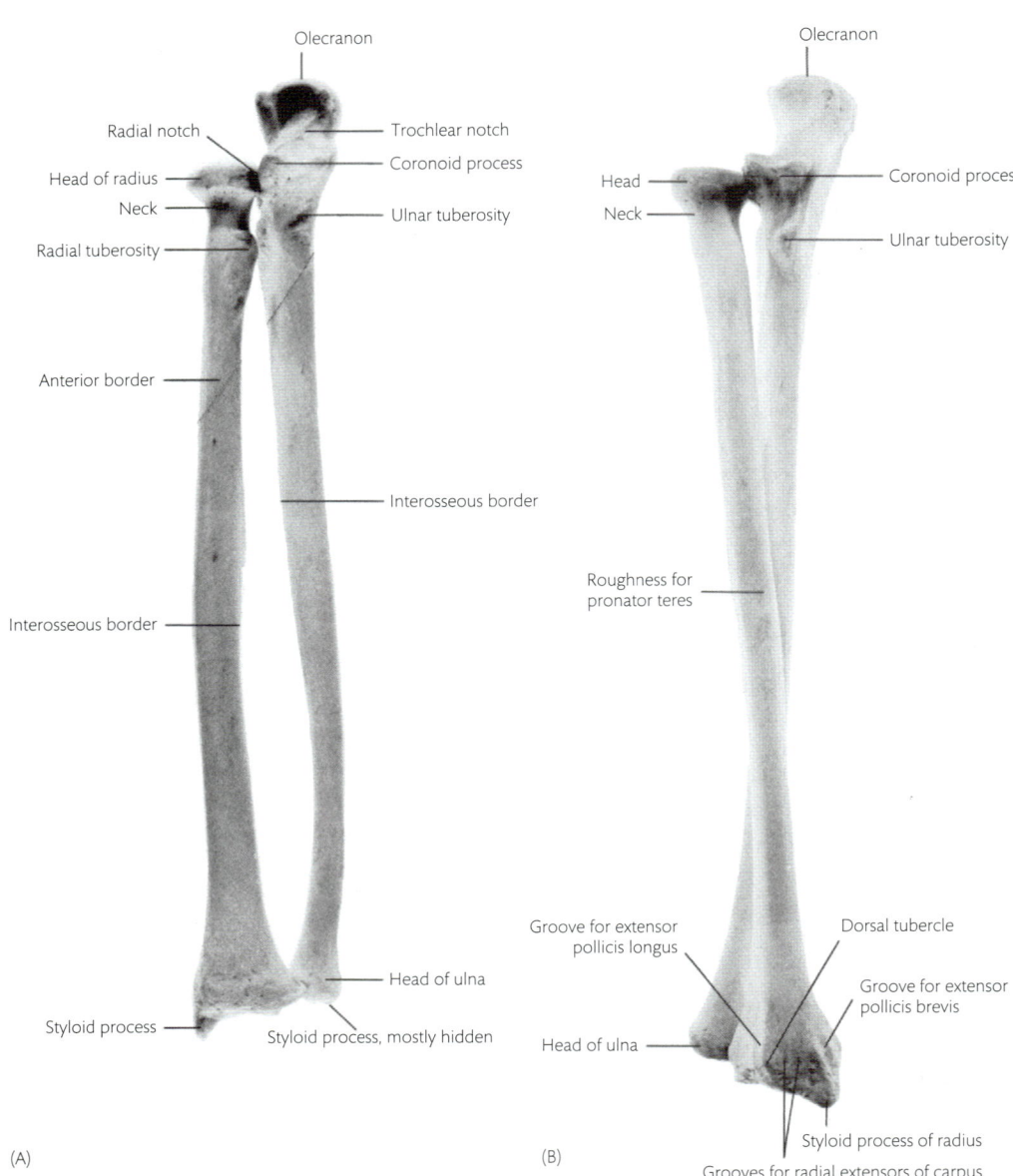

Olecranon

Radial notch

Head of radius
Neck
Radial tuberosity

Trochlear notch
Coronoid process

Ulnar tuberosity

Anterior border

Interosseous border

Interosseous border

Head of ulna

Styloid process

Styloid process, mostly hidden

(A)

Olecranon

Head
Neck

Coronoid process

Ulnar tuberosity

Roughness for
pronator teres

Groove for extensor
pollicis longus

Dorsal tubercle

Groove for extensor
pollicis brevis

Head of ulna

Styloid process of radius

Grooves for radial extensors of carpus

(B)

Fig. 4.11 Right radius and ulna (anterior surface). (A) Supinated position. (B) Prone position.

uous with the posterior margin of the medial surface of the inferior end. The distal three-quarters of this border give attachment to the interosseous membrane which connects the radius to the ulna and forms the middle radioulnar joint. The **anterior** and **posterior** borders are rounded and well defined only in the middle third. The anterior border curves medially in the upper part towards the radial tuberosity.

The **posterior surface** of the shaft lies between the interosseous and posterior border. The abduc-

tor pollicis longus and extensor pollicis brevis take attachment from the posterior surface of the radius and the adjoining part of the interosseous membrane.

The **lateral surface** lies between the anterior and posterior borders. The supinator curves around the head of the radius and is attached to the upper part of the posterior and lateral surfaces. The pronator teres is attached on the mid shaft at the point of maximum convexity. The brachioradialis is attached to the lateral surface just above the styloid process.

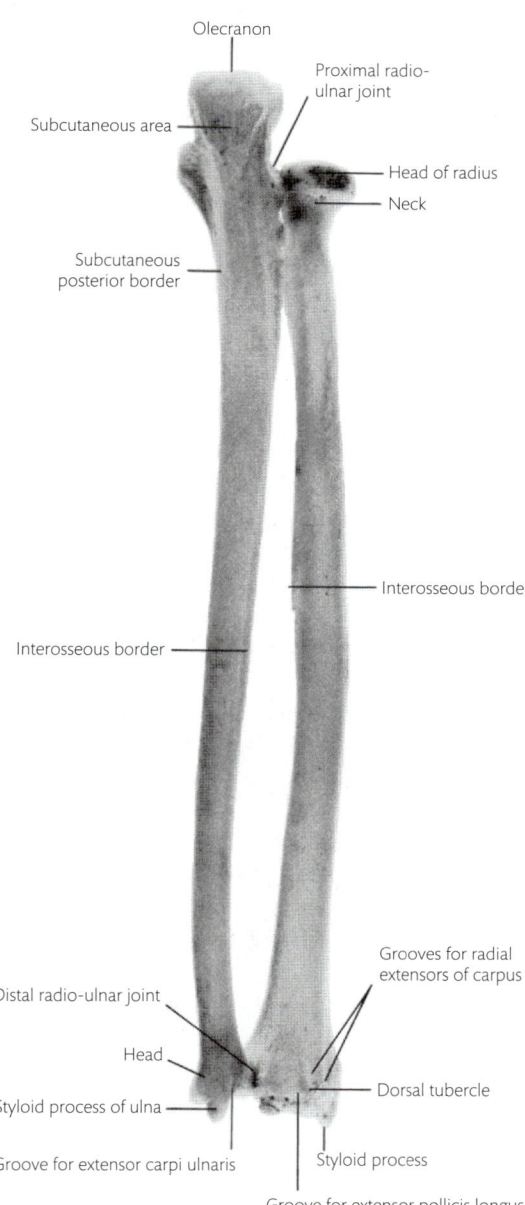

Fig. 4.12 Right radius and ulna (posterior surface).

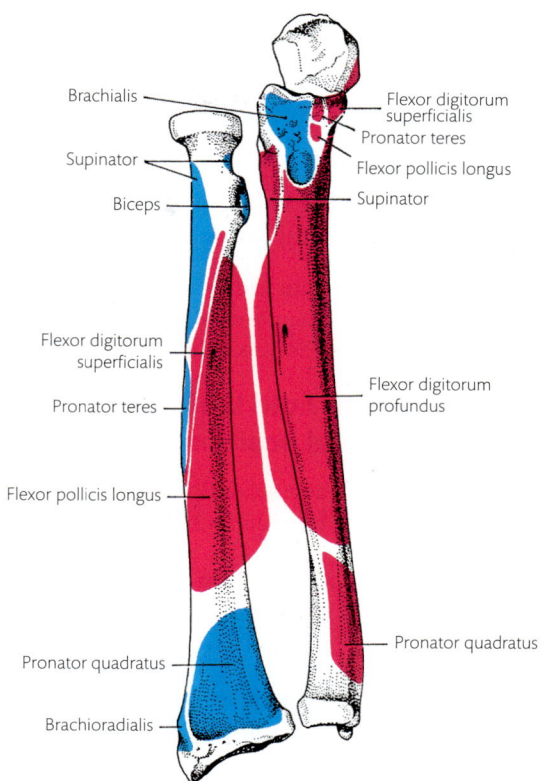

Fig. 4.13 Muscle attachments to the anterior surface of the right radius and ulna.

The **anterior surface** lies between the anterior and interosseous borders and is gently concave. It is crossed obliquely in its proximal half by the **anterior border**. The radial head of the flexor digitorum superficialis is attached to a long, thin area in the upper part. Most of the middle of the anterior surface and the adjoining part of the interosseous membrane give attachment to the flexor pollicis longus. The lower one-fourth of the anterior surface gives attachment to the pronator quadratus.

Lower end

The **anterior** or **ventral surface** of the lower end of the radius is smooth and concave.

The **posterior** or **dorsal** surface is convex and has a prominent **dorsal tubercle**. A well-defined groove lies medial to the dorsal tubercle and lodges the extensor pollicis longus tendon. A similar, but shallow, depression lateral to the dorsal tubercle is related to the extensor carpi radialis longus and brevis. The **lateral surface** of the lower end of the radius projects distally as the **styloid process of the radius**. The **medial surface** of the lower end of the radius has an articular facet for articulation with the head of the ulna at the inferior radioulnar joint. Superior to the facet, a roughened triangular area gives attachment to the lower part of the interosseous membrane. The **inferior surface** of the radius is concave and covered with articular cartilage. It articulates with the carpal bones to form the wrist joint. This surface is divided by a ridge into a medial and lateral area. The medial area is quadrilateral and articulates with the lunate. The lateral area is triangular and continuous with the medial aspect of the styloid process. It articulates with the scaphoid.

Exercise 4.6A: to determine the side of the radius

Identify the side of the radius by using the following instructions: hold the radius such that:

1. The head of the radius is directed superiorly
2. The sharp interosseous border is directed medially
3. The dorsal tubercle is directed posteriorly.

Exercise 4.6B: to hold the radius in the anatomical position

Hold the radius in the anatomical position and in the hand of the side to which it belongs.

Exercise 4.7: to articulate the radius with the humerus

Articulate the humerus and radius together by bringing the facets on the capitulum of the humerus and the head of the radius together. (Make sure the two bones you are using belong to the same side.)

Ulna

General features of the ulna

The **ulna** is one of the two forearm bones. It lies medial to the radius.

Like the other long bones, it consists of an **upper end**, **shaft**, and **lower end** [Figs. 4.11, 4.12].

The upper end has the olecranon process directed superiorly, and the coronoid process directed anteriorly. The anterior surface of the olecranon process and the superior surface of the coronoid process together form a continuous notch, the trochlear notch for articulation with the trochlea of the humerus at the elbow joint. The **shaft** has three surfaces—**anterior**, **posterior**, and **medial**; and three borders—**anterior**, **posterior**, and **lateral**. The lateral border is sharp and is the **interosseous border**. The posterior aspect of the olecranon and the posterior border of the shaft form a continuous margin that is subcutaneous throughout its length. The lower end of the ulna is rounded to form the head of the ulna. A small, downward projection,

the **styloid process** of the ulna, projects down on the medial aspect.

Bony landmarks and muscle attachments

By using Figs. 4.11, 4.12, 4.13, and 4.14, and dry bones, identify the following bony landmarks and muscle attachments on the ulna.

Upper end

The trochlear notch is covered by articular cartilage and is divided by a vertical ridge into medial and lateral parts, for articulating with the corresponding parts of the trochlea of the humerus. The **olecranon process** projects upwards from the shaft. Its tip is curved and fits into the olecranon fossa of the humerus on full elbow extension. The posterior surface of the olecranon process is smooth and subcutaneous. The upper aspect gives insertion to the triceps brachii. The anterior aspect of the olecranon process is covered by articular cartilage and forms the upper part of the trochlear notch. The lateral part of the olecranon (and posterior surface of the shaft of the ulna) gives insertion to the anconeus.

The **coronoid process** projects forwards from the anterior surface and fits into the coronoid fossa of the humerus during full flexion of the elbow. The superior aspect forms part of the trochlear notch. It is divided by a shallow ridge onto a medial and lateral part which fit into the medial and lateral flanges of the trochlear of the humerus. The inferior sloping surface of the coronoid process is triangular and is the **ulnar tuberosity**. It gives insertion to the brachialis. On the lateral aspect of the coronoid process is the **radial notch**. It is covered with articular cartilage and articulates with the head of the radius at the superior radioulnar joint. Inferior to the radial facet is the **supinator crest** which gives attachment to the supinator. The anterior aspect of the coronoid gives secondary attachment to the pronator teres, flexor pollicis longus, and flexor digitorum superficialis [Fig. 4.13].

Shaft

The **lateral border** of the ulna is the sharp **interosseous border**. The interosseous membrane is attached to it. Proximally, this border continues with the supinator crest. The anterior and posterior borders are rounded. The **anterior surface** lies

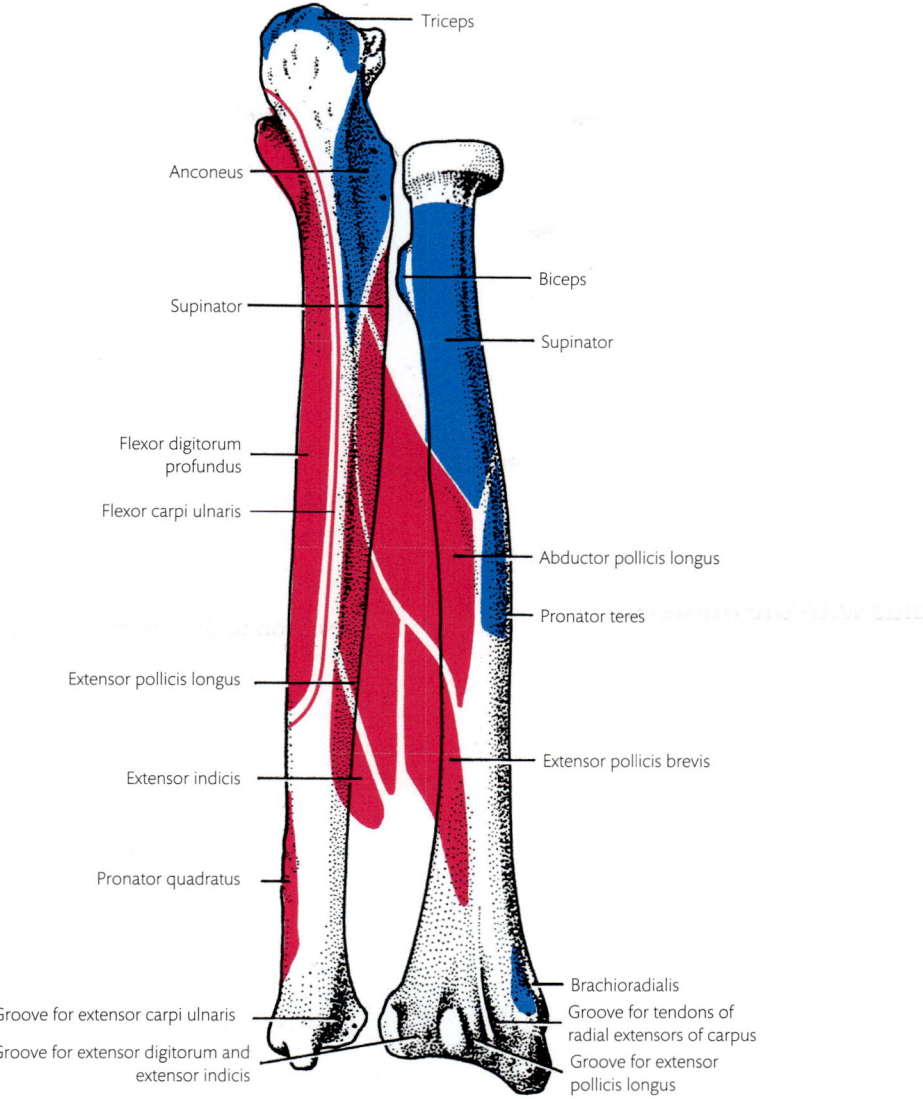

Fig. 4.14 Muscle attachments to the posterior surface of the right radius and ulna.

between the anterior and sharp lateral border. It is slightly concave. The flexor digitorum profundus has an extensive origin from the upper three-fourths of the anterior and medial surfaces of the shaft. The pronator quadratus takes origin from the distal one-fourth of the anterior surface. The **posterior border** extends from the posterior aspect of the olecranon to the styloid process. The deep fascia of the forearm is attached to this border. The flexor carpi ulnaris has a thin, linear origin from the back of the olecranon and the posterior border. The attachment of the flexor digitorum profundus on the medial surface also extends up to the posterior border. The anconeus is attached to the upper part of the **posterior surface**,

just below the olecranon. Below this, the abductor pollicis longus, extensor pollicis longus, and extensor digitorum arise from the posterior surface and adjoining interosseous membrane. The **medial surface** between the anterior and posterior borders is curved and smooth. It gives attachment to the flexor digitorum profundus.

Lower end

The head of the ulna is rounded and covered by articular cartilage laterally where it articulates with the ulnar notch on the medial side of the radius to form the inferior radioulnar joint. The inferior aspect of the head of the ulna is rounded and smooth. It is covered by articular cartilage and articulates with the articular disc of the wrist joint. The styloid process projects down from the posterolateral aspect of the ulna.

Exercise 4.8A: to determine the side of the ulna

Identify the side of the radius by using the following instructions; hold the radius such that:

1. The olecranon process is directed superiorly
2. The sharp interosseous border is directed laterally
3. The coronoid process is directed anteriorly.

Exercise 4.8B: to hold the ulna in the anatomical position

Hold the ulna in the anatomical position and in the hand of the side to which it belongs.

Exercise 4.9: to articulate the radius with the ulna

Articulate the radius and ulna together at the upper end, by bringing the facets on the head of the radius and the radial notch on the ulna together. At the same time, at the lower end, bring the head of the ulna and the ulnar notch of the radius together. (Make sure the two bones you are using belong to the same side.)

Exercise 4.10: to articulate the radius and ulna with the humerus

Articulate the lower end of the humerus with the upper end of the radius and ulna. The trochlear notch of the ulna should be in apposition with the trochlea of the humerus, and the head of the radius with the capitulum of the humerus. (Make sure the three bones you are using belong to the same side.)

Articulated hand

General features of the hand

The skeleton of the hand is made up of eight carpal bones, five metacarpals, and the phalanges [Figs. 4.15, 4.16].

The **carpals**, or wrist bones, are arranged in two rows of four bones each. The bones in the proximal row, except for the pisiform, articulate with the inferior surface of the radius and the articular disc on the lower end of the ulna. They are the **scaphoid**, **lunate**, **triquetrum**, and **pisiform**, from lateral to medial. The bones of the distal row are the **trapezium**, **trapezoid**, **capitate**, and **hamate**. The five **metacarpals** are numbered one to five, from lateral to medial. They articulate with the distal row of carpal bones. Each metacarpus has a **base**, a **shaft**, and a **head** [Fig. 4.17]. The thumb has two **phalanges**—proximal and distal. The medial four fingers have three phalanges each—proximal, middle, and distal. The heads of the metacarpals articular with the base of the proximal phalanx at the metacarpophalangeal joints. The phalanges of each digit articulate with each other at the proximal and distal interphalangeal joints.

Bony landmarks and muscle attachments on the bones of the hand

By using Figs. 4.15, 4.16, 4.17, 4.18, and 4.19, and dry bones, identify the following bony landmarks and muscle attachments on the bones of the hand.

The eight carpal bones fit together closely to form a semi-flexible unit that is concave anterior-

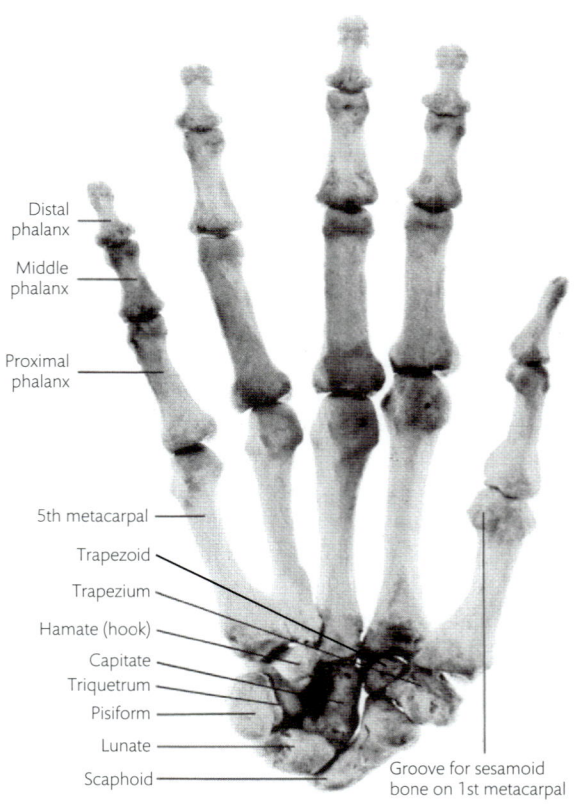

Distal phalanx

Middle phalanx

Proximal phalanx

5th metacarpal

Trapezoid

Trapezium

Hamate (hook)

Capitate

Triquetrum

Pisiform

Lunate

Scaphoid

Groove for sesamoid bone on 1st metacarpal

Fig. 4.15 Palmar aspect of bones of the right hand.

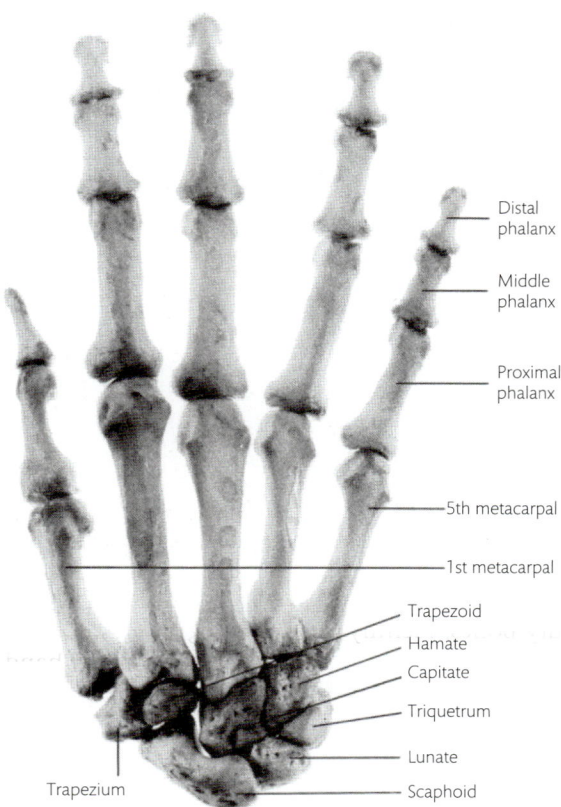

Distal
phalanx

Middle
phalanx

Proximal
phalanx

5th metacarpal

1st metacarpal

Trapezoid

Hamate

Capitate

Triquetrum

Lunate

Trapezium

Scaphoid

Fig. 4.16 Dorsal aspect of bones of the right hand.

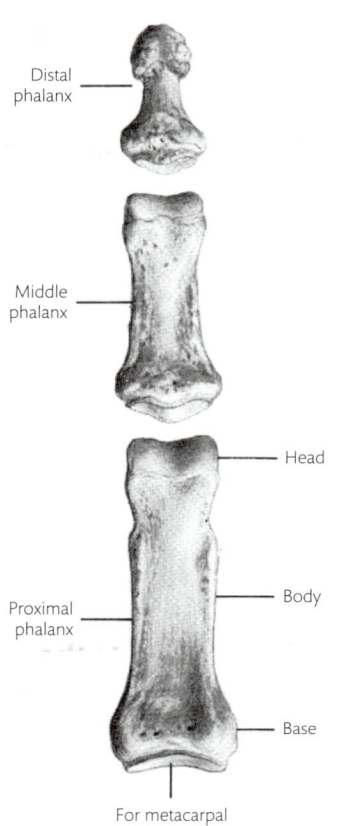

Distal
phalanx

Middle
phalanx

Head

Proximal
phalanx

Body

Base

For metacarpal

Fig. 4.17 Phalanges of a finger, palmar view. By comparison, in the foot, the base of the proximal phalanx is large and the bodies are much thinner.

ly. The scaphoid, lunate, and triquetrum articulate with the lower end of the radius and the articular disc on the ulna. The pisiform lies on the palmar surface of the triquetrum. The bones of the distal row articulate with those of the proximal row and with the base of the metacarpals. The first metacarpal articulates with the trapezium at the first carpometacarpal joint. The second and third metacarpals articulate with the trapezoid and capitate. The fourth and fifth metacarpals articulate with the hamate. The hamate has a prominent **hook of the hamate** which is directed anteriorly. The pisiform is a sesamoid bone developed in the tendon of the flexor carpi ulnaris. The tendon continues to the base of the fifth metacarpal and hook of the hamate as the pisometacarpal and pisohamate ligaments, respectively. The trapezium is grooved on the palmar aspect by the tendon of the flexor carpi radialis.

The muscles of the thenar eminence have partial attachment to the scaphoid and trapezoid. The muscles of the hypothenar eminence have partial attachment to the pisiform and hook of the hamate. The flexor carpi radialis is attached to the base of the second and third metacarpals, on the palmar aspect. The extensor carpi radialis longus and brevis are attached to the base of the second and third metacarpals on the dorsal aspect. The extensor carpi ulnaris is attached to the base of the fifth metacarpal on the dorsal side. The palmar and dorsal interossei take origin from the body of the metacarpals.

Applied anatomy

➲ The distal end of the radius articulates distally with the lateral two **carpal bones**—the scaphoid and lunate in the proximal row. This forms the only direct articulation of the carpal bones with the bones of the forearm. Forces transmitted from the hand to the forearm pass through the scaphoid and lunate to the radius—a feature which accounts for the fracture of the scaphoid or radius (but not of the ulna) from falling onto an outstretched hand.

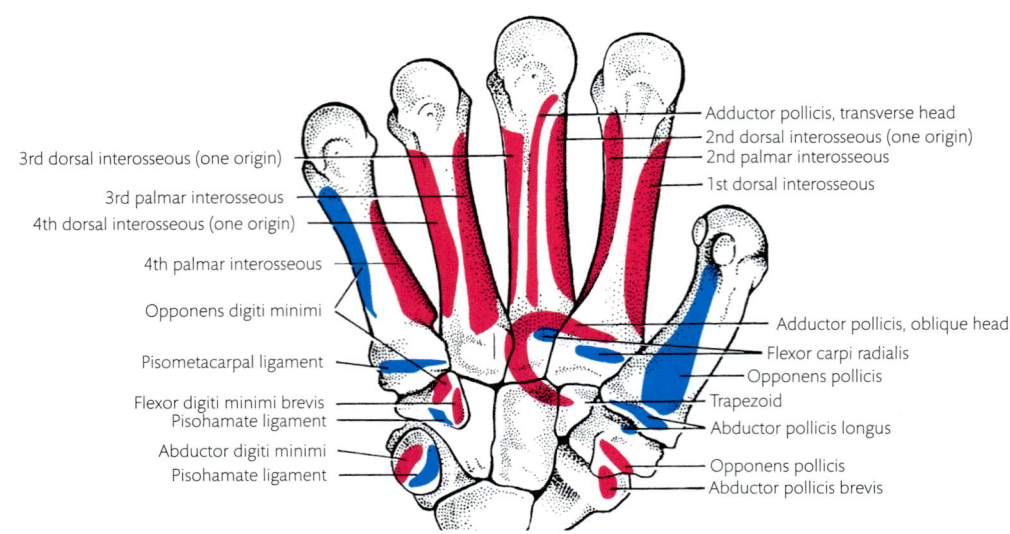

3rd dorsal interosseous (one origin)
3rd palmar interosseous
4th dorsal interosseous (one origin)
4th palmar interosseous
Opponens digiti minimi
Pisometacarpal ligament
Flexor digiti minimi brevis
Pisohamate ligament
Abductor digiti minimi
Pisohamate ligament

Adductor pollicis, transverse head
2nd dorsal interosseous (one origin)
2nd palmar interosseous
1st dorsal interosseous

Adductor pollicis, oblique head
Flexor carpi radialis
Opponens pollicis
Trapezoid
Abductor pollicis longus
Opponens pollicis
Abductor pollicis brevis

Fig. 4.18 Muscle attachments to the palmar surfaces of the carpus and metacarpus.

The body of the scaphoid is constricted to form a narrow 'waist' that is prone to fractures, especially when a person falls onto an outstretched hand. The proximal part of the bone is devoid of periosteum and gets its arterial supply through the distal segment. In the event of a fracture, the arterial supply to the proximal segment is compromised and can lead to avascular necrosis of that segment. (Avascular=lack of blood supply; necrosis=death of tissue.)

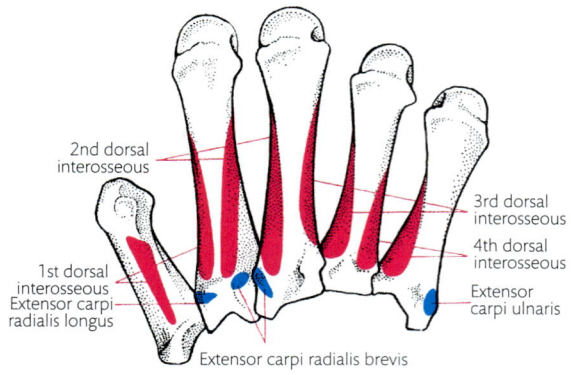

2nd dorsal interosseous
1st dorsal interosseous
Extensor carpi radialis longus
Extensor carpi radialis brevis

3rd dorsal interosseous
4th dorsal interosseous
Extensor carpi ulnaris

Fig. 4.19 Muscle attachments to the dorsal aspect of the right metacarpus.

Exercise 4.11A: to determine the side of the articulated hand

Identify the side of the articulated hand by using the following instructions: hold the hand such that:

1. The carpal bones are directed superiorly
2. The metacarpal of the thumb is directed laterally
3. The hook of the hamate is directed anteriorly.

Exercise 4.11B: to hold the articulated hand in the anatomical position

Hold the skeleton of the hand in the anatomical position and in the hand of the side to which it belongs.

Exercise 4.12: to articulate the hand with the radius

Articulate the hand with the distal end of the radius by bringing the scaphoid and lunate and the distal end of the radius together. (Make sure the two bones you are using belong to the same side.)

CHAPTER 5
The pectoral region and axilla

Introduction

Muscles covering the front of the chest and holding the free upper limb to the torso, and their vessels and nerves constitute the **pectoral region**. The pyramidal space between the upper part of the thorax and the arm is the **axilla**.

Overview of the axilla

The axilla is a four-sided pyramidal space between: (1) the upper limb; (2) the muscles connecting the upper limb to the front of the thorax; (3) the muscles connecting the upper limb to the back of the thorax; and (4) the lateral wall of the thorax. When the arm is by the side, the axilla is a narrow space. When the arm is abducted, the volume of the axilla increases and its floor (base) rises, forming a definite 'armpit'. Also, when the arm is abducted, the muscular inferior margin of its anterior wall stands out as the **anterior axillary fold**, and the inferior margin of the posterior wall stands out as the **posterior axillary fold** [Fig. 5.1].

The superior part of the axilla—the apex—lies lateral to the first rib and is continuous over its superior surface, with the superior aperture of the thorax below and the root of the neck above. This continuity permits blood vessels from the thorax and nerves from the neck to enter the axilla on their way to the upper limb. (These vessels and nerves pass over the superior surface of the first rib behind the clavicle [Fig. 5.2].)

Surface anatomy of the pectoral region and axilla

All points mentioned in this section should be confirmed on the living body and on specimens of the bones.

The **clavicle** (collar bone) is palpable throughout its length. It follows a slight curve which is convex forwards in its medial two-thirds and concave forwards in its lateral one-third [see Fig. 4.1]. Draw a finger along your clavicle, and note that its ends project above the acromion of the scapula laterally and the manubrium of the sternum medially. Thus, the positions of these joints are easily identified, though the medial end of the clavicle is somewhat obscured by the attachment of the sternocleido-mastoid muscle.

Between the medial ends of the clavicles, feel the **jugular notch** on the superior margin of the manubrium [Fig. 5.3]. Draw a finger downwards

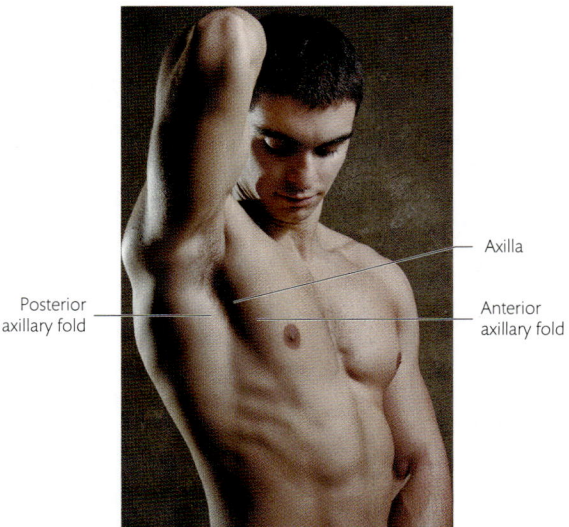

Axilla

Posterior axillary fold

Anterior axillary fold

Fig. 5.1 A pyramidal-shaped hollow—the axilla—is clearly seen between the chest wall and the arm. The anterior and posterior axillary folds are seen bounding the floor of the axilla.

FotoAndalucia/Shutterstock.com

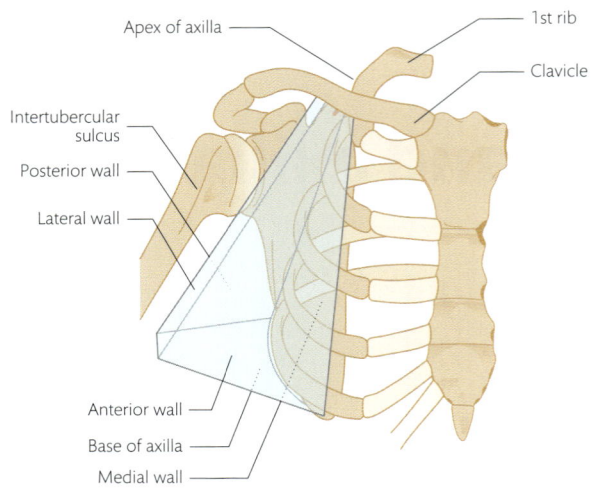

Fig. 5.2 Schematic drawing of the axilla, showing the base, apex, and four walls in relation to the bones of the thorax, pectoral girdle, and arm.

from this notch in the median plane till a blunt transverse ridge is felt on the sternum. This bony landmark is the **sternal angle**, a joint between the manubrium and the body of the sternum. At this level, the **cartilage of the second rib** articulates with the side of the sternum. The second rib may be identified in this way, even in obese subjects, as the sternal angle is always readily palpable. The other ribs are identified by counting down from the second rib. The anterior part of the first rib is hid-

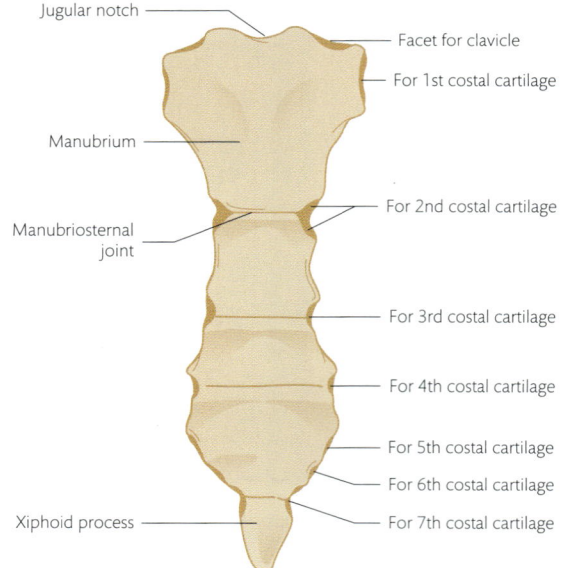

Fig. 5.3 Sternum (anterior view).

den by the medial part of the clavicle. Immediately inferior to the lower end of the body of the sternum is a small median depression—the **epigastric fossa**—which overlies the **xiphoid process**, the lowest piece of the sternum. The **cartilages of the seventh ribs** lie on either side of this fossa.

The **nipple** is very variable in position, even in the male, but usually lies over the fourth intercostal space, near the junction of the ribs with their cartilages. It is just medial to a vertical line passing through the middle of the clavicle (the **midclavicular line**).

The **infraclavicular fossa** is a depression inferior to the junction of the lateral and middle thirds of the clavicle. The pectoralis major muscle on the front of the chest lies medial to the fossa, and the deltoid muscle, which clasps the shoulder, is lateral to it. The **coracoid process** of the scapula can be felt just lateral to the fossa and under cover of the deltoid muscle, 2–3 cm below the clavicle.

Follow the clavicle laterally to its articulation with the **acromion**—a subcutaneous, flattened piece of bone about 2.5 cm wide, on the top of the shoulder. The **acromioclavicular joint** can be felt as a slight dip, as the clavicle projects slightly above the level of the acromion (*acron*=summit; *omos*=shoulder).

Raise the arm from the side (i.e. abduct it), and identify the hollow of the **axilla**, the anterior axillary fold (containing the pectoralis major muscle), and the posterior axillary fold (containing the latissimus dorsi and teres major muscles) [Fig. 5.1]. The teres major is a thick, rounded muscle which connects the inferior angle of the scapula to the humerus, and can be felt in the posterior axillary fold when the arm is raised above the head. The latissimus dorsi muscle extends from the lower part of the back to the humerus. It can be made to stand out by depressing the horizontal arm against resistance.

With the arm by the side, push your fingers into the axilla. The anterior and posterior walls are soft and fleshy, but the lateral margin of the scapula can be felt in the posterior wall. The medial wall is formed by the ribs covered by a sheet-like muscle—the serratus anterior. In the lateral angle, the biceps brachii and coracobrachialis muscles lie parallel to the humerus. Some of the large nerves in the axilla can be rolled between the fingers and the humerus, and the axillary artery can be felt pulsating. By pushing the fingers up into the axilla, the head of the humerus can be felt laterally and the lateral border of the first rib medially.

Pectoral region

Cutaneous nerves

The skin on the anterior and lateral surfaces of the thorax is supplied by:

1. The supraclavicular nerves from the cervical plexus—principally the fourth cervical ventral ramus
2. The anterior and lateral cutaneous branches of the ventral rami of the second to eleventh thoracic nerves (intercostal nerves) [Fig. 5.4].

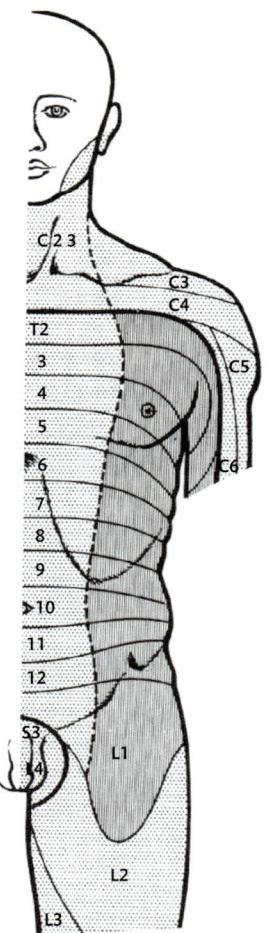

Fig. 5.4 Dermatomal pattern on the front of the trunk. The areas of skin supplied by the ventral rami are illustrated.

The **supraclavicular nerves** [Fig. 5.5] arise in the neck from the third and fourth cervical nerves (C. 3, C. 4). Diverging as they descend, the nerves pierce the deep fascia in the neck. They cross the clavicle to supply the skin on the front of the chest and shoulder [see Figs. 7.1 and 7.2] down to a horizontal line at the level of the second costal cartilage. They are named, according to their positions: medial, intermediate, and lateral.

The **anterior cutaneous branches of the intercostal nerves** (except the first and occasionally the second) emerge from the intercostal spaces near the lateral border of the sternum, pierce the pectoralis major, and supply the skin from the anterior median line almost to a vertical line through the middle of the clavicle (mid-clavicular line). They are accompanied by perforating branches of the internal thoracic artery, an artery which lies immediately deep to the costal cartilages. In the female, these arterial branches are enlarged in the second to fourth spaces to supply the mammary gland. The arteries have lymph vessels running with them from the skin of the anterior thoracic wall and the medial part of the mammary gland (breast) to **parasternal** nodes which lie beside the internal thoracic artery.

The **lateral cutaneous branches of the intercostal nerves** pierce the deep fascia along the mid-axillary line. Each nerve divides and enters the superficial fascia as anterior and posterior branches. The nerves pierce, or pass between, the digitations of the serratus anterior but play no part in supplying this muscle, the pectoral muscles, or the latissimus dorsi over which they run. They supply the part of the skin between the parts supplied by the anterior cutaneous branches (midline in front to the mid-clavicular line) and the dorsal ramus (midline of the back to approximately 10 cm from the midline).

There are usually no lateral or anterior cutaneous branches from the first intercostal nerve. The lateral cutaneous branch of the second intercostal nerve is the **intercostobrachial nerve**. It emerges as a large single branch and communicates with the medial cutaneous nerve of the arm and the lateral cutaneous branch of the third intercostal nerve. Together, these three nerves supply the skin of the medial side of the arm and the floor of the axilla.

Dissection 5.1 describes how to reflect the skin of the front and side of chest. [See also Fig. 5.6.]

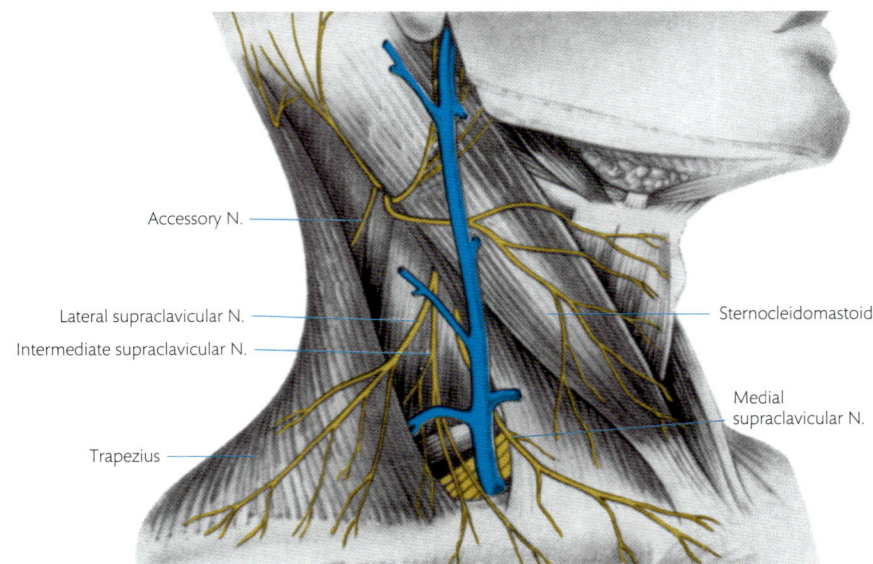

Fig. 5.5 Course and distribution of the supraclavicular nerves.

Labels in figure: Accessory N., Lateral supraclavicular N., Intermediate supraclavicular N., Trapezius, Sternocleidomastoid, Medial supraclavicular N.

The breast

Location

The mamma, or breast, lies in the superficial fascia of the pectoral region.

Structure

It is made up of: (1) the mammary gland; (2) the fatty superficial fascia in which it is embedded; and (3) the overlying skin with the nipple and the surrounding pigmented skin—the **areola** [Fig. 5.7].

In the male, the mammary gland is rudimentary; the nipple is small, and the areola is commonly surrounded by fine hairs. In the non-lactating female, the breast consists mainly of fatty tissue of the superficial fascia, in which are enclosed 15–20 lobes of rudimentary glandular tissue. These glands

---

DISSECTION 5.1 Skin reflection of the front and side of the chest

Objectives

I. To reflect the skin on the front and side of the chest. II. To examine the superficial fascia. III. To find the cutaneous vessels and nerves.

Instructions

1. Make the skin incisions 1–4, shown in Fig. 5.6. Make sure to carry incision 4 backwards as far as the posterior axillary fold.
2. Cut through the superficial fascia in incisions 1, 3, and 4.
3. Start from the midline (incision 1). Reflect the flaps of skin and the superficial fascia laterally by blunt dissection. Do not detach them. Leave the nipple and the surrounding skin in position as a landmark.
4. As the flap is separated from the skin of the neck along the clavicle, split the superficial fascia with a blunt instrument. Avoid cutting through the thin sheet of muscle (platysma) and the supraclavicular nerves [Fig. 5.5]. The supraclavicular nerves pass anterior to the clavicle to supply the skin of the upper part of the anterior thoracic wall and the shoulder.
5. Identify the medial, intermediate, and lateral branches of the supraclavicular nerves.
6. Note the fibrous strands connecting the deep fascia to the skin, especially deep to the breast in the female.
7. Find the anterior cutaneous nerves and vessels which emerge from the anterior ends of the intercostal spaces. Follow the branches of one of these nerves medially and laterally as far as possible.
8. Find the lateral cutaneous branches which pierce the chest wall in the mid-axillary line. They emerge through the deep fascia, one inferior to the other, in a vertical line. Follow the branches of one of them anteriorly and posteriorly as far as possible.

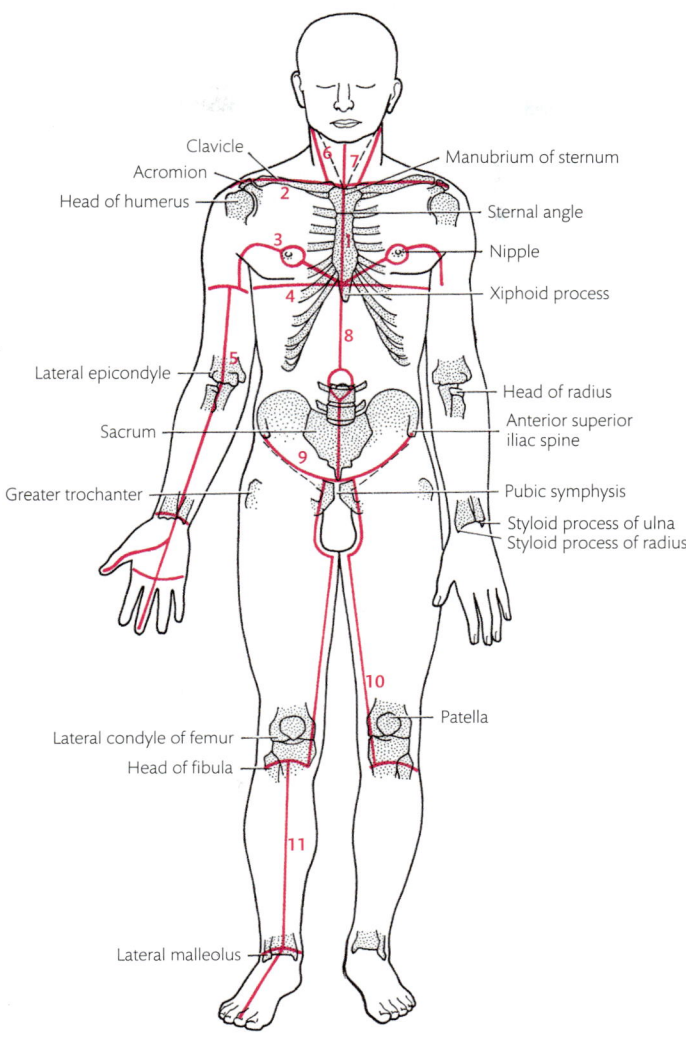

Clavicle
Acromion
Head of humerus
Manubrium of sternum
Sternal angle
Nipple
Xiphoid process
Lateral epicondyle
Head of radius
Sacrum
Anterior superior iliac spine
Greater trochanter
Pubic symphysis
Styloid process of ulna
Styloid process of radius
Patella
Lateral condyle of femur
Head of fibula
Lateral malleolus

Fig. 5.6 Landmarks and incisions. Left forearm pronated, and right forearm supinated.

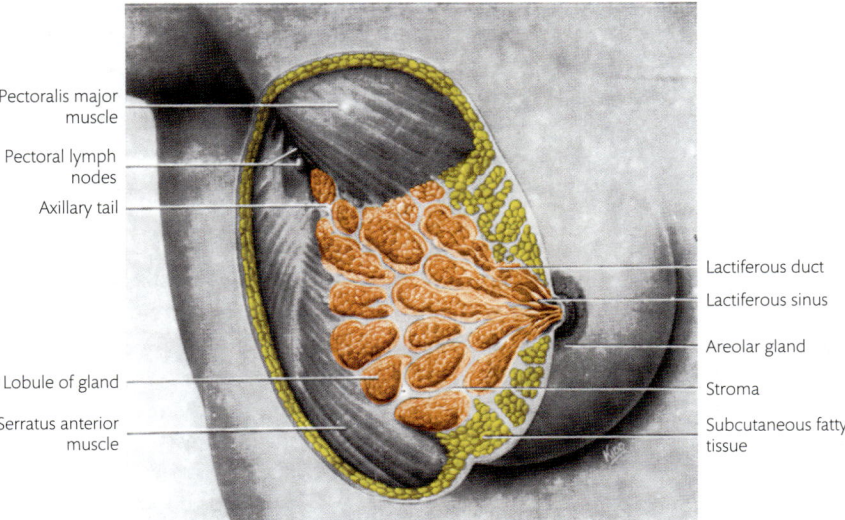

Pectoralis major muscle
Pectoral lymph nodes
Axillary tail
Lactiferous duct
Lactiferous sinus
Areolar gland
Lobule of gland
Stroma
Serratus anterior muscle
Subcutaneous fatty tissue

Fig. 5.7 Dissection of the right mammary gland.

radiate outwards from the nipple, giving the gland the shape of a flattened cone. Each lobe has a main **lactiferous duct** which passes to open separately on the nipple. At the base of the nipple, the duct is dilated to form a **lactiferous sinus**. The gland has no capsule, but its lobes are separated by fibrous strands of the superficial fascia which pass from the skin to the deep fascia. These fibrous strands are attached to the gland and anchor it both to the skin and to the underlying deep fascia.

Extent and deep relations

The base of the mammary gland extends from the margin of the sternum to almost the mid-axillary line, and from the second to sixth ribs. It lies largely on the pectoralis major muscle. Inferolaterally, it extends on to the costal origins of the serratus anterior and the external oblique muscle of the abdomen. The '**axillary tail**' arises from the superolateral quadrant of the breast and passes into the axilla, up to the level of the third rib [Fig. 5.7]. The apex of the gland—the **nipple**—lies a little below the midpoint of the gland, approximately at the fourth intercostal space in the nulliparous woman. The nipple is free of fat but contains circular and longitudinal smooth muscle fibres which can erect or flatten it. The skin of the nipple and areola contains modified sweat and sebaceous glands, particularly at the outer margin of the areola. These sebaceous glands tend to enlarge in the early stages of pregnancy, and shortly thereafter there is an increase in pigmentation in both the nipple and areola which never return to their original colour. In the later stages of pregnancy, the greater part of the fat in the gland is replaced by the proliferation of its ducts and the growth of many secretory alveoli from their branching ends.

Age changes in the breast

In infancy and early childhood, the breast consists solely of lactiferous ducts. No alveoli or precursors of alveoli are present. Breast changes occur in the female child at puberty under the influence of ovarian hormones. The ducts become branched, and masses of polyhedral cells develop at their ends. The areola becomes larger and more deeply pigmented. Adipose tissue is laid down in the stroma and the breasts enlarge and become rounded. The nipple becomes more raised. After menopause, ducts and alveoli atrophy and fatty tissue replaces the glandular tissue.

During the later stages of pregnancy, the ducts proliferate and alveoli are formed at the ends of the ducts.

Blood supply

The gland receives its **blood supply** from perforating branches of the intercostal and internal thoracic arteries medially and from the lateral thoracic artery laterally.

Lymphatic drainage

Lymph vessels drain principally: (1) to the axillary lymph nodes—(a) along the axillary tail to the **pectoral lymph nodes**, and (b) through the pectoralis major and clavipectoral fascia to the **apical axillary nodes** via the **infraclavicular nodes**; (2) to the **parasternal nodes** along the internal thoracic artery by passing along the branches of that artery which supply the gland; and (3) some lymph also drains to the posterior intercostal nodes. Since there is communication of lymph vessels across the median plane, there may be drainage to the opposite side, especially when some of the pathways are blocked by disease [Fig. 5.8].

➲ The most common breast conditions are painful breasts or mastitis (inflammation of the breast), cysts, benign tumours, and cancer. Clinical Application 5.1 discusses some of the clinical features of breast cancer.

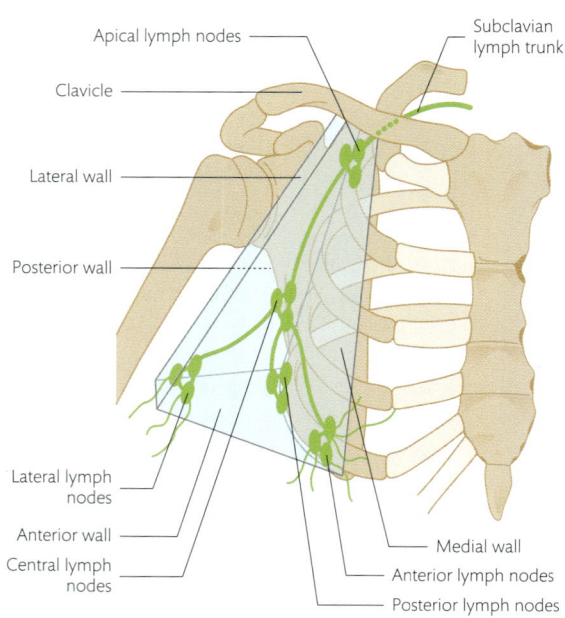

Fig. 5.8 Lymph nodes and lymph vessels of the axilla.

Dissection 5.2 describes the dissection of the breast.

This is not usually very successful in the elderly female and should not be attempted in the male.

Deep fascia of the pectoral region

The deep fascia covering the pectoralis major is continuous with the periosteum of the clavicle and sternum, and passes over the infraclavicular fossa and deltopectoral groove (between the pectoralis major and the deltoid) to become continuous with the fascia covering the deltoid. It curves over the inferolateral border of the pectoralis major to become continuous with the fascia of the axillary floor (**axillary fascia**). The axillary fascia stretches between the pectoralis major and the latissimus dorsi. When the arm is abducted, the axillary fascia rises into the axilla to form the armpit.

The **clavipectoral fascia** lies in the anterior wall of the axilla, deep to the pectoralis major. It extends from the clavicle to the axillary fascia, and encloses the pectoralis minor and subclavius muscles [Fig. 5.9].

See Dissection 5.3 for instructions on dissecting the pectoral region. [See also Fig 5.10.]

Pectoralis major

Origin: this powerful, fan-shaped muscle takes origin from: (1) the medial half of the front of the clavicle; (2) the anterior surfaces of the sternum; (3) the upper six costal cartilages; and (4) the aponeurosis of the external oblique muscle of the abdomen [Fig. 5.11]. **Insertion**: it is inserted into the lateral lip of the intertubercular sulcus or crest of the greater tubercle of the humerus [see Fig. 4.8]. At the insertion, the abdominal part twists under the sternocostal part to form a U-shaped tendon with it. The lowest abdominal fibres are inserted deep to the upper sternocostal fibres, while the intermediate fibres form the base of the U in the anterior axillary fold.

The clavicular part passes inferolaterally, fuses with the anterior layer of the U-shaped tendon, and extends further inferiorly on the humerus. The clavicular part lies at right angles to the abdominal and lower sternocostal parts and has different actions [Fig. 5.11].

Nerve supply: medial and lateral pectoral nerves. **Actions**: the clavicular part of the pectoralis major adducts and medially rotates the humerus. With the arm above the head, the lowest fibres act with the latissimus dorsi to pull down the arm or raise the body, as in climbing a rope. The muscle can also return the extended humerus to the anatomical position, then continue to flex the shoulder joint with its clavicular part which passes in front of the shoulder.

Pectoralis minor

Origin: this triangular muscle originates from the third to fifth ribs, near their cartilages. **Insertion**: it passes superolaterally to the tip of the coracoid process [Fig. 5.11]. **Nerve supply**: medial pectoral nerve. **Actions**: it pulls the scapula (and hence the shoulder) downwards and forwards. It raises the ribs in inspiration when the scapula is fixed.

Dissection 5.4 continues the dissection of the pectoral region.

Subclavius

Origin: this small muscle arises from the adjacent parts of the upper surfaces of the first costal cartilage and rib. **Insertion**: it passes parallel to the clavicle and is inserted into the groove on the inferior surface of the clavicle [Figs. 5.9 and 4.2]. **Nerve supply**: nerve to the subclavius from the upper trunk of the brachial plexus. **Actions**: it holds the medial end of the clavicle against the articular disc of the sternoclavicular joint during movements of the shoulder girdle.

Sternoclavicular joint

The sternoclavicular joint is a synovial joint between the shallow notch at the superolateral angle of the manubrium of the sternum and the larger medial end of the clavicle. A complete artic-

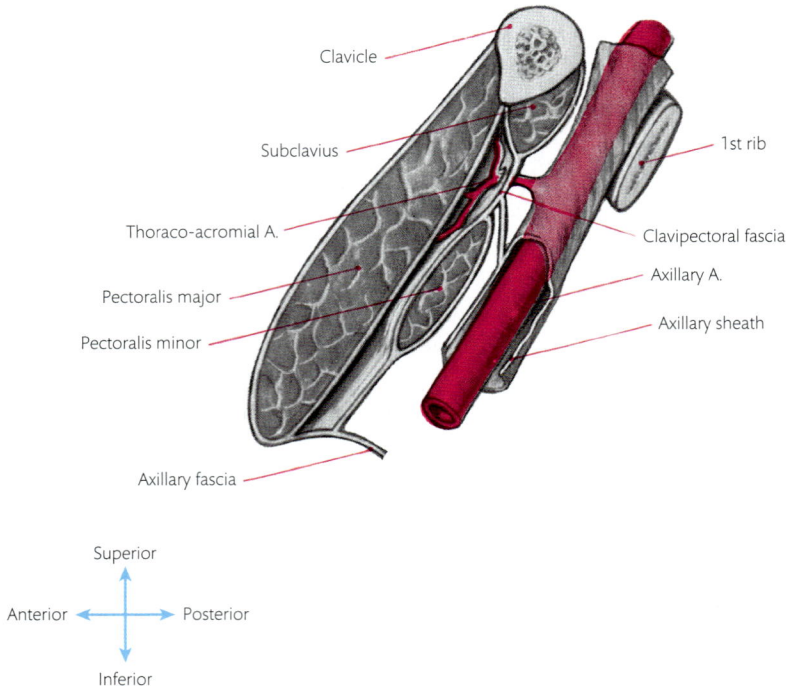

Clavicle

Subclavius

Thoraco-acromial A.

Pectoralis major

Pectoralis minor

1st rib

Clavipectoral fascia

Axillary A.

Axillary sheath

Axillary fascia

Superior

Anterior ←→ Posterior

Inferior

Fig. 5.9 Diagram of the clavipectoral fascia shown enclosing the subclavius and pectoralis minor and extending to the axillary fascia.

ular disc intervenes between these two articular surfaces. The joint also extends on to the superior surface of the first costal cartilage. This is the only articulation of the upper limb bones with the axial skeleton. Thus, the clavicle forms a support which maintains the scapula in position and transmits forces from the upper limb to the trunk

---

DISSECTION 5.3 Pectoral region-1

Objectives

I. To study the pectoralis major and the deltoid.
II. To identify the cephalic vein.

Instructions

1. Cut the deep fascia in the deltopectoral groove to uncover the **cephalic vein** passing to the infraclavicular fossa [Fig. 5.11].
2. Occasional lymph nodes found beside the vein receive lymph from the adjacent superficial tissues and transmit it through the infraclavicular fossa to the apical nodes of the axilla [Fig. 5.8].
3. Remove the fascia from the anterior parts of the pectoralis major and the deltoid, and define the attachments of these muscles.

(e.g. forces generated in falling on an outstretched hand). Functionally, the joint behaves like a ball-and-socket joint with a wide range of movements, as it has to move with each change in scapular position. It is subject to considerable force, but the bony surfaces give little intrinsic stability. For this reason, it is strengthened by powerful ligaments which are designed to prevent dislocation of the medial end of the clavicle from the shallow fossa on the sternum. The **articular capsule** is attached close to the articular margins of the bones. It is thickened anteriorly and posteriorly to form the anterior and posterior **sternoclavicular ligaments** [Fig. 5.12].

The **articular disc** is a nearly circular plate of fibrocartilage attached at its margins to the articular capsule. It divides the joint into two separate synovial cavities. Its strongest attachments are to the upper surface of the medial end of the clavicle and to the junction of the sternum and first costal cartilage. It assists the costoclavicular ligament in preventing the upward displacement of the medial end of the clavicle and acts as a shock absorber of compression forces applied from the upper limb.

The **costoclavicular ligament** is a powerful band which passes upwards and laterally from the

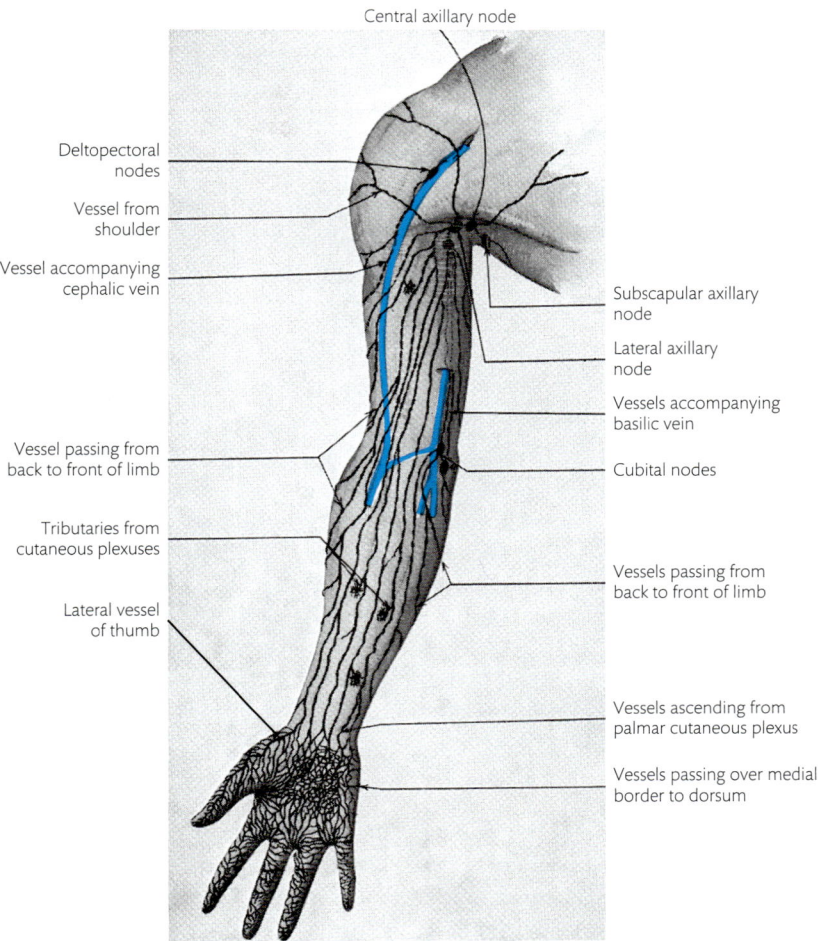

Central axillary node

Deltopectoral nodes

Vessel from shoulder

Vessel accompanying cephalic vein

Vessel passing from back to front of limb

Tributaries from cutaneous plexuses

Lateral vessel of thumb

Subscapular axillary node

Lateral axillary node

Vessels accompanying basilic vein

Cubital nodes

Vessels passing from back to front of limb

Vessels ascending from palmar cutaneous plexus

Vessels passing over medial border to dorsum

Fig. 5.10 Superficial lymph vessels and lymph nodes of the front of the upper limb.

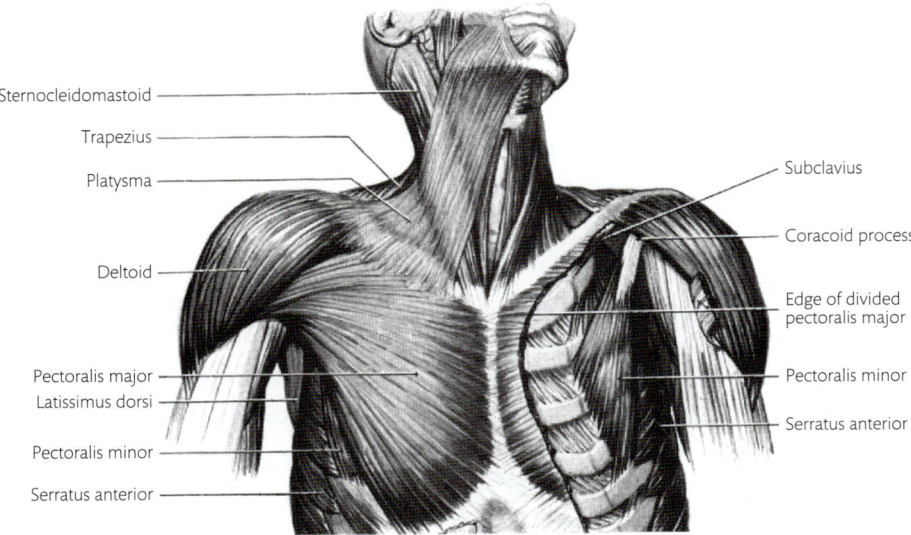

Sternocleidomastoid

Trapezius

Platysma

Deltoid

Pectoralis major

Latissimus dorsi

Pectoralis minor

Serratus anterior

Subclavius

Coracoid process

Edge of divided pectoralis major

Pectoralis minor

Serratus anterior

Fig. 5.11 Muscles of the anterior wall of the trunk. On the left side, the pectoralis major is removed to show the pectoralis minor and subclavius.

DISSECTION 5.4 Pectoral region-2

Objectives

I. To study the pectoralis minor, subclavius, and clavipectoral fascia. II. To identify and trace the cephalic vein, thoracoacromial artery, and medial and lateral pectoral nerves. III. To explore the continuity of the axillary vessels with the subclavian vessels.

Instructions

1. Cut across the clavicular head of the **pectoralis major** below the clavicle, and reflect it towards its insertion. Identify the branches of the lateral pectoral nerve and thoracoacromial artery that pierce the clavipectoral fascia to enter the pectoralis major.
2. Cut across the remainder of the pectoralis major about 5 cm from the sternum. Reflect it laterally. Identify the branch of the medial pectoral nerve which pierces the pectoralis minor to enter the pectoralis major.
3. Note the entire sheet of the **clavipectoral fascia** deep to the pectoralis major, and then remove it from the pectoralis minor.

4. Trace the **pectoralis minor** to its attachments.
5. Follow the cephalic vein through the upper part of the clavipectoral fascia to the axillary vein, and the thoracoacromial artery and the lateral pectoral nerve to their origins.
6. Expose the vessels and nerves superior to the pectoralis minor.
7. Cut through the anterior layer of the clavipectoral fascia immediately inferior to, and parallel with, the clavicle to expose the subclavius muscle.
8. Gently push a finger, inferior to that muscle, along the line of the axillary vessels. It will pass over the first rib, deep to the clavicle, into the root of the neck. If the finger is pressed medially between the axillary artery and vein, a firm resistance of the scalenus anterior muscle can be felt on the upper surface of the first rib between the artery and the vein. (Note that the vessels felt on the first rib are the subclavian vessels.)
9. Pass a finger deep to the pectoralis minor through the lower part of the axilla. Lift it from the subjacent structures, but preserve the medial pectoral nerve which enters its deep surface.

junction of the first rib and its cartilage to a rough area on the inferior surface of the clavicle near its medial end. The **interclavicular ligament** passes between the medial ends of the two clavicles and is fused with the articular capsules and the jugular notch of the sternum [Fig. 5.12].

Dissection 5.5 explains the dissection of the sternoclavicular joint.

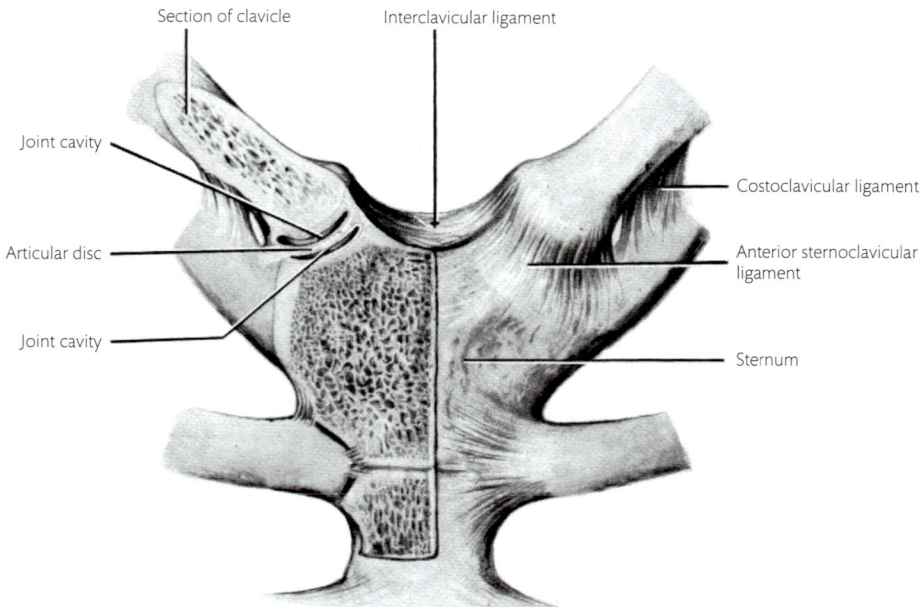

Section of clavicle — Interclavicular ligament

Joint cavity

Articular disc

Joint cavity

Costoclavicular ligament

Anterior sternoclavicular ligament

Sternum

Fig. 5.12 Sternoclavicular joint. A coronal section has been made through the anterior surface of the sternum and clavicle on the right side, opening the right sternoclavicular joint.

Acromioclavicular joint

The acromioclavicular joint is a **plane synovial** joint between the facets on the acromial (lateral) end of the clavicle and the acromion process of the scapula [see Fig. 4.3]. Both surfaces are covered by fibrocartilage. An intra-articular fibrocartilaginous disc may be present. The fibrous capsule surrounds the articular margin and is lined by synovial membrane. It is strengthened by superior and inferior acromioclavicular ligaments, and allows sliding movement between the bones when the shoulder girdle is moved.

Axilla

Boundaries and contents

Start by reviewing the overview of the axilla on p. 55 and in Fig. 5.2. The **anterior wall** of the axilla extends from the clavicle to the **anterior axillary fold**. It consists of the pectoralis major, the pectoralis minor, the subclavius, and the fascia enclosing them [Fig. 5.11]. The **posterior wall** consists of the lateral part of the costal surface of the scapula, covered by the subscapularis superiorly, and the teres major muscle with the latissimus dorsi winding round its lower border inferiorly. Identify these muscles by using Fig. 5.13. The **anterior** and **posterior axillary folds** are formed by the lower borders of the pectoralis major and the latissimus dorsi, respectively. The convex **medial wall** is formed by the lateral wall of the thorax (the first five ribs and intercostal spaces) covered by the serratus anterior. The narrow **lateral boundary** is formed by the humerus covered by the upper parts of the biceps and coracobrachialis muscles.

The **apex** of the axilla is bounded by the clavicle, first rib, and upper border of the scapula. It is continuous medially with the superior aperture of the thorax and the root of the neck. Through the apex, vessels from the thorax and the nerves of the brachial plexus from the neck enter the axilla [Fig. 5.2]. These vessels and nerves descend through the axilla to the arm and form the contents of the axilla, together with the axillary lymph nodes and loose fatty tissue. The **floor** of the axilla is formed by the axillary fascia which stretches between the anterior and posterior axillary folds. It is pierced by the axillary tail of the breast.

See Dissection 5.6 which begins the dissection of the axilla.

Serratus anterior

Origin: the **serratus anterior** arises from the outer surface of the upper eight ribs. **Insertion**: its fibres pass posteriorly around the lateral surface of the chest wall, forming the medial wall of the axilla. On the back, the fibres run deep to the scapula and are inserted into the costal surface of the scapula along the medial border. **Nerve supply**: long thoracic nerve [Figs. 5.13, 5.14, 4.6]. **Actions**: (1) it holds the scapula against the ribs—this action is lost when the long thoracic nerve is injured, resulting in winging of the scapula; (2) it also protracts the scapula; and (3) the lower fibres are powerful lateral rotators of the scapula.

Axillary artery

The **axillary artery** is the main artery of the upper limb. **Origin**: it is a continuation of the subclavian artery at the outer border of the first rib. It passes through the apex and lateral part of the axilla. **Termination**: it becomes the brachial artery at the lower border of the teres major, close to the humerus. For the purpose of description, it is divided into three parts by the pectoralis minor. The first part lies superior to the pectoralis minor, the second part deep to it, and the third part inferior to it. **Relations**: the cords of the brachial plexus lie posterior to the first part. They are arranged around the second part according to their names.

DISSECTION 5.6 Axilla-1

Objectives

I. To remove the loose connective tissue and fat of the axilla. **II.** To remove the fascia overlying the lateral wall of the axilla. **III.** To identify and trace the axillary artery, axillary vein, musculocutaneous nerve, median nerve, ulnar nerve, medial cutaneous nerve of the arm and forearm, intercostobrachial nerve, lateral cutaneous branches of the upper intercostal nerves, lateral thoracic artery, and long thoracic nerve.

Instructions

1. Remove the loose connective tissue, fat, and lymph nodes in the axilla to expose its contents. Only a few of the large number of lymph nodes will be seen, unless they are enlarged by disease, so it is not worth trying to dissect them, though they will be felt as slightly firmer structures among the fat. Lymph vessels will not be seen.
2. Expose the coracobrachialis and short head of the biceps muscles which arise from the tip of the coracoid process.
3. Find the **axillary artery** and the **median nerve** medial to these muscles, and the **musculocutaneous nerve** entering the deep surface of the cora-

cobrachialis. Follow this nerve upwards, and find its branch to the muscle.
4. Identify the **axillary vein** medial to the axillary artery.
5. Between the axillary artery and vein, identify the **medial cutaneous nerve of the forearm**, and more posteriorly the larger **ulnar nerve**.
6. Find the **medial cutaneous nerve of the arm** medial to the vein. Trace it superiorly. A branch from the **intercostobrachial nerve** usually joins it. Follow this branch upwards to the emergence of the intercostobrachial nerve from the second intercostal space in the medial axillary wall. Trace that nerve down to the axillary floor and the medial side of the arm.
7. Note again the series of lateral cutaneous branches of the third, fourth, fifth, and sixth intercostal nerves, as they emerge in a vertical line inferior to the point of emergence of the intercostobrachial nerve. Note that the nerves emerge posterior to the pectoralis major and between the digitations (attachments) of the serratus anterior on each rib.
8. Find the lateral thoracic artery and the long thoracic nerve descending on the lateral surface of the serratus anterior muscle which they supply [Fig. 5.13].

The main nerves arising from the cords surround the third part.

The axillary artery supplies the structures in, and surrounding, the axilla [Fig. 5.15]: (1) the **thoracoacromial artery** arises from the second part of the axillary artery and supplies the anterior axillary wall of the axilla, including the clavicle, acromion, and anterior part of the deltoid; (2) the **superior thoracic artery** (a branch of the first part) and the **lateral thoracic artery** (a branch of the second part) supply the medial axillary wall, the lateral part of the mammary gland, and surrounding structures; (3) the **subscapular artery** (a branch of the third part) supplies the posterior axillary wall, including the scapula and muscles covering its posterior aspect; it gives off two branches—the **circumflex scapular** and **thoracodorsal** arteries—which anastomose with branches from the subclavian artery; and (4) the **anterior** and **posterior circumflex humeral arteries** (branches of the third part) supply the proximal

part of the humerus, the muscles covering it, and the shoulder joint.

Axillary vein

The axillary vein lies on the anteromedial aspect of the axillary artery and extends from the lower border of the teres major to the outer border of the first rib. It is a continuation of the **brachial vein** and receives tributaries corresponding to the branches of the axillary artery. In addition, the axillary vein also receives the venae comitantes of the brachial artery inferiorly, and the cephalic vein superiorly. It continues as the subclavian vein at the outer border of the first rib [Fig. 5.13].

Dissection 5.7 continues the dissection of the axilla.

Axillary lymph nodes

The axillary lymph nodes drain the lymph vessels of the: (1) upper limb; and (2) superficial vessels of the trunk above the level of the umbilicus and iliac

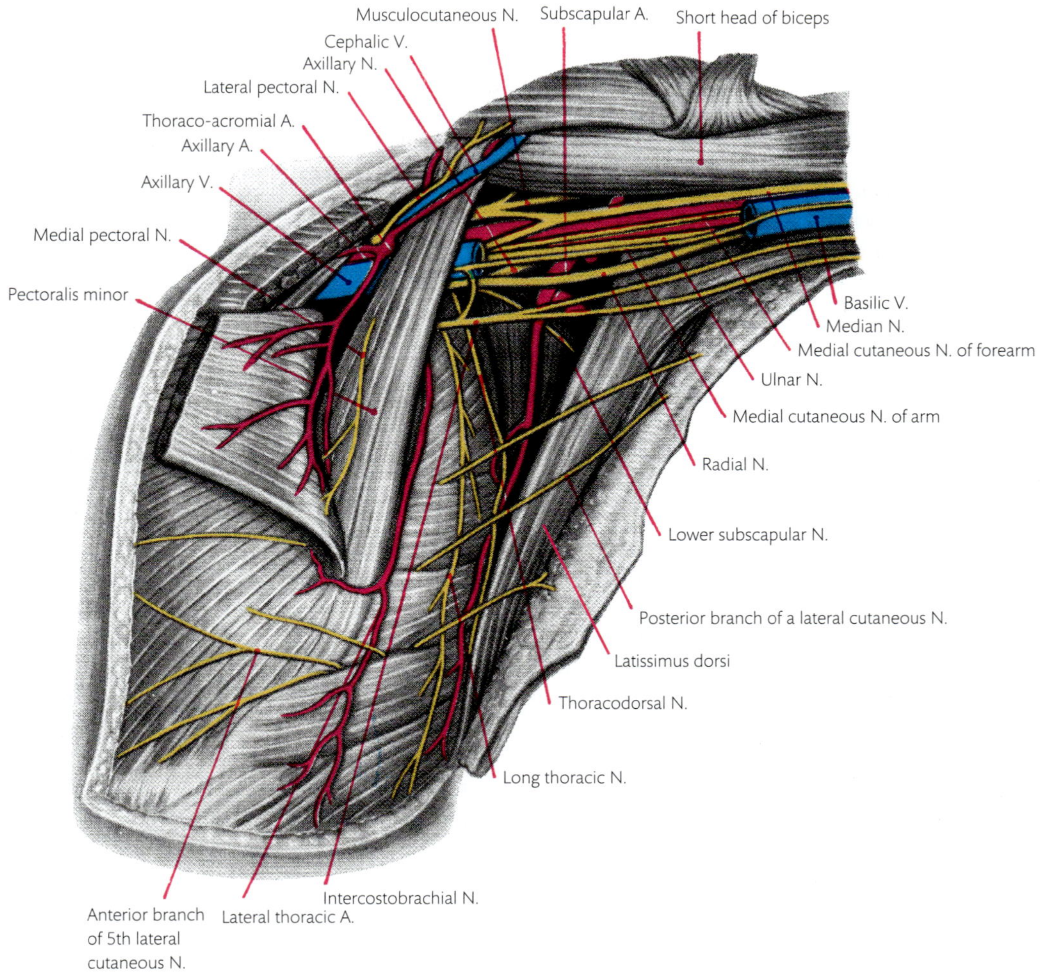

Musculocutaneous N. Subscapular A. Short head of biceps
Cephalic V.
Axillary N.
Lateral pectoral N.
Thoraco-acromial A.
Axillary A.
Axillary V.
Medial pectoral N.
Pectoralis minor

Basilic V.
Median N.
Medial cutaneous N. of forearm
Ulnar N.
Medial cutaneous N. of arm
Radial N.
Lower subscapular N.
Posterior branch of a lateral cutaneous N.
Latissimus dorsi
Thoracodorsal N.
Long thoracic N.
Intercostobrachial N.
Anterior branch   Lateral thoracic A.
of 5th lateral
cutaneous N.

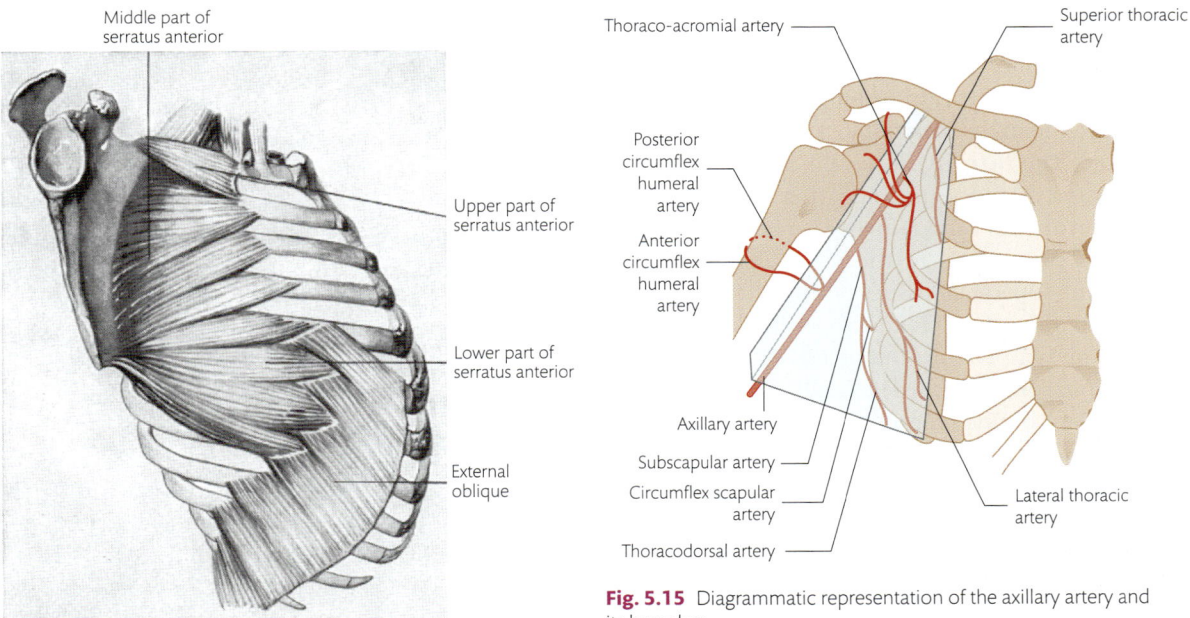

Fig. 5.13 Contents of the axilla exposed by reflection of the pectoralis major and removal of fat, the fascia, and lymph nodes. Part of the axillary vein has been removed to display the medial cutaneous nerve of the forearm and the ulnar nerve.

Middle part of
serratus anterior

Upper part of
serratus anterior

Lower part of
serratus anterior

External
oblique

Thoraco-acromial artery

Superior thoracic
artery

Posterior
circumflex
humeral
artery

Anterior
circumflex
humeral
artery

Axillary artery

Subscapular artery

Circumflex scapular
artery

Thoracodorsal artery

Lateral thoracic
artery

Fig. 5.15 Diagrammatic representation of the axillary artery and
its branches.

Fig. 5.14 Serratus anterior. The scapula is drawn away from the
side of the chest to show the insertion into the scapula.

Axilla

67

DISSECTION 5.7 Axilla-2

Objectives

I. To clean the connective tissue and fascia over the axillary artery, ulnar nerve, radial nerve and its branches, axillary nerve, and subscapular artery and its branches. II. To examine the relationship of the cords of the brachial plexus to the axillary artery. III. To remove the fascia and define the posterior wall of the axilla.

Instructions

1. Expose the axillary artery and vein, and the large nerves surrounding them. If necessary, remove the smaller tributaries of the vein, in order to get a clear view of the nerves. (Since the veins follow the branches of the artery, their loss is of little significance.)
2. Identify and follow the **ulnar nerve**. It lies behind and between the axillary artery and vein.
3. Find the **median nerve** lateral to the axillary artery. Follow its lateral root to the lateral cord of the brachial plexus, and its medial root to the medial cord of the brachial plexus.
4. Identify the **radial nerve** which lies behind the artery. Trace the radial nerve proximally and distally to the lower border of the subscapularis.
5. Find the **axillary nerve** which passes posteriorly along with the posterior humeral circumflex artery.
6. Find the posterior cutaneous nerve of the arm.
7. Find the muscular branches of the radial nerve to the long and medial heads of the triceps muscle.
8. Find the **subscapular artery** as it arises from the axillary artery close to the axillary nerve. Trace it and its major branches—the circumflex scapular and thoracodorsal arteries. The thoracodorsal artery runs along the chest wall parallel to the margin of the latissimus dorsi, together with the thoracodorsal nerve, to that muscle. (You will study the latissimus dorsi in further detail later.) The **circumflex scapular artery** lies close to the nerve (lower subscapular nerve), entering the teres major.
9. Cut across the pectoralis minor, and follow the axillary vessels to the outer border of the first rib. Note that the medial, lateral, and posterior **cords of the brachial plexus** lie around the artery posterior to the pectoralis minor. Above the level of the pectoralis minor, all three cords of the brachial plexus lie posterior to the artery.
10. Expose the anterior surface of the subscapularis, and identify the **upper subscapular nerve(s)** entering it. Follow the **upper** and **lower subscapular** and **thoracodorsal nerves** to their origin from the posterior cord of the brachial plexus [Fig. 5.13].

---

crest. The nodes are scattered throughout the fascia of the axilla and, for the most part, transfer lymph towards the nodes at its apex (**apical nodes**). For descriptive purposes, the axillary lymph nodes are divided into five groups, four of which lie in one angle of the axillary pyramid and drain a specific territory. The **lateral nodes** lie along the axillary vessels and drain the greater part of the upper limb. The **pectoral** or **anterior group** lies in the anteromedial angle, deep to the pectoralis major, and drains the superficial tissues of the anterior and lateral parts of the thoracic and upper abdominal walls, including lymph from the breast. The **subscapular** or **posterior group** lies along the subscapular vessels and drains lymph from the corresponding region on the back. All these nodes communicate with the more centrally placed **central nodes**. The efferent vessels of all these nodes pass to the apical group [Fig. 5.8], which also receives vessels from nodes on the cephalic vein and in the infraclavicular fossa. The efferent vessels of the apical

nodes form the **subclavian lymph trunk** which usually drains into the subclavian vein.

Brachial plexus

The brachial plexus is an important nerve plexus that supplies sensory and motor innervation to the upper limb. The plexus begins in the lower part of the neck (**supraclavicular part**: roots and trunks) and passes as divisions behind the middle third of the clavicle into the apex of the axilla [Figs. 5.16, 5.17, 5.18]. The plexus is formed successively by roots, trunks, divisions, and cords. Nerves supplying the muscles and skin of the upper limb arise from the roots, trunks, and cords of the plexus.

Roots of the brachial plexus

The five **roots** of the brachial plexus are formed by the ventral rami of the lower four cervical nerves, the greater part of the ventral ramus of the first thoracic nerve (C. 5 to T. 1). Small twigs from the ventral rami of the fourth cervical and second tho-

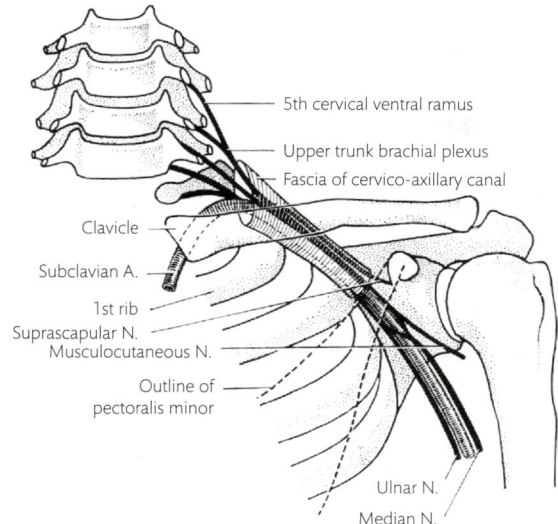

Fig. 5.16 Diagram showing the route of entry of the nerves and subclavian artery into the upper limb. The fascial sheath which binds these structures into a narrow bundle is the axillary sheath.

racic nerves may join the plexus. When the contribution to the brachial plexus from C. 4 is large and that from T. 1 is small, it is considered as a **prefixed brachial plexus**. Similarly, when the contribution from C. 5 is small and that from T. 2 is large, it is considered as a **post-fixed brachial plexus**.

Trunks and divisions of the brachial plexus

Three trunks—the superior, middle, and inferior trunks—arise from the roots. The ventral rami of the fifth and sixth cervical nerves unite to form the **upper trunk** [Fig. 5.18]. The seventh cervical ventral ramus continues as the **middle trunk**. The eighth cervical and first thoracic ventral rami unite to form the **lower trunk**.

Just above the clavicle, each of these trunks splits into an anterior and a posterior division, in this way giving rise to six divisions. The three **posterior divisions** supply the extensor muscles and the skin on the back of the limb. The three **anterior divisions** supply the flexor muscles and the skin on the front of the limb.

Cords of the brachial plexus

The three posterior divisions unite to form the **posterior cord** of the plexus. The anterior divisions of the upper and middle trunks unite to form the **lateral cord** of the plexus, and the anterior division of the lower trunk forms the **medial cord**.

In the axilla (**infraclavicular part**), the cords first lie posterior to the first part of the axillary artery, but lower down posterior to the pectoralis minor, they surround the second part of the axillary artery in positions which correspond to their names. The plexus ends at the lower border of

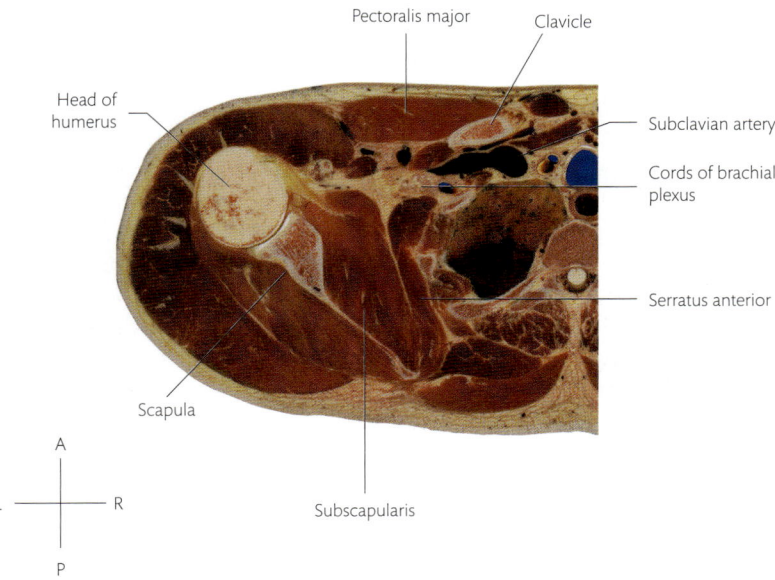

Fig. 5.17 Horizontal section at the level of the shoulder joint. The chief structures in the axilla and its walls are shown. A = anterior; P = posterior; L = left; R = right.

Image courtesy of the Visible Human Project of the US National Library of Medicine.

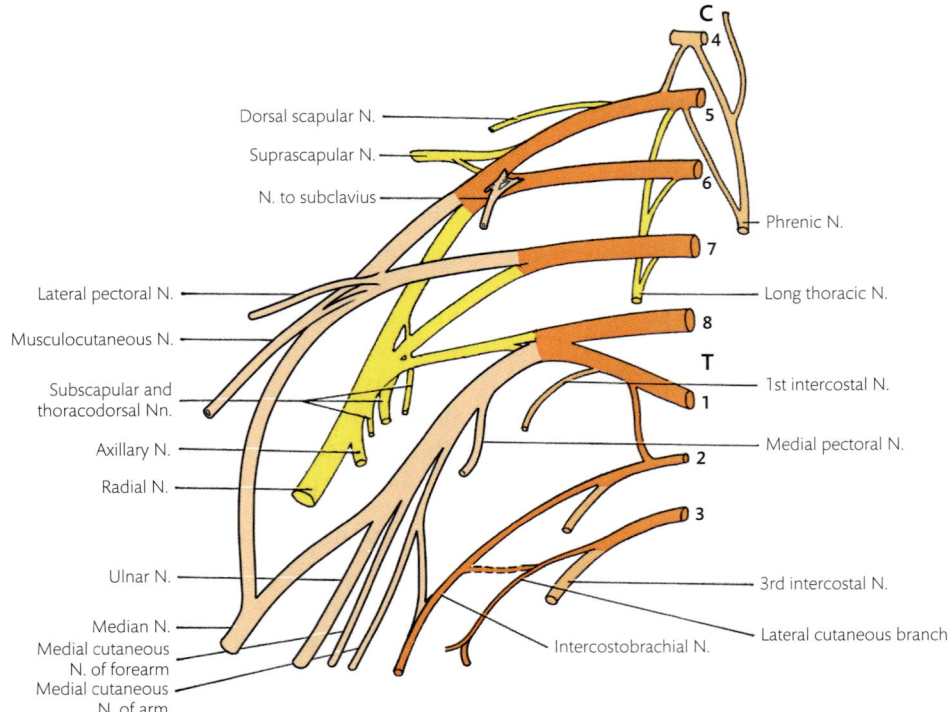

Fig. 5.18 Diagram of the right brachial plexus. Ventral divisions, light orange; dorsal divisions, yellow. C = cervical; T = thoracic.

the pectoralis minor by dividing into a number of branches.

Each cord of the plexus (and in consequence the nerves which arise from that cord) contains nerve fibres from more than one spinal (segmental) nerve. The lateral cord contains nerve fibres from the cervical (C.) 5 to 7 [Fig. 5.18]. The medial cord contains nerve fibres from C. 8 and thoracic (T.) 1; and the posterior cord from C. 5 to T. 1. The root value of a nerve is commonly indicated in brackets, with the name of the nerve. ➲ A knowledge of these segmental, or root values, helps in accurate localization of the injury to the spinal nerves, or the section of the spinal cord from which they arise.

Branches of the brachial plexus

Branches from the roots

The **dorsal scapular nerve** (C. 5) supplies the rhomboid major and minor, and the levator scapulae. It will be seen later on the deep surface of the rhomboid muscles.

The **long thoracic nerve** (C. 5, 6, 7) arises from the posterior aspect of these ventral rami. It descends behind the brachial plexus and axillary artery, and then on the lateral surface of the serratus anterior muscle which it supplies.

Branches from the trunk

The **suprascapular nerve** (C. 5, 6) arises from the upper trunk, and supplies the supraspinatus and infraspinatus muscles. It runs inferolaterally behind the clavicle and crosses the superior border of the scapula to its posterior surface [Fig. 5.19].

The **nerve to the subclavius** (C. 5, 6) also arises from the upper trunk and descends in front of the plexus to supply the subclavius.

Branches from the cords

The **lateral cord** gives rise to: (1) the **lateral pectoral nerve** (C. 5, 6, 7); (2) the **musculocutaneous nerve** (C. 5, 6); and (3) the **lateral root of the median nerve**, which joins the medial root to form the **median nerve**.

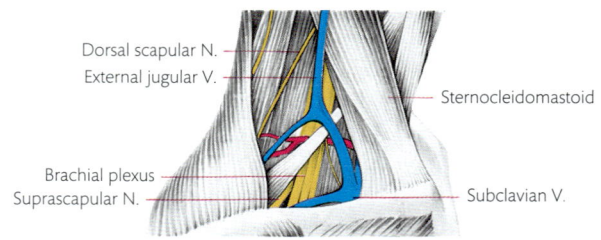

Fig. 5.19 Dissection of the lower part of the posterior triangle of the neck showing the supraclavicular part of the brachial plexus.

The **medial cord** gives rise to: (1) the **medial pectoral nerve** (C. 8, T. 1); (2) the **medial cutaneous nerve of the arm**; (3) the **medial cutaneous nerve of the forearm**; (4) the **medial root of the median nerve**; and (5) the **ulnar nerve** (C. 8; T. 1).

The **posterior cord** gives rise to: (1) the **upper** and (2) **lower subscapular nerves** (C. 5, 6); (3) the **thoracodorsal nerve**; (4) the **axillary nerve** (C. 5, 6); and (5) the **radial nerve** (C. 5, 6, 7, 8; occasionally T. 1).

The **median nerve** (C. 5, 6, 7, 8; T. 1) is formed by one root, each from the medial and lateral cords of the brachial plexus.

Injuries to the brachial plexus

Injuries to any part of the brachial plexus, or to its branches will lead to sensory and motor loss in the area supplied by it. Injuries of the brachial plexus are best understood after you have completed the study of the upper limb, and are described in Chapter 12.

See Clinical Applications 5.1 and 5.2.

CLINICAL APPLICATION 5.1 Breast cancer

The following observations were made during breast examination of a 36-year-old woman with breast cancer. The right breast was firmly adherent to the underlying tissue.

Study question 1: name the tissue which lies immediately deep to the breast. What does this immobility/tethering of the breast tell you about the disease process? (Answer: deep fascia, pectoralis major, and serratus anterior; the breast being fixed to the underlying tissue means that the cancer has invaded the underlying muscle.) The skin over the upper lateral quadrant of the breast is thick and pitted, resulting in an orange-peel appearance. This appearance is caused by two factors: (1) blockage of lymph vessels by cancer cells, resulting in lymphoedema;

and (2) the fact that the subcutaneous tissue is prevented from swelling uniformly by the shortened suspensory ligaments which are also invaded by disease.

Study question 2: to what structures are the suspensory ligaments attached? (Answer: the suspensory ligaments run from the glands to the underlying deep fascia and to the overlying skin.)

Study question 3: on examination of the axilla, hard and 3 cm-sized masses were felt immediately deep to the anterior axillary fold. What are these masses likely to be, and how are they related to the disease process? (Answer: the masses are most likely enlarged anterior axillary lymph nodes, to which the cancer cells from the breast have spread.)

CLINICAL APPLICATION 5.2 Axillary lymph node dissection

Axillary lymph node dissection is a surgical procedure that is used for staging breast cancer. The surgeon explores the axilla to identify, examine, and remove lymph nodes. Axillary lymph node status on whether or not they are invaded by cancer cells, and to what extent they are involved, gives valuable information for planning treatment. Lymph drainage of the upper limb may be impeded after removal of the axillary nodes.

Study question 1: why is it common for patients who have undergone this procedure to have swelling of the upper limb? What name is given to swelling due to this cause? (Answer: the upper limb drains into the axillary

lymph nodes, which have been removed during surgery. As such, the lymph collects in the limb tissue. Such swelling is called 'lymphoedema'.) The long thoracic nerve and the thoracodorsal nerve have a long course in the axilla and may become infiltrated by cancer cells. These nerves may also be damaged during the surgery. The thoracodorsal nerve lies on the posterior wall of the axilla and enters the latissimus dorsi near its medial border. The axillary tail of the breast lies close to it.

Study question 2: what would be the result of damage to the thoracodorsal nerve? (Answer: weakened medial rotation and adduction of the arm.)

CHAPTER 6
The back

Turn the body face downwards, and examine the structures which connect the upper limb to the back of the trunk.

Surface anatomy of the back

Begin by reviewing the structure of the scapula [Fig. 6.1].

The **scapula** is placed against the posterolateral wall of the thorax. It lies over the second to seventh ribs and extends into the posterior wall of the axilla. It is thickly covered with muscles, but most of its outline can be felt in the living subject. Find the acromion at the top of the shoulder. Draw your finger along the crest of the spine of the scapula which runs medially and slightly downwards from the acromion to the medial border of the scapula [Fig. 4.2]. Trace the medial border to the inferior angle and, if possible, to the superior angle, palpating it through the muscles that cover it. The scapula is held in position by muscles and the clavicle. It is very mobile—the scapulae move apart when the arms are folded across the chest. When the shoulders are drawn back, the medial borders of the scapulae are brought close to each other and the posterior median line.

The rib felt immediately inferior to the inferior angle of the scapula is usually the **eighth rib**, and the lower ribs may be counted from it. The twelfth rib is not palpable, unless it projects beyond the lateral margin of the back muscle—the erector spinae [Fig. 6.1].

The **iliac crest** is the curved bony ridge felt below the waist. Trace it forwards to the **anterior superior iliac spine** and backwards to the **posterior superior iliac spine**. The posterior superior iliac spine is felt in a shallow dimple in the skin above the buttock and about 5 cm from the median plane.

Between the left and right dimples is the back of the **sacrum**. Usually, three sacral spines can be palpated in the median plane. The **coccyx** is the slightly mobile bone felt deep between the buttocks in the median plane.

Feel the tips of the **spines of the vertebrae** in the median furrow of the back. These are the only parts of the vertebral column which are easily felt. It is difficult to identify individual spines directly, but the seventh cervical spine (**vertebra prominens**) is the uppermost spine which can be readily felt at the root of your neck. Below this, the approximate levels of other spines are as described in Table 6.1.

Above the vertebra prominens, only the **second cervical spine** can be felt easily. It is about 5 cm below the **external occipital protuberance** which is on the lower part of the back of the head where the median furrow of the neck (**nuchal groove**) meets the skull. The short cervical spines (compare with C. 7) are separated from the skin by a median fibrous partition—the **ligamentum nuchae**. The posterior edge of the ligamentum nuchae stretches from the external occipital protuberance to the seventh cervical spine.

The **superior nuchal line** is a curved ridge on the occipital bone of the skull, extending laterally from a midline bony elevation—the external occipital protuberance.

Deep fascia

On the dorsal surface of the scapula, a dense layer of fascia extends from the spine of the scapula to the margins of the infraspinous fossa. Superolaterally, it splits to enclose the deltoid muscle.

See Dissection 6.1 and Figure 6.2 for instructions on skin reflection and cutaneous nerves of the back.

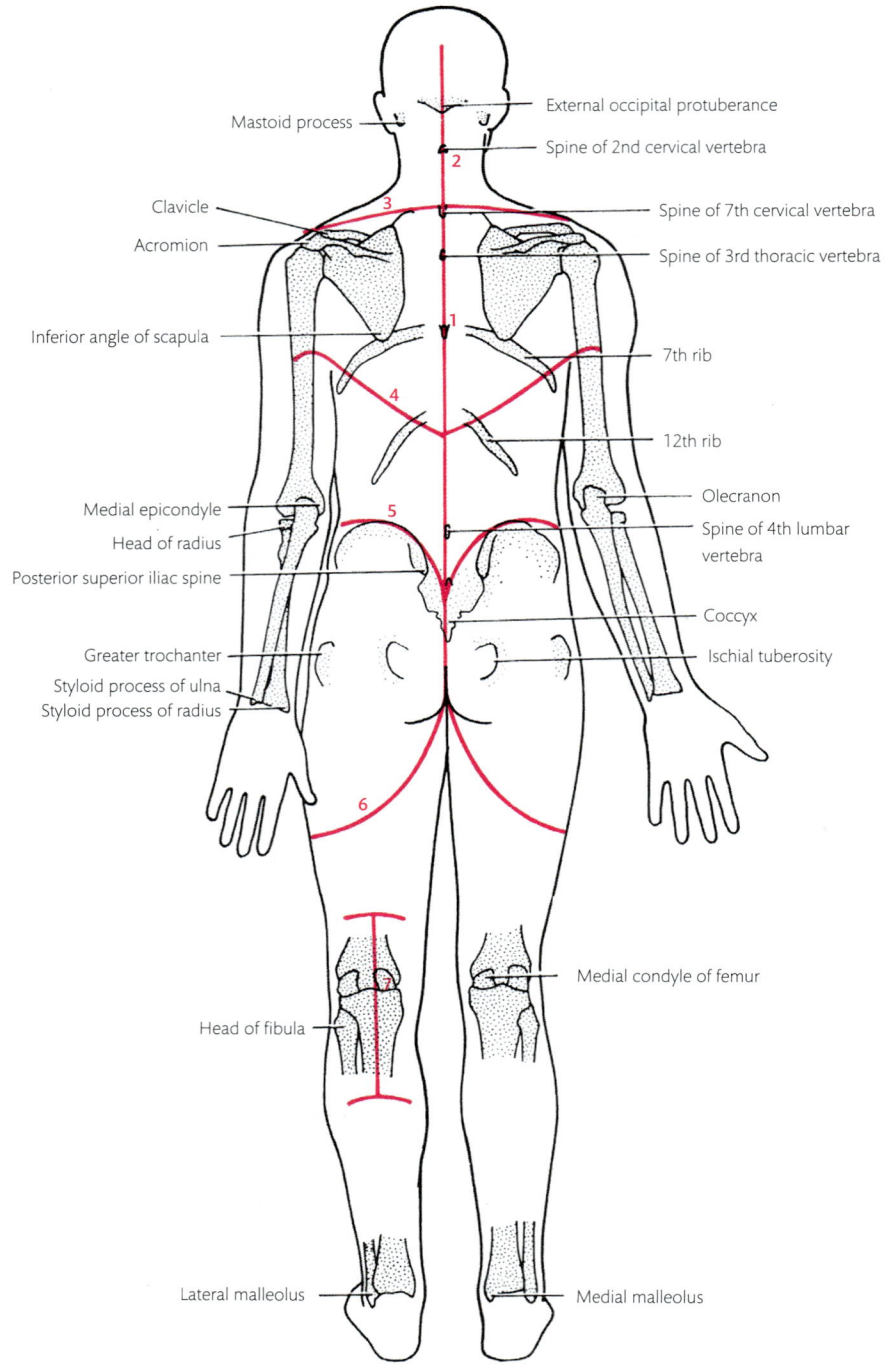

Fig. 6.1 Landmarks and incisions. Left forearm pronated, right forearm supinated.

Cutaneous nerves and arteries

The cutaneous nerves of the back are branches of the **dorsal rami of the spinal nerves**. Each dorsal ramus divides into a medial and a lateral branch [see Fig. 1.22]. Both of these enter and supply the erector spinae muscles. But only one continues through the erector spinae to supply the overlying skin. In the cervical and upper six or seven thoracic segments, the medial branches from the dorsal rami form the **cutaneous nerves**. They pierce the deep

Table 6.1 Vertebral levels for bony surface landmarks

Vertebral spine	Level
Third thoracic	Junction of the scapular spine and the medial border
Seventh thoracic	Inferior angle of the scapula
Fourth lumbar	Highest point of the iliac crest
Second sacral	Posterior superior iliac spine

fascia close to the median plane. Below this level, the lateral branches form the cutaneous nerves and emerge in line with the lateral edge of the erector spinae, piercing either the latissimus dorsi (upper nerves) or the dense deep fascia (thoracolumbar fascia) of the small of the back (lower nerves). Each of these cutaneous nerves divides into a smaller medial and a larger lateral branch. In the thoracic and lumbar nerves, branches of the dorsal rami descend before entering the skin. Thus, the area of skin supplied by the dorsal ramus of each nerve lies at a lower level than that at which the spinal nerve emerges. This makes the **dermatomes** of the trunk more nearly horizontal than would be expected from the oblique course of the ventral rami.

The cutaneous branches of the dorsal rami of the upper three lumbar nerves pierce the deep fascia a short distance superior to the iliac crest and turn down to supply the skin of the gluteal region.

The **arteries** which accompany the cutaneous nerves of the back arise from the dorsal branches of the posterior intercostal and lumbar arteries.

Muscles attaching the scapula to the trunk

Two of these muscles—the **pectoralis minor** and the **serratus anterior**—have been seen already.

Other muscles attaching the scapula to the trunk are the **trapezius**, **rhomboid major**, **rhomboid minor**, and **levator scapulae** [see Fig. 6.2].

The latissimus dorsi will also be seen in the dissection of the back. The latissimus dorsi and pectoralis major are the only two muscles which attach the humerus to the trunk.

See Dissection 6.2.

Trapezius

The trapezius is a broad muscle placed superficially in the upper part of the back. The right and left trapezius together are shaped like a trapezium, which gives the muscle its name. **Origin**: it has an extensive origin from: (1) the external occipital protuberance; (2) the superior nuchal line on the occiput; (3) the ligamentum nuchae; (4) the seventh cervical vertebra in the neck; (5) the spines of all 12 thoracic vertebrae; and (6) the supraspinous ligaments. **Insertion**: from this almost completely midline origin, the fibres run laterally and come together to be inserted into the: (1) lateral third of the clavicle; (2) the acromion; and (3) the crest of the spine of the scapula [see Fig. 4.3]. **Nerve supply**: cranial nerve XI—the spinal accessory nerve [see Fig. 6.2]. **Actions**: elevation of the shoulder, and retraction and lateral rotation of the scapula.

Latissimus dorsi

Origin: this broad sheet of muscle takes origin from: (1) the lower six thoracic spines and (2) the supraspinous ligaments between them, deep to the trapezius; (3) the thoracolumbar fascia [see Fig. 6.2]; (4) the posterior part of the iliac crest; and (5) the lower three or four ribs. **Insertion**: the fibres come together and wind around the inferior border of the teres major to reach its anterior surface and are inserted into the intertubercular sulcus of the humerus [see Fig. 4.8A]. **Nerve supply**: thoracodorsal nerve [see Fig. 5.13]. **Actions**: it is a powerful adductor of the

---

DISSECTION 6.1 Skin reflection of the back

Objectives

I. To reflect the skin of the back. **II.** To identify the cutaneous nerves of the back.

Instructions

1. Make the skin incisions 1, 3, 4, and 5 [Fig. 6.1].

2. Reflect the two skin flaps laterally, stripping the skin and the superficial fascia from the deep fascia by blunt dissection.

3. Find the cutaneous nerves as they pierce the deep fascia [see Fig. 6.2]. This is more difficult than on the flexor surface, because of the denser connections of the superficial fascia to the deep fascia on the back.

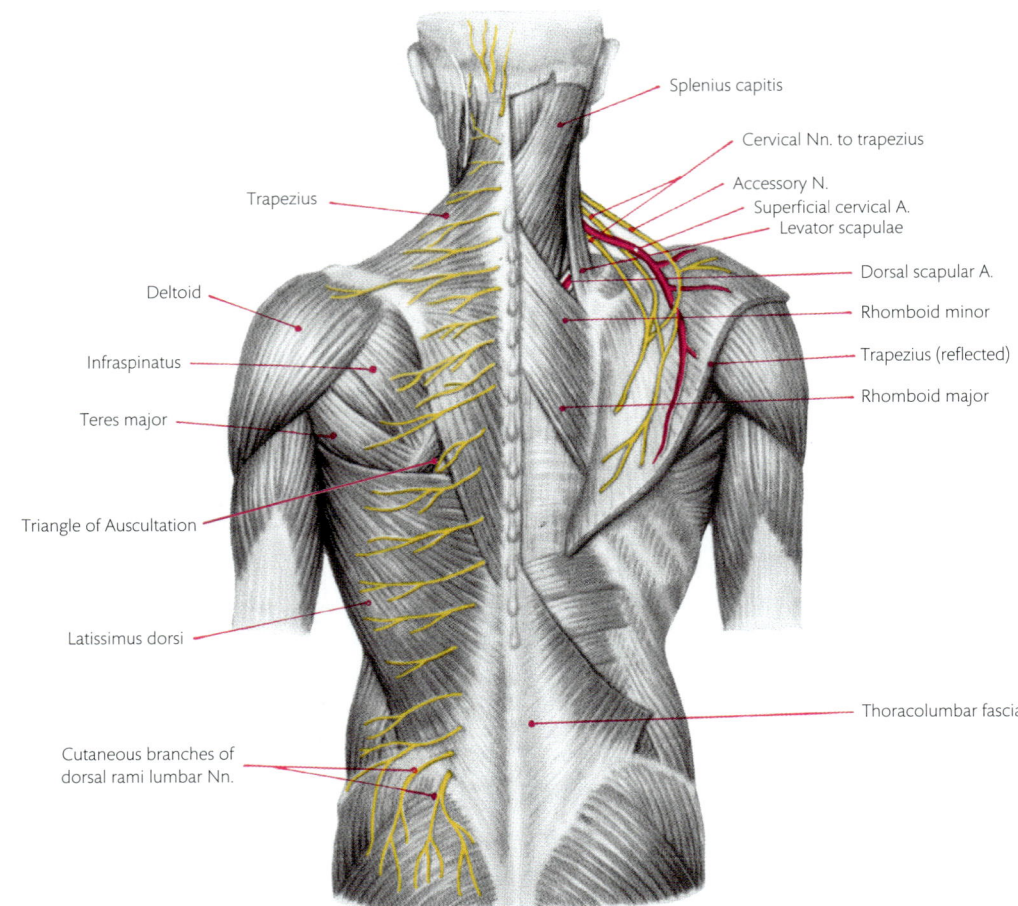

Fig. 6.2 Dissection of muscles and nerves of the back.

humerus and a depressor of the shoulder. It is used to pull the arm down from its fully abducted position above the head, as in rope climbing. When the shoulder is flexed, it acts as an extensor of that joint. When the shoulder is fixed, it helps with retraction and medial rotation of the scapula.

Levator scapulae

Origin: the levator scapulae takes origin from the transverse processes of cervical vertebrae 1 to 4. **Insertion**: it is inserted into the medial border of the scapula, between the superior angle and the root of the spinous process. **Nerve supply**: dorsal scapular nerve [see Fig. 6.2]. **Actions**: elevation and medial rotation of the scapula.

Triangle of auscultation

The rib cage is uncovered by muscles in a small area bounded by the lateral border of the trapezius, the medial border of the latissimus dorsi, and

the lateral margin of the scapula. This area is called the 'triangle of auscultation', as breath sounds are heard more distinctly over it [see Fig. 6.2].

Rhomboid minor

Origin: the rhomboid minor arises from the lower part of the ligamentum nuchae and the spine of the first thoracic vertebra. **Insertion**: it is inserted into the medial border of the scapula, at the root of the spine. **Nerve supply**: dorsal scapular nerve [see Fig. 6.2]. **Actions**: retraction and elevation of the scapula.

Rhomboid major

Origin: the rhomboid major arises from the spines of the second to fifth thoracic vertebrae. **Insertion**: it is inserted into the medial border of the scapula, inferior to the root of the spine. **Nerve supply**: dorsal scapular nerve [see Fig. 6.2]. **Actions**: retraction, medial rotation, and elevation of the scapula.

DISSECTION 6.2 The back

Objectives

I. To study the muscles of the back—latissimus dorsi, trapezius, levator scapulae, rhomboid minor, rhomboid major, serratus anterior, and inferior belly of the omohyoid. **II.** To identify and trace the dorsal scapular nerve, the deep branch of the transverse scapular artery, and the suprascapular vessels and nerves.

Instructions

1. Remove the deep fascia from the surface of the trapezius below the spine of the seventh cervical vertebra. The upper part of the muscle will be dissected with the head and neck.
2. Expose the **latissimus dorsi** [see Fig. 6.2]. This is difficult because of the varying direction of its fibres and the thinness of its upper part, which has an ill-defined margin. Define its attachment to the thoracolumbar fascia and to the iliac crest. Expose the lateral border of the muscle and its slips from the lowest three or four ribs. These lower fibres interdigitate with the external oblique muscle of the abdomen.
3. Identify a small triangular area between the lateral border of the trapezius, the medial border of the latissimus dorsi, and the lateral border of the scapula, where the rib cage is exposed. This is the **triangle of auscultation**.
4. Cut the **trapezius** horizontally halfway between the clavicle and the spine of the scapula, and vertically 5 cm lateral to the median plane. In the latter cut, take care not to injure the underlying rhomboid muscles, as the trapezius is thin near its origin. Reflect the lower part of the **trapezius**, and define its attachments to the thoracic vertebral spines, the medial border of the acromion, and the superior margin of the crest of the spine of the scapula.
5. Note the superficial branch of the transverse cervical vessels and the accessory nerve on the deep surface of the lateral part of the trapezius. Expose the upper part of the muscle at its attachments to the clavicle and the acromion [see Fig. 4.3].
6. Define the **levator scapulae, rhomboid minor,** and **rhomboid major** muscles, which are attached from above downwards, in that order, to the medial border of the scapula, deep to the trapezius [see Fig. 4.7]. The rhomboid minor is attached where the spine of the scapula meets the medial border. Only the lower part of the levator scapulae can be seen at this time. Free this part of the muscle from the underlying fat, and identify the **dorsal scapular nerve** and the **deep branch of the transverse cervical artery**. Follow them to the deep surface of the rhomboid muscles.
7. Divide both rhomboid muscles midway between the vertebral spines and the medial border of the scapula. Reflect these muscles, and define their attachments.
8. Trace the dorsal scapular nerve and the deep branch of the transverse cervical artery deep to them.
9. Lift the medial border of the scapula from the thoracic wall. Note how easily the scapula and the underlying **serratus anterior muscle** are lifted clear because of the loose connective tissue deep to them.
10. Pass one hand between the thoracic wall and the serratus anterior. Then place the other hand in the axilla from the front, and slide it backwards between the subscapularis and the serratus anterior, which is now between your hands. Define the attachment of the serratus anterior to the scapula [see Fig. 5.14].
11. Turn the latissimus dorsi downwards and note how the inferior digitations of the serratus anterior converge on the anterior surface of the inferior angle of the scapula [see Fig. 5.14].
12. Remove the fat from the suprascapular region deep to the cut edge of the trapezius, and find the inferior belly of the **omohyoid muscle**, and the **suprascapular vessels** and **nerve** passing to the superior border of the scapula at the scapular notch [Figs 4.4, 4.5].

Spinal accessory nerve

The accessory, or eleventh, cranial nerve consists of cranial and spinal parts. The **spinal part** arises from the cervical spinal cord (C. 1 to 5), enters the skull, has a short course in the cranium, and re-emerges in the neck. It passes posteroinferiorly through the sternocleidomastoid [see Fig. 5.5] to the deep surface of the trapezius supplying both muscles.

Dorsal scapular nerve

The dorsal scapular nerve arises in the lower part of the neck from the ventral ramus of the fifth cervical nerve. It passes posteroinferiorly and runs deep to the lower part of the levator scapulae and the rhomboid muscles, and supplies them. It is accompanied by the deep branch of the transverse cervical artery [see Fig. 6.2].

Transverse cervical artery

The **transverse cervical artery** is a branch of the thyrocervical trunk from the first part of the subclavian artery. It divides into a **superficial cervical artery**, which accompanies the spinal accessory nerve deep to the trapezius, and the **dorsal scapular artery**, which accompanies the dorsal scapular nerve deep to the levator scapulae and the rhomboids [see Fig. 6.2]. Both branches take part in the anastomosis around the scapula.

Movements of the scapula

The scapula is able to slide freely over the chest wall because of the loose connective tissue deep to the serratus anterior. The scapular movements are protraction, retraction, elevation, depression, and medial and lateral rotation. These movements are produced by the muscles which attach the scapula to the trunk and indirectly by the muscles passing from the trunk to the humerus when the shoulder joint is fixed. All these movements take place around the **sternoclavicular joint**, with minor adjustments at the **acromioclavicular joint**.

Protraction

This forward movement of the scapula on the chest wall is produced by the **serratus anterior** and is assisted by the pectoral muscles [see Fig. 6.3B]. During protraction, all eight digitations of the serratus anterior contract with the **pectoralis minor** and the sternocostal fibres of the **pectoralis major** [see Fig. 5.14]. This movement is used in reaching forwards, punching, and pushing.

Retraction

Retraction is the reverse of protraction. It draws the scapulae back towards the median plane and braces back the shoulders. It is produced by contraction of the middle fibres of the **trapezius**, which pass

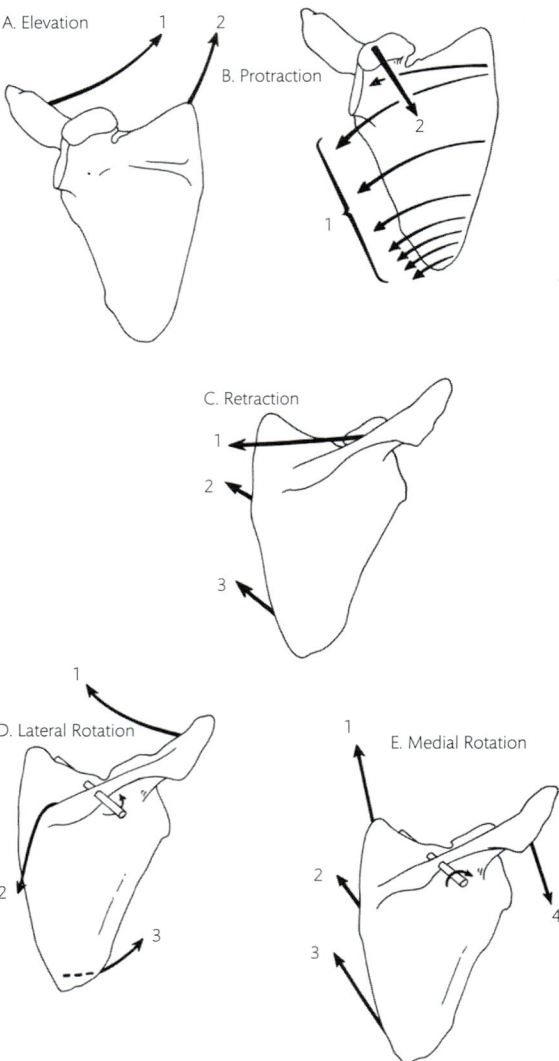

Fig. 6.3 Diagrams to show the direction of pull of muscles acting on the scapula. (A) Elevation of the scapula: 1. upper fibres of the trapezius; 2. levator scapulae. (B) Protraction of the scapula, as in punching or stretching forwards: 1. the serratus anterior pulls the scapula forwards; 2. the pectoralis minor assists. (C) Retraction of the scapula, as in drawing the shoulders back: 1. middle fibres of the trapezius; 2. rhomboid minor; 3. rhomboid major. (D) and (E) Rotation of the scapula. In these figures, the axis of movement is shown as a rod piercing the scapula. The arrow shows the direction of rotation. Lateral rotation: 1. upper fibres of the trapezius; 2. lower fibres of the trapezius; 3. lower part of the serratus anterior. Medial rotation: 1. levator scapulae; 2. rhomboid minor; 3. rhomboid major; 4. the weight of the upper limb.

horizontally from the ligamentum nuchae, and the seventh cervical and upper thoracic spines to the acromion and the lateral part of the spine of the scapula, and also by the **rhomboid muscles** passing from a similar origin to the medial border of the scapula [see Fig. 6.3C].

Elevation

Elevation, as in shrugging the shoulders, is achieved by the simultaneous contraction of the **levator scapulae** and the upper fibres of the **trapezius**, which descend from the skull and ligamentum nuchae to the clavicle and acromion [see Fig. 6.3A].

Depression

This movement is achieved by gravity and the contraction of the pectoralis minor, the lower fibres of the pectoralis major and trapezius, and the latissimus dorsi.

Lateral rotation

Rotation takes place around a horizontal axis passing through the middle of the scapular spine and the sternoclavicular joint [see Fig. 6.3D]. In **lateral rotation**, the upper fibres of the **trapezius** raise the acromion and lateral part of the clavicle, while its lower fibres depress the medial end of the spine of the scapula. Together, they laterally rotate the scapula. The lower five digitations of the ser-

Table 6.2 Trunk muscles acting on the shoulder girdle

Muscle	Origin	Insertion	Action	Nerve supply
Trapezius	Upper fibres: occiput, external occipital protuberance, and superior nuchal line Ligamentum nuchae Seventh cervical spine	Clavicle, lateral third Scapula, acromion, medial edge Scapula, superior margin of spine	Elevates shoulder	Spinal accessory
	Lower fibres: seventh cervical spine All thoracic vertebral spines	Scapula, superior margin of spine	1. Retracts scapula 2. Depresses medial part of spine of scapula	Spinal accessory
	All fibres: occiput, external occipital protruberance, and superior nuchal line Ligamentum nuchae Seventh cervical spine All thoracic vertebral spines	Clavicle, lateral third Scapula, acromion, medial edge Scapula, superior margin of spine	Rotates scapula laterally	Spinal accessory
Serratus anterior	Upper eight ribs, anterolateral surface	Scapula, medial border	1. Protracts scapula 2. Holds scapula against ribs 3. Lower five digitations laterally rotate scapula	Long thoracic nerve
Pectoralis minor	Ribs 3–5, anterolateral surface	Scapula, coracoid process	1. Protracts scapula 2. Depresses shoulder	Medial pectoral nerve
Rhomboid major	Second to fifth thoracic vertebral spines	Scapula, medial border inferior to spine	1. Retracts scapula 2. Medially rotates scapula 3. Elevates scapula	Dorsal scapular nerve
Rhomboid minor	Lower ligamentum nuchae First thoracic vertebral spine	Scapula, medial border at spine	1. Retracts scapula 2. Elevates scapula	Dorsal scapular nerve
Levator scapulae	Cervical transverse processes 1–4	Scapula, medial border between superior angle and rhomboid minor	1. Elevates scapula 2. Medially rotates scapula	Dorsal scapular nerve
Subclavius	First costal cartilage	Clavicle, inferior surface	Holds clavicle to sternum	Nerve to subclavius
Pectoralis major, sternocostal*	Sternum Costal cartilages 1–6	Humerus, lateral lip of intertubercular sulcus	1. Protracts scapula 2. Depresses shoulder	Medial and lateral pectoral nerves
Latissimus dorsi*	Lower ribs Thoracolumbar fascia Iliac crest	Humerus, intertubercular sulcus	1. Depresses shoulder 2. Medially rotates scapula 3. Retracts scapula	Thoracodorsal nerve

* The sternocostal part of the pectoralis major and latissimus dorsi can act on the scapula only when the shoulder is fixed.

Table 6.3 Movements of the shoulder girdle

Movement	Muscles	Nerve supply (motor)
Elevation	Levator scapulae	Dorsal scapular, and C. 3 and 4
	Trapezius, upper part	Accessory
	Rhomboids	Dorsal scapular
Depression	Pectoralis minor	Medial pectoral
	Trapezius, lower part	Accessory
	Latissimus dorsi*	Thoracodorsal
	Pectoralis major, lower sternocostal part*	Medial pectoral
Protraction	Serratus anterior	Long thoracic
	Pectoralis minor	Medial pectoral
	Pectoralis major, sternocostal part*	Medial and lateral pectoral
Retraction	Rhomboid major and minor	Dorsal scapular
	Trapezius, middle part	Accessory
	Latissimus dorsi*	Thoracodorsal
Lateral rotation (e.g. in abduction of arm)	Serratus anterior, lower five digitations	Long thoracic
	Trapezius, upper and lower parts	Accessory
Medial rotation	Rhomboid major and minor	Dorsal scapular
	Levator scapulae	Dorsal scapular, and C. 3 and 4
	Pectoralis major, lower sternocostal part*	Medial pectoral
	Pectoralis minor	Medial pectoral
	Latissimus dorsi*	Thoracodorsal

* These muscles can only act on the scapula through a fixed shoulder joint.

ratus anterior converge on the inferior angle of the scapula and play a powerful part in this movement by pulling that angle laterally and forwards. Lateral rotation tilts the glenoid cavity upwards [see Fig. 4.5] and is important in abduction of the upper limb above the horizontal. ➲ Normally, scapular and shoulder joint movements occur together, but if the scapular movement is paralysed, abduction of the limb to the horizontal cannot be achieved because the weight of the limb forces the scapula into medial rotation.

Medial rotation

This is the opposite movement to lateral rotation. Gravity plays a large part in this movement, as in depression of the scapula. In addition, combined contraction of the levator scapulae, rhomboids, and latissimus dorsi produces an active movement which is assisted by the pectoral muscles [see Fig. 6.3E].

When all the muscles attaching the scapula to the trunk are contracted, the scapula is fixed to form a stable base on which upper limb movements can take place. It is also used in transmitting forces from the trunk to the upper limbs, as in lifting heavy weights in the hands by straightening the flexed legs.

Muscles acting on, and movements of, the scapula

Having studied the muscles acting on the scapula and its movements, review the details of the muscles by using Table 6.2. This table lists the origin, insertion, nerve supply, and action of muscles acting on the shoulder girdle. In Table 6.3, the muscles are grouped according to their actions (and the nerve supply of each muscle is repeated). This allows an easy assessment of the degree of loss of a particular movement, following the destruction of a particular nerve. (These tables show only the actions of a particular muscle when it actively shortens.)

Clinical Applications 6.1 and 6.2 demonstrate how this anatomy relates to your clinical practice.

CLINICAL APPLICATION 6.1 Winging of the scapula

The long thoracic nerve descends from the brachial plexus and runs on the lateral surface of the serratus anterior before supplying that muscle. It may be affected in metastatic cancer of axillary lymph nodes and their surgical resection. It is also vulnerable to damage by direct trauma, such as a blow to the ribs, because of its superficial location when the arm is raised in sport or combat.

Study question 1: what change would be observed in the position of the scapula at rest, when the serratus anterior is paralysed? (Answer: the medial border of the scapula would project posteriorly away from the thoracic wall.) When the arm is raised, the medial border and inferior angle will move sharply away from the chest wall.

Study question 2: what movements would be lost when the serratus anterior is paralysed? (Answer: lateral rotation of the scapula. Also the patient will not be able to abduct the arm above the horizontal, as this movement is dependent on lateral rotation of the scapula.)

CLINICAL APPLICATION 6.2 Testing cranial nerve XI

Functional assessment of cranial nerves is an essential part of the general physical examination. The distal part of the spinal accessory nerve (cranial nerve XI) is susceptible to injury in the neck due to its superficial position between the muscles. Injury to the spinal accessory nerve can cause diminished or absent function of the upper portion of the trapezius muscle.

Study question 1: using your knowledge of movements of the shoulder girdle, name the movement most severely affected by injury to the spinal accessory nerve. (Answer: elevation of, or shrugging, the shoulder.)

Study question 2: how does the physician test the function of cranial nerve XI? (Answer: the patient is asked to shrug the shoulder with and without resistance. One-sided weakness of the trapezius indicates injury to the spinal accessory nerve of that side.)

CHAPTER 7
The free upper limb

Introduction

Regions of the free upper limb will be described and dissected in the subsequent chapters. This chapter gives an overview of the cutaneous nerves, and which are best studied in continuity from the shoulder down to the hand, as well as superficial veins and lymphatics which are described in continuity from the hand to the shoulder.

Superficial veins

The main superficial veins are extremely variable. They begin in the **dorsal venous network** of the hand—an irregular arrangement of veins on the dorsum of the hand, usually with a transverse element 2–3 cm proximal to the heads of the metacarpals. This network receives the dorsal digital veins through the **intercapital veins**, and also communicating veins from the palm which pass through the intermetacarpal spaces. This network drains into a number of veins, of which one or more on the medial side form the **basilic vein** and on the lateral side the **cephalic vein** [Fig. 7.1]. Both the basilic and cephalic veins turn round the corresponding border of the forearm to reach the anterior surface and ascend on the anteromedial and anterolateral surfaces of the forearm [Figs 7.1, 7.2]. At the elbow, the arrangement of the superficial veins is variable, but they are usually united anterior to the cubital fossa by a **median cubital vein**. Much of the blood in the cephalic vein is transferred to the basilic vein. Another common arrangement is shown in Fig. 7.3. The veins then ascend on the corresponding sides of

the biceps muscle. The cephalic vein pierces the deep fascia and runs between the pectoralis major and the deltoid in the deltopectoral groove to the infraclavicular fossa. Here it passes deep to join the axillary vein near the apex of the axilla. The basilic vein pierces the deep fascia at about the middle of the arm. It unites with the brachial veins and runs with the brachial artery to become the axillary vein at the lower border of the axilla [see Fig. 5.10].

Lightly compress your right arm with your left hand, and contract your right forearm muscles by clenching and unclenching your fist. This distends the **superficial veins** and makes them visible. Compare your superficial veins with those of the other students, and note their variability and the presence of a cephalic and a basilic vein in most cases.

The median cubital vein is most frequently used for venepuncture (taking of a blood sample, administering intravenous drugs, and transfusing blood), because the vein is large and superficially located, and therefore easily seen and felt. For more details on venepuncture of the median cubital vein, see Clinical Application 7.1.

Lymph vessels and nodes of the upper limb

It is not possible in an ordinary dissection to display the lymph vessels in any part of the body, and lymph nodes are difficult to find. It is necessary, however, to know the arrangement of the vessels and nodes, because they form a common route for the spread of infection and cancer. The primary source of disease can often be deduced from the nodes which are involved.

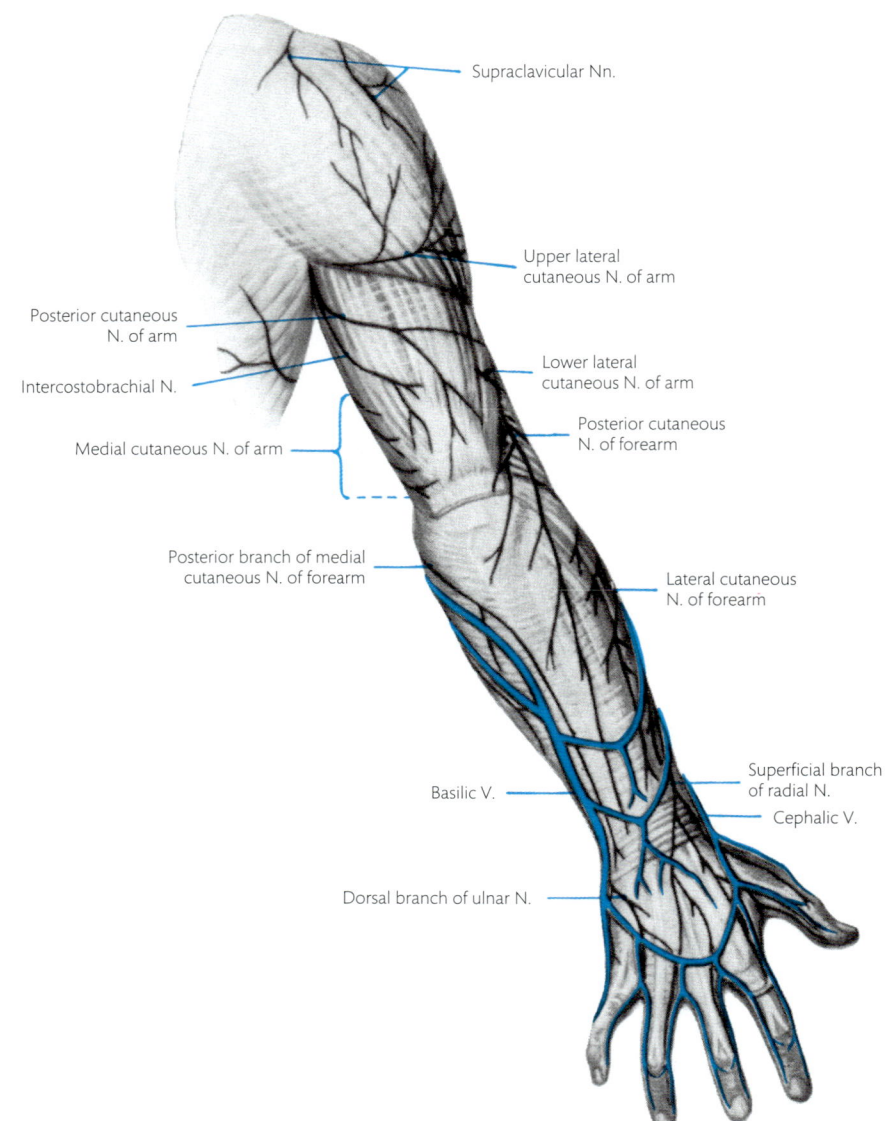

Fig. 7.1 Superficial veins and nerves of the back of the upper limb.

Labels on figure: Supraclavicular Nn.; Upper lateral cutaneous N. of arm; Posterior cutaneous N. of arm; Intercostobrachial N.; Lower lateral cutaneous N. of arm; Medial cutaneous N. of arm; Posterior cutaneous N. of forearm; Posterior branch of medial cutaneous N. of forearm; Lateral cutaneous N. of forearm; Basilic V.; Superficial branch of radial N.; Cephalic V.; Dorsal branch of ulnar N.

In the upper limb, as elsewhere, the lymph vessels and nodes are divided into two groups by the deep fascia: the superficial vessels and nodes, and the deep vessels and nodes.

Deep lymph vessels, which are much less numerous than superficial vessels, drain structures which are deep to the deep fascia. They accompany the main blood vessels and drain into the axillary lymph nodes. Some of the lymph they contain may have passed through a small number of **deep lymph nodes** which are occasionally found on the arteries of the forearm, in the cubital fossa, and on the brachial artery.

The **superficial lymph nodes** of the upper limb are few in number: (1) one or two lie a little superior to the medial epicondyle near the basilic vein; and (2) a few are scattered along the upper part of the cephalic vein [see Fig. 5.10].

Superficial lymph vessels of the upper limb drain the skin and subcutaneous tissues, and most end in the lateral group of axillary nodes. The larger lymph vessels, unlike the veins, do not unite into larger trunks but run individual courses directly towards the axilla. The general arrangement of superficial lymph vessels is shown in Figs 5.10 and 7.4. Note the following: (1) the dense

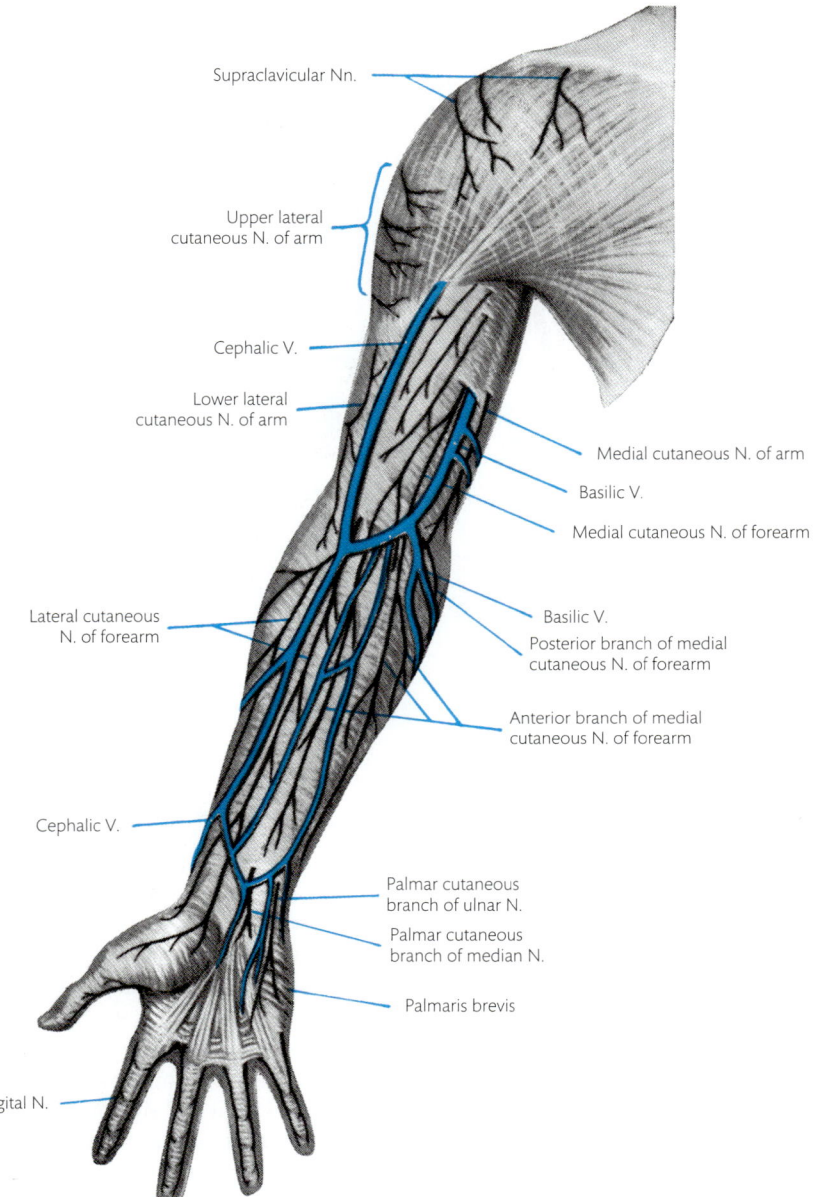

Supraclavicular Nn.

Upper lateral
cutaneous N. of arm

Cephalic V.

Lower lateral
cutaneous N. of arm

Medial cutaneous N. of arm

Basilic V.

Medial cutaneous N. of forearm

Lateral cutaneous
N. of forearm

Basilic V.

Posterior branch of medial
cutaneous N. of forearm

Anterior branch of medial
cutaneous N. of forearm

Cephalic V.

Palmar cutaneous
branch of ulnar N.

Palmar cutaneous
branch of median N.

Palmaris brevis

Digital N.

Fig. 7.2 Superficial veins and nerves of the front of the upper limb.

palmar plexus drains mainly to the dorsum of
the hand to join the posterior vessels of the forearm; (2) the vessels on the posterior surfaces of
the forearm and arm spiral upwards round the
medial and lateral surfaces to reach the anteri-
or surface and approach the floor of the axilla;
most vessels remain superficial, until they pierce
the fascial floor of the axilla; however, some lymphatics on the medial aspect enter the superficial
cubital nodes, and efferents from these nodes pass

through the deep fascia with the basilic vein; (3)
a few vessels from the lateral side of the arm and
shoulder run with the cephalic vein into the apical
axillary nodes; and (4) vessels from the shoulder
and upper parts of the thoracic wall curve round
the anterior and posterior axillary folds to enter
the pectoral and subscapular groups of **axillary
nodes**; those from the walls of the lower parts of
the thorax and upper abdomen converge directly
on the axilla.

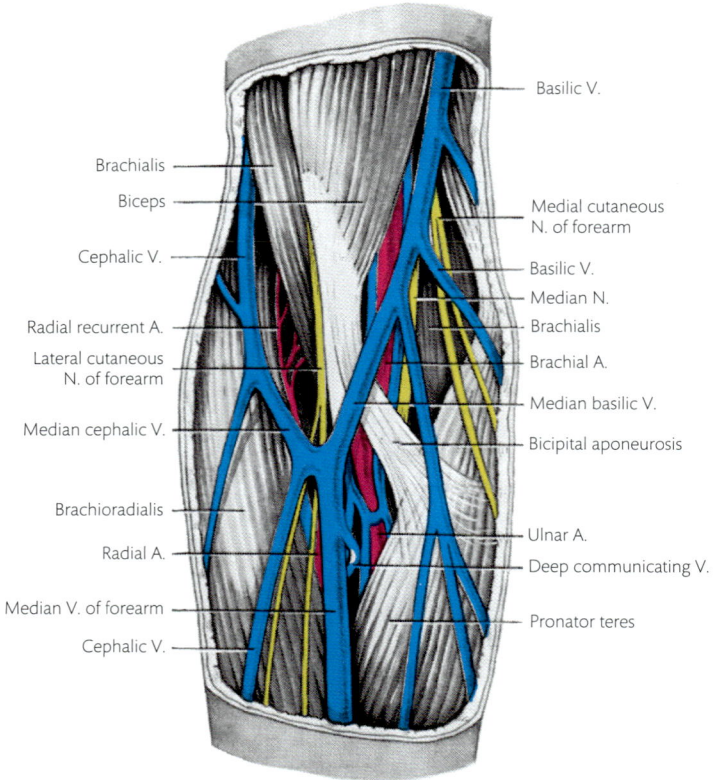

Fig. 7.3 Commonly seen arrangement of superficial veins at the bend of the elbow.

General features of cutaneous nerves

There are certain general principles about the distribution of nerves to the skin which are of clinical importance.

1. Each nerve which passes to the skin is distributed to a circumscribed area. The area of the upper limb skin supplied by spinal nerves C. 4 to T. 2 is shown in Fig. 7.5.

2. Limb plexuses are formed by the plaiting together of the ventral rami of several spinal nerves. As a result of this: (1) each part of the plexus and each branch of the brachial plexus contain nerve fibres from more than one ventral ramus—examples: the upper trunk contains fibres from C. 5 and C. 6; the ulnar nerve contains fibres from C. 8 and T. 1; and (b) several of these branches contain some nerve fibres from the same ventral ramus—example: the suprascapular nerve and the axillary nerve contain fibres from C. 5 and C. 6 [see Fig. 5.18]. Thus, the area of distribution of a cutaneous nerve, and consequently the area of sensory

loss, when it is destroyed is different from the loss resulting from the destruction of a ventral ramus.

3. Cutaneous nerves supplying adjacent areas of the skin overlap with each other to a considerable degree. Thus, the destruction of a single cutaneous nerve leads to total loss of sensation only in a small area within the area of distribution of that nerve. Surrounding this anaesthetized area, there will be an area of altered sensation due to the presence of nerve fibres from adjacent uninjured nerves.

Each thoracic spinal nerve (except for T. 1 which is involved in the formation of a plexus) supplies a strip of skin (**dermatome**) which overlaps those of adjacent nerves, so that destruction of a single thoracic spinal nerve produces altered sensation only within its dermatome. In the plexus regions, each ventral ramus supplies a circumscribed area of the skin in sequence with, and overlapped by, the areas of adjacent ventral rami. The overlap of these dermatomes is accounted for by the presence of nerve fibres from multiple ventral rami in every branch of the plexus.

4. The major branches of the plexus (i.e. the main nerves of a limb) give rise to several cutaneous

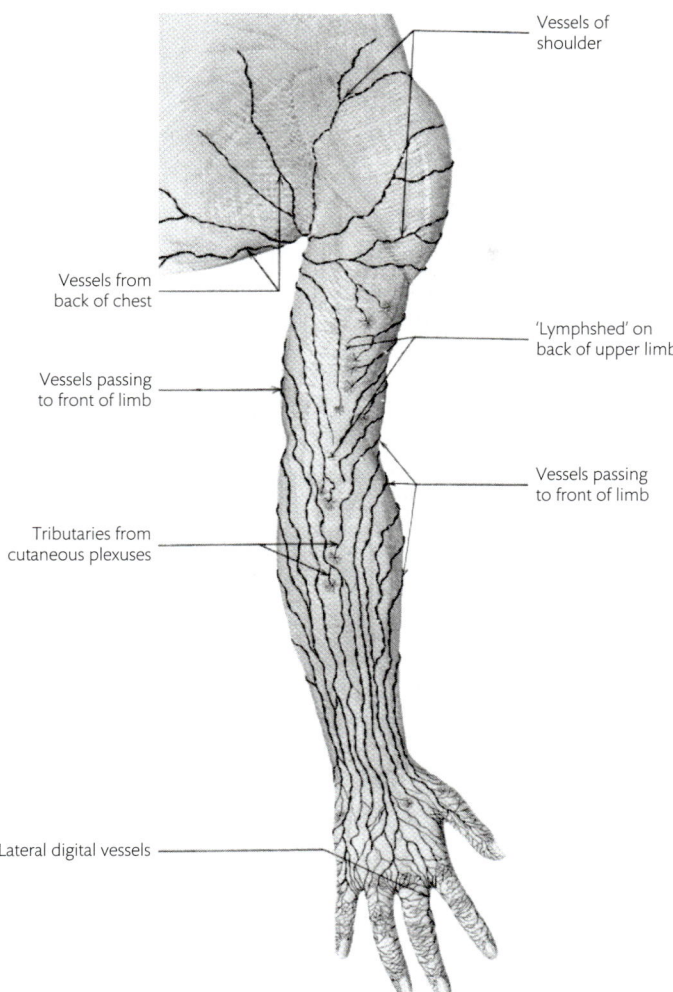

Fig. 7.4 Superficial lymph vessels and lymph nodes of the back of the upper limb.

Vessels of shoulder

Vessels from back of chest

'Lymphshed' on back of upper limb

Vessels passing to front of limb

Vessels passing to front of limb

Tributaries from cutaneous plexuses

Lateral digital vessels

branches which leave them at different points. Destruction of a nerve before it has given off any branches will produce a different distribution of sensory loss to that which occurs when the nerve is destroyed after giving off one or more branches. It is more important to know the total distribution of these major nerves than that of their individual cutaneous branches. But the ability to make a detailed diagnosis of the site of an injury depends on knowing the distribution of the individual cutaneous nerves and the approximate site of origin of these nerves from the parent trunk [see Fig. 12.1]. The diagrams of nerve distribution in this book take no account either of the overlap or of the fact that nerve fibres may sometimes pass to their destinations by unusual routes and hence modify the expected clinical effects of destruction of a particular nerve.

5. In both upper and lower limbs, the nerves which pass to the anterior surface supply a greater area of the skin than those which pass to the posterior surface. In the upper limb, this means that a greater part of the skin is supplied by nerves arising from the medial and lateral cords of the brachial plexus, which are formed from the anterior divisions of the trunks of the plexus.

Cutaneous nerves of the upper limb

Cutaneous nerves from the spinal nerves adjacent to the brachial plexus

1. The **supraclavicular nerves** (C. 3, 4) descend from the neck, cross the superficial surface of

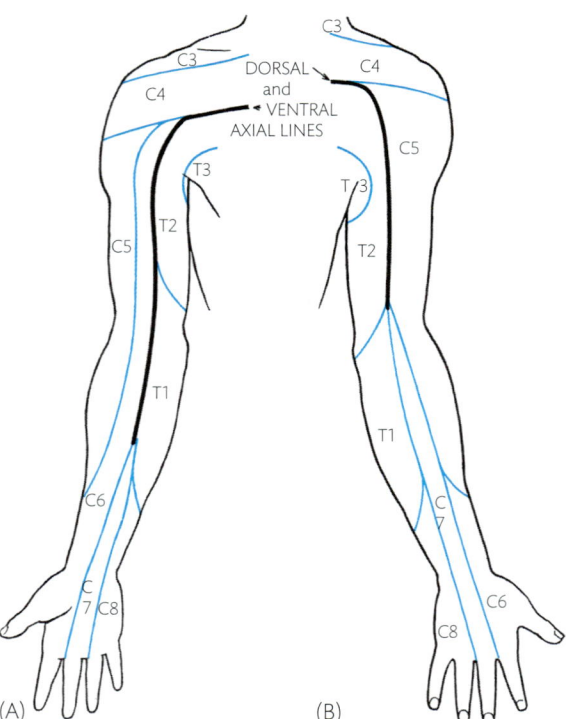

Fig. 7.5 Dermatomes of the upper limb. (A) Front of the upper limb. (B) Back of the upper limb.

After Head, 1893, and Foerster, 1933.

the clavicle and acromion, and supply the skin over the upper part of the front of the chest and deltoid muscle to the level of the sternal angle [see Fig. 5.5].

2. The **intercostobrachial nerve** (T. 2)—this lateral cutaneous branch of the second intercostal nerve enters the axilla from the second intercostal space. It descends obliquely across the axilla, communicates with the medial cutaneous nerve of the arm, and sends branches to the floor of the axilla. It pierces the deep fascia below the axilla to supply the skin on the upper posteromedial part of the arm [Fig. 7.1].

Cutaneous nerves from the posterior cord

These supply the posterior surface of the limb. They arise from either the axillary or the radial nerve.

1. Cutaneous branches of the axillary nerve [Figs 7.1, 7.2]: the **upper lateral cutaneous nerve of the arm** (C. 5, 6) pierces the deep fascia at a variable point posterior to the lower half of the deltoid muscle. It curves round the posterior bor-

der of that muscle and supplies the skin over its lower half.

2. Cutaneous branches of the radial nerve [Figs 7.1, 7.2; see also Fig. 12.1]:

    a. The **posterior cutaneous nerve of the arm** (C. 5) arises in the axilla, pierces the deep fascia below the posterior axillary fold, and supplies the skin of the back of the arm from the insertion of the deltoid to the olecranon.

    b. The **lower lateral cutaneous nerve of the arm** (C. 5, 6) arises posterior to the humerus and pierces the deep fascia below the insertion of the deltoid. It supplies the skin of the lateral side of the arm below that insertion.

    c. The **posterior cutaneous nerve of the forearm** (C. 6, 7, 8) arises with the lower lateral cutaneous nerve of the arm and pierces the deep fascia below it. It gives some branches to the lateral side of the arm, descends posterior to the lateral epicondyle, and lies in the middle of the back of the forearm. It supplies the skin on the back of the forearm to the level of the wrist or occasionally on to the dorsum of the hand.

    d. The **superficial branch of the radial nerve** (C. 6, 7, 8) is a terminal branch of the radial nerve. It descends in the forearm between the extensor and flexor muscle groups, passes posteriorly in the distal half of the forearm, and pierces the deep fascia 5 cm superior to the styloid process of the radius. It supplies the lateral two-thirds of the dorsum of the hand, the dorsal surfaces of the thumb, and the lateral two-and-a-half fingers through five **dorsal digital nerves**. These do not supply the terminal parts of the fingers. The area supplied by the nerve varies reciprocally with the other nerves with which it communicates on the dorsum of the hand (ulnar, posterior cutaneous nerve of the forearm, and median).

Cutaneous nerves from the lateral cord

They arise from the musculocutaneous nerve [Figs 7.1, 7.2; see also Fig. 12.1].

1. The **lateral cutaneous nerve of the forearm** (C. 5, 6) pierces the deep fascia just lateral to the biceps, 2–3 cm proximal to the bend of the elbow. It divides into anterior and posterior branches which supply the anterolateral and posterolateral surfaces of the forearm, with the anterior branch extending on to the ball of the thumb.

Note that the lateral cord also contributes to the formation of the median nerve, which is described below.

Cutaneous nerves from the medial cord

They arise from either the medial cord or the ulnar nerve.

1. Cutaneous branches of the medial cord [Figs 7.1, 7.2; see also Fig. 12.1]:
   a. The **medial cutaneous nerve of the arm** (T. 1, 2) pierces the deep fascia on the medial side of the middle of the arm. It supplies the skin on the medial side of the inferior half of the arm, posterior to the basilic vein.
   b. The **medial cutaneous nerve of the forearm** (C. 8, T. 1) pierces the deep fascia with the basilic vein. As it descends, it divides into anterior and posterior branches which supply the anteromedial and posteromedial surfaces of the forearm to the wrist. The anterior branch supplies the skin of the distal part of the front

of the arm, and together they supply the skin of the medial half of the forearm.

2. Cutaneous branches of the ulnar nerve (C. 8, T. 1, with fibres of C. 7 received from the median nerve in the axilla) [Figs 7.1, 7.2; see also Fig. 12.1]: this nerve supplies only the skin of the hand and fingers.
   a. The **dorsal branch of the ulnar nerve** (C. 7, 8) arises from the ulnar nerve in the middle of the forearm and descends with it almost to the pisiform bone. It then passes obliquely backwards across the medial surface of the carpus to divide into two **dorsal digital nerves**. These supply the skin of the medial third of the back of the hand and the dorsal surfaces of the little finger and the medial half of the ring finger, except for the terminal part of the fingers which is supplied by the palmar digital branches of the ulnar nerve.
   b. The **palmar (cutaneous) branch of the ulnar nerve** (C. 7, 8) arises in the distal half of the forearm, pierces the deep fascia anterior to the wrist, and supplies the medial third of the palmar skin.
   c. Two **palmar digital nerves** (C. 7, 8) arise from the superficial branch of the ulnar nerve, distal to the pisiform bone. The medial nerve is the proper palmar digital nerve to the medial side of the little finger. The lateral nerve is a common palmar digital nerve, which divides near the cleft between the little and ring fingers to give a proper palmar digital nerve to the contiguous sides of each. (Palmar and plantar digital nerves and arteries are called '**proper**' when each is distributed only to one finger or toe. The term '**common**' indicates that the nerve or artery is distributed to two adjacent fingers or toes as two proper digital branches.)

Note that the medial cord also contributes to the formation of the median nerve, which is described below.

Cutaneous nerves from the median nerve (medial and lateral cords)

1. Cutaneous branches of the median nerve (C. 5, 6, 7 from the lateral cord; C. 8, T. 1 from the medial cord) which supply the skin on the hand [Figs 7.1, 7.2; see also Fig. 12.1]:
   a. The **palmar (cutaneous) branch** (C. 6, 7, 8) arises a little superior to the wrist, pierces the deep fascia just above it, and descends to supply the lateral two-thirds of the palmar skin.

b. Five **palmar digital branches** (C. 6, 7, 8) arise in the palm. The medial two are common palmar digital nerves. Each divides near the cleft on the medial and lateral aspects of the middle finger [see Fig. 10.4] to form a proper palmar digital nerve to each side of that cleft. The common palmar digital nerve to the ring and middle fingers communicates with the ulnar nerve. The next palmar digital nerve is a proper palmar digital nerve to the lateral side of the index finger. The lateral two nerves curve round the distal border of the thenar eminence as proper palmar digital nerves to the sides of the thumb. The proper palmar digital nerves not only supply the corresponding parts of the fingers, but also send branches to the dorsal skin on the middle and terminal phalanges.

See Clinical Applications 7.1 and 7.2 for the practical implications of the anatomy described in this chapter.

CLINICAL APPLICATION 7.1 Intravenous injections

Understanding the anatomy of the median cubital vein is an essential prerequisite for performing a successful venepuncture.

Study question 1: where is the median cubital vein located? (Answer: cubital fossa—superficial fascia.)

Study question 2: which two superficial veins does this vein connect? (Answer: cephalic and basilic.) Distension of these veins is accomplished by applying a tourniquet mid arm and by asking the patient to clench and unclench the fist.

Study question 3: which veins are compressed by the tourniquet at mid arm? (Answer: basilic vein and brachial veins.)

Known complications: thrombosis formation with occlusion of the injected vein is a common consequence of intravenous injection. This obliteration may be asymptomatic or may result in inflammation of the vein—phlebitis. Leakage of the injected drug could lead to injury of the lateral or medial cutaneous nerve of the forearm. Inadvertent puncture of the brachial artery is known to occur.

Study question 4: what firm structure normally separates the median cubital vein from the brachial artery? (Answer: bicipital aponeurosis.)

CLINICAL APPLICATION 7.2 Paraesthesiae over the shoulder

A 10-year-old boy was given an armpit crutch to help him walk after he fractured his right femur. A few days later, he complained of a sensation of numbness and tingling over the outer part of his right shoulder and the lateral side of his right upper arm.

Study question 1: what is the sensory innervation to the affected area? (Answer: the upper lateral cutaneous nerve of the arm, a branch of the axillary nerve.)

Study question 2: what nerve is likely to be compressed? (Answer: axillary nerve.)

Study question 3: what is the probable cause of the injury? (Answer: pressure on the axillary nerve due to improper use of the armpit crutch.)

CHAPTER 8
The shoulder

Surface anatomy

Begin by reviewing the scapula and proximal part of the humerus [see Figs 4.4, 4.5, 4.8].

The **humerus** [see Fig. 4.8] is almost entirely covered by the muscles of the arm, so that its outlines can only be felt indistinctly. In the distal part, the lateral and medial supracondylar ridges become readily palpable. The medial epicondyle is more prominent than the lateral epicondyle. In the anatomical position, the epicondyles are in the positions suggested by their names, and the hemispherical head of the humerus faces medially. When the palms of the hands face medially (the arm is semi-pronated), the lateral epicondyle is more anterior and the head of the humerus is directed posteromedially.

The upper half of the humerus is covered on its anterior, lateral, and posterior surfaces by the deltoid muscle. Inferiorly, the apex of that muscle is attached to the lateral side of the middle of the humerus—the **deltoid tubercle**.

Deep fascia

In the arm, the deep fascia is strongest posteriorly where it covers the triceps muscle. Below the insertion of the deltoid, it is thickened on each side and sends strong **medial** and **lateral intermuscular septa** to the corresponding supracondylar lines and epicondyles of the humerus. These septa lie between the triceps muscle posteriorly and the muscles attached to the anterior surface of the distal half of the humerus.

At the elbow, the deep fascia is thickened by extensions from the triceps and biceps muscles, and by the origin of the forearm muscles from its deep surface. The **bicipital aponeurosis** is a strong slip which extends medially from the tendon of the biceps to the deep fascia and the subcutaneous posterior border of the ulna. The aponeurosis is readily felt by sliding a finger down the medial side of the taut biceps tendon.

See Dissection 8.1 for instructions on skin reflection of the front of the arm and forearm.

DISSECTION 8.1 Skin reflection of the front of the arm and forearm-1

Objectives

I. To reflect the skin on the front of the arm and forearm.
II. To find the cutaneous vessels and nerves.

Instructions

1. Make incision 5, as shown in Fig. 5.6. Extend the incision along the anterior surfaces of all the fingers, leaving them covered with skin at present to avoid drying of the tissues.
2. Strip the skin and superficial fascia from the deep fascia by blunt dissection.
3. Note and follow the large cutaneous veins and the cutaneous nerves, as they pierce the deep fascia.

Muscles attaching the humerus to the scapula

Muscles attaching the humerus to the scapula are best described in two groups:

1. Those which have considerable mechanical advantage over the shoulder joint by being attached at some distance from it—deltoid, teres major, coracobrachialis, and biceps brachii (short head)
2. Those which lie close to the shoulder joint and have a smaller mechanical advantage over it. They help to stabilize the joint in any position and act as the main ligaments of the joint—subscapularis, supraspinatus, infraspinatus, teres minor (together forming the rotator cuff), and the long heads of the biceps and triceps brachii.

Deltoid

This large muscle forms the bulge of the shoulder. **Origin**: it takes origin from the lateral third of the clavicle, acromion, and spine of the scapula [see Fig. 4.3]. **Insertion**: it is inserted into the deltoid tubercle of the humerus [see Figs 4.9, 4.10, 8.1]. **Nerve supply**: axillary nerve. **Actions**: (1) the anterior fibres flex and medially rotate the shoulder; (2) the middle fibres abduct the shoulder; and (3) the posterior fibres extend and laterally rotate the shoulder.

Subscapularis

Origin: the subscapularis originates from a wide area on the costal surface of the scapula [see Fig. 4.5]. **Insertion**: it is inserted on the lesser tubercle of the humerus [see Figs. 4.9, 8.2]. The fibres fuse with the capsule of the shoulder joint and are separated from it and the lateral part of the costal surface of the scapula by the subscapular bursa. The subscapular bursa frequently communicates with the cavity of the shoulder joint. **Nerve supply**: upper and lower subscapular nerves. **Actions**: medial rotation of the shoulder; plays an important role in stabilizing the shoulder joint.

Supraspinatus

Origin: the supraspinatus arises from the supraspinous fossa of the scapula. **Insertion**: it is

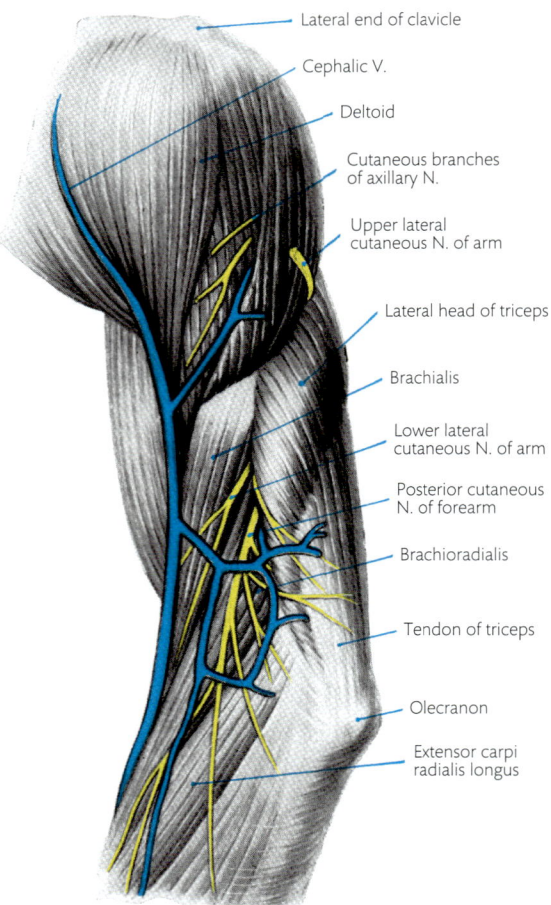

Fig. 8.1 Deltoid muscle and lateral aspect of the arm.

inserted on the superior facet on the greater tubercle of the humerus [see Figs 4.7, 4.9]. Its fibres pass under the coracoacromial ligament. **Nerve supply**: suprascapular nerve. **Actions**: abduction of the shoulder joint.

Infraspinatus

Origin: the infraspinatus arises from the infraspinous fossa of the scapula. **Insertion**: it is inserted on the middle facet on the greater tubercle of the humerus [see Figs 4.7, 4.10]. **Nerve supply**: suprascapular nerve. **Actions**: lateral rotation of the shoulder joint.

Teres major

Origin: the teres major takes origin from the lateral one-third of the lateral margin of the scapula. **Insertion**: its fibres pass to the anterior surface of the humerus and are inserted into the medial lip

Labels on figure:
- Lateral end of clavicle
- Cephalic V.
- Deltoid
- Cutaneous branches of axillary N.
- Upper lateral cutaneous N. of arm
- Lateral head of triceps
- Brachialis
- Lower lateral cutaneous N. of arm
- Posterior cutaneous N. of forearm
- Brachioradialis
- Tendon of triceps
- Olecranon
- Extensor carpi radialis longus

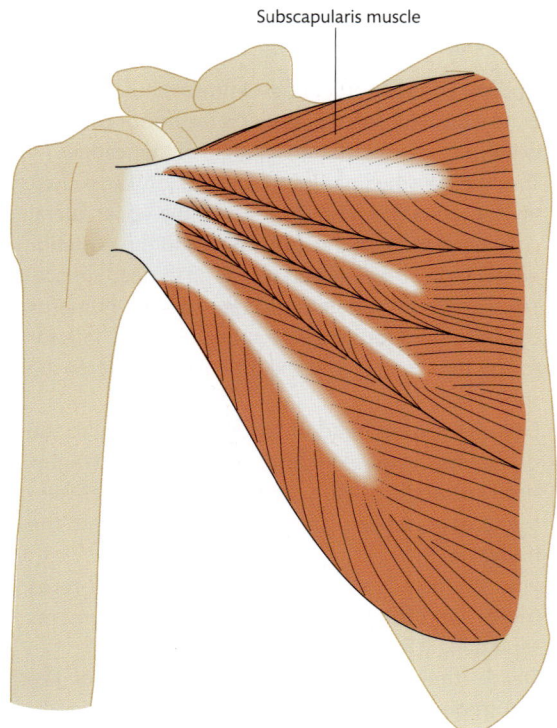

Subscapularis muscle

Fig. 8.2 Diagram showing the attachments of the subscapularis.

of the intertubercular sulcus [see Figs 4.7, 4.9]. The lower margin of the teres major is closely related to the latissimus dorsi and forms the posterior axillary fold. **Nerve supply**: lower subscapular nerve. **Actions**: adduction and medial rotation of the shoulder joint.

Teres minor

Origin: the teres minor takes origin from the superior two-thirds of the lateral margin of the scapula. **Insertion**: it is inserted into the posterior facet on the greater tubercle of the humerus [see Figs 4.7, 4.10]. **Nerve supply**: axillary nerve. **Actions**: lateral rotation and stabilization of the shoulder joint.

Rotator cuff muscles

The supraspinatus, infraspinatus, teres minor, and subscapularis together are known as the rotator cuff muscles. They take origin from the scapula, lie in close contact with the capsule of the shoulder joint, and are inserted into the greater and lesser tubercles of the humerus. They stabilize the joint and act

as expansile ligaments. For rotator cuff injuries, see Clinical Application 8.1.

Dissection 8.2 (and Figs 8.3 and 8.4) explores the shoulder region.

Dissection 8.3 continues to explore the shoulder region.

Axillary nerve (C. 5, 6)

The axillary nerve is a terminal branch of the posterior cord of the brachial plexus and is formed near the lower border of the subscapularis. It curves back on the lower border of the subscapularis and passes through the quadrangular space with the posterior circumflex humeral artery. It lies medial to the surgical neck of the humerus, immediately inferior to the capsule of the shoulder joint. The nerve gives a branch to the **shoulder joint** and then divides into anterior and posterior branches.

The **posterior branch** supplies the teres minor and the posterior part of the deltoid. It then descends over the posterior border of the deltoid and supplies the skin over the lower half of that muscle as the **upper lateral cutaneous nerve of the arm** [see Fig. 7.1].

The **anterior branch** continues horizontally between the deltoid muscle and the surgical neck of the humerus. It supplies the deltoid and sends a few branches through it to the overlying skin [Fig. 8.4].

⊃ The axillary nerve is at risk of damage in a number of clinical conditions: (1) in downward dislocation of the head of the humerus and in fractures of the surgical neck of the humerus because of its close relation to the joint and bone [Fig. 8.5]; and (2) it can also be injured during a poorly administered intramuscular injection into the deltoid. When the nerve is damaged, the deltoid and teres minor muscles are paralysed.

⊃ Axillary nerve injury and the resultant paralysis of the deltoid and teres minor severely affect shoulder abduction, extension, and lateral rotation. In such an injury, the arm is held in a position of adduction, medial rotation, and flexion. Sensory loss is experienced in the skin over the lateral part of the upper arm—regimental patch anaesthesia.

DISSECTION 8.2 Shoulder region-1

Objectives

I. To study the deltoid, infraspinatus, teres minor, and long head of the triceps, and to identify the boundaries of the quadrangular and triangular spaces. **II.** To identify and trace the axillary and radial nerves.

Instructions

1. Remove the fascia from the surface of the **deltoid muscle**, and study its attachments. It has a V-shaped origin from the lateral third of the clavicle, the acromion, and the crest of the spine of the scapula. It is inserted into the deltoid tuberosity of the humerus. Note that the long anterior and posterior fibres run parallel to each other on the corresponding surfaces of the shoulder joint. The lateral fibres are short and multipennate to increase the power of this part.

2. Separate the muscle from the spine of the scapula, and turn this part downwards to expose the underlying **infraspinatus muscle**. Remove the dense deep fascia from the surface of the **infraspinatus**. Define its attachments to the infraspinous fossa and the greater tubercle of the humerus.

3. Find the inferior border of the infraspinatus, and separate it from the **teres major** and the **minor muscles** which arise from the lateral margin of the scapula.

4. Turn the detached part of the deltoid forwards; identify the **axillary nerve** (from which the upper lateral cutaneous nerve of the arm arises) and the posterior humeral circumflex vessels supplying its deep surface.

5. Trace these on the surgical neck of the humerus through the **quadrangular space** [Fig. 8.3], inferior to the teres minor and the articular capsule of the shoulder joint.

6. Expose the **long head of the triceps** medial to the quadrangular space. It descends from the **infraglenoid tubercle** of the scapula and passes between the teres minor and the major muscles close to the humerus.

7. Find the branch of the axillary nerve to the **teres minor**, and follow this muscle to its attachments separating it from the teres major.

8. Divide the remainder of the **deltoid** from the acromion and clavicle, and turn it downwards. It lies on the proximal end and surgical neck of the humerus, superficial to the anastomosis of the circumflex humeral vessels [Fig. 8.4].

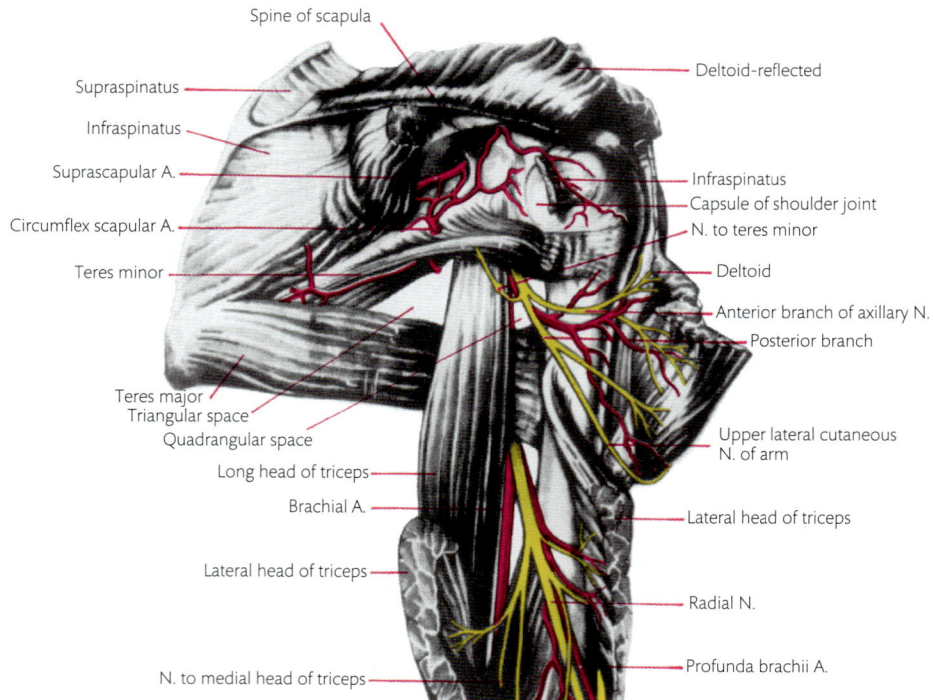

Spine of scapula

Supraspinatus

Infraspinatus

Suprascapular A.

Circumflex scapular A.

Teres minor

Teres major
Triangular space
Quadrangular space

Long head of triceps

Brachial A.

Lateral head of triceps

N. to medial head of triceps

Deltoid-reflected

Infraspinatus
Capsule of shoulder joint
N. to teres minor

Deltoid

Anterior branch of axillary N.
Posterior branch

Upper lateral cutaneous N. of arm

Lateral head of triceps

Radial N.

Profunda brachii A.

Fig. 8.3 Dissection of the scapular region and back of the arm showing the quadrangular and triangular spaces. The lateral head of the triceps has been divided and turned aside to expose the spiral groove on the humerus.

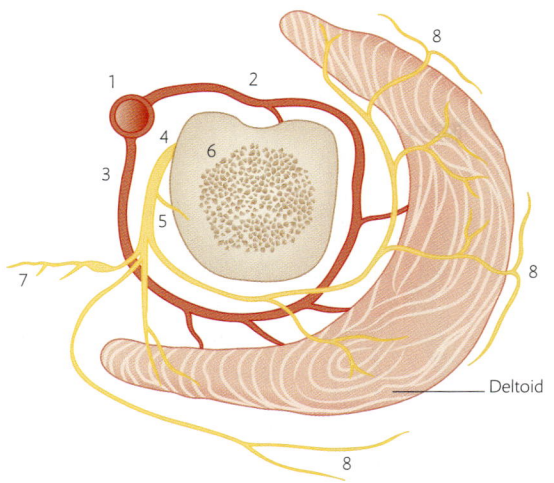

Deltoid

Fig. 8.4 Diagram of the circumflex humeral arteries and axillary nerve. 1. Axillary artery. 2. Anterior circumflex humeral artery. 3. Posterior circumflex humeral artery. 4. Axillary nerve. 5. Articular branch. 6. Humerus. 7. Branch to teres minor. 8. Cutaneous branches.

Circumflex humeral arteries

The **anterior** and **posterior circumflex humeral arteries** are branches of the third part of the axillary artery. Together they form a circular anastomosis at the surgical neck of the humerus [Fig. 8.4]. They supply the surrounding muscles, shoulder joint, and upper end of the humerus. They also anastomose with the profunda brachii artery via a descending branch.

Dissection 8.4 (and Figs 8.6, 8.7, and 8.8) continues to explore the shoulder region.

Suprascapular artery

The suprascapular artery arises from the first branch of the subclavian artery. It enters the supraspinous

---

DISSECTION 8.3 Shoulder region-2

Objectives

I. To study the teres major, coracobrachialis, short head of the biceps, and subscapularis.   **II.** To identify and trace the axillary and radial nerves.   **III.** To explore the subacromial and subscapular bursae.

Instructions

1. On the anterior surface of the scapula, follow the **subscapularis muscle** from the subscapular fossa to the lesser tubercle [Fig. 8.2].
2. Separate it inferiorly from the **teres major**, and trace that muscle to its insertion on the medial lip of the intertubercular crest of the humerus immediately behind the insertion of the **latissimus dorsi**. Both these insertions are lateral to the coracobrachialis and the short head of the biceps which descend in front of them from the coracoid process.
3. Find the **axillary nerve** arising from the posterior cord of the brachial plexus, and follow it to the quadrangular space which, when seen from the front, is between the subscapularis and the teres major.
4. Identify the **radial nerve** as it arises from the posterior cord. It descends between the parts of the triceps muscle, after giving branches to the long and medial heads [Fig. 8.3].

5. Remove the fascia covering the **coracobrachialis** and the **short head of the biceps** from the coracoid process to the insertion of the coracobrachialis on the medial aspect of the body of the humerus.
6. Follow the **musculocutaneous nerve** from the lateral cord of the brachial plexus into the medial aspect of the coracobrachialis, and find the branch which it gives to that muscle.
7. Pull these muscles medially, and identify the tendon of the **long head of the biceps**. It lies in the intertubercular sulcus posterolateral to the short head. Follow the long head upwards to the lower border of the lesser tubercle where it disappears deep to the articular capsule of the shoulder joint.
8. Move the fascia covering the superior surface of the greater tubercle of the humerus. It slides easily on the tubercle because of the **subacromial bursa** deep to the fascia.
9. Make a small incision into the bursa, and explore its limits with a blunt seeker. Then open it widely. The bursa separates the superior surface of the humerus and the capsule of the shoulder joint from the acromion and coracoacromial ligament, and makes a secondary synovial socket for the humerus with the coracoacromial arch.

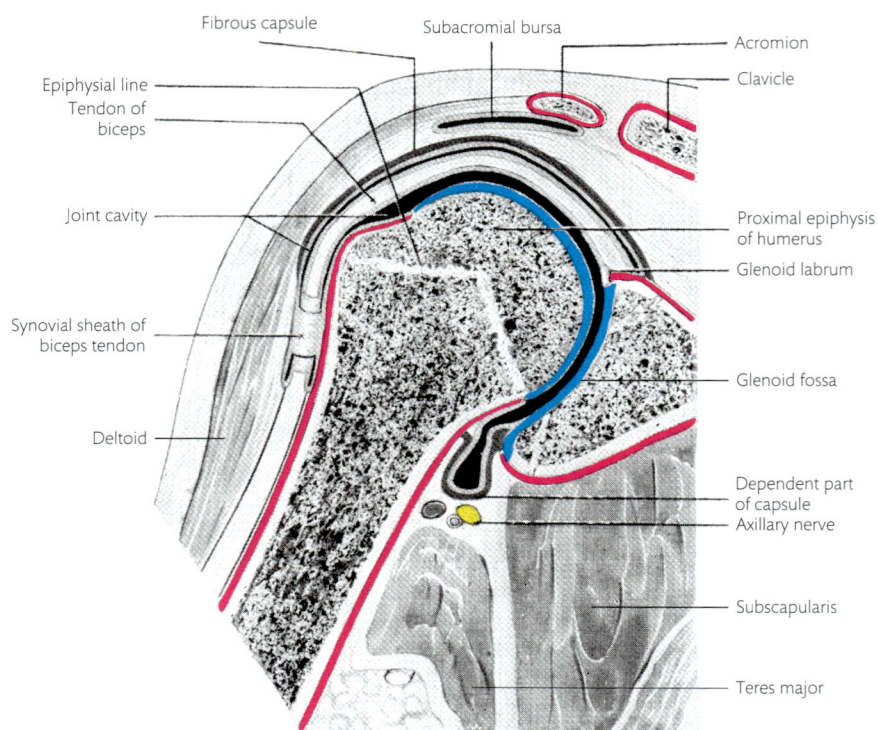

Fibrous capsule — Subacromial bursa — Acromion
— Clavicle
Epiphysial line —
Tendon of biceps —
Joint cavity —
— Proximal epiphysis of humerus
— Glenoid labrum
Synovial sheath of biceps tendon —
— Glenoid fossa
Deltoid —
— Dependent part of capsule
— Axillary nerve
— Subscapularis
— Teres major

Fig. 8.5 Coronal section through the right shoulder joint. The synovial sheath of the biceps tendon has been partially left in place. Periosteum, red. Articular cartilage, blue.

fossa of the scapula above the scapular ligament, and runs deep to, and supplies, the supraspinatus. It then passes through the spinoglenoid notch and supplies the infraspinatus. It anastomoses with the circumflex scapular artery and with branches of the transverse cervical artery.

third part of the axillary artery when the main stem is blocked between these two points. Note that slow occlusion will allow time for collateral circulation to be established. Sudden occlusion is more dangerous, as the collateral circulation may not be ready to take over the increased supply.

Anastomosis around the scapula

➲ Arteries supplying the muscles on the dorsal surface of the scapula come from two distinct sources. The dorsal scapular artery and the suprascapular artery are branches of the thyrocervical trunk of the subclavian artery. The posterior circumflex humeral artery and the circumflex scapular arteries are branches of the axillary artery. The anastomosis around the scapula [Fig. 8.7] ensures that the mobile scapula is not dependent on only one source. The anastomosis forms an alternate route through which blood from the first part of the subclavian artery can reach the

Suprascapular nerve (C. 5, 6)

The suprascapular nerve arises from the upper trunk of the brachial plexus [see Fig. 5.18]. It passes downwards and backwards, superior to the plexus, to join the suprascapular vessels. It enters the supraspinous fossa through the scapular notch. Here it supplies the supraspinatus and gives branches to the acromioclavicular and shoulder joints. The nerve then descends immediately lateral to the root of the spine of the scapula to enter the infraspinous fossa and supply the infraspinatus and shoulder joint.

DISSECTION 8.4 Shoulder region-3

Objectives

I. To study the supraspinatus and infraspinatus. **II.** To identify and trace the suprascapular nerve, suprascapular artery, and circumflex scapular artery. **III.** To study the coracoacromial ligament, and the subacromial and subscapular bursae. **IV.** To study the acromioclavicular joint.

Instructions

1. Remove the subacromial bursa, and expose the **supraspinatus** passing from the supraspinous fossa, beneath the coracoacromial arch and bursa, to the superior surface of the greater tubercle. The tendon of the supraspinatus is firmly fused to this part of the capsule of the shoulder joint [Fig. 8.6].

2. Cut across the infraspinatus and teres minor muscles at the level of the neck of the scapula. Turn their parts medially and laterally [Fig. 8.7].

3. Find the **suprascapular** and **circumflex scapular arteries**, passing deep to the muscles and anastomosing on the posterior surface of the scapula [Fig. 8.7].

4. Find the **suprascapular nerve** as it enters the deep surface of the infraspinatus from the supraspinatus. The tendons of the teres minor and infraspinatus fuse with the capsule of the shoulder joint as they pass to the greater tubercle of the humerus. There

may be a bursa deep to the tendon of the infraspinatus, and this may communicate with the cavity of the shoulder joint [Fig. 8.6].

5. Follow the tendon of the **long head of the triceps** to its attachment on the infraglenoid tubercle, and find the dependent part of the capsule of the shoulder joint lateral to it. The **axillary nerve** passes posteriorly on the surgical neck of the humerus, immediately inferior to this part of the capsule, and sends a branch to it.

6. Cut across the **subscapularis** at the neck of the scapula, and reflect its parts. Laterally, it is fused with the capsule of the shoulder joint. Medially, it is separated from the neck of the scapula and the root of the coracoid process by the **subscapular bursa**. This bursa facilitates movement of the muscle over these bony structures. It communicates with the cavity of the shoulder joint through an aperture in its capsule [Fig. 8.8].

7. Remove the trapezius and the remainder of the deltoid from the capsule of the acromioclavicular joint.

8. Cut away the superior part of the capsule (acromioclavicular ligament) of this joint, and note the articular disc separating the two bones.

9. Identify the **suprascapular artery** and **nerve** crossing the superior margin of the scapula at the scapular notch [Fig. 8.7].

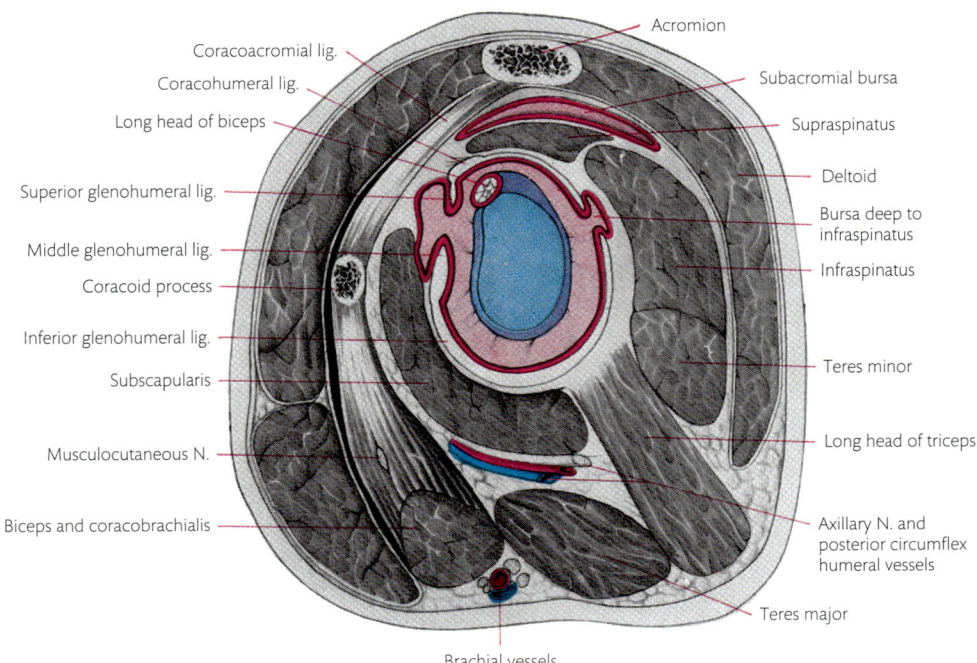

Fig. 8.6 Diagrammatic sagittal section through the left shoulder. The subscapular bursa protrudes between the superior and middle glenohumeral ligaments.

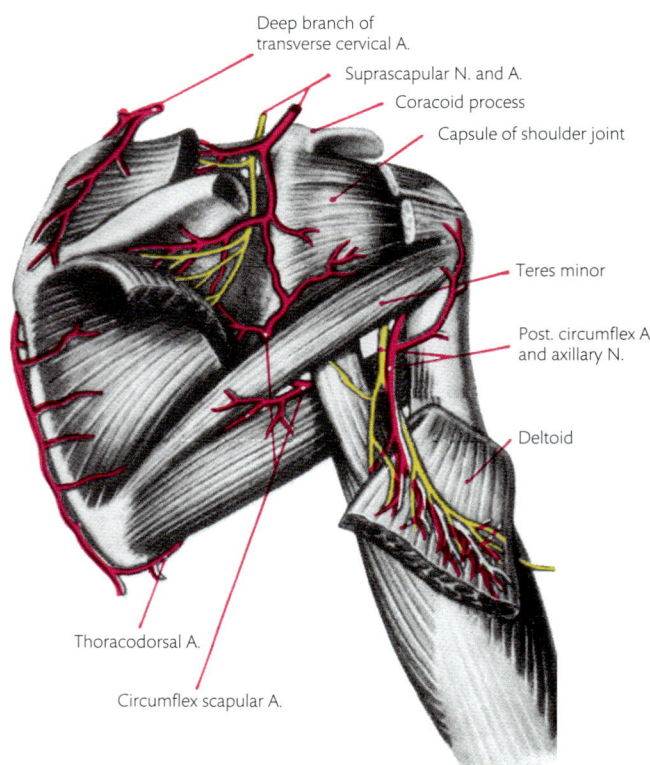

Deep branch of
transverse cervical A.

Suprascapular N. and A.

Coracoid process

Capsule of shoulder joint

Teres minor

Post. circumflex A.
and axillary N.

Deltoid

Thoracodorsal A.

Circumflex scapular A.

Fig. 8.7 Anastomosing arteries around the scapula. The acromion has been removed to expose the suprascapular artery and nerve.

Coracoacromial ligament

The **coracoacromial ligament** is a strong triangular band. Its base is attached to the lateral bor-

der of the coracoid process, and its apex to the tip of the acromion. It lies between the subacromial bursa inferiorly and the deltoid muscle superiorly [Fig. 8.8].

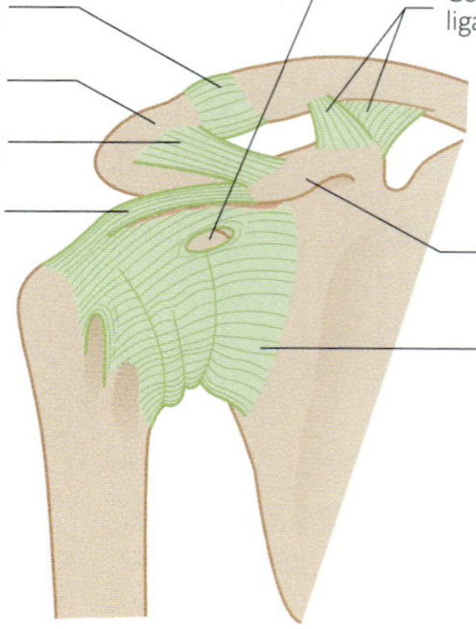

Acromioclavicular
ligament

Acromion

Coracoacromial
ligament

Coracohumeral
ligament

Communication with
subscapularis bursa

Coracoclavicular
ligaments

Coracoid
process

Fibrous
capsule

Fig. 8.8 Diagram of the anterior aspect of the right shoulder joint, showing the fibrous capsule, coracoacromial arch, and coracoclavicular ligaments.

Coracoclavicular ligament

The powerful **coracoclavicular ligament** passes between the upper surface of the coracoid process and the clavicle. It has two parts. The posteromedial part is the **conoid ligament**. It is shaped like an inverted cone and is attached above to the **conoid tubercle** of the clavicle [see Figs 4.1, 8.8]. The anterolateral triangular part is the **trapezoid ligament**. It passes superolaterally to the trapezoid line on the clavicle. The coracoclavicular ligament is the main structure suspending the scapula, and hence the upper limb from the clavicle. It supports the acromioclavicular joint and helps to transmit forces from the upper limb to the trunk. ➲ If the clavicle is broken medial to this ligament, the upper limb sags—a sign characteristic of this fracture.

Shoulder joint

Type

The shoulder joint is a synovial joint of the ball-and-socket type.

Articular surfaces

The **head of the humerus**, which is disproportionately large, articulates with the small, shallow **glenoid cavity**. **Articular cartilage** covers the articular surfaces [Fig. 8.5].

Articular capsule

The outer **fibrous membrane** of the articular capsule is a thin, but relatively strong, tubular structure. It is attached to the margin of the glenoid cavity and to the anatomical neck of the humerus, except inferiorly where it extends downwards 1.5–2.0 cm on the surgical neck of the bone [Figs 8.5, 8.8]. With the arm by the side, this inferior part of the membrane hangs down in a redundant fold between the teres major and the minor muscles. When the arm is abducted to a right angle, this fold is tensed. In the latter position, the lower part of the articular surface of the humeral head lies on this part of the articular capsule, with the long head of the triceps and teres major muscles supporting it below. Anteriorly, the attachment of the fibrous membrane extends inferiorly (from the anatomical neck) between the tubercles of the humerus, bridging over the upper part of the intertubercular sulcus. Deep to it is a synovial-lined tunnel, through which the tendon of the long head of the biceps muscle leaves the joint. This is one of three apertures in the fibrous membrane; the other two are extensions of the synovial membrane through the fibrous membrane to form the subscapular and infraspinatus bursae. The **subscapular bursa** is large and more constant. It lies close to the root of the coracoid process and occasionally allows dislocation of the head of the humerus at this point.

Synovial membrane

The **synovial membrane** lines the fibrous capsule and covers the intracapsular part of the surgical neck of the humerus. It also forms a sheath around the tendon of the long head of the biceps and is continuous with the lining of the bursae which communicate with the joint.

Ligaments

There are four slight thickenings of the articular capsule: (1) the **coracohumeral ligament** lies obliquely across the upper surface of the joint from the base of the coracoid process to the superior surface of the greater tubercle of the humerus; and (2) two or three **glenohumeral ligaments** [Fig. 8.9] may be visible as thickenings of the anterior part of the membrane, when viewed from the interior of the joint.

Relations

The important relations of the shoulder joint are shown in Fig. 8.6. Anteriorly, the shoulder joint is covered by the anterior fibres of the deltoid and pectoralis major; superiorly by the infraspinatus bursa, infraspinatus, acromion process, and middle fibres of the deltoid; posteriorly by the posterior fibres of the deltoid and infraspinatus; and inferiorly by the long head of the triceps, axillary nerve, and posterior circumflex humeral arteries and teres major muscle.

Blood supply

The shoulder joint receives its blood supply from the surrounding arteries, and the thoracoacromial and anterior and posterior circumflex humeral arteries.

Nerve supply

The **nerve supply** of the shoulder joint is from the axillary, suprascapular, and lateral pectoral nerves.

Dissection 8.5 studies the intracapsular parts of the shoulder joint.

Movements of the shoulder joint and muscles involved

(Note that movements of the scapula described in Chapter 4 are different from movements of the shoulder joint. The two are associated but should not be confused.)

The wide range of movements which is possible at the shoulder joint is the result of: (1) the nature of the articular surfaces (the large hemispherical

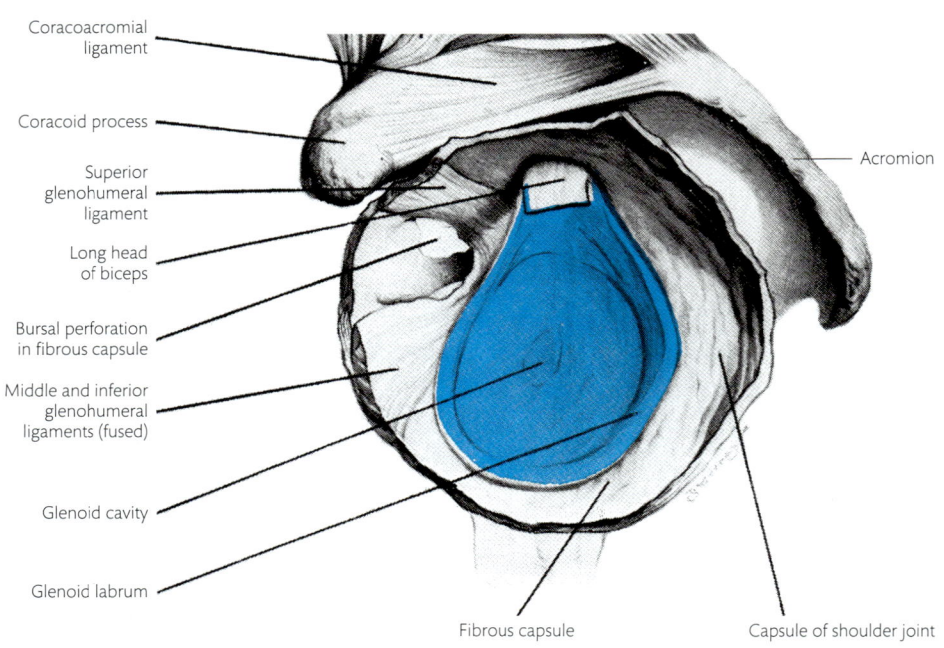

Coracoacromial ligament

Coracoid process

Superior glenohumeral ligament

Long head of biceps

Bursal perforation in fibrous capsule

Middle and inferior glenohumeral ligaments (fused)

Glenoid cavity

Glenoid labrum

Acromion

Fibrous capsule

Capsule of shoulder joint

Fig. 8.9 Left shoulder joint. The articular capsule has been cut across and the humerus removed, together with the surrounding muscles. Articular cartilage and glenoid labrum, blue.

head of the humerus fitted to the small, shallow glenoid cavity); (2) a loose-fitting articular capsule; and (3) the replacement of ligaments by a group of muscles. The **glenoid cavity** faces anterolaterally at rest, and its plane is parallel to the axis around which the scapula is rotated in movements of the shoulder girdle. Movement of the shoulder joint can take place independently but is usually accompanied by movements of the shoulder girdle. Even

when the scapula is not moved, the muscles would be in tension to maintain a stable scapula on which the limb may be moved.

Flexion

Flexion is carried out by muscles which pass anterior to the shoulder joint (the short head of the biceps, the coracobrachialis, and the clavicular parts of the deltoid and pectoralis major), but the

---

DISSECTION 8.5 Shoulder joint

Objective

I. To study the intracapsular parts of the shoulder joint.

Instructions

1. Most of the surface of the thin capsule has already been exposed by removal of the muscles which closely surround it. These muscles are partly fused to the fibrous membrane, so that they prevent it from passing between the joint surfaces when they contract.

2. Make a vertical incision through the posterior part of the articular capsule of the shoulder joint. Rotate the arm medially, and dislocate the head of the humerus through the cut in the capsule.

3. Identify the **tendon of the long head of the biceps** passing over the superior surface of the head of the humerus within the capsule to reach the supraglenoid tubercle [Fig. 8.9]. Note that it becomes continuous here with a fibrocartilaginous ring—the **glenoid labrum**. Identify the glenoid labrum which is attached to the margin of the glenoid cavity and the internal surface of the articular capsule. By surrounding the glenoid cavity, the labrum deepens it.

4. Try to expose the glenohumeral ligaments on the deep surface of the anterior part of the capsule, and note the aperture of the subscapular bursa between them. If necessary, cut through the tendon of the long head of the biceps to increase the mobility of the humerus.

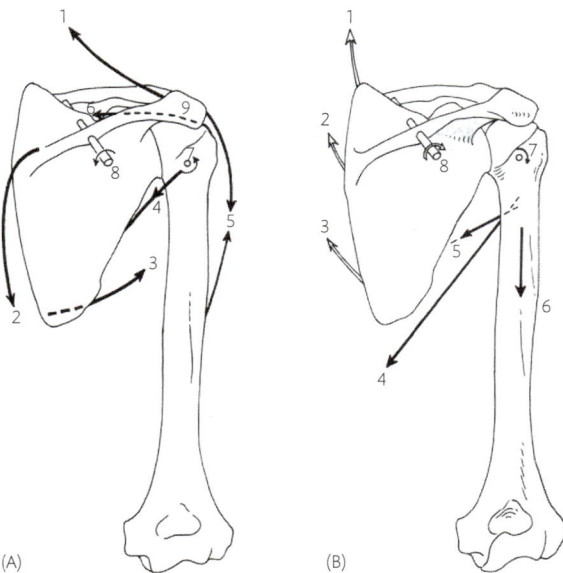

(A)   (B)

Fig. 8.10 Diagrams to show the direction of pull of muscles producing (A) abduction and (B) adduction of the arm. 7 = axis of abduction of the humerus. 8 = axis of scapular rotation. (A) Abduction of the arm. This action requires two separate movements: (i) lateral rotation of the scapula by the (1) upper and (2) lower fibres of the trapezius, and (3) lower fibres of the serratus anterior; (ii) abduction at the shoulder joint by (5) the deltoid and (6) the supraspinatus. When the deltoid contracts, it tends to pull the humerus upwards against the acromion (9). This action is counteracted by the teres minor (4). (B) Adduction of the arm. This movement is normally produced by gravity (6). In this case, the arm is lowered to the side by the muscles which produce abduction. When adduction is produced against resistance, two separate movements are involved: (i) medial rotation of the scapula by the (1) levator scapulae, (2) rhomboid minor, and (3) rhomboid major; (ii) adduction of the humerus by the (5) teres major, (4) latissimus dorsi, and pectoralis major (not shown).

arm can be flexed to the horizontal position only if the inferior angle of the scapula is also pulled forwards on the chest wall (lateral rotation) by the **serratus anterior**, thus turning the glenoid cavity upwards.

Extension

Two sets of muscles extend the shoulder. Extension of the shoulder from the anatomical position is restricted and produced by the posterior fibres of the **deltoid**, assisted initially by the latissimus dorsi. Further extension is completed by elevation of the scapula on the convex thoracic wall by the trapezius and levator scapulae. Extension of the flexed shoulder joint back to the anatomical position against resistance is produced by the latissimus dorsi, the teres major, and the sternocostal part

of the pectoralis major, assisted by the rhomboid major and pectoralis minor, both of which rotate the scapula medially.

Abduction

Abduction [Fig. 8.10A] is produced by the middle fibres of the **deltoid** and by the **supraspinatus**, both of which pass superior to the joint. The supraspinatus is responsible for initiating the movement. While the deltoid is contracting, simultaneous contraction of the teres minor and the lower fibres of the subscapularis prevents the humerus from being pulled up against the coracoacromial arch [Fig. 8.9]. The deltoid can abduct the humerus on the scapula to the horizontal, but this movement is associated from the beginning with **lateral rotation of the scapula** [Fig. 8.10A]. Lateral rotation of the scapula (produced by the serratus anterior and trapezius) permits the humerus to be carried upwards to the vertical position, by turning the glenoid cavity to face superiorly. To confirm this, note the elevation of the shoulder and the lateral projection of the inferior angle of the scapula in full abduction in the living.

Adduction

Adduction against resistance of the arm abducted above the head is first produced by the **latissimus dorsi** and the lowest sternocostal fibres of the **pectoralis major**, and is assisted by the teres major and the medial rotators of the scapula [Fig. 8.10B]. Once the horizontal position has been passed, progressively higher fibres of the pectoralis major are involved. When this movement is not resisted, those muscles which are active in abduction act eccentrically to control the pull of gravity on the limb. This situation is common to every movement where gravity is the driving force. It can be demonstrated in the shoulder by the continuing firmness (contraction) of the deltoid as the arm is lowered to the side and its immediate flaccidity when the movement encounters resistance.

Medial and lateral rotation of the humerus

Medial and lateral rotation of the shoulder may occur in any position but are best demonstrated with the arm by the side and the elbow flexed at right angle. The hand can then be swung laterally (lateral rotation of the humerus) or medially (medial rotation

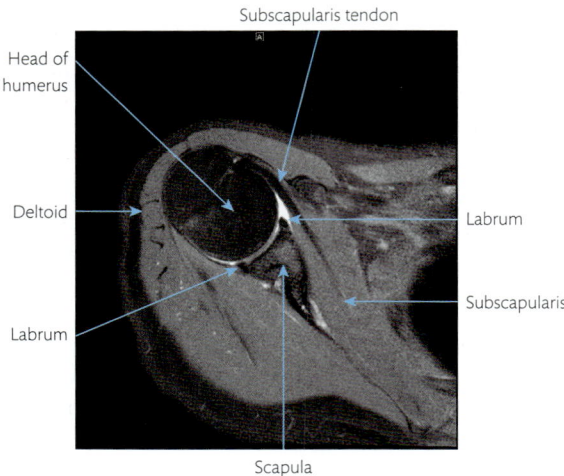

Head of humerus

Subscapularis tendon

Deltoid

Labrum

Labrum

Subscapularis

Scapula

Fig. 8.11 Plain radiographs of the shoulder. (A) Arm is adducted at the shoulder. (B) Superoinferior (axial) view.

of the humerus). In this position, medial rotation is produced by muscles passing to the front of the humerus from the trunk (pectoralis major, latissimus dorsi, subscapularis, teres major, anterior fibres of the deltoid). Lateral rotation results from contraction of muscles passing to the back of the humerus from the trunk (infraspinatus, teres minor, and posterior fibres of the deltoid).

Muscles, movements, and nerves of the shoulder joint

The action of a muscle crossing the shoulder joint can be predicted not only from its origin and insertion, but also from its relation to the shoulder joint. These details, along with the nerve supply to the muscle, are shown in Table 8.1. In Table 8.2, the muscles are grouped according to their action on the shoulder joint, and the nerve supply of each muscle is repeated. This allows an easy assessment of the degree of paralysis of a particular movement following destruction of a particular nerve. (The tables show only the actions of a particular muscle when it actively shortens.)

➲ Axillary nerve injury and the resultant paralysis of the deltoid and teres minor severely affect shoulder abduction, extension, and lateral rotation. In such an injury, the arm is held in a position of adduction, medial rotation, and flexion.

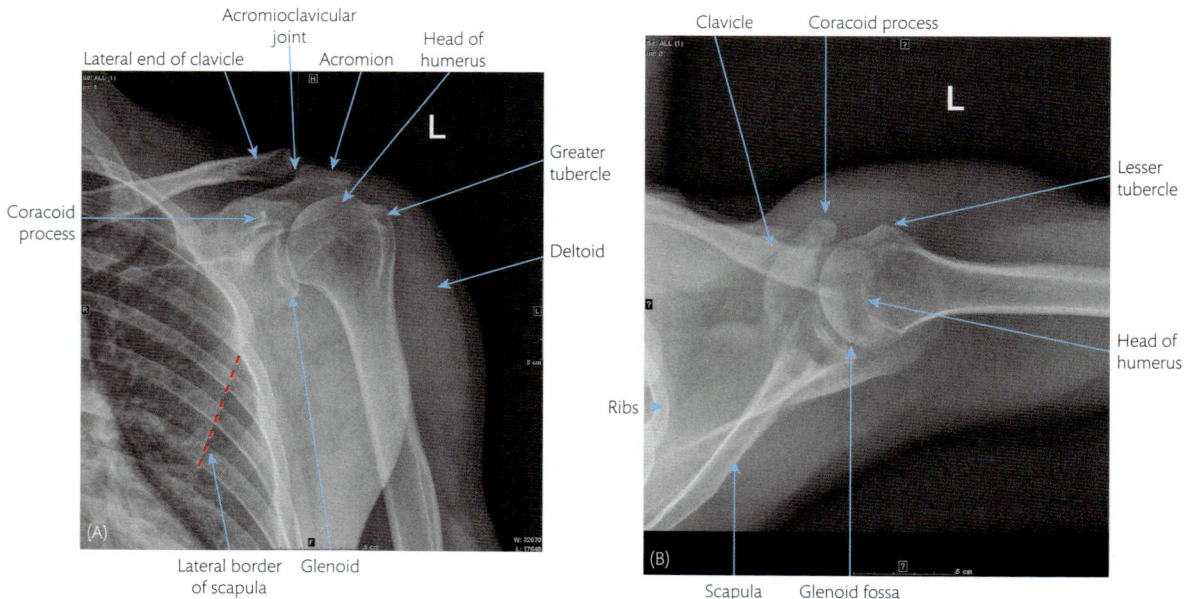

Fig. 8.12 Magnetic resonance axial image of the shoulder.

Table 8.1 Muscles acting on the shoulder joint

Muscle	Origin	Insertion	Relation to joint	Action	Nerve supply
Pectoralis major	Clavicle, medial two-thirds	Humerus, lateral lip of intertubercular sulcus	Anterior	Flexion and medial rotation	Medial and lateral pectoral nerves
	Sternum Costal cartilages 1–6	Humerus, lateral lip of intertubercular sulcus	Anterior	Adduction and medial rotation	Medial and lateral pectoral nerves
Latissimus dorsi	Lower ribs Thoracolumbar fascia Iliac crest	Humerus, intertubercular sulcus	Inferior	Adduction, medial rotation, extension if flexed	Thoracodorsal nerve
Deltoid	Clavicle, lateral one-third	Humerus, deltoid tuberosity	Anterior	Flexion and medial rotation	Axillary nerve
	Scapula, acromion, spine	Humerus, deltoid tuberosity	Superior Posterior	Abduction, extension, lateral rotation	Axillary nerve
Biceps brachii, short head	Scapula, coracoid process	Radius	Anterior	Flexion Stabilization	Musculocutaneous nerve
Biceps brachii, long head	Scapula, supraglenoid tubercle	Radius	Anterior	Flexion Stabilization	Musculocutaneous nerve
Coracobrachialis	Scapula, coracoid process	Humerus, middle of body medially	Anterior	Flexion	Musculocutaneous nerve
Teres major	Scapula, lateral margin inferior one-third	Humerus, medial lip of intertubercular sulcus	Inferior	Adduction, medial rotation	Lower subscapular nerve
Supraspinatus	Scapula, supraspinous fossa	Humerus, greater tubercle superior surface	Superior	Abduction, stabilization	Suprascapular nerve
Infraspinatus	Scapula, infraspinous fossa	Humerus, greater tubercle posterosuperior surface	Posterior	Lateral rotation, stabilization	Suprascapular nerve
Teres minor	Scapula, lateral margin superior two-thirds	Humerus, greater tubercle posterior surface	Posterior	Lateral rotation, stabilization*	Axillary nerve
Subscapularis	Scapula, subscapular fossa	Humerus, lesser tubercle	Anterior	Medial rotation, stabilization*	Upper and lower subscapular nerves
Triceps, long head	Scapula, infraglenoid tubercle	Ulna, olecranon process	Inferior	Stabilization	Radial nerve

* These muscles stabilize the shoulder joint in abduction and prevent the head of the humerus from rising in the glenoid and hitting on the acromion when the deltoid contracts.

Applied anatomy

The disproportionately large head of the humerus, the small and shallow glenoid cavity, and the loose articular capsule give the shoulder joint a wide range of movements but make the joint inherently unstable [Fig. 8.11]. This instability is overcome by the powerful muscles which closely surround the joint [Figs 8.6, 8.12]. These muscles support the joint in any position, without restricting movement which ligaments would do.

Dislocation

There is an increased risk of displacement of the head of the humerus from the glenoid cavity (dislocation) when the joint is suddenly pulled upon. This displacement frequently occurs through the lower part of the joint capsule which is inadequately supported by the long head of the triceps. This inferior dislocation of the humeral head can result in damage to the adjacent axillary nerve.

See Clinical Application 8.1 for the practical implications of the anatomy in this chapter.

Table 8.2 Movements at the shoulder joint

Movement	Muscles	Nerve supply
Flexion	Pectoralis major, clavicular part	Pectoral nerves
	Deltoid, clavicular part	Axillary
	Biceps, short head	Musculocutaneous
	Coracobrachialis	Musculocutaneous
Extension	Deltoid, posterior part	Axillary
	Latissimus dorsi (if shoulder flexed)	Thoracodorsal
	Teres major (if shoulder flexed)	Subscapular
Abduction	Deltoid, acromial part	Axillary
	Supraspinatus	Suprascapular
Adduction	Pectoralis major, sternocostal part	Pectoral
	Latissimus dorsi	Thoracodorsal
	Teres major	Subscapular
Lateral rotation of humerus	Deltoid, posterior part	Axillary
	Infraspinatus	Suprascapular
	Teres minor	Axillary
Medial rotation of humerus	Pectoralis major	Pectoral
	Latissimus dorsi	Thoracodorsal
	Deltoid, clavicular part	Axillary
	Teres major	Subscapular
	Subscapularis	Subscapular
Stabilization*	Subscapularis	Subscapular
	Supraspinatus	Suprascapular
	Infraspinatus	Suprascapular
	Teres minor	Axillary
	Triceps, long head	Radial
	Biceps, long head	Musculocutaneous

The actions of muscles shown in this table presuppose a fixed scapula.

* All muscles of stabilization are attached close to the shoulder joint, have a poor mechanical advantage over it, and are more effective in holding the joint surfaces together than in moving it.

CLINICAL APPLICATION 8.1 Rotator cuff injury

A young swimmer came to the doctor with a history of sudden pain in the right shoulder and inability to sleep on the injured side. After examining the patient, a diagnosis of rotator cuff injury was made.

Study question 1: what are the muscles which form the 'rotator cuff'? (Answer: supraspinatus, infraspinatus, teres minor, and subscapularis.)

Study question 2: apart from the movement these muscles bring about at the shoulder, what is the other important function they perform? (Answer: they act as expansile ligaments of the shoulder joint and steady it.) The doctor suspects injury to the supraspinatus.

Study question 3: where is the supraspinatus inserted? What is its action? Under which bony projection does the tendon pass? (Answer: it is inserted into the upper facet on the greater tubercle and produces abduction of the shoulder. The tendon passes under the acromion process of the scapula.)

CHAPTER 9
The arm

Surface anatomy of the arm

The surface anatomy of the upper part of the arm is covered in Chapter 8. At the midpoint of the anterior surface of the bend of the elbow, find the **tendon of the biceps**. Medial to this, feel the pulsations of the **brachial artery**, and trace it superiorly on the medial side of the arm. The **median nerve** may be palpable, posteromedial to the artery.

Deep fascia of the arm

In the arm, the deep fascia is strongest posteriorly where it covers the triceps muscle. The deep fascia

enclosing the arm sends septa between the groups of muscles to allow them to slide on each other and to give an increased area for origin. Two of these septa—the **lateral** and **medial intermuscular septa**—pass to the corresponding supracondylar lines and epicondyles of the humerus, thus dividing the distal part of the arm into anterior and posterior compartments [Figs 9.1, 9.2]. These septa lie between the triceps muscle posteriorly and the muscles attached to the anterior surface of the distal half of the humerus.

At the elbow, the deep fascia is thickened by extensions from the triceps and biceps muscles, and by the origin of the forearm muscles from its deep surface. The **bicipital aponeurosis** is a strong slip which extends medially from the tendon of the

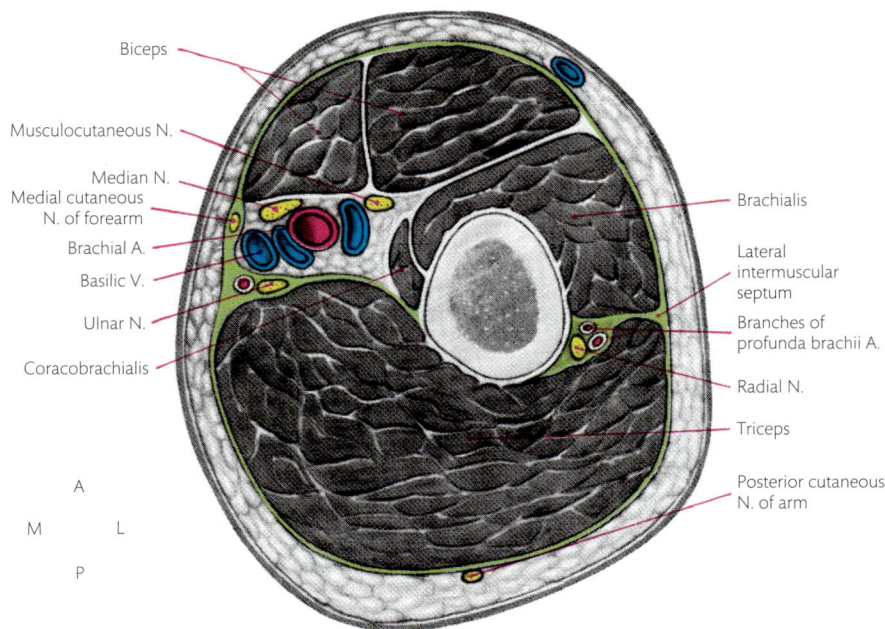

Fig. 9.1 Section through the middle of the right arm showing positions of neurovascular bundles and intermuscular septa. A = anterior; P = posterior; M = medial; L = lateral.

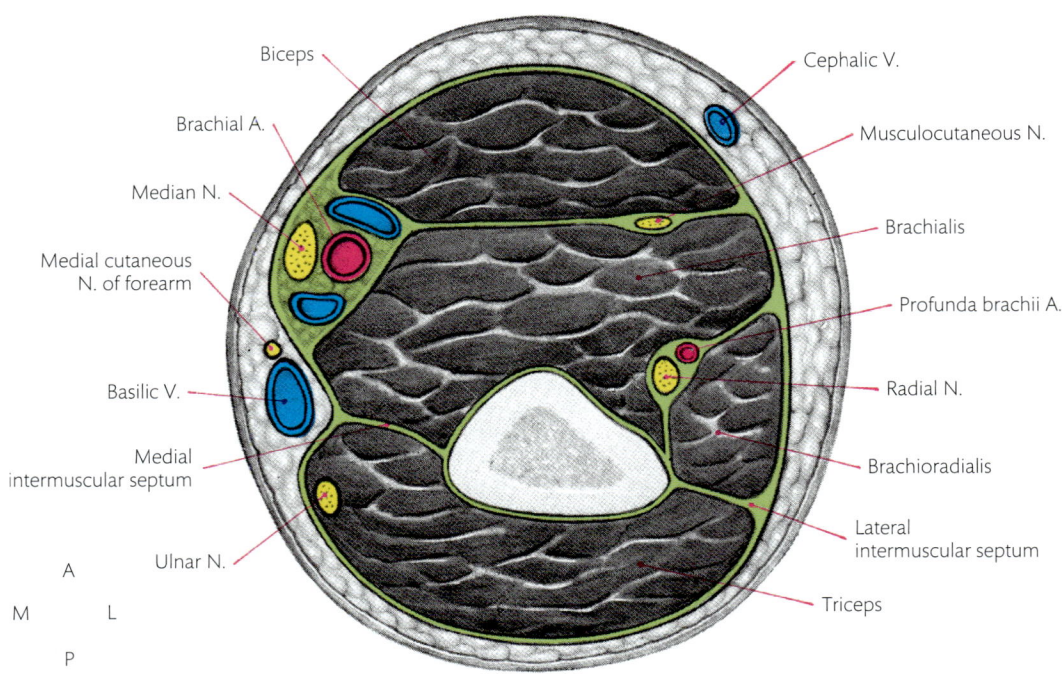

Biceps

Brachial A.

Median N.

Medial cutaneous
N. of forearm

Basilic V.

Medial
intermuscular septum

Ulnar N.

Cephalic V.

Musculocutaneous N.

Brachialis

Profunda brachii A.

Radial N.

Brachioradialis

Lateral
intermuscular septum

Triceps

A

M   L

P

Fig. 9.2 Section through the distal third of the right arm. A = anterior; P = posterior; M = medial; L = lateral.

biceps to the deep fascia and the subcutaneous posterior border of the ulna. The aponeurosis is readily felt by sliding a finger down the medial side of the taut biceps tendon.

Anterior compartment of the arm

See Dissection 8.1 for instructions on skin reflection of the front of the arm and forearm.

See Dissection 9.1 and Fig. 9.3 for an exploration of the anatomy of the front of the arm.

Biceps brachii

Origin: the biceps brachii muscle arises from the scapula by two heads—the long and short heads. The **short head** arises with the coracobrachialis from the coracoid process. The **long head** arises from the supraglenoid tubercle within the shoulder joint. **Insertion**: its tendon runs on the superior surface to the head of the humerus and emerges from the joint through the intertubercular groove. The two heads of the biceps fuse in the distal third of the arm and form a short tendon which passes to the posterior surface of the tuberosity of the radius [Fig. 9.4]. The biceps tendon also gives off the **bicipital aponeurosis**. The

bicipital aponeurosis forms a secondary insertion of the biceps to the posterior border of the ulna. **Nerve supply**: musculocutaneous nerve. The short head of the **biceps** flexes the shoulder joint. The long head of the biceps helps to hold the head of the humerus against the glenoid cavity, especially when the arm is abducted. The biceps also supinates the forearm because of its attachment to the posterior surface of the radial tuberosity.

Coracobrachialis

Origin: the coracobrachialis originates from the coracoid process of the scapula, along with the short head. **Insertion**: it is inserted into the middle of the medial surface of the humerus. **Nerve supply**: musculocutaneous nerve. **Actions**: the **coracobrachialis** flexes the shoulder joint.

Brachialis

Origin: this muscle arises from the anterior surface of the distal half of the humerus [see Fig. 4.9]. **Insertion**: it descends across the anterior surface of the elbow joint to be inserted into the coronoid process of the ulna. **Nerve supply**: musculocutaneous nerve, and some nerve fibres from the radial nerve. **Actions**: the **brachialis** is a pure flexor of the elbow joint.

DISSECTION 9.1 Front of the arm

Objectives

I. To study the muscles of the front of the arm. **II.** To identify and trace the musculocutaneous, median, and ulnar nerves in the front of the arm. **III.** To identify the muscles arising from the lateral supracondylar ridge, radial nerve, and profunda brachii artery on the lateral side of the elbow.

Instructions

1. Cut vertically through the deep fascia on the anterior surface of the arm, as far as the elbow. Cut transversely through it at the elbow. Reflect the flaps to uncover the biceps brachii.
2. Lift the biceps brachii, and find the musculocutaneous nerve. Separate the biceps brachii from the brachialis muscle posteriorly. Follow the nerve, and the biceps and coracobrachialis muscles proximally and distally. Neither the tendon of origin of the long head of the biceps nor its tendon of insertion can be followed to the end.
3. Remove the fascia from the brachialis, and define the forearm muscles—the brachioradialis and extensor carpi radialis longus. They arise from the lateral supracondylar line and the lateral intermuscular septum, and appear to be part of the brachialis but are separated from it by a thin septum which contains the radial nerve and a terminal branch of the profunda brachii artery [Figs 9.1, 9.2].
4. Find the radial nerve in this situation and its branches to the brachioradialis, extensor carpi radialis longus, and brachialis [Fig. 9.3].
5. Trace the radial nerve proximally to the point where it passes posterior to the humerus. Identify the lower lateral cutaneous nerve of the arm and the posterior cutaneous nerve of the forearm that arise from the radial nerve.
6. Find the principal neurovascular bundle of the arm immediately deep to the deep fascia, medial to the biceps. This bundle includes the median nerve, ulnar nerve, and brachial artery and veins. Trace these structures proximally in continuity with the structures in the axilla, and distally to the level of the elbow.

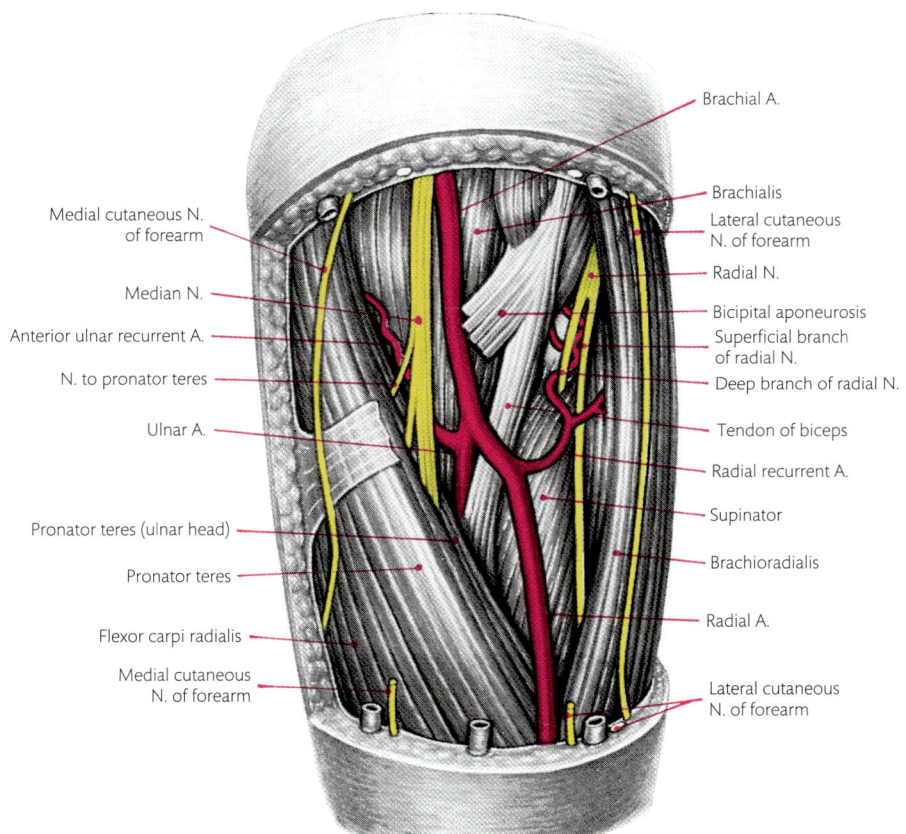

Fig. 9.3 Dissection of the left cubital fossa. The bicipital aponeurosis has been cut.

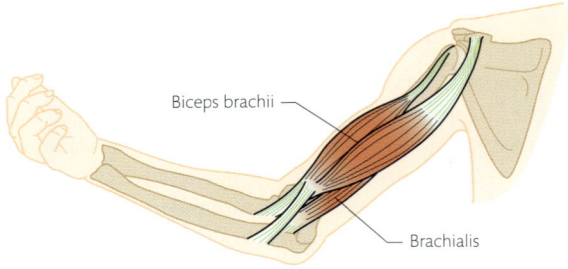

Fig. 9.4 Diagram of the biceps brachii showing the two heads of origin and the insertions.

Musculocutaneous nerve in the arm (C. 5, 6)

This nerve arises in the axilla from the lateral cord of the brachial plexus. It passes inferolaterally to supply, and then pierce, the coracobrachialis. It then descends between the biceps and brachialis, supplying both muscles, and emerges at the lateral border of the biceps [see Fig. 7.3] as the lateral cutaneous nerve of the forearm.

Nerve injury

⊃ The musculocutaneous nerve is the motor supply to all of the muscles in the front of the arm. Thus, damage to this nerve interferes with flexion at the shoulder and elbow joints, especially the elbow joint, and weakens supination.

Principal neurovascular bundle of the arm

Above the insertion of the coracobrachialis, this bundle consists of: (1) the brachial artery; (2) the basilic and brachial veins; (3) the median nerve; (4) the ulnar nerve; (5) the radial nerve; and (6) the medial cutaneous nerve of the forearm. The **radial nerve** is the first to leave the bundle. It passes inferolaterally to the groove for the radial nerve on the posterior surface of the humerus. In the groove, the nerve is accompanied by the **profunda brachii artery** [Fig. 9.5]. Slight traction on that nerve will confirm its continuity with the nerve already exposed on the lateral side of the arm. The **ulnar**

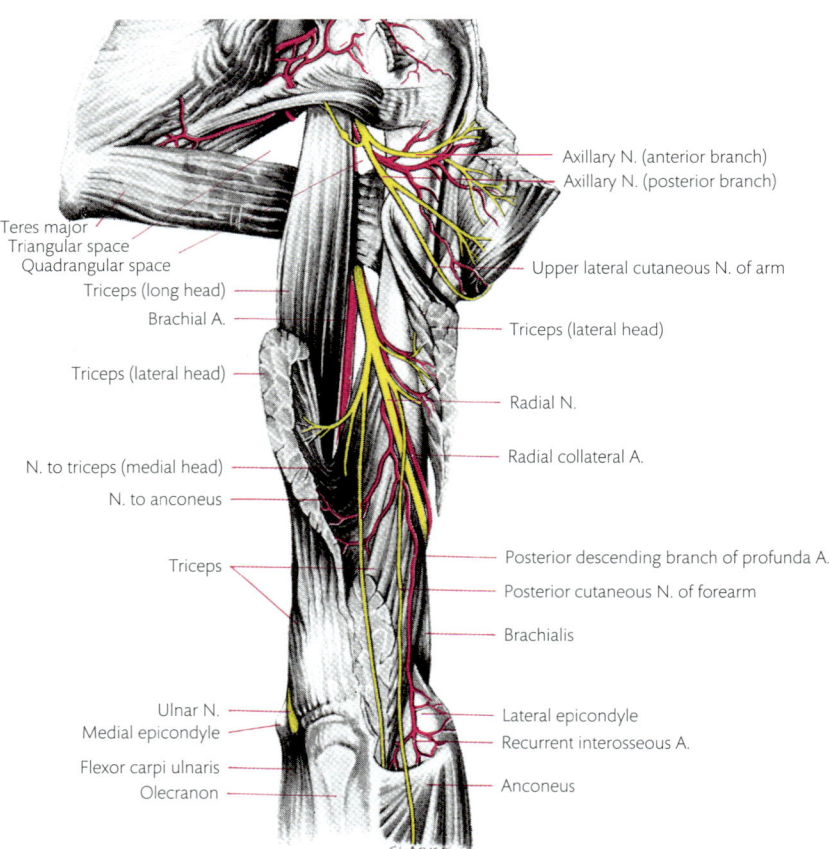

Teres major
Triangular space
Quadrangular space
Triceps (long head)
Brachial A.

Triceps (lateral head)

N. to triceps (medial head)
N. to anconeus

Triceps

Ulnar N.
Medial epicondyle

Flexor carpi ulnaris
Olecranon

Axillary N. (anterior branch)
Axillary N. (posterior branch)

Upper lateral cutaneous N. of arm

Triceps (lateral head)

Radial N.

Radial collateral A.

Posterior descending branch of profunda A.
Posterior cutaneous N. of forearm

Brachialis

Lateral epicondyle
Recurrent interosseous A.

Anconeus

Fig. 9.5 Dissection of the back of the shoulder and arm. The lateral head of the triceps has been divided and turned aside to expose the radial nerve.

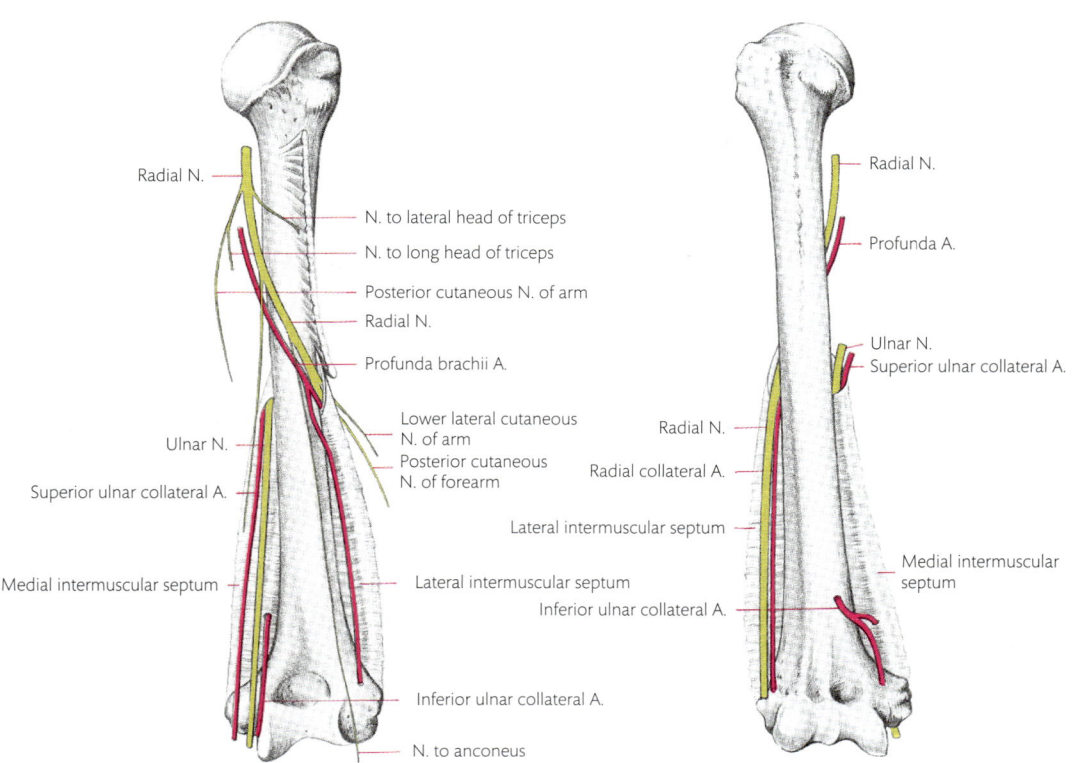

Fig. 9.6 Diagram to show the relation of the radial and ulnar nerves, and the profunda brachii and ulnar collateral arteries to the humerus. Posterior surface = left; anterior surface = right.

nerve is the next to leave the bundle. It passes back through the medial intermuscular septum into the posterior compartment [Fig. 9.6]. Below this level, the **basilic vein** and the **medial cutaneous nerve of the forearm** pierce the deep fascia and enter the superficial fascia. The bundle then consists of the **median nerve**, and the **brachial artery** and veins in the lower third of the arm. These structures incline forwards, in front of the brachialis, to lie just medial to the tendon of the biceps at the elbow [Fig. 9.3].

Median nerve in the arm (C. 5, 6, 7, 8; T. 1)

This nerve is formed in the axilla by one root, each from the medial and lateral cords of the brachial plexus. It descends anterior to the axillary and upper part of the brachial arteries to the medial aspect of the brachial artery in the distal half of the arm. The median nerve does not supply any muscle in the arm but supplies the sympathetic post-ganglionic fibres to the axillary and brachial arteries. It supplies most of the flexor muscles in the anterior aspect of the forearm. In the hand, it

supplies the thenar muscles and two lumbricals. It also supplies the skin on the hand and fingers.

Ulnar nerve in the arm (C. 8; T. 1)

The ulnar nerve arises from the medial cord of the brachial plexus in the axilla, and descends medial to the distal part of the axillary artery and the proximal part of the brachial artery. In the middle of the arm, it pierces the medial intermuscular septum and runs distally in the posterior compartment of the arm. It enters the forearm by passing over the posterior surface of the medial epicondyle [Fig. 9.6]. This nerve is concerned principally with the supply of the small muscles of the hand, some muscles of the forearm, and the skin on the hand and fingers. It does not supply any structures in the axilla or the arm.

Brachial artery

The **brachial artery** begins at the lower border of the teres major as a continuation of the axillary artery. It ends in the cubital fossa, at the neck of the radius by dividing into the radial and ulnar arteries. It supplies the structures of the arm via branches which accompany the major nerves and via smaller branches which pass directly to the muscles.

Named branches

The **profunda brachii artery** arises from the brachial artery and accompanies the radial nerve. It ends by dividing into **radial collateral** and **middle collateral arteries**. It gives a descending branch on each side of the lateral intermuscular septum [Figs 9.6, 9.7]. The **superior ulnar collateral branch** of the brachial artery accompanies the ulnar nerve. The **inferior ulnar collateral branch** arises 5 cm above the elbow and sends descending branches on each side of the medial intermuscular septum. These vessels supply adjacent muscles and take part in anastomosis around the elbow joint. The **nutrient artery to the humerus** arises from the brachial artery near the middle of the humerus. The brachial artery ends by dividing into the **radial** and **ulnar arteries** in the cubital fossa (see below).

Anastomosis around the elbow

All the blood to the forearm and hand reached it through the brachial artery. Should there be an obstruction to the blood flow in the brachial artery (e.g. blockage of the artery by a blood clot), the blood supply to the forearm and hand will be compromised. However, the anastomosis of arteries around the elbow joint provides an alternate route for blood flow to the forearm in such a case.

Four to five arteries from the arm—**collateral arteries**—anastomose with a similar number of arteries from the forearm—**recurrent arteries**. On the lateral side of the elbow: (1) the radial collateral branch of the profunda brachii anastomoses with the radial recurrent branch of radial artery, and (2) the middle collateral branch of the profunda brachii anastomoses with the interosseous recurrent artery—a branch of the common interosseous artery from the ulnar artery; on the medial side of the elbow, (3) the superior ulnar collateral branch of the brachial artery anastomoses with the posterior ulnar recurrent branch of the ulnar artery, and (4) the inferior ulnar collateral branch of the brachial artery anastomoses with the anterior ulnar recurrent artery. The superior ulnar collateral artery may give an additional branch which passes anterior the medial epicondyle and anastomoses with a branch of the anterior ulnar recurrent artery [Fig. 9.7].

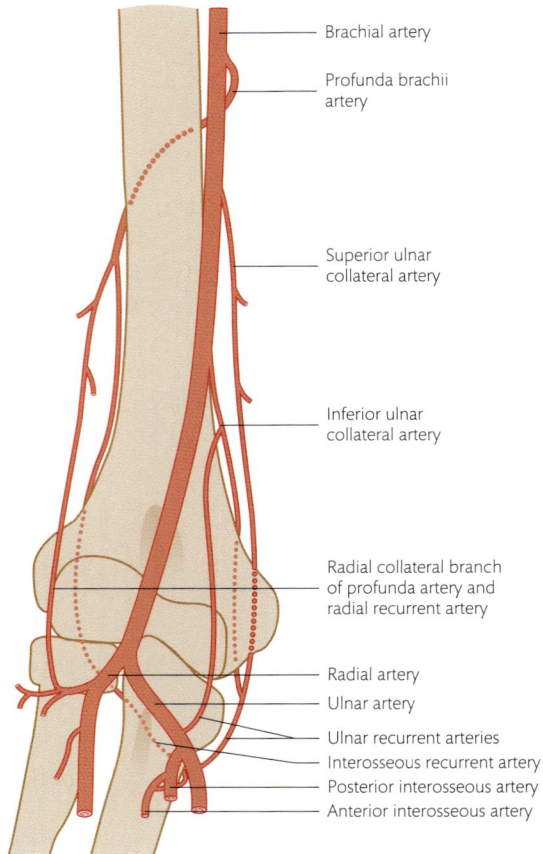

Fig. 9.7 Diagram showing the brachial artery and the anastomosis around the elbow joint.

Brachial veins

Brachial veins are formed in the cubital fossa by the union of small veins—venae comitantes—that run along the radial and ulnar arteries. They run alongside the brachial artery and continues as the axillary artery at the lower border of the teres major. Brachial veins receive small veins draining the arm, and the basilic vein in the middle or the arm.

Cubital fossa

The cubital fossa is the triangular intermuscular depression at the front of the elbow [Fig. 9.3]. The **base** of the fossa is an imaginary line drawn between the two epicondyles of the humerus. The **medial border** is the pronator teres muscle. The **lateral border** is the brachioradialis muscle, up to a point where it overlaps with the pronator teres. The **floor** is formed by the brachialis and supinator. The **roof** is the deep fascia of the forearm, with the cutaneous vessels and nerves superficial to it. The **contents** of the cubital fossa include the median nerve, brachial artery, and tendon of the biceps, which enter the fossa from above. In the cubital fossa, the brachial artery divides into the **radial** and **ulnar arteries**. The radial artery leaves the fossa at the apex; the ulnar artery leaves by passing deep to the pronator teres. The median nerve supplies the muscles medial to it and leaves the fossa through the pronator teres. The tendon of the biceps passes between the forearm bones to reach the radial tuberosity. If the elbow is flexed and the margins are pulled apart, the contents of the fossa are seen after the deep fascia covering it has been removed. The radial nerve lies just deep to the brachioradialis in the lateral border and is considered as a content of the cubital fossa. (The cubital fossa will be dissected with the front of the forearm.)

Back of the arm

Dissection 9.2 explores the anatomy of the back of the arm.

Triceps brachii

The triceps brachii has three heads of origin: the long head, medial head, and lateral head. **Origin**: the **long head** of the triceps brachii takes origin from the infraglenoid tubercle of the scapula. The **lateral head** arises from the upper third of the posterior surface of the humerus [see Fig. 4.10]. The **medial head** arises from the posterior surface of the humerus, distal to the groove for the radial nerve. These long and lateral heads descend posterior to the groove for the radial nerve and are joined on their deep surfaces by the medial head of the triceps. **Insertion**: the three heads form a common tendon which is inserted into the superior surface of the olecranon [see Fig. 4.14] and the surrounding deep fascia. Some of the deep fibres pass to the articular capsule of the elbow joint and pull up the redundant part of the capsule to avoid it being caught between the olecranon and the humerus in extension of the joint. **Nerve supply**: radial nerve. **Actions**: all three heads of the triceps are extensors of the elbow joint. The long head also acts on the shoulder joint. It is mainly responsible for steadying the humerus in the glenoid cavity, especially when it is stretched across the inferior surface of

---

DISSECTION 9.2 Back of the arm

Objectives

I. To study the muscles of the back of the arm. **II.** To identify and trace the profunda brachii artery, and the radial nerve and its branches. **III.** To follow the ulnar nerve to the back of the medial epicondyle.

Instructions

1. Remove the deep fascia from the back of the arm to expose the triceps muscle which fills the posterior compartment.
2. Superiorly, separate the medially placed long head of the triceps, which arises from the infraglenoid tubercle of the scapula [see Fig. 4.7], from the lateral head, which has a linear origin from the posterior surface of the humerus between the insertions of the teres minor and the deltoid [see Fig. 4.10].
3. Find the radial nerve in the axilla, posterior to the axillary artery. Trace the nerve as far as the triceps, and separate the parts of the triceps by passing a blunt seeker along the nerve in that muscle. Divide and reflect the parts of the lateral head of the triceps

where it covers the radial nerve to expose the radial nerve and the profunda brachii artery in the groove on the back of the humerus. The medial head of the triceps takes origin from the humerus, inferior to the groove [Fig. 9.5].
4. Follow the branches of the radial nerve, and check its continuity with the part of the radial nerve already seen between the brachialis and the brachioradialis. Follow this part of the nerve distally. Check its branches to the brachioradialis, extensor carpi radialis longus, and brachialis. Identify its division into superficial and deep branches at the level of the elbow joint [Fig. 9.3].
5. Follow the ulnar nerve into the posterior compartment of the arm with the superior ulnar collateral artery and the branch of the radial nerve to the medial head of the triceps. Trace the ulnar nerve to the back of the medial epicondyle.
6. Remove the connective tissue from the posterior surface of the medial intermuscular septum, and find the posterior branch of the inferior ulnar collateral artery [Figs 9.5, 9.6].

Biceps brachii

Biceps brachii

Triceps brachii

(A)

(B)

Fig. 9.8 Muscles of the arm. (A) Anterior view. (B) Posterior view.

achernishev/Shutterstock.com. Fesus Robert/Shutterstock.com.

the joint in abduction of the arm. In abduction of the arm, the long heads of the triceps and biceps lie, like supporting cables, inferior and superior to the joint, and the long head of the triceps is the main support inferiorly. (For the surface anatomy of the muscles of the arm, see Fig. 9.8.)

Radial nerve in the arm

The **radial nerve** arises as a terminal branch of the posterior cord of the brachial plexus in the axilla. In the axilla, it gives off the **nerve to the long head of the triceps** and then pass into the arm, posterior to the brachial artery. Almost immediately, it gives off the **posterior cutaneous nerve of the arm** and a branch to the **medial head of the triceps** (which accompanies the ulnar nerve in the posterior compartment). It then passes inferolaterally into the groove for the radial nerve on the posterior surface of the humerus [see Fig. 4.8B], and winds around this surface of the humerus with the profunda brachii artery. In the groove, the nerve gives off branches to the **lateral head**

of the triceps and a long slender branch which descends through the medial head of the triceps of the **anconeus** muscle [Fig. 9.5]. Two cutaneous branches—the **lower lateral cutaneous nerve of the arm** and the **posterior cutaneous nerve of the forearm**—are also given off here.

At the lower end of the arm, the nerve pierces the lateral intermuscular septum and descends in the anterior compartment between the brachialis (medially) and the brachioradialis and extensor carpi radialis longus (laterally). It divides into superficial and deep branches. The **superficial branch** is a sensory nerve to the back of the fingers and hand [see Fig. 7.1]. The **deep branch** supplies the muscles of the back of the forearm and the joints at the wrist. The radial nerve is sensory to the elbow joint, principally through its branches to the anconeus and the medial head of the triceps, and to the superior and inferior radioulnar joints.

See Clinical Applications 9.1 and 9.2 which demonstrate the practical implications of the anatomy explored in this chapter.

CLINICAL APPLICATION 9.1 Fracture of the shaft of the humerus

Mid-shaft fractures of the humerus are common and may damage the radial nerve, as it traverses the groove on the back of the humerus. Knowledge of the muscular and sensory branches given off by the radial nerve, along with the point of origin of these branches, will help you predict what lesions arise from such a fracture.

Study question 1: name the muscles denervated by such an injury. (Answer: triceps, lateral head; anconeus; brachioradialis; extensor carpi radialis longus and brevis. The medial and long heads are supplied before the nerve enters the groove.) The extensor muscles of the forearm and wrist are also paralysed. You will learn about these in Chapter 10.

Study question 2: name the movements likely to be lost or weakened because of this injury. (Answer: loss of extension of the wrist, hand, and fingers; weakness of supination.)

Study question 3: what sensory nerves will be affected? (Answer: lower lateral cutaneous nerve of the arm, posterior cutaneous nerve of the forearm, superficial branch of the radial nerve.)

Study question 4: what sensory loss will result from this nerve injury? (Answer: loss of sensation over the lateral part of the lower arm, the posterior forearm, the lateral part of the dorsum of the hand, and the dorsal aspect of the lateral three-and-a-half digits, excluding their nail beds.)

Study question 5: name the artery which could be damaged in a fracture of the mid shaft of the humerus. (Answer: profunda brachii artery.)

CLINICAL APPLICATION 9.2 Measurement of blood pressure

Blood pressure is the force exerted by circulating blood upon the walls of blood vessels. It is one of the vital signs of physical well-being. It is measured by using a sphygmomanometer, an inflatable rubber cuff connected to a manometer. The cuff is tied around the arm, and the pressure in the cuff inflated till the blood flow through the brachial artery is cut off. The cuff is then deflated slowly, allowing blood flow to resume. A stethoscope placed over the brachial artery at the cubital fossa monitors the start of blood flow in the artery. When the pressure exerted by the cuff is above the systolic pressure, all blood flow through the artery is stopped.

No sounds are audible, as there is no blood flow. As the cuff pressure falls below the systolic, the sound of spurts of blood flowing past the cuff is heard. When the cuff pressure falls below the diastolic, the blood flow through the brachial artery becomes streamlined (no longer turbulent) and no sound is audible through the stethoscope. The physician measuring the blood pressure makes a note of two pressure readings: (1) when the blood flow first becomes audible, with a drop in cuff pressure; and (2) when the blood flow is no longer audible. These two readings give the patient's systolic and diastolic pressures.

CHAPTER 10
The forearm and hand

Introduction

Surface anatomy of the forearm

Begin by revising the palpable parts of the **radius** and **ulna** in your own forearm, and compare these with the bones themselves. Articulate a radius and an ulna parallel to each other, with the head of the radius lying in the radial notch of the ulna and the head of the ulna in the ulnar notch of the radius [see Figs 4.11, 4.12]. Note the following points:

1. The ulna extends further proximally than the radius. This extension of the ulna is formed by the **trochlear notch** and the **olecranon**.
2. The radius projects further distally than the ulna. The distal end of the radius has markings for the lateral two carpal bones of the proximal row—**scaphoid** and **lunate**. This is the only articulation between a forearm bone and the carpal bones. Because of this: (1) the carpal bones (and hence the hand) move with the radius in movements of pronation and supination; and (2) forces applied to the hand are transmitted to the radius through these two carpal bones, which are therefore vulnerable to injury.
3. The radius is markedly convex laterally. The radius and ulna are held together by an interosseous membrane attached to the sharp **interosseous borders** on their adjacent surfaces.
4. The distal end of the anterior surface of the radius curves forwards to form a flat surface against which the pulsations of the radial artery may be felt (**radial pulse**).
5. Palpate the head of the radius just distal to the lateral epicondyle of the humerus, and feel it rotating within the annular ligament when you pronate and supinate your forearm. When this is done, the distal end of the radius turns around the head of the ulna and comes to lie medial to the distal end of the ulna. In this position, the posterior surface of the radius is directed anteriorly and the shaft of the radius crosses the ulna. This movement is **pronation**, and the reverse movement is **supination**. These are the movements at the radioulnar joints. Note the following points about pronation and supination: (1) the axis for pronation and supination passes through the centre of the head of the radius proximally and the centre of the head of the ulna distally; (2) any muscle that can rotate the radius medially (e.g. pronator teres) produces pronation, whereas muscles that rotate it laterally (e.g. biceps brachii) produce supination; (3) in pronation and supination, only the radius moves (there is no rotation of the forearm on the arm); (4) the hand is carried with the radius and the palm faces posteriorly in pronation (elbow extended); (5) if medial rotation of the humerus is combined with pronation, the palm is brought to face posterolaterally; and (6) testing of the range of pronation is carried out with the elbow flexed.
6. The proximal part of the olecranon is readily palpable. On its posterior surface is a triangular subcutaneous area which is continuous distally with the **posterior margin** (border) **of the ulna**. The entire length of this posterior margin is palpable and ends distally in the **styloid process** which projects from the posteromedial aspect of the **head** of the bone [see Figs 4.11, 4.12]. This palpable margin not only allows the entire length of the ulna to be examined for fractures, but also forms the line of separation between the anteromedial flexor group of muscles of the forearm (supplied by the median and ulnar nerves) and the posterolateral extensor group (supplied by the

Extensor
compartment
of forearm

Flexor compartment of forearm

Fig. 10.1 Muscles on the back of the forearm. The subcutaneous posterior border of the ulna (dotted line) separates the flexor muscles medially from the extensor muscles laterally.

Dean Drobot/Shutterstock.com.

radial nerve) [Fig. 10.1]. These two muscle groups abut on each other anteriorly along a line from the lateral side of the tendon of the biceps to the styloid process of the radius.

Surface anatomy of the wrist

The lower end of the radius is expanded into a cuboidal mass. The lateral surface of this mass extends distally to form the blunt **styloid process**. When the thumb is fully extended (i.e. is moved laterally), a surface depression—'the ana-

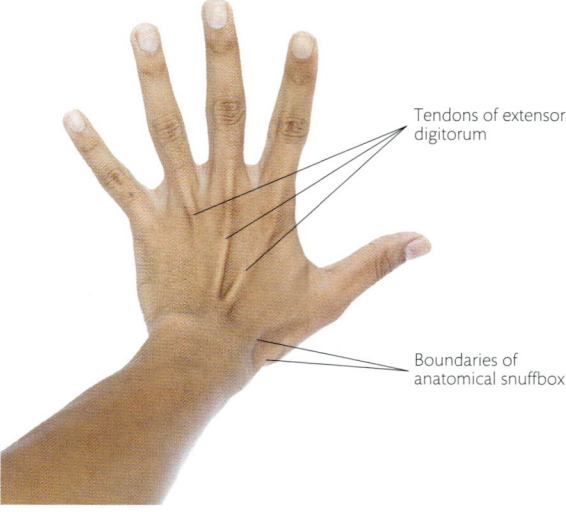

Tendons of extensor digitorum

Boundaries of anatomical snuffbox

Fig. 10.2 Boundaries of the anatomical snuffbox.

Only background/Shutterstock.com

tomical snuffbox'—appears on the lateral side of the back of the wrist between the tendons passing to the thumb [Fig. 10.2]. The styloid process of the radius lies in the proximal part of the depression and is covered by tendons of the short extensor and long abductor muscles of the thumb. The bones at the lateral ends of the proximal and distal rows of the carpal bones (**scaphoid** and **trapezium**) and the base of the metacarpal of the thumb [see Figs 4.14, 4.15] can be felt in the 'snuffbox', distal to the styloid process. Light pressure applied over the trapezium reveals the pulsations of the **radial artery**. The 'radial pulse' is more easily felt where the artery crosses the anterior surface of the distal end of the radius (medial to the styloid process).

Feel your radial and ulnar styloid processes simultaneously. Note that the radial styloid projects further distally than the ulnar styloid—a situation which is altered when the radius is fractured. If the tendon forming the posterior boundary of the 'snuffbox' (extensor pollicis longus) is followed proximally, it curves round the medial surface of the **dorsal tubercle** of the radius which is readily palpable.

The dorsal surfaces of both rows of carpal bones can be felt through the tendons which cover them, though the individual bones cannot be defined. On the palmar surface of the wrist, only the **pisiform bone** (medially) and the **tubercle of the scaphoid** (laterally) can be felt at the level of the distal transverse skin crease (junction of the forearm and wrist) when the wrist is fully extended. If the wrist is passively flexed, the pisiform may be gripped between the finger and the thumb and moved on the triquetrum bone with which it articulates, and the **tendon of the flexor carpi ulnaris** can be felt passing to its proximal surface [Fig. 10.3]. The **hook of the hamate** can be felt deeply through the proximal parts of the muscles forming the ball of the little finger (**hypothenar eminence**), and the **tubercle of the trapezium** can be felt deeply through the proximal parts of the muscles forming the ball of the thumb (**thenar eminence**). These four palpable bony points lying at the ends of the two rows of the carpal bones—tubercle of the scaphoid, pisiform, tubercle of the trapezium, and hook of the hamate—give attachment to the **flexor retinaculum**.

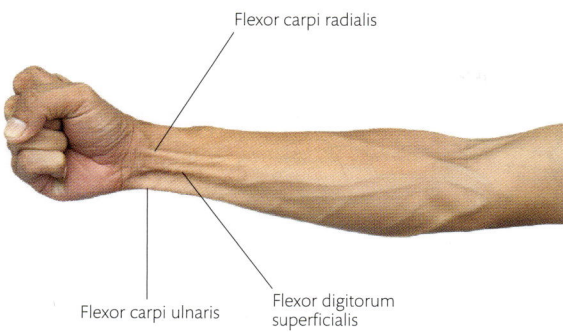

Flexor carpi radialis

Flexor carpi ulnaris

Flexor digitorum superficialis

Fig. 10.3 Flexor tendons at the wrist.

Only background/Shutterstock.com

Surface anatomy of the palm

The skin of the central region of the palm is firmly bound to the thickened underlying deep fascia—the **palmar aponeurosis**. The palmar aponeurosis is continuous distally with the deep fascia of the fingers, and proximally with the flexor retinaculum and the **tendon of the palmaris longus**. If present, this tendon enters the palm, superficial to the retinaculum, and is the most superficial tendon immediately proximal to the wrist.

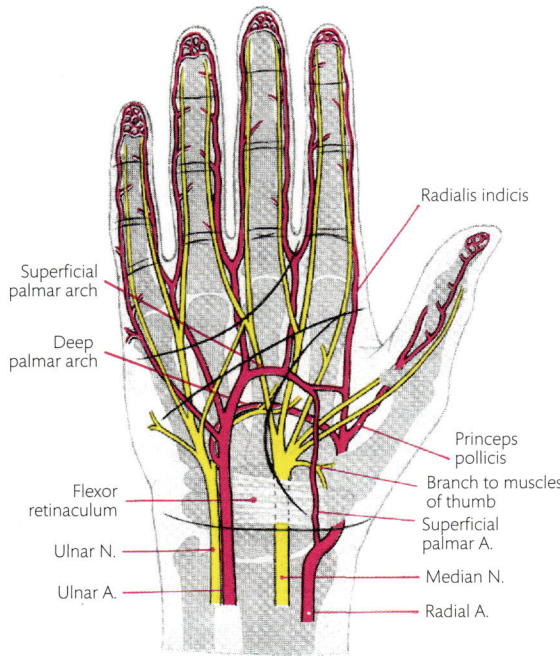

Radialis indicis

Superficial palmar arch

Deep palmar arch

Flexor retinaculum

Ulnar N.

Ulnar A.

Princeps pollicis

Branch to muscles of thumb

Superficial palmar A.

Median N.

Radial A.

Fig. 10.4 Diagram of nerves and vessels of the hand in relation to bones and skin creases.

Contraction of the palmaris longus and extension of the fingers tighten the palmar aponeurosis and stabilize the palmar skin to maintain a firm grip. The distal **skin crease** of the palm lies just proximal to the metacarpophalangeal joints, whereas that at the roots of the fingers lies approximately 3 cm distal to them [Fig. 10.4]. The middle skin crease of each finger lies at the level of the proximal interphalangeal joint, and the distal skin crease lies proximal to the distal interphalangeal joint. The **metacarpals** (hand bones) are readily palpated on their dorsal surfaces. When a fist is formed, the **knuckles** are the distal ends of the **heads of the metacarpal bones**, uncovered by movement of the proximal phalanges onto the palmar surfaces of these heads. The heads of the proximal and middle phalanges of the fingers are similarly exposed.

Surface anatomy of the digits

The thumb lies at right angles to the fingers, with its nail facing laterally, and not posteriorly. Thus, **flexion** moves the tip of the thumb medially across the palm, whereas **extension** moves it laterally. **Abduction** swings the tip of the thumb anteriorly; **adduction** moves it to the index finger. The first carpometacarpal joint, between the metacarpal of the thumb and the trapezium, allows for greater freedom of movement, when compared with the carpometacarpal joints of the fingers. To test this, grip the head of each metacarpal in turn, with the thumb and index finger of your other hand on the palmar and dorsal surfaces, and attempt to move it. The metacarpal of the middle finger scarcely moves, whereas those of the index, ring, and little fingers, in that order, have an increasing, but small, range of flexion and extension and slight rotation only in the little finger.

Flexion of the **metacarpophalangeal joint** of the thumb is less than 90 degrees. It is 90 degrees in the index and middle fingers, and exceeds this in the ring and little fingers. There is some degree of rotation in the ring and little fingers. Check the bones, and note that there is nothing in their contours to prevent rotation. Rotation is limited by ligaments. Also there is very little abduction or adduction at the metacarpophalangeal joint in the thumb, though it is relatively free in the fingers.

The **interphalangeal joint** of the thumb flexes to approximately 90 degrees. The proximal interphalangeal joints of the fingers flex to more than 90 degrees, and the distal interphalangeal joints slightly less than 90 degrees.

The fingers separate on extension and come together in flexion. This is the result of the curve in which the heads of the metacarpals lie and a slight obliquity of the interphalangeal joints. Note also that the tips of all four fingers meet the palm at the same time, despite the differences in their lengths. Check your interphalangeal joints for rotation. This is minimal because of the shape of the articular surfaces of the phalanges.

Cup the palm of your hand by spreading the fingers as though to grasp a large ball. The hollow between the proximal parts of the thenar and hypothenar eminences marks the position of the **flexor retinaculum**. The tightening of the skin on the medial side of the hypothenar eminence is produced by contraction of the **palmaris brevis** which heaps up the hypothenar skin to form a pad for grasping. Note too that the thumb is slightly flexed at the carpometacarpal joint and rotated medially, so that its palmar surface faces that of the little finger (opposition), which is also rotated laterally. The other three fingers are not parallel to each other but, on flexion, converge to meet the tip of the thumb.

Superficial fascia

There are no special features about this fascia of the forearm, but in the palm, it contains dense bundles of fibrous tissue connecting the skin to the palmar aponeurosis. It is thickened transversely in the webs of the fingers to form the **superficial transverse metacarpal ligament**. On the proximal part of the hypothenar eminence, it contains fibres of the skeletal muscle—**palmaris brevis**—which passes from the skin of the medial border of the hand to the palmar aponeurosis.

Introduction to fascial thickening in the forearm and hand

The deep fascia of the forearm and hand is thickened at the wrist to form the flexor and extensor retinacula, in the palm of the hand to form the palmar aponeurosis, and on the palmar aspect of the digits to form the fibrous flexor sheaths. These specializations are briefly introduced here but are discussed in detail after the dissections.

Retinacula

The wrist has thickened bands of deep fascia attached to all the subcutaneous bony points (retinacula). The flexor retinaculum binds down the flexor tendons and median nerve anteriorly. The extensor retinaculum binds down the extensor tendons posteriorly and laterally. The **flexor retinaculum** is a strong band (approximately 2×2 cm) which is continuous with the deep fascia of the forearm at the distal flexor skin crease of the wrist. The **extensor retinaculum** extends from the lateral aspect and the styloid process of the radius to the ulna. Its deep surface is attached to the subcutaneous parts of these bones. Both retinacula act as pulleys for the corresponding tendons which slide freely deep to them in synovial sheaths when the wrist is moved.

Palmar aponeurosis

This thick, triangular deep fascia lies in the central part of the palm, with its apex at the flexor retinaculum and its base at the level of the heads of the metacarpals [Fig. 10.5]. It stabilizes the palmar skin which is firmly adherent to it. Medially and laterally, the aponeurosis is continuous posteriorly with the layer of fascia which covers the anterior surfaces of the metacarpals, and the muscles overlying them. Distally, the palmar aponeurosis splits into slips passing to each finger. The edges of these slips turn back around the long flexor tendons of the finger, forming a tunnel around them.

Fibrous flexor sheaths

On the ventral aspect of the digits, the deep fascia is thickened to form fibrous tunnels called **fibrous flexor sheaths**. The posterior margins of the fibrous flexor sheath are attached to the edges of the palmar surfaces of the phalanges. The sheath, along with the bone, produces a fibro-osseous tube, through which the long flexor tendons of the fingers pass to the phalanges.

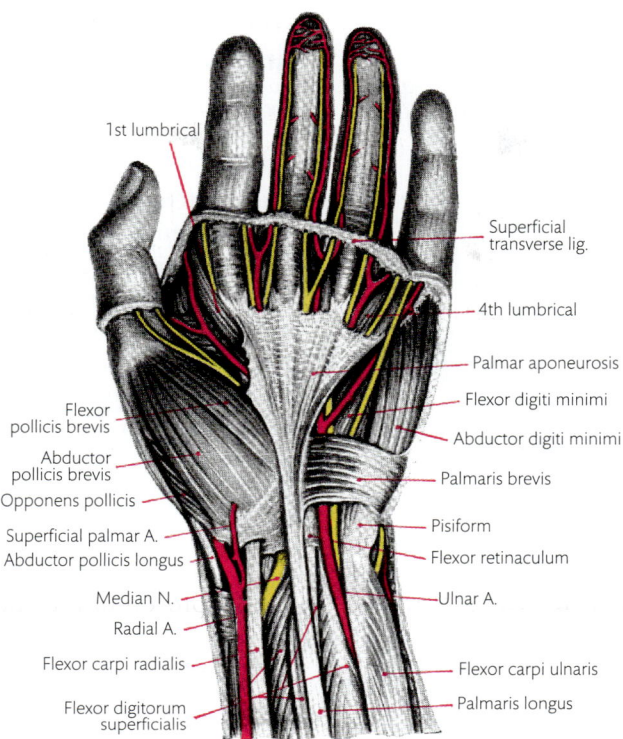

1st lumbrical

Superficial transverse lig.

4th lumbrical

Palmar aponeurosis

Flexor digiti minimi

Flexor pollicis brevis

Abductor pollicis brevis

Opponens pollicis

Superficial palmar A.

Abductor pollicis longus

Median N.

Radial A.

Flexor carpi radialis

Flexor digitorum superficialis

Abductor digiti minimi

Palmaris brevis

Pisiform

Flexor retinaculum

Ulnar A.

Flexor carpi ulnaris

Palmaris longus

Fig. 10.5 Superficial dissection of the palm to show the palmar aponeurosis. The deep fascia has been removed from the thenar and hypothenar eminences.

Introduction to synovial sheaths of long tendons

Synovial sheaths surround tendons wherever they pass through fascial or osteofascial tunnels. In the hand, synovial sheaths are found around tendons, as they pass deep to the retinacula, and the osteofascial tunnel formed by the phalanges and fibrous flexor sheaths [Fig. 10.6A]. These sheaths consist of two concentric tubes or layers of smooth synovial membrane joined together at their ends. The layers are separated from each other by a capillary interval (the cavity of the sheath) which contains synovial fluid to lubricate the opposed surfaces as they slide on each other. The inner layer surrounds, and is adherent to, the tendon; the outer layer lines the fibrous or bony canal and is fused with it [Fig. 10.6B]. More details of the synovial sheaths of the tendons are given after these have been dissected.

Dissection 10.1 continues the study of the cutaneous nerves of the forearm.

Introduction to muscles of the forearm

The arrangement of these muscles is complex. In general, both the flexor and extensor groups are arranged in two layers—superficial and deep. In addition, muscles are grouped together, based on their attachments and actions: (1) muscles that arise from the humerus and pass to the hand (these act on the elbow, the wrist, and, in some instances, the joints of the digits); (2) muscles that arise from the forearm bones and pass to the hand (acting on the wrist and digital joints—but not on the elbow); (3) muscles that arise from both the humerus and the forearm bones and pass to the hand (acting on the elbow, wrist, and digital joints); (4) muscles that pass from the humerus to the forearm bones (acting on the elbow and proximal radioulnar joint); and (5) muscles that pass between the two forearm bones (acting on the radioulnar joints).

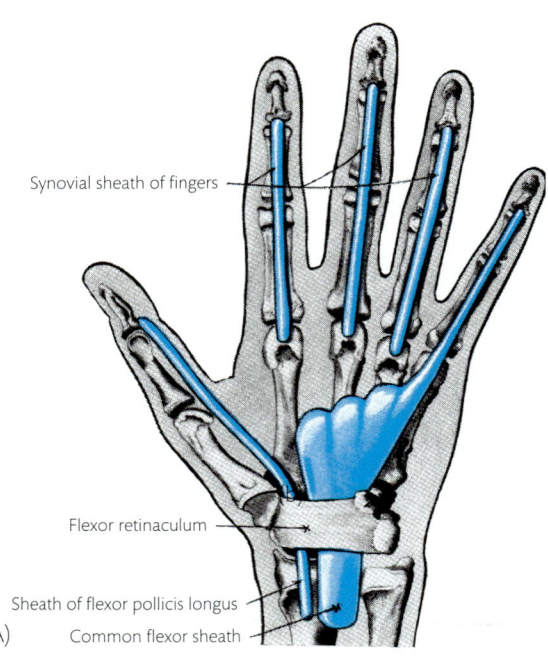

Synovial sheath of fingers

Flexor retinaculum

Sheath of flexor pollicis longus

(A) Common flexor sheath

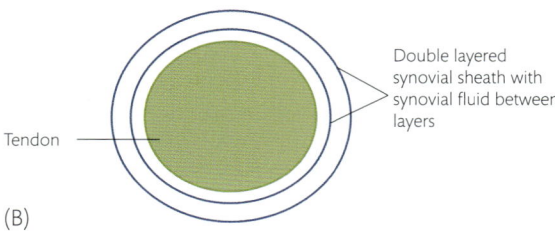

Tendon

Double layered synovial sheath with synovial fluid between layers

(B)

Fig. 10.6 (A) Synovial sheaths of flexor tendons of digits. (B) Schematic drawing to show the relationship of a tendon to the inner and outer layers of the synovial sheath.

Front of the forearm and hand

Dissection 10.2 looks at the superficial muscles, vessels, and nerves of the front of the forearm [see also Figs 10.7, 10.8, 10.9, 10.10, 10.11, 10.12].

Muscles of the front of the forearm and hand

Extensor muscles

Brachioradialis

The **brachioradialis** belongs to the extensor group of muscles but is seen in the front of the forearm. **Origin**: it takes origin from high on the lateral supracondylar line of the humerus. It descends in front of the elbow joint, with the extensor carpi radialis longus which arises inferior to it [see Fig. 4.9]. Both flex the elbow joint, though they belong to the extensor group of muscles. **Insertion**: the brachioradialis is inserted into the lateral surface of the distal end of the radius [see Fig. 4.14]. **Nerve supply**: radial nerve. **Actions**: in addition to flexing the elbow joint, it also partially supinates the fully pronated forearm because of its spiral course through the forearm in this position.

Flexor muscles—superficial group

The flexor muscles of the forearm are arranged into three groups—superficial, intermediate, and deep groups.

The superficial group consists of the pronator teres, flexor carpi radialis, palmaris longus, and flexor carpi ulnaris. The flexor digitorum superficialis is closely associated with the superficial muscles but forms the intermediate layer. All these muscles take origin from the medial epicondyle of the humerus [see Fig. 4.9] and the fascia between them. Some muscles have a subsidiary attachment to the coronoid process of the ulna [see Fig. 4.13]. All these muscles pass anterior to the elbow joint and are weak flexors, though each has other specific actions. All of them are supplied by the median nerve in the cubital fossa, except the flexor carpi ulnaris which is supplied by the ulnar nerve.

DISSECTION 10.1 Cutaneous nerves of the front of the forearm

Objective

I. To continue the study of the cutaneous nerves of the forearm.

Instructions

1. Many of the cutaneous nerves will have been found when the skin and superficial fascia were removed from the upper limb in Dissection 3.1. Their distribution is best determined by reference to Figs 7.1 and 7.2, as they are very difficult to follow through the superficial fascia.

2. Retain nerves that have been found piercing the deep fascia, so that they can later be traced to their parent nerves.

DISSECTION 10.2 Front of the forearm–superficial dissection

Objectives

I. To study the superficial muscles of the front of the forearm.   II. To identify and trace the brachial, radial, and ulnar arteries, and the median and ulnar nerves.   III. To study the palmaris brevis and palmar aponeurosis.

Instructions

1. Divide the deep fascia of the forearm vertically from the cubital fossa to the proximal margin of the flexor retinaculum. Make a transverse incision just proximal to the retinaculum. Reflect the flaps of the fascia, but avoid cutting the structures deep to it.
2. The muscles uncovered consist of the flexor group medially and the extensor group laterally. Separate the most superficial muscle—the **brachioradialis** [Fig. 10.7]—on the lateral side of the front of the forearm. Follow it to its insertion on the lateral surface of the distal end of the radius.
3. Push aside the tendons of the abductor pollicis longus and extensor pollicis brevis which overlie the insertion of the brachioradialis, as they pass to the base of the thumb [Fig. 10.8]. Avoid injury to the superficial branch of the radial nerve which crosses their superficial surface.
4. Pull the brachioradialis laterally. This exposes the extensor carpi radialis longus and separates the extensor and flexor groups of muscles.
5. In the groove between the flexors and extensors, identify the **radial artery** and the **superficial branch of the radial nerve**. Follow them distally and proximally [Fig. 10.9]. In this way, find the origin of the artery from the brachial artery in the cubital fossa (occasionally high in the arm) and the branch from the **deep branch of the radial nerve** which supplies the extensor carpi radialis brevis—a muscle deep to the extensor carpi radialis longus.
6. Deep to the extensor carpi radialis brevis is the supinator muscle.
7. Separate the **superficial group of flexor muscles**. These arise from the distal part of the medial supracondylar line and the medial epicondyle of the humerus. Most also have a minor attachment to the coronoid process of the ulna [see Figs 4.9, 4.13]. These superficial muscles spread out from the medial epicondyle as a narrow fan with a vertical medial edge—the **flexor carpi ulnaris** [Fig. 10.7].

8. The most lateral muscle—the **pronator teres**—arises furthest superiorly and passes obliquely across the proximal half of the forearm to the point of maximum convexity of the radius [see Fig. 4.13]. It passes deep to the radial artery, the superficial branch of the radial nerve, and the extensor muscles overlying the anterolateral surface of the radius [Fig. 10.7].
9. Medial to the pronator teres is the **flexor carpi radialis** (the radial flexor of the wrist), which also takes an oblique course across the forearm. Its tendon disappears deep to the lateral part of the flexor retinaculum, medial to the **radial artery**. Identify this tendon in your own wrist, and feel the pulsation of the radial artery (**radial pulse**) lateral to it on the distal margin of the radius [see Fig. 10.3].
10. Medial to the flexor carpi radialis is the **palmaris longus** (if present). Follow its tendon superficial to the flexor retinaculum to join the palmar aponeurosis.
11. Beneath the tendon of the palmaris longus and between it and the flexor carpi radialis, the median nerve becomes superficial, just proximal to the flexor retinaculum [Fig. 10.7].
12. The **flexor carpi ulnaris** is immediately medial to the palmaris longus in the proximal third of the forearm. Further distally, the two muscles separate and the flexor digitorum superficialis appears between them [Fig. 10.7]. Follow the tendon of the ulnaris to its insertion on the **pisiform** bone.
13. Find the **ulnar artery and nerve** which become superficial between the tendons of the flexor carpi ulnaris and those of the flexor digitorum superficialis, just proximal to the flexor retinaculum. Here the artery and nerve pierce the deep fascia and enter the hand, superficial to the retinaculum and immediately lateral to the insertion of the flexor carpi ulnaris. They then pass deep to the palmaris brevis and divide into their terminal branches [Figs 10.5, 10.7].
14. The **palmaris brevis** is a thin cutaneous muscle which arises from the palmar aponeurosis and flexor retinaculum. It passes transversely across the proximal 2–3 cm of the hypothenar eminence and is inserted into the skin on the medial side [Fig. 10.5]. It is supplied by the ulnar nerve. ➲ When it contracts, it bunches up the skin over the eminence, thus deepening the concavity of the palm and producing a cushion of skin against which the handle of a tool can be held steadily.

15. Expose the superficial surface of the **flexor carpi ulnaris**. Note its origin from the medial epicondyle, the olecranon, and the proximal two-thirds of the posterior (subcutaneous) border of the ulna [see Fig. 4.14].

16. Follow the **ulnar nerve** from the arm, posterior to the medial epicondyle, then between the humeral and ulnar attachments of the flexor carpi ulnaris on its deep surface [see Figs 9.6, 10.10]. Pull the flexor carpi ulnaris medially, and expose the ulnar nerve on its deep surface lying on the flexor digitorum profundus which covers the ulna. Trace the nerve and ulnar artery into continuity with the parts exposed at the wrist. Note the branches of the nerve to the flexor carpi ulnaris and flexor digitorum profundus, the **dorsal branch** arising near the middle of the forearm, and the small **palmar branch** [see Fig. 7.2] which arises in the distal half of the forearm. Trace the dorsal branch to the medial side of the forearm.

17. Cut through the middle of the flexor carpi radialis, palmaris longus, and pronator teres muscles, and reflect their parts. Identify the branches of the median nerve entering them proximally [Fig. 10.9].

18. The **flexor digitorum superficialis** is now exposed. Identify the radial head of this muscle attached to the anterior border of the radius, deep to the distal part of the pronator teres [see Figs 10.9, 4.13].

19. Follow the **median nerve** from the cubital fossa between the radial and humero-ulnar heads of the flexor digitorum superficialis. Lift the medial edge of the muscle, and find the nerve on its deep surface. Note its branches to the muscle. A short distance proximal to the wrist, the nerve becomes superficial between the tendons of the flexor carpi radialis and the palmaris longus. It gives off the **palmar branch** of the median nerve which enters the palm, superficial to the flexor retinaculum. The median nerve enters the hand, deep to the flexor retinaculum [Fig. 10.5]. ➲ Identify this position in your own wrist, and note how easily the nerve could be injured by a relatively superficial cut at this point.

20. Turn the tendon of the palmaris longus distally. Note its attachment to the flexor retinaculum and its continuity with the apex of the palmar aponeurosis.

21. Complete the exposure of the superficial surface of the **palmar aponeurosis** and the slips which pass from its distal margin to each of the fingers. Note that the edges of each slip turn posteriorly into the palm, leaving spaces through which the digital vessels and nerves (and lumbrical muscles) pass from beneath the palmar aponeurosis into the fingers [Fig. 10.5]. Each slip of the aponeurosis is attached: (1) to the phalanges by fusing with the fibrous flexor sheath; and (2) with the deep transverse metacarpal ligament (described later) [Fig. 10.5]. Confirm this arrangement by pushing a blunt instrument proximally between any two fingers—it passes readily into the palm, deep to the aponeurosis, without having to pierce it.

22. Separate the palmaris longus tendon and the aponeurosis from the surface of the flexor retinaculum. Turn the aponeurosis distally. You will need to separate the edges of the aponeurosis from the thinner deep fascia covering the thenar and hypothenar muscles and divide the septum which passes backwards from each edge. The septa fuse with the fascia anterior to the interosseous muscles and the adductor pollicis [Fig. 10.11]. Avoid injury to the vessels and nerves immediately deep to the aponeurosis.

23. Remove the fat from the interdigital region, and expose the **digital nerves and vessels**, with a **lumbrical muscle** posterior to them [Fig. 10.12].

Pronator teres

The **pronator teres** is the most lateral and most oblique of these muscles. **Origin**: it takes origin from the medial epicondyle (few fibres arise from the coronoid process of the ulna). **Insertion**: it is inserted on the most lateral part of the convex shaft of the radius [see Fig. 4.13]. **Nerve supply**: median nerve. **Actions**: it is a pronator (medial rotator of the radius on the ulna) and has an increased mechanical advantage by being attached to the most lateral part of the convex radius.

Flexor carpi radialis

Origin: the **flexor carpi radialis** arises from the medial epicondyle and runs obliquely across the anterior surface of the forearm to the anterolateral surface of the wrist. **Insertion**: it passes deep to the lateral end of the flexor retinaculum, grooves the trapezium, and is inserted into the base of the second and third metacarpals (that of the thumb is the first). **Nerve supply**: median nerve. **Actions**: it is a flexor of the wrist and abductor (radial deviator) of the hand. Acting with the flexor carpi ulnaris, it

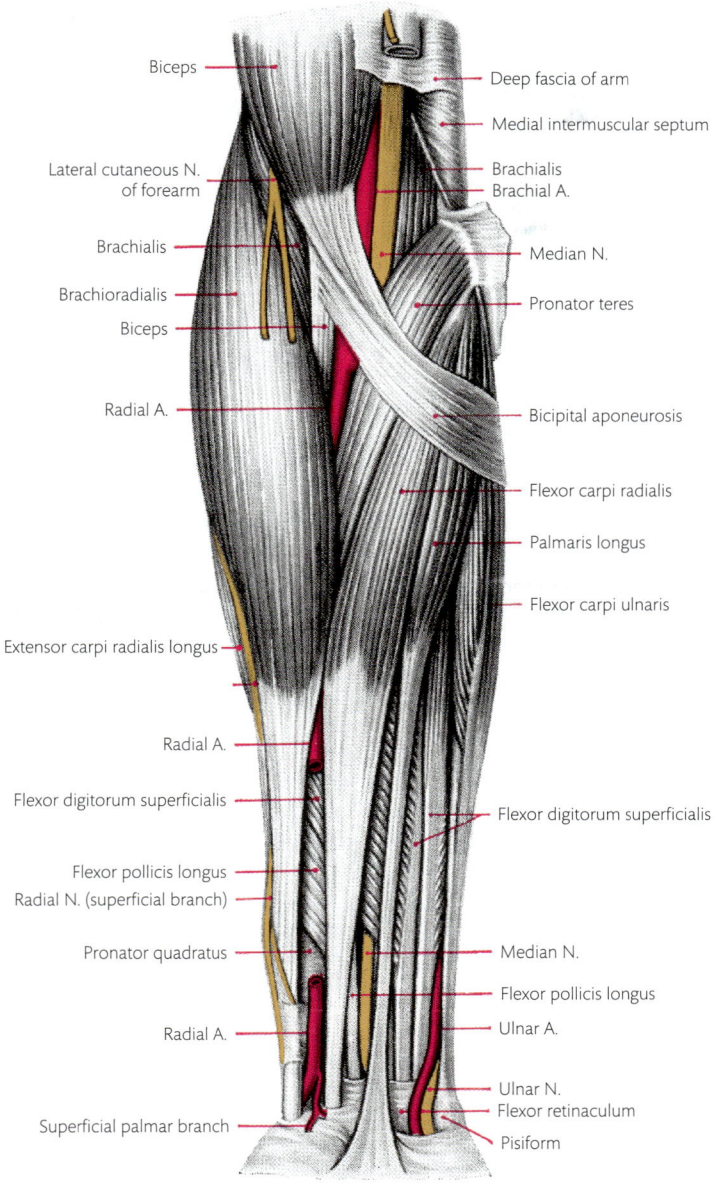

Biceps

Deep fascia of arm

Medial intermuscular septum

Lateral cutaneous N. of forearm

Brachialis
Brachial A.

Brachialis

Median N.

Brachioradialis

Pronator teres

Biceps

Radial A.

Bicipital aponeurosis

Flexor carpi radialis

Palmaris longus

Flexor carpi ulnaris

Extensor carpi radialis longus

Radial A.

Flexor digitorum superficialis

Flexor digitorum superficialis

Flexor pollicis longus
Radial N. (superficial branch)

Pronator quadratus

Median N.

Flexor pollicis longus

Ulnar A.

Radial A.

Ulnar N.
Flexor retinaculum

Superficial palmar branch

Pisiform

Fig. 10.7 Dissection of superficial muscles, arteries, and nerves of the front of the forearm. Part of the radial artery was removed to show the muscles deep to it.

produces pure flexion of the wrist joint. Because of its oblique course in the forearm, it tends to rotate the hand and radius medially, and so assists in pronation.

Palmaris longus

Origin: the **palmaris longus** takes origin from the medial epicondyle. **Insertion**: at the wrist, it passes superficial to the flexor retinaculum and expands into the palmar aponeurosis. **Nerve supply**: median nerve. **Actions**: it tenses the palmar aponeurosis and stabilizes the palmar skin in grasping an object. As it passes anterior to the wrist joint, it may assist in its flexion. The muscle is frequently absent.

Flexor carpi ulnaris

Origin: the **flexor carpi ulnaris** arises from the olecranon and posterior border of the ulna, in addition to the humeral origin from the medial epicondyle. **Insertion**: it passes vertically down and is inserted into the pisiform, and to the hook

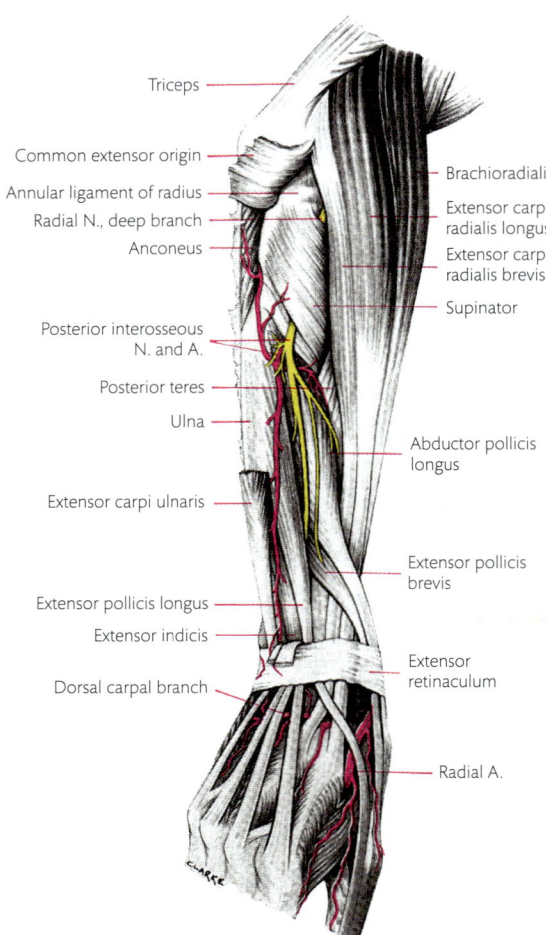

Triceps

Common extensor origin

Annular ligament of radius

Radial N., deep branch

Anconeus

Posterior interosseous N. and A.

Posterior teres

Ulna

Extensor carpi ulnaris

Extensor pollicis longus

Extensor indicis

Dorsal carpal branch

Brachioradialis

Extensor carpi radialis longus

Extensor carpi radialis brevis

Supinator

Abductor pollicis longus

Extensor pollicis brevis

Extensor retinaculum

Radial A.

Fig. 10.8 Deep dissection of the back of the forearm.

of the hamate and the base of the fifth metacarpal through the **pisohamate** and **pisometacarpal ligament**. Both ligaments are essentially continuations of the tendon, as the pisiform is a sesamoid bone developed in it. **Nerve supply**: ulnar nerve. **Actions**: it is a flexor of the wrist and an adductor (ulnar deviator) of the hand.

Flexor digitorum superficialis

Origin: the **flexor digitorum superficialis** lies deep to the other muscles of this group. The humero-ulnar head arises with the other muscles from the medial epicondyle of the humerus and the coronoid process of the ulna. An additional thin **radial head** [see Figs 10.9, 4.13] arises from the anterior border of the radius. **Course**: the muscle gives rise to four tendons, one to each of the medial four digits. In the hand, the tendons of the flexor digitorum superficialis lie superficial to, and in close apposition with, the tendons of the

flexor digitorum profundus (the deep flexor). The four tendons of the flexor digitorum superficialis enter the palm through the carpal tunnel (the space deep to the flexor retinaculum) in the same synovial sheath as those of the flexor digitorum profundus. (The tendons to the middle and ring fingers are anterior to those of the index and little fingers.) **Insertion**: one tendon passes to each of the medial four digits. At the proximal phalanx, each tendon splits into two bands or slips which curve posteriorly on each side of the underlying tendon of the flexor digitorum profundus and are inserted on the palmar surface of the middle phalanx [Fig. 10.13]. The most posterior fibres of each band pass behind the profundus tendon to intermingle with the corresponding fibres of the other band (**chiasma tendinum**) and end in its insertion. Thus, they form an oblique sling around the profundus tendon. **Nerve supply**: median nerve. **Actions**: the tendons of the flexor digitorum superficialis cross the anterior surfaces of all the joints from the wrist to the proximal interphalangeal joint, and so may flex the wrist, and metacarpophalangeal and proximal interphalangeal joints. (The range of movement at the carpometacarpal joints is so small that it may be ignored.) Usually, they act on the metacarpophalangeal and proximal interphalangeal joints, with the wrist being fixed by the extensors. The tendon to the little finger is occasionally missing. The radial head of the flexor digitorum superficialis straightens the pull of the muscle and aligns it to the fingers.

The deep group of flexor muscles arise from the ulna and radius [see Fig. 4.13], and consist of the flexor digitorum profundus, flexor pollicis longus, and pronator quadratus. They are described after deep dissection of that region.

The deep group of flexor muscles arise from the ulna and radius [see Fig. 4.13], and consist of the flexor digitorum profundus, flexor pollicis longus, and pronator quadratus. They are described after dissection of that region.

Dissection 10.3 examines the palmar aponeurosis and cutaneous nerves of the palm.

Arteries of the flexor compartment of the forearm

The brachial artery divides into the radial and ulnar arteries at the level of the neck of the radius, in the cubital fossa. The ulnar artery is the principal source of supply to the forearm; the radial

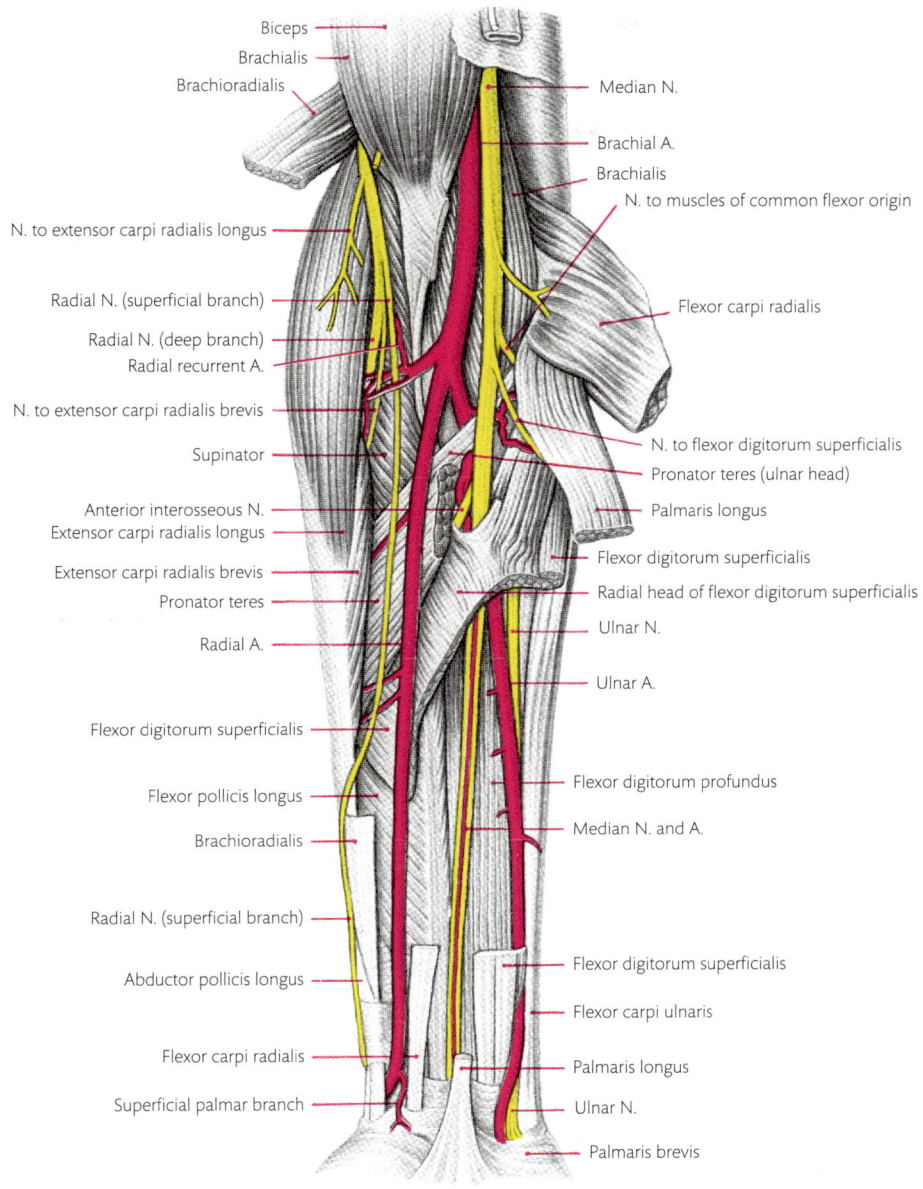

Biceps
Brachialis
Brachioradialis
N. to extensor carpi radialis longus
Radial N. (superficial branch)
Radial N. (deep branch)
Radial recurrent A.
N. to extensor carpi radialis brevis
Supinator
Anterior interosseous N.
Extensor carpi radialis longus
Extensor carpi radialis brevis
Pronator teres
Radial A.
Flexor digitorum superficialis
Flexor pollicis longus
Brachioradialis
Radial N. (superficial branch)
Abductor pollicis longus
Flexor carpi radialis
Superficial palmar branch

Median N.
Brachial A.
Brachialis
N. to muscles of common flexor origin
Flexor carpi radialis
N. to flexor digitorum superficialis
Pronator teres (ulnar head)
Palmaris longus
Flexor digitorum superficialis
Radial head of flexor digitorum superficialis
Ulnar N.
Ulnar A.
Flexor digitorum profundus
Median N. and A.
Flexor digitorum superficialis
Flexor carpi ulnaris
Palmaris longus
Ulnar N.
Palmaris brevis

Fig. 10.9 Deep dissection of muscles and nerves of the front of the forearm. The division of the brachial artery is slightly lower than usual.

artery is usually the main artery of the hand. Both arteries send **recurrent branches** to anastomose around the elbow joint with branches of the brachial (**ulnar collateral arteries**) and profunda brachii (**radial collateral arteries**) arteries [see Fig. 9.7]. Both arteries supply the muscles adjacent to them and give the **palmar** and **dorsal carpal arteries** at the wrist. These arteries anastomose on, and supply, the carpal bones, the surrounding joint tissues, and the adjacent parts of the hand.

Radial artery

The **radial artery** is the smaller of the two terminal branches. It passes inferolaterally between the brachioradialis and the flexor carpi radialis to reach the anterior surface of the distal end of the radius between the tendons of these muscles. Here the artery can be felt readily (radial pulse) against the bone. Except in the initial part of its course, the artery is immediately deep to the deep fascia. It crosses the superficial surface of the pronator teres with the superficial branch of the radial nerve

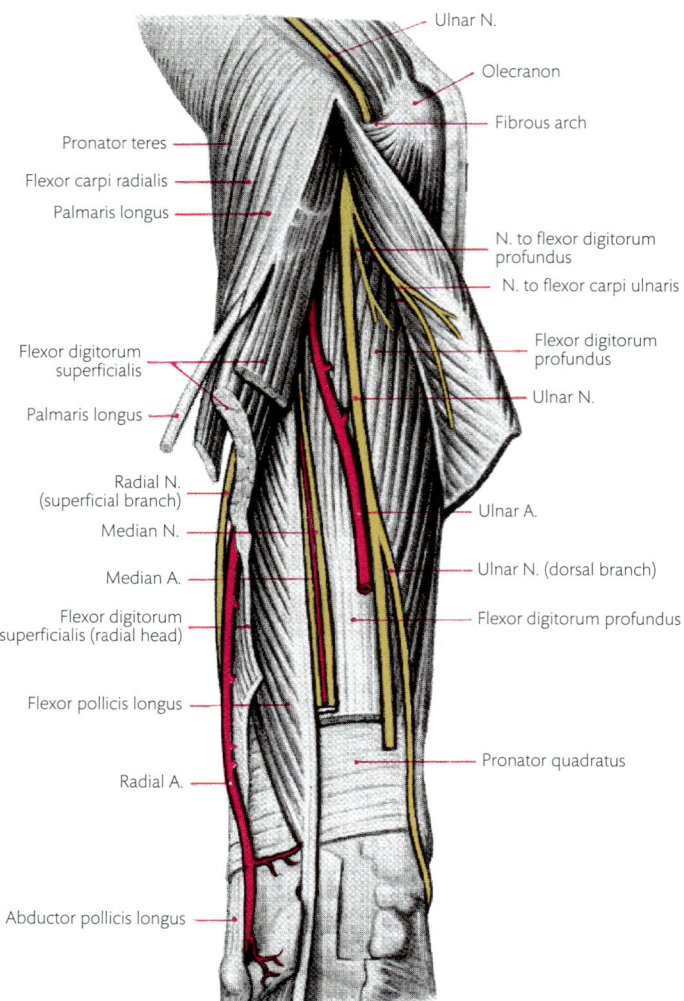

Fig. 10.10 Deep dissection of the front of the forearm. The elbow is partially flexed, and the forearm semi-pronated. The superficial muscles are cut and reflected. The flexor digitorum superficialis and flexor carpi ulnaris are separated to display deeper structures.

Labels for Fig. 10.10:
Ulnar N.
Olecranon
Fibrous arch
Pronator teres
Flexor carpi radialis
Palmaris longus
N. to flexor digitorum profundus
N. to flexor carpi ulnaris
Flexor digitorum superficialis
Flexor digitorum profundus
Palmaris longus
Ulnar N.
Radial N. (superficial branch)
Median N.
Ulnar A.
Median A.
Ulnar N. (dorsal branch)
Flexor digitorum superficialis (radial head)
Flexor digitorum profundus
Flexor pollicis longus
Radial A.
Pronator quadratus
Abductor pollicis longus

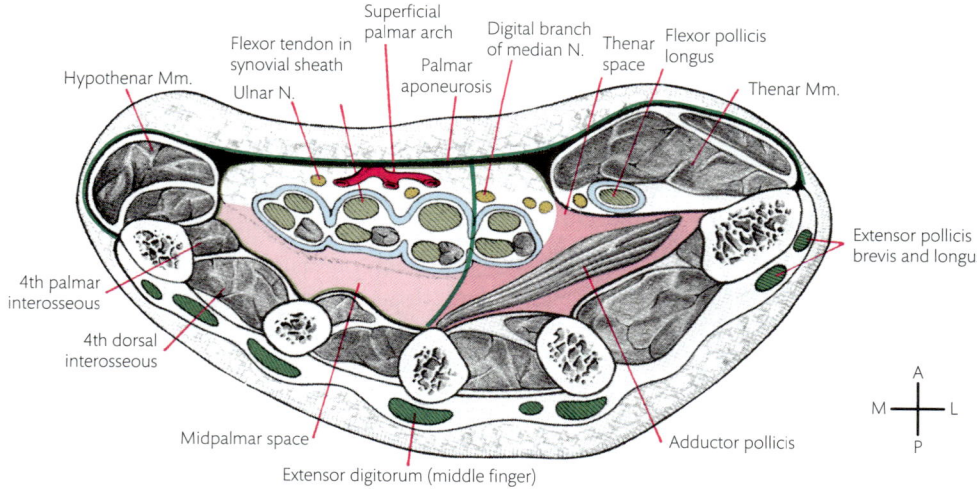

Fig. 10.11 Section through the hand to show the fascial layers and spaces of the palm. A = anterior; P = posterior; M = medial; L = lateral.

Labels for Fig. 10.11:
Superficial palmar arch
Flexor tendon in synovial sheath
Digital branch of median N.
Thenar space
Flexor pollicis longus
Hypothenar Mm.
Ulnar N.
Palmar aponeurosis
Thenar Mm.
4th palmar interosseous
Extensor pollicis brevis and longus
4th dorsal interosseous
Adductor pollicis
Midpalmar space
Extensor digitorum (middle finger)
A
M — L
P

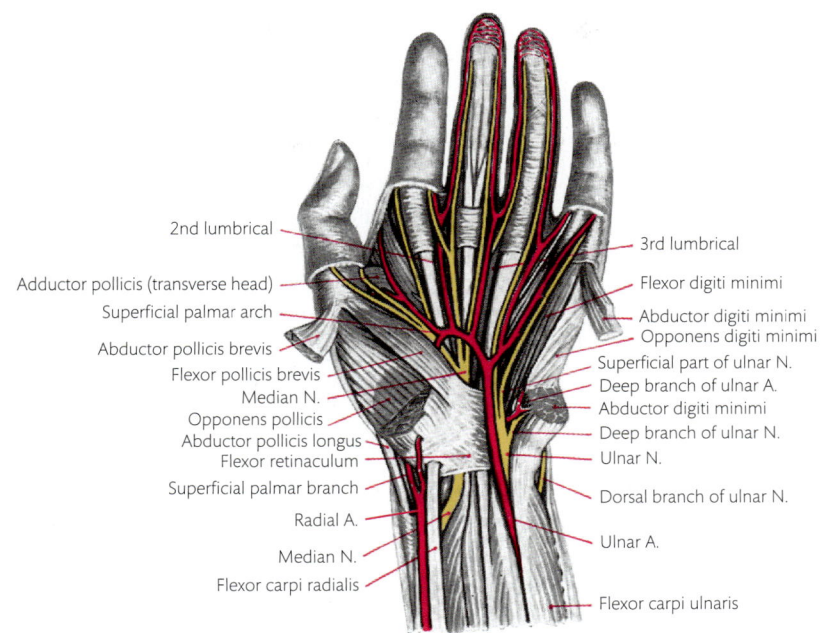

Fig. 10.12 Palmar structures displayed by removing the palmar aponeurosis.

[Fig. 10.9]. At the wrist, the radial artery sends a **superficial palmar branch** of variable size into the thenar muscles. It lies superficial to the flexor retinaculum and may anastomose with the superficial palmar arch of the ulnar artery [Fig. 10.12], occasionally forming a large or separate part of it.

Ulnar artery

The **ulnar artery** passes inferomedially, deep to the median nerve and the muscles arising from

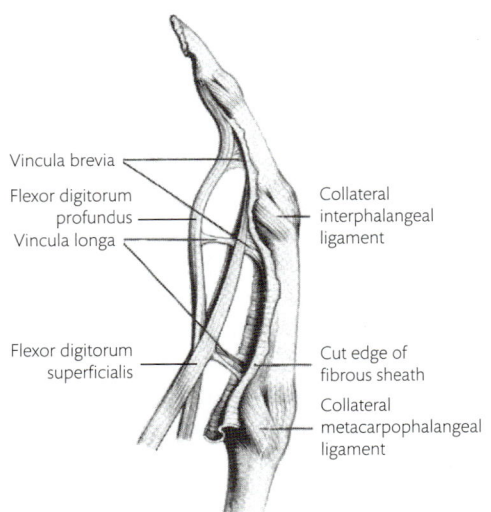

Fig. 10.13 Flexor tendons of the finger with the vincula tendinum.

the medial epicondyle of the humerus and the coronoid process of the ulna. It lies on the brachialis and flexor digitorum profundus. It meets the ulnar nerve above the middle of the forearm and descends vertically with it to pierce the deep fascia just proximal to the flexor retinaculum, between the tendons of the flexor carpi ulnaris and flexor digitorum superficialis [Fig. 10.9].

The **common interosseous artery** is the principal branch of the ulnar artery in the forearm [see Fig. 9.7]. It passes to the upper border of the interosseous membrane and divides into the **anterior** and **posterior interosseous arteries**. The posterior interosseous artery passes above the membrane to the back of the forearm. The anterior interosseous artery gives off the long, slender median artery, which runs with the median nerve, and then descends on the front of the interosseous membrane between the flexor pollicis longus and the flexor digitorum profundus to the pronator quadratus [Fig. 10.9]. It supplies these muscles, sends nutrient arteries to the radius and ulna, and pierces the interosseous membrane at the upper border of the pronator quadratus to enter the back of the forearm.

Superficial palmar arch

The superficial palmar arch begins as a terminal branch of the ulnar artery on the flexor retinacu-

DISSECTION 10.3 Palmar aponeurosis and cutaneous nerves of the palm

Objectives

I. To examine the palmar aponeurosis. II. To identify and trace the common and proper palmar digital nerves, the dorsal branch of the ulnar nerve, and the superficial branch of the radial nerve.

Instructions

1. Expose the palmar aponeurosis by removing the superficial fascia from the palm. Follow it proximally into continuity with the deep fascia of the forearm and the palmaris longus tendon. Follow it distally to the slips it sends to the palmar surface of each finger.
2. Look between these slips for the three common palmar digital nerves, and follow them and their branches (proper palmar digital nerves) into the fingers.
3. Identify and follow the proper palmar digital arteries which accompany the nerves.
4. Find the proper palmar digital nerve to the medial side of the little finger on the hypothenar eminence [Fig. 10.5], and follow it distally.
5. Find the corresponding nerve to the lateral side of the index finger, immediately lateral to the slip of the palmar aponeurosis to that finger.
6. Find the digital nerves of the thumb at the distal margin of the thenar eminence [Fig. 10.5], and trace them into the thumb.
7. Remove the fat from the medial side of the wrist, distal to the ulnar styloid process, and expose the dorsal branch of the ulnar nerve. Follow it and its dorsal digital branches, thus confirming the presence of four digital nerves (two palmar and two dorsal) in the little finger, as in each of the digits.
8. Expose the deep fascia on the lateral surface of the lower end of the radius, and find the superficial branch of the radial nerve. Trace its branches to the thumb and fingers, and note its communication with the dorsal branch of the ulnar nerve.

lum, distal to the pisiform bone. It runs forwards, medial to the hook of the hamate, and turns laterally, deep to the palmar aponeurosis, to join one of the branches of the radial artery [Fig. 10.4]. The radial branch may form a significant portion of an incomplete arch. The distal point of the arch lies at the same level as the distal border of the thenar eminence when the thumb is fully extended.

Branches

The principal branches of the superficial palmar arch are the four palmar digital arteries. The most medial is the proper palmar digital artery to the medial side of the little finger. The other three branches are the **common palmar digital arteries** to adjacent sides of two fingers. In the interdigital clefts, the common palmar digital artery receives the corresponding palmar metacarpal artery from the deep palmar arch. It then divides into two **proper palmar digital arteries** to the adjacent sides of two fingers [Figs 10.4, 10.12]. The proper palmar digital arteries to each finger form a rich anastomosis in the pulp of the finger and in the nail bed.

Nerves of the flexor compartment of the forearm and hand

Median nerve

The median nerve lies medial to the brachial artery at the elbow. It gives branches to the pronator teres, flexor carpi radialis, palmaris longus, flexor digitorum superficialis, and elbow joint, as it lies on the brachialis in the cubital fossa. As it leaves the fossa on the anterior surface of the ulnar artery, it gives off the **anterior interosseous nerve**. This nerve runs with the corresponding artery and supplies the flexor digitorum profundus (lateral half), flexor pollicis longus, and pronator quadratus. It passes posterior to the pronator quadratus to end on the front of the wrist and distal radioulnar joints [see Figs 9.3, 10.9].

The median nerve descends on the deep surface of the flexor digitorum superficialis. Near the wrist, the nerve lies between the tendons of the flexor digitorum superficialis and flexor carpi radialis, deep to the tendon of the palmaris longus. It gives off the palmar cutaneous branch, traverses the carpal tunnel, and divides into six branches near the distal border of the retinaculum. The most lateral branch turns proximally on the retinaculum and enters the thenar muscles to supply the abductor pollicis brevis, flexor pollicis brevis, and opponens pollicis. Medial to this, the mode of division is variable, but commonly three **proper palmar digital nerves** pass, one to each side of the thumb and the radial side of the index finger [Fig. 10.4]. The nerve to the index finger supplies the first lumbrical muscle (a muscle of the hand). A **common palmar digital nerve** passes towards each side of the middle finger. The lateral one supplies the second lumbrical muscle, and the medial one communicates with the common palmar digital nerve from the ulnar nerve. Proximal to the division of the corresponding common palmar digital arteries, each of these nerves divides into two **proper palmar digital nerves**. These nerves pass between the slips of the palmar aponeurosis to enter the subcutaneous tissue of the index, middle, and ring fingers [Fig. 10.12]. They lie anterior to the corresponding arteries.

The proper palmar digital nerves supply the skin on the palmar aspects of the finger and on the dorsal surface of the distal one-and-a-half to two phalanges through the dorsal branches. In the thumb, they frequently supply the skin on the dorsal surface of the terminal phalanx [Fig. 12.1]. The areas supplied by the digital nerve overlap to a large extent.

Ulnar nerve

The ulnar nerve enters the forearm on the posterior surface of the medial epicondyle of the humerus. It gives a branch to the elbow joint, passes deep to the flexor carpi ulnaris [Fig. 10.10], and gives a branch to it and to the medial half of the flexor digitorum profundus. It descends between these two muscles in the medial part of the front of the forearm and is joined by the ulnar artery. It gives off the **dorsal branch of the ulnar nerve** above the middle of the forearm and becomes superficial on the lateral side of the tendon of the flexor carpi ulnaris, close to the pisiform bone. It pierces the deep fascia with its palmar cutaneous branch and the ulnar artery, passes on to the flexor retinaculum, and divides into deep and superficial branches.

The **superficial branch** supplies the palmaris brevis, passes deep to it, and divides into a **proper palmar digital nerve** to the medial side of the little finger and a **common palmar digital nerve** to the fourth interdigital cleft. The latter communicates with the adjacent common palmar digital nerve from the median nerve and divides into proper palmar digital nerves to the adjacent sides of the ring and little fingers [Fig. 10.4].

The ulnar nerve may transmit nerve fibres from the seventh cervical ventral ramus to the skin of the hand. It receives these C. 7 fibres either through a communication with the median nerve in the forearm or through a branch from the lateral cord of the brachial plexus in the axilla.

Dissection 10.4 describes the dissection of the structures immediately under the palmar aponeurosis and explores the fibrous flexor sheaths.

Dissection 10.5 looks at the long flexor tendons, deep muscles of the front of the forearm, and nerves of the anterior compartment.

Deep muscles of the front of the forearm

The deep group of flexor muscles arise from the ulna and radius [see Fig. 4.13], and consist of the flexor digitorum profundus, flexor pollicis longus, and pronator quadratus.

Flexor digitorum profundus

Origin: the **flexor digitorum profundus** arises distal to the coronoid process from the medial and anterior surfaces of the proximal parts of the ulna and the adjacent interosseous membrane. When the fist is clenched, it may be felt contracting through the aponeurotic origin of the flexor carpi ulnaris by placing your fingers medial to the posterior border of the ulna. **Course**: the muscle gives rise to four tendons which pass through the carpal tunnel and separate in the palm. In the hand, these tendons give origin to the **lumbrical muscles** from their radial or adjacent sides. Each tendon enters the fibrous flexor sheath of the corresponding finger, deep to the tendon of the flexor digitorum superficialis. **Insertion**: it traverses the aperture between the slips of the flexor digitorum superficialis tendon, and is inserted into the palmar surface of the base of the distal phalanx [Fig. 10.13]. **Nerve supply**: the lateral half, which passes to the index and middle fingers, is supplied by the anterior interosseous branch of the median nerve; the medial half, which passes to the ring and little fingers, is supplied by the ulnar nerve. **Actions**: this muscle flexes the wrist and fingers, as does the flexor digitorum superficialis. In addition, it also flexes the distal interphalangeal joints and forms the structure from which the lumbricals act. ➲ When the muscle is paralysed, the obvious effect is absence of flexion of the terminal phalanges.

Flexor pollicis longus

Origin: the **flexor pollicis longus** arises from the anterior surface of the radius between the attachments of the flexor digitorum superficialis (radial head) and pronator quadratus, and from the interosseous membrane [see Fig. 4.13]. The single tendon traverses the lateral part of the carpal tunnel and the palmar surface of the thumb between the muscles of the thenar eminence and adductor pollicis. **Insertion**: it enters the fibrous flexor sheath

DISSECTION 10.4 Palm of the hand-1

Objectives

I. To identify and trace the superficial palmar arch and its branches, the superficial branch of the ulnar nerve, and branches of the median nerve. II. To study the digital fibrous flexor sheath.

Instructions

1. Remove any remnants of the palmaris brevis, and follow the **ulnar nerve** and **artery** distally, superficial to the flexor retinaculum. Note, but do not follow, the deep branch of the nerve (motor) and the deep palmar branch of the artery entering the hypothenar muscles.
2. Follow the superficial palmar arch, which the ulnar artery forms, and trace its branches. Trace the branches of the ulnar nerve (sensory and sympathetic) to the fingers [Fig. 10.12]. The superficial palmar arch is immediately deep to the palmar aponeurosis. Deep to the arch are the branches of the median and ulnar nerves, and further deeper are the long flexor tendons [Fig. 10.11].
3. Pull gently on the **median nerve**, proximal to the flexor retinaculum. This demonstrates its position at the distal edge of the retinaculum. Carefully follow the branches of the median nerve, distal to the retinaculum, but do not disturb the flexor tendons. The nerve gives a short, thick recurrent (motor) branch into the thenar muscles, close to the distal edge of the retinaculum.
4. The median nerve divides into the **common** and **proper palmar digital nerves** (mainly sensory and sympathetic) to the thumb and lateral two-and-a-half fingers [Fig. 10.4]. The most medial digital branch communicates with the most lateral digital branch of the ulnar nerve.
5. The branch to the lateral side of the index finger supplies the first lumbrical muscle, and the branch to the medial side supplies the second lumbrical.
6. Follow the digital branches at least to the roots of the fingers, and trace them to the tip of one finger. Note the branches which pass dorsally in the distal part of the finger. Through these, the skin on the dorsal surfaces of the distal two phalanges is supplied by the palmar digital nerves.
7. Expose the surface of the **fibrous flexor sheath** in one finger at least, but avoid damage to the tissue surrounding the flexor tendons at its proximal end. Note that the sheath is thick where it lies opposite the bodies of the phalanges but is thinner at the level of the interphalangeal joint.

at the base of the proximal phalanx of the thumb and is inserted into the base of the distal phalanx. **Nerve supply**: the anterior interosseous branch of the median nerve. **Actions**: it flexes all the joints of the thumb (including the carpometacarpal joint) and the wrist.

Pronator quadratus

The **pronator quadratus** [see Figs 4.13, 10.10] is placed transversely. **Origin**: it takes origin from the anterior surface of the distal quarter of the ulna. **Insertion**: it is inserted on the distal quarter of the radius. **Nerve supply**: anterior interosseous branch of the median nerve. **Actions**: pronation.

Synovial sheaths, and radial and ulnar bursae

Where tendons pass deep to retinacula or through osteofascial tunnels, they are protected from the wear and tear of friction by double-layered synovial sheaths [Fig. 10.6]. In the carpal tunnel, the tendons of the flexor digitorum superficialis and flexor digitorum profundus are enclosed in a common synovial sheath, sometimes referred to as the **ulnar bursa**. More laterally, the tendon of the flexor pollicis longus is enclosed in its own synovial sheath—the **radial bursa** [Fig. 10.14]. These two sheaths may communicate with each other, and both extend 2–3 cm proximally into the forearm, posterior to the tendons. Distally, the sheath for the flexor pollicis longus extends to the terminal phalanx of the thumb. The sheath for the flexor digitorum extends to the middle of the palm for the lateral three digits and to the terminal phalanx of the little finger. The synovial sheaths within the fibrous flexor sheaths of the middle three digits are separate from the common sheath. The synovial sheath of the little finger is incomplete to allow the lumbrical muscles (which are attached to the tendons of the flexor digitorum profundus) to emerge from the sheath. The tendon of the flexor carpi radialis has its own synovial sheath.

Tendons may be completely surrounded by the synovial space or suspended in the sheath by a

DISSECTION 10.5 Front of the forearm (deep dissection) and palm of the hand

Objectives

I. To study the long flexor tendons in the hand. **II.** To study the deep muscles on the front of the forearm. **III.** To identify and trace the branches of the median nerve, ulnar nerve, and anterior interosseous nerve in the forearm.

Instructions

1. Attempt to demonstrate the synovial sheath around the flexor digitorum superficialis tendon. Clean the loose connective tissue (external layer of the sheath) around the tendons, close to the proximal margin of the flexor retinaculum, and note the smooth, slippery internal surface of the sheath and external surface of the tendon. Introduce a blunt probe into the sheath, and try to define its extent. Note that it allows the probe to pass easily into the hand, deep to the flexor retinaculum, but also that the structures here are tightly packed in the carpal tunnel.

2. In one finger, divide the fibrous flexor sheath longitudinally. Note it is thick at the level of the bodies of the phalanges and relatively thin opposite the joints. Examine the extent of the digital synovial sheath.

3. Lift the tendons of the flexor digitorum superficialis and profundus within the digital sheath, and note the vinculae [Fig. 10.13] passing between them and the outer layer of the synovial sheath on the phalanges.

4. Divide the flexor retinaculum by a vertical cut between the thenar and hypothenar muscles. Take care not to damage the median nerve. Establish the continuity of the trunk of the median nerve with the branches found in the hand.

5. Follow the tendons of the flexor digitorum superficialis distally to the fingers. Note how each tendon divides to enclose the corresponding tendon of the flexor digitorum profundus. It is then inserted on the palmar aspect of the middle phalanx, dorsal to the profundus tendon [Fig. 10.13].

6. In the forearm, cut transversely through the humero-ulnar head of the flexor digitorum superficialis. Reflect the distal part of the muscle laterally to uncover the deep muscles of the forearm, the median nerve, and the ulnar artery.

7. Separate the **median nerve** from the deep surface of the flexor digitorum superficialis, and trace it proximally. Find its branches to the muscle and the **anterior interosseous nerve** arising from it in the cubital fossa.

8. Follow the **ulnar artery** to its origin from the brachial artery, and trace its principal branches [see Figs 9.7, 10.9]. Do not follow the posterior interosseous artery at this stage. Trace the anterior interosseous artery and nerve on the interosseous membrane between the deep flexor muscles of the forearm.

9. Identify the deep muscles of the forearm. These are the **flexor digitorum profundus** medially, the **flexor pollicis longus** laterally, and the **pronator quadratus**. The flexor pollicis longus and flexor digitorum profundus arise from the proximal two-thirds of the ulna and radius [see Figs 4.13, 4.14]. The pronator quadratus is a rectangular muscle which passes transversely between the anterior surfaces of the distal quarters of the radius and ulna.

10. Find the branches of the anterior interosseous artery and nerve to the deep muscles. Note that the artery passes posteriorly through the interosseous membrane at the proximal border of the pronator quadratus.

11. Cut across the flexor digitorum superficialis, distal to the origin of its radial head, and turn its tendons distally. Follow the tendons of the flexor digitorum profundus and flexor pollicis longus through the carpal tunnel into the palm.

12. Note the separation of the tendons of the flexor digitorum profundus in the palm and the four small muscles (**lumbricals**) which arise from them. Follow the lumbrical muscles to their tendons, as they pass with the proper digital vessels and nerves to the lateral side of the base of each finger [Fig. 10.12]. Their tendons will be traced later.

13. Reconfirm the nerve supply to the medial part of the flexor digitorum profundus from the ulnar nerve and the origin of that muscle from the medial and anterior surfaces of the ulna.

14. If possible, follow the anterior interosseous nerve through the pronator quadratus to the anterior surface of the **wrist joint**. It supplies this joint and the **distal radioulnar joint**.

fold of the synovial membrane (**mesotendon**). The mesotendon allows blood vessels to reach the tendon at any point along the sheath. The **vincula tendinum** [Fig. 10.13] are thin fibrous slips enclosed in the mesotendon which transmit blood vessels to the tendons within the digital synovial sheaths. The vincula brevia are triangular and attached to the tendons immediately proximal to

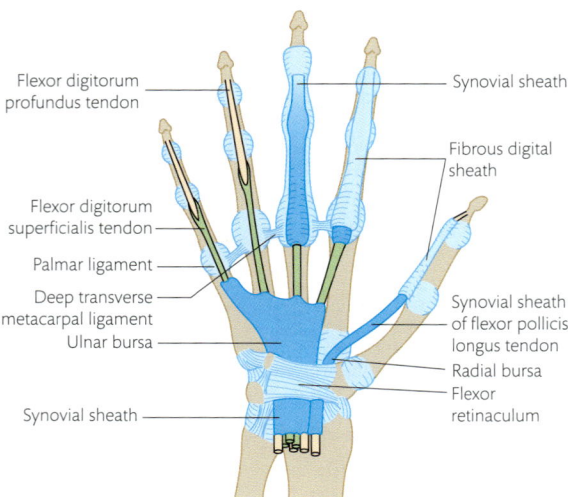

Flexor digitorum
profundus tendon

Flexor digitorum
superficialis tendon

Palmar ligament

Deep transverse
metacarpal ligament

Ulnar bursa

Synovial sheath

Synovial sheath

Fibrous digital
sheath

Synovial sheath
of flexor pollicis
longus tendon

Radial bursa

Flexor
retinaculum

Fig. 10.14 Relationship between the fibrous flexor sheath, synovial sheath, and long flexor tendons.

their insertion. The vincula longa are more slender and lie nearer the root of the finger.

➲ The cavity of a synovial sheath forms a route for rapid spread of infection along its length. When inflamed, the cavity becomes distended with fluid, tense, and painful, and the increased pressure within the sheath may interfere with the blood supply of the tendon. This condition is known as tenosynovitis. See Clinical Application 10.1.

Flexor retinaculum and carpal tunnel

This dense, fibrous band unites the ends of the concavity of the carpal bones and thus converts the space within the arch into an osteofibrous **carpal tunnel**. The median nerve, the long flexor tendons of the fingers and thumb, and the tendon of the flexor carpi radialis pass through this tunnel. The flexor retinaculum acts as a pulley for these tendons in flexion of the wrist. The retinaculum is continuous with the deep fascia of the forearm and the palmar aponeurosis in the hand.

Proximally, the flexor retinaculum is attached to the tuberosity of the scaphoid and the pisiform bone. Distally, it is attached to the tubercle of the trapezium and the hook of the hamate bone, and is continuous with the palmar aponeurosis. The attachment to the trapezium is to either side of the groove which lodges the tendon of the flexor carpi radialis, thus forming a separate tunnel for this tendon. *The flexor retinaculum forms the anterior limit of the carpal tunnel.* The superficial surface of the retinaculum gives partial origin to the thenar

and hypothenar muscles and has the ulnar nerve and vessels, the palmar cutaneous branches of the median and ulnar nerves, and the tendon of the palmaris longus on it. The tendons of the flexor carpi radialis, flexor pollicis longus, flexor digitorum longus, and flexor digitorum superficialis, and the median nerve lie in it [Fig. 10.15A and B].

Carpal tunnel syndrome

Carpal tunnel syndrome is pain, numbness, and tingling of the lateral part of the hand, combined with weakness of the muscles of the thumb. It is due to compression of the median nerve deep to the flexor retinaculum.

Fibrous flexor sheaths

Fibrous flexor sheaths are found on the flexor surface of the digits, deep to the superficial fascia. Together with the synovial sheath which lies within them, they form coverings for the long flexor tendons of the fingers [Fig. 10.16]. These thickened sheaths are attached to the sides of the palmar surfaces of the phalanges and of the palmar ligaments of the metacarpophalangeal and interphalangeal joints. It is thick over the surface of the bones, but thinner at the joints to allow flexion. Together with the phalanges, each sheath forms an osteofascial tube and prevents the tendons from springing away from the bone during contraction.

Dissection 10.6 continues to dissect the palm of the hand.

Fascial compartments of the palm

The deep fascia of the palm covers the muscles in the hand and forms intermuscular septa between them [Fig. 10.11]. The fascial compartments of the palm are potential spaces enclosed by septa. Anterior to the central region of the palm is the palmar aponeurosis. Two septa pass posteriorly from the medial and lateral margins of the aponeurosis. On the lateral side, the septum becomes continuous with the fascia covering the palmar aspect of the adductor pollicis. On the medial side, it becomes continuous with the fascia over the medial two interosseous spaces. The thenar and hypothenar muscles lie in separate fascial compartments. The adductor pollicis lies in a separate compartment, posterior to the lateral part of the central sheath.

The central region contains the long flexor tendons of the fingers, the lumbricals, branches of the median and ulnar nerves, and the arteries in the palm. The space is continuous proximally with the carpal tunnel and distally with the superficial tissues of the fingers. It is not directly continuous with the cavities of the fibrous flexor sheaths of the fingers, because these are filled by the digital synovial sheaths. Distal to the common synovial sheath for the flexor tendons, several septa pass between the tendons from the palmar aponeurosis to the fascia covering the metacarpals.

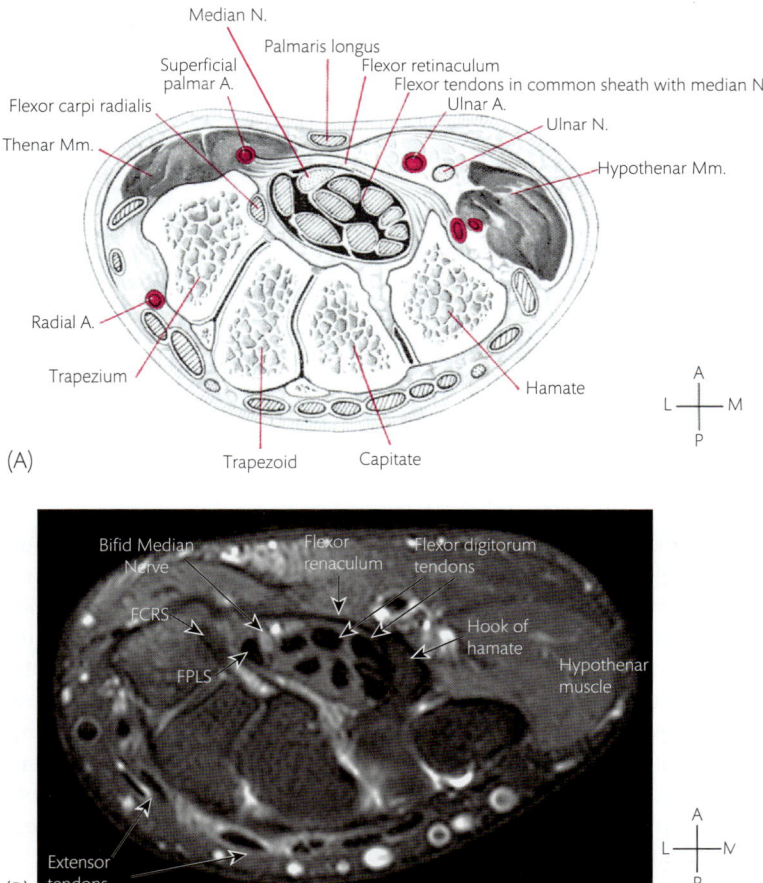

Fig. 10.15 (A) Transverse section at the level of the distal row of the carpal bones. The flexor pollicis longus, the median nerve, and the tendons of the two flexors (superficial and deep) of the digits are seen in the carpal tunnel. A = anterior; P = posterior; M = medial; L = lateral. (B) Magnetic resonance image of the hand at the same level. FCRS = flexor carpi radialis; FPLS = flexor pollicis longus.

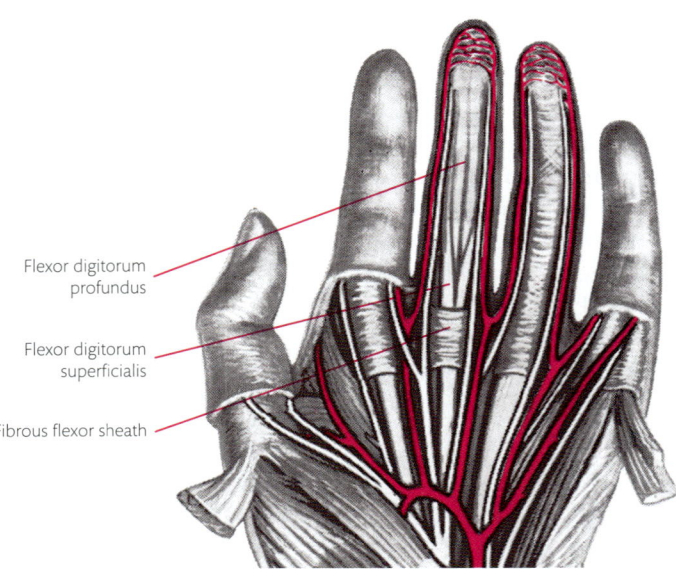

Flexor digitorum
profundus

Flexor digitorum
superficialis

Fibrous flexor sheath

Fig. 10.16 Palmar surface of the middle finger. A window has been cut in the fibrous flexor sheath to show the long flexor tendons.

DISSECTION 10.6 Palm of the hand-2

Objectives

I. To study the muscles of the thenar eminence, hypothenar eminence, and adductor pollicis. **II.** To identify the deep palmar arch and trace its branches.

Instructions

1. Remove the deep fascia from the surface of the thenar and hypothenar eminences. Avoid damage to the branch of the median nerve and the deep branch of the ulnar nerve which supply these muscles.

2. In the **thenar eminence**, the most anterior muscle is the **abductor pollicis brevis**. It covers the **opponens pollicis** completely and the **flexor pollicis brevis** partially [Fig. 10.12]. Pass the handle of a scalpel behind the lateral border of the abductor, and lift it from the underlying muscles. Define its attachments, and cut across it to expose the opponens. Separate the flexor brevis from the opponens; the flexor brevis is inserted, with the abductor brevis, into the anterior aspect of the base of the first phalanx. The opponens pollicis is inserted into the anterior surface of the first metacarpal [see Fig. 4.18]. (The use of the terms anterior and posterior in the thumb may be misleading, but necessary because the palmar surface of the thumb faces medially.)

3. All three muscles of the thenar eminence arise from the flexor retinaculum and the tubercle of the trapezium. The flexor has a deep head from the trapezoid and the capitate as well.

4. Cut across the middle of the flexor pollicis brevis, and reflect its parts. This exposes the tendon of the flexor pollicis longus and, behind it, the adductor pollicis passing to the posterior surface of the base of the proximal phalanx.

5. Examine the synovial sheath of the long flexor tendon; then expose and divide the fibrous flexor sheath in the thumb to follow the tendon and its synovial sheath to the terminal phalanx.

6. In the **hypothenar eminence**, separate the **abductor digiti minimi** from the medial side of the **flexor digiti minimi brevis** [Figs 10.5, 10.12]. Identify and follow the **deep branch of the ulnar nerve** and the deep palmar branch of the ulnar artery between these muscles. Define the attachments of these muscles, and then cut through the middle of the abductor and reflect its parts to expose the **opponens digiti minimi.** Follow this muscle to its attachments.

7. Cut through the flexor digitorum profundus in the forearm and both ends of the superficial palmar arch in the hand. Turn the distal parts of the flexor digitorum superficialis and profundus towards the fingers. Note the nerves to the lumbricals, as they are reflected with the tendons.

8. Study the remainder of the adductor pollicis, the deep palmar arterial arch, and the deep branch of the ulnar nerve in the palm. Establish the continuity of the last two with the ulnar artery and nerve by tracing them over the medial surface of the hook of the hamate bone. Divide the flexor digiti minimi brevis, if necessary [Fig. 10.12].

9. Follow the artery and nerve laterally on the proximal parts of the bodies of the metacarpals, deep to the long tendons. Find the branches of the nerve which pass to the interosseous muscles between the metacarpals, and trace its branch to the adductor pollicis.
10. Define the attachments of the adductor pollicis [see Fig. 4.18]. Cut across the muscle midway between its origins and insertion, and follow the branch of the nerve and the artery between the two parts of the muscle.
11. The deep branch of the ulnar nerve passes to the first dorsal interosseous muscle which is now exposed in the palm.
12. The deep branch of the ulnar artery completes the deep palmar arch by meeting the radial artery entering the palm between the two heads of the first dorsal interosseous muscle. Find the branch of the radial artery which passes to the palmar surface of the thumb (**princeps pollicis**) and the **radialis indicis artery** to the lateral side of the palmar surface of the index finger [Fig. 10.4].

Palmar spaces

Posterior to the structures in the common synovial sheath is a zone of loose connective tissue known as the '**mid-palmar space**', and surrounding the adductor pollicis is the '**thenar space**' [Fig. 10.11].

➲ Puncture wounds and infections of the hand can lead to collection or pus in one or more of the palmar spaces. Both the mid-palmar space and thenar space may become distended with fluid in infections of the hand. If pressure from such infections develops, it must be relieved by opening the sheath and draining the fluid.

➲ Infected puncture wounds of the palm may also lead to a collection of fluid between the palmar aponeurosis and the tendons. Fluid in this space cannot escape to the surface, because of the thick aponeurosis, and may involve the tendon sheaths if not drained by surgical incision.

➲ Minor injuries of the fingers, when infected, can lead to considerable effusion of fluid and development of pressure in the dense tissues of the fingers. Such finger infection may be in the pulp of the finger, distal to the end of the fibrous flexor sheaths, or within the fibrous flexor sheath or the synovial sheath. If untreated, the pressure may interfere with the blood supply of local tissues and lead to the death of the terminal phalanx if the pulp is involved or to necrosis of the flexor tendons if the flexor sheath is involved.

➲ Infection in the finger may also track to the mid-palmar space along the lumbrical muscles or from the digital synovial sheaths.

Short muscles of the thumb

The short muscles of the thumb include the **abductor pollicis brevis**, **flexor pollicis brevis**, **opponens pollicis**, and **adductor pollicis**. The first three together form the **thenar eminence**.

Opponens pollicis

Origin: from the lateral part of the flexor retinaculum and from the tubercles of the scaphoid and trapezium [see Fig. 4.18], along with the flexor pollicis brevis and abductor pollicis brevis. **Insertion**: it fans out to be inserted on the anterior surface of the first metacarpal. **Nerve supply**: median nerve. **Actions**: the opponens pollicis acts only on the carpometacarpal joint. It moves the metacarpal towards the centre of the palm and rotates it medially, so that the palmar surface of the thumb is turned to face the palmar surfaces of the fingers.

Abductor pollicis brevis

Origin: from the lateral part of the flexor retinaculum and from the tubercles of the scaphoid and trapezium [see Fig. 4.18]. **Insertion**: the abductor pollicis brevis is inserted into the anteromedial surface of the base of the proximal phalanx through a tendon which contains a small sesamoid bone. It sends part of its tendon into the extensor expansion on the dorsal aspect of the thumb, through which it is inserted into the dorsal surface of the base of the distal phalanx and can produce extension of the interphalangeal joint. **Nerve supply**: median nerve. **Actions**: the abductor pollicis brevis produces abduction mainly at the carpometacarpal joint. It also causes weak flexion at the metacarpophalangeal joint of the thumb.

Flexor pollicis brevis

Origin: from the lateral part of the flexor retinaculum and from the tubercles of the scaphoid and trapezium [see Fig. 4.18]. **Insertion**: it is inserted into the anteromedial surface of the base of the proximal phalanx of the thumb, along with the abductor pollicis brevis. **Nerve supply**: median nerve. **Actions**: the flexor pollicis brevis crosses the anteromedial surfaces of the first carpometacarpal and metacarpophalangeal joints, and so produces flexion at both joints.

Adductor pollicis

Origin: the **adductor pollicis** lies deep in the palm and takes origin by two heads—the oblique and transverse heads [see Fig. 4.18]. The transverse head arises from the anterior surface of the shaft of the third metacarpal. The oblique head arises from the base of the second and third metacarpals, the capitate, and trapezoid. **Insertion**: the muscles converge on the posteromedial surface of the base of the proximal phalanx of the thumb in a tendon containing a small sesamoid bone. **Nerve supply**: deep branch of the ulnar nerve. **Actions**: the adductor draws the thumb posteriorly towards the palm at the carpometacarpal joint. It flexes the metacarpophalangeal joint in the fully opposed thumb.

Sesamoid bones of the thumb

The **sesamoid bones** of the thumb are small, ovoid bones lying in the tendons and adherent to the capsule of the metacarpophalangeal joint. One flattened surface of each bone is covered with cartilage and slides directly on the cartilage of the palmar surface of the head of the first metacarpal. In a firm grip, sesamoid bones prevent compression of the tendons against the bone and facilitate their movements on the bone.

Opposition of the thumb

Opposition is a complex movement which brings the palmar surface of the distal segment of the thumb into contact with the corresponding surface of a finger. The movement consists of abduction, followed by combined flexion and medial rotation at the carpometacarpal joint, with or without flexion of the other joints of the thumb. Opposition of the thumb is usually (but not necessarily) accompanied by flexion at the metacarpophalangeal and interphalangeal joints of the thumb. **Opposition** is produced by combined action of the abductor pollicis longus and brevis, followed by action of the opponens pollicis (medial rotation) synchronously with the flexor pollicis brevis. The reverse movement is brought about by extension and lateral rotation of the thumb occurring together.

Opposition of the thumb must not be confused with simple flexion of all of its joints, which can be produced by the flexor pollicis longus acting alone. Simple flexion brings the tip of the thumb into contact with the lateral side of the terminal segment of any flexed finger.

➲ Simple flexion of the thumb can be carried out by the flexor pollicis longus alone when true opposition is lost due to paralysis of the short muscles of the thumb. This condition is seen when the median nerve is injured, distal to the origin of its anterior interosseous branch supplying the long flexor of the thumb.

Short muscles of the little finger

The **abductor digiti minimi**, **flexor digiti minimi brevis**, and **opponens digiti minimi** form the hypothenar eminence. The **abductor digiti minimi** and **flexor digiti minimi brevis** lie side by side, superficial to the **opponens digiti minimi**.

Abductor digiti minimi

Origin: the abductor digiti minimi, along with the other two muscles, arises from the flexor retinaculum, the pisiform bone, and the hook of the hamate bone [see Fig. 4.18]. **Insertion**: it is inserted into the anteromedial surface of the base of the proximal phalanx of the little finger. **Nerve supply**: deep branch of the ulnar nerve. **Actions**: abduction of the little finger.

Flexor digiti minimi brevis

Origin: the flexor digiti minimi brevis also takes origin from the flexor retinaculum, the pisiform bone, and the hook of the hamate bone [see Fig. 4.18]. **Insertion**: it is inserted into the anteromedial surface of the base of the proximal phalanx of the little finger. **Nerve supply**: deep branch of the ulnar nerve. **Actions**: it flexes the metacarpophalangeal joint of the fifth digit, along with the flexor digitorum superficialis and profundus.

Opponens digiti minimi

Origin: the opponens digiti minimi also takes origin from the flexor retinaculum, the pisiform bone, and the hook of the hamate bone [see Fig. 4.18]. The opponens is inserted along the anteromedial surface of the fifth metacarpal. **Nerve supply**:

deep branch of the ulnar nerve. **Actions**: limited opposition (lateral rotation) of the fifth metacarpal.

Lumbricals and interossei

The lumbricals and interossei are small muscles of the hand which act on the digits, mainly the four fingers.

Lumbrical muscles

Origin: the four lumbricals arise from the lateral side of the four tendons of the flexor digitorum profundus in the palm of the hand [Fig. 10.16]. **Insertion**: each passes to the radial side of the corresponding finger, anterior to the metacarpophalangeal joint and the deep transverse metacarpal ligament (except the first), to join the lateral edge of the extensor expansion—an aponeurotic expansion on the dorsal aspect of each digit [Fig. 10.17]. **Nerve supply**: the lateral two lumbricals are innervated by the median nerve, and the medial two by the ulnar nerve. **Actions**: these muscles flex the metacarpophalangeal joints of the fingers and extend the interphalangeal joints—the position of the fingers in writing.

Interosseous muscles

Four dorsal interossei and four palmar interossei muscles lie between the metacarpal bones [Figs 10.18, 10.19]. They are numbered sequentially from lateral to medial.

Dorsal interossei

The dorsal interossei are larger than the palmar interossei. **Origin**: they arise from two adjacent metacarpals—the metacarpal bones [see Figs 4.18, 4.19]. **Course of the tendon**: the first dorsal interosseus passes on the lateral side of the index finger, the second on the lateral side of the middle finger, the third on the medial side of the middle finger, and the fourth on the medial side of the ring finger. They cross the metacarpophalangeal joints, posterior to the deep transverse metacarpal ligament (except for the first dorsal interosseus). **Insertion**: on the dorsal aspect of the digit, part of the tendon passes deep to the extensor expansion to the base of the proximal phalanx, and the remainder joins the corresponding margin of the extensor expansion [Figs 10.18, 10.20]. **Nerve supply**: deep branch of the ulnar nerve. **Actions**: dorsal interossei (1) abduct the fingers from the line of the middle finger; (2) extend the interphalangeal joints; and (3) play a part in flexion of the metacarpophalangeal joints (especially the first dorsal interosseous).

Palmar interossei

Origin: the smaller palmar interossei arise from the metacarpal bone of the finger on which they

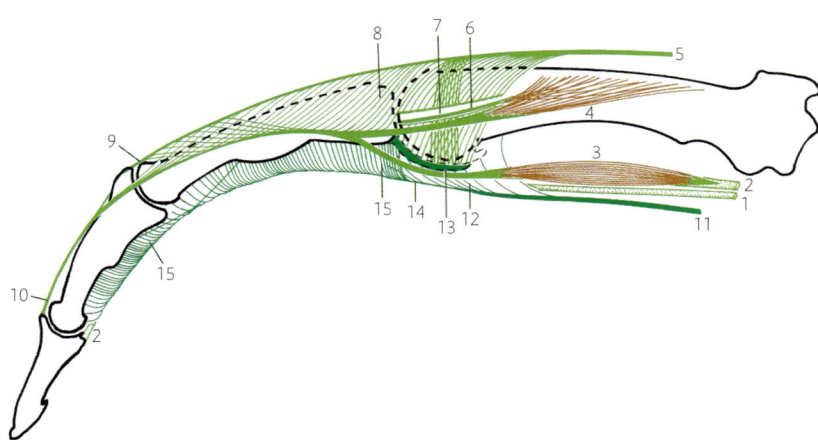

Fig. 10.17 Diagram showing the relationship of the extensor expansion of a finger to the palmar aponeurosis, palmar ligament of the metacarpophalangeal joint, fibrous flexor sheath of the long flexor tendons, and lumbrical and interosseous muscles. 1. Tendon of the flexor digitorum superficialis. 2. Tendon of the flexor digitorum profundus. 3. Lumbrical muscle. 4. Interosseous muscle. 5. Tendon of the extensor digitorum. 6. Deep part of the interosseous tendon passing to the base of the proximal phalanx. 7. Window cut in the extensor expansion. 8. Extensor expansion. 9. Central part of the extensor expansion passing to the base of the middle phalanx. 10. Conjoined lateral and medial parts of the extensor expansion inserting into the terminal phalanx. 11. Palmar aponeurosis with fibres (12) passing to the palmar ligament of (13) the metacarpophalangeal joint, and (14) fibres passing to (15) the fibrous flexor sheath.

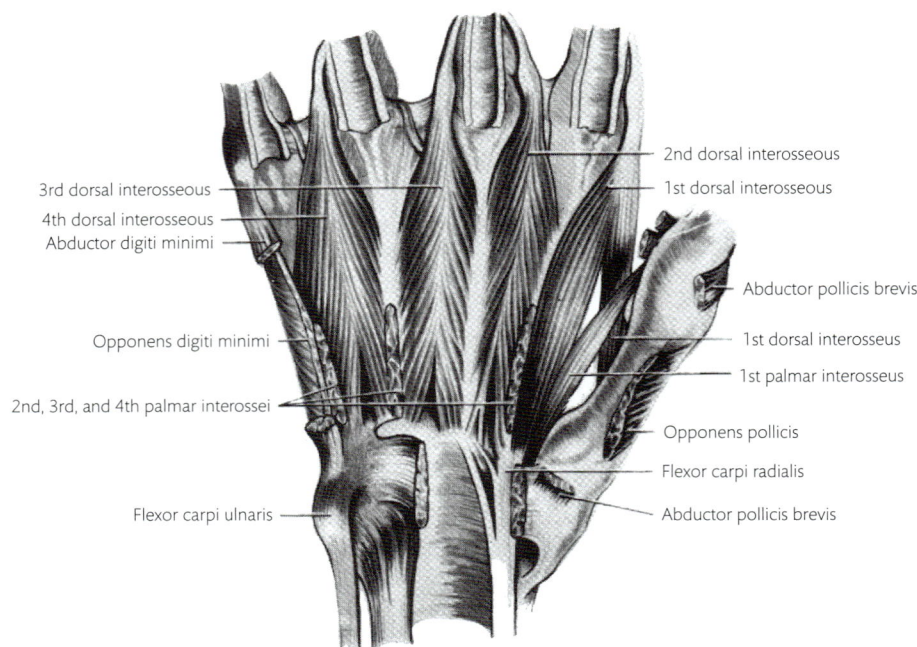

Fig. 10.18 Dorsal interosseous muscles of the right hand (seen from the palmar aspect).

act. **Course of the tendon**: the first palmar interosseus passes on the medial side of the thumb, the second on the medial side of the index finger, the third on the lateral side of the ring finger, and the fourth on the lateral side of the little finger. They cross the metacarpophalangeal joints, posterior to the deep transverse metacarpal ligament (except for those in the first intermetacarpal space). **Insertion**: they are inserted into the corresponding margin of the extensor expansion and into the base of the proximal phalanx [Fig. 10.19]. **Actions**: palmar interossei (1) adduct the fingers and thumb to the line of the middle finger; (2) extend the interphalangeal joints; and (3) play a part in flexion of the metacarpophalangeal joints.

Abduction and adduction at the metacarpophalangeal joints are markedly reduced when these joints are flexed. This is because of tightening of the collateral ligaments and the reduced mechanical advantage of the interossei in the flexed position. Table 10.1 shows the position of the interossei tendons in relation to the metacarpophalangeal joints, and their action on the digits.

Abduction and adduction at the metacarpophalangeal joints are markedly reduced when these joints are flexed. This is because of tightening of the collateral ligaments and the reduced mechanical advantage of the interossei in the flexed position.

Palmar interossei {
4th
3rd
2nd
1st
}

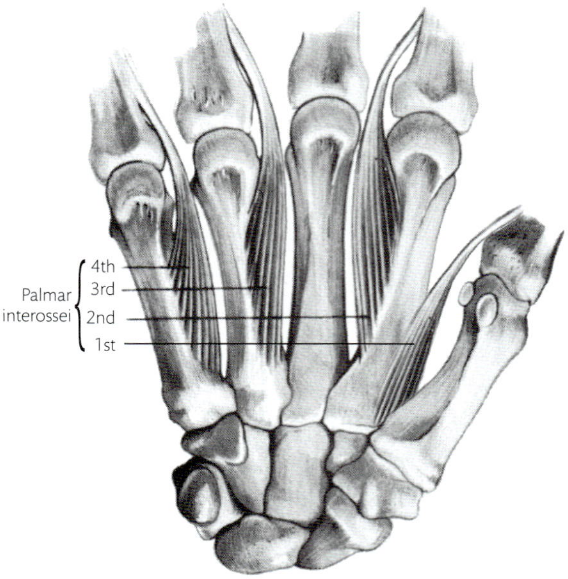

Fig. 10.19 Palmar interosseous muscles of the right hand.

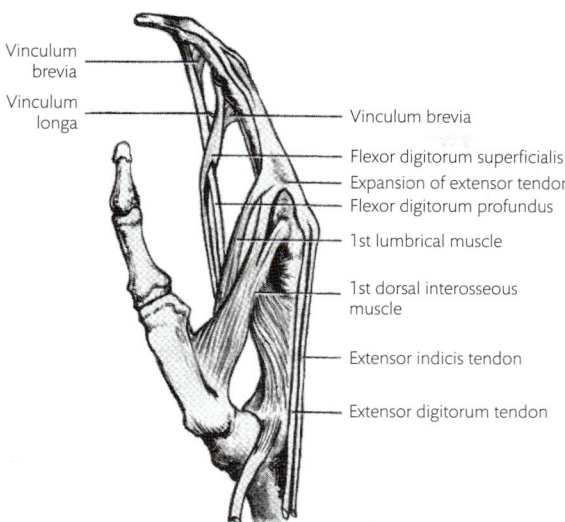

Vinculum brevia

Vinculum longa

Vinculum brevia

Flexor digitorum superficialis

Expansion of extensor tendon

Flexor digitorum profundus

1st lumbrical muscle

1st dorsal interosseous muscle

Extensor indicis tendon

Extensor digitorum tendon

Fig. 10.20 Tendons attached to the index finger.

Deep branch of the ulnar nerve

The deep branch of the ulnar nerve arises from the ulnar nerve on the flexor retinaculum [Fig. 10.12]. It supplies the muscles of the hypothenar eminence and passes deep between the abductor digiti minimi and the flexor digiti minimi brevis with the deep **palmar branch of the ulnar artery**. The nerve and artery turn laterally on the hook of the hamate and cross the palm on the proximal parts of the metacarpal bones, deep to the long flexor tendons. The nerve supplies all interossei muscles, the medial two lumbricals, and the adductor pollicis.

Table 10.1 Position of interossei tendons and their action on the digits

Interossei	Position of tendon	Action of muscle
Dorsal interossei		
First	Index finger—lateral side	Abduction—index finger
Second	Middle finger—lateral side	Abduction*—middle finger
Third	Middle finger—medial side	Abduction**—middle finger
Fourth	Ring finger—medial side	Abduction—ring finger
Palmar interossei		
First	Thumb—medial side	Adduction—thumb
Second	Index finger—medial side	Adduction—index finger
Third	Ring finger—lateral side	Adduction—ring finger
Fourth	Little finger—lateral side	Adduction—little finger

* Moves the middle finger laterally.
** Moves the middle finger medially.

In summary, the deep branch of the ulnar nerve supplies all the muscles in the palm, except the three muscles of the thenar eminence and the lateral two lumbricals. These muscles are supplied by the median nerve. The flexor pollicis brevis frequently receives a branch from the ulnar nerve, in addition to that from the median nerve, and occasionally the ulnar nerve is its only source of supply.

Deep palmar arch

The deep palmar arch is formed by the **radial artery** which enters the palm between the two heads of the first dorsal interosseous muscle. It gives off the **princeps pollicis** and the **radialis indicis arteries** [Fig. 10.4], passes between the heads of the adductor pollicis, runs medially, and unites with the deep palmar branch of the ulnar artery to complete the arch. The deep palmar arch lies a finger's breadth proximal to the superficial palmar arch.

Branches

The arch gives a **palmar metacarpal artery** in each of the medial three interosseous spaces. These join the distal ends of the corresponding common palmar digital arteries of the superficial palmar arch and may sometimes replace them.

The **princeps pollicis** passes to the metacarpal of the thumb and divides into the two palmar digital arteries of the thumb, one on each side of the tendon of the flexor pollicis longus.

The **radialis indicis** passes distally between the first dorsal interosseous and the adductor pollicis to become the proper palmar digital artery on the lateral side of the index finger.

Extensor compartment of the forearm and hand

Begin by revising the surface anatomy and cutaneous nerves of the region.

The **posterior cutaneous nerve of the forearm** (a branch of the radial nerve) is the main cutaneous nerve in this region. The **medial** and **lateral cutaneous nerves of the forearm** also spread onto the posterior surface. In the hand, the main supply is from the **superficial branch of** the **radial nerve** and the **dorsal branch of the ulnar nerve**, often in almost equal proportions. These nerves give rise to the **dorsal digital nerves** to the dorsal aspect of the proximal part of the fingers up to the proximal interphalangeal joints. The cutaneous supply of the dorsal surfaces of the hand and fingers shows considerable variation.

Occasionally, the lateral and/or posterior cutaneous nerves of the forearm also supply the dorsum of the hand, and the ulnar nerve regularly supplies a greater area on the dorsum of the hand than on the palmar aspect [see Fig. 7.1].

Dissection 10.7 looks at the back of the forearm (see also Figs 10.21, 10.22, 10.23).

DISSECTION 10.7 Back of the forearm

Objectives

I. To study the superficial muscles of the back of the forearm. II. To define the extensor retinaculum. III. To identify and trace the branches of the posterior interosseous nerve and artery.

Instructions

1. Find the superficial branch of the radial nerve and the dorsal branch of the ulnar nerve on the front of the forearm. Trace both to their distribution on the dorsal surfaces of the hand and fingers.
2. Remove the deep fascia from the back of the forearm, but leave the thickened part at the wrist—the **extensor retinaculum** [Fig. 10.21].
3. As the proximal edge of the retinaculum is defined, try to demonstrate the synovial sheaths of the tendons of the extensor muscles which pass deep to it [Figs 10.22, 10.23].
4. Separate the superficial muscles from each other, starting with the tendons at the wrist. Because of their limited bony origin, these muscles arise mainly from extensive tendinous sheets between them, so that separation proximally is artificial, but necessary if the deeper structures are to be seen.
5. Completely separate the three anterolateral muscles (brachioradialis, extensor carpi radialis longus, and extensor carpi radialis brevis) from the extensor digitorum [Fig. 10.21], and expose the **supinator** lying deep to them [Fig. 10.8].
6. Expose the **posterior interosseous nerve** emerging from the supinator near its distal border. Follow the branches of this nerve to the extensor digitorum, extensor digiti minimi, extensor carpi ulnaris, and the muscles deep to it. (This nerve is the continuation of the deep branch of the radial nerve.)
7. Pull the brachioradialis and the extensor carpi radialis longus and brevis laterally to expose the radial nerve at the elbow.

8. Complete the exposure of this nerve and its deep branch which gives branches to the supinator and then pierces it. Find the branch to the extensor carpi radialis brevis which arises here. Pull gently on the deep branch to establish its continuity with the posterior interosseous nerve.
9. Find the posterior interosseous artery on the back of the forearm. It emerges between the radius and the ulna, immediately distal to the supinator and close to the posterior interosseous nerve. Trace the main branches of the artery.
10. Identify the remaining superficial muscles. The extensor digiti minimi is the separated medial part of the extensor digitorum and leaves it at the extensor retinaculum to pass through a separate compartment, medial to the extensor digitorum. Medial to this is the extensor carpi ulnaris. It is attached to the posterior border of the ulna by the thick deep fascia. Divide the fascia over the extensor carpi ulnaris in the proximal third of the forearm to demonstrate the anconeus [Fig. 10.21].
11. Lift the extensor digitorum, and expose the deep layer of the extensor muscles. Immediately distal to the supinator is the abductor pollicis longus. The other three muscles (extensor pollicis brevis, extensor pollicis longus, and extensor indicis) arise distal to this [see Fig. 4.14]. Trace the tendons of these muscles to the extensor retinaculum.
12. Expose the tendons on the back and lateral side of the hand. Note that the thumb, index finger, and little finger each has two extensor tendons. Identify all these tendons at your own wrist, and note which of them become taut in movements of the wrist and in movements of the fingers and thumb [see Fig. 10.2]. At the same time, review the position of the flexor tendons on the anterior aspect of the distal part of your forearm [see Fig. 10.3].

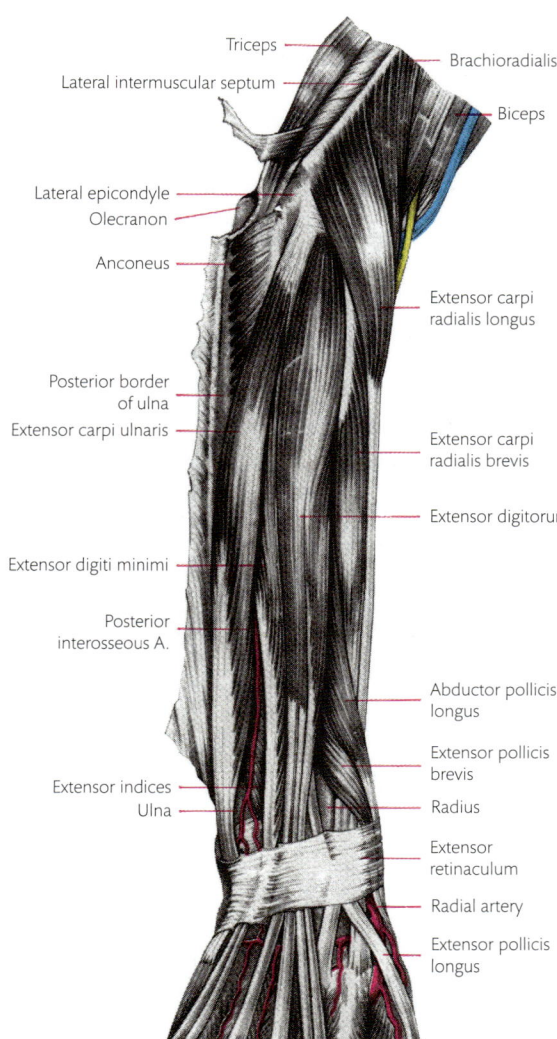

Fig. 10.21 Superficial dissection of the back of the forearm.

Labels on figure:
Triceps
Lateral intermuscular septum
Brachioradialis
Biceps
Lateral epicondyle
Olecranon
Anconeus
Extensor carpi radialis longus
Posterior border of ulna
Extensor carpi ulnaris
Extensor carpi radialis brevis
Extensor digitorum
Extensor digiti minimi
Posterior interosseous A.
Abductor pollicis longus
Extensor pollicis brevis
Extensor indices
Ulna
Radius
Extensor retinaculum
Radial artery
Extensor pollicis longus

Muscles of the back of the forearm

These are divided into two groups—superficial and deep.

Superficial muscles of the back of the forearm

There are seven muscles in the superficial layer; from lateral to medial, they are the brachioradialis, extensor carpi radialis longus, extensor carpi radialis brevis, extensor digitorum, extensor digiti minimi, extensor carpi ulnaris, and anconeus. They have a common origin from the lateral supracondylar line and lateral epicondyle.

Brachioradialis

Origin: the **brachioradialis** takes origin from high on the lateral supracondylar line of the humerus and descends in front of the elbow joint [see Fig. 4.9]. **Insertion**: into the lateral surface of the distal end of the radius [see Fig. 4.14]. **Nerve supply**: radial nerve. **Actions**: it flexes the elbow joint, though it belongs to the extensor group of muscles. In addition to flexing the elbow joint, it also partially supinates the fully pronated forearm because of its spiral course through the forearm in this position.

Extensor carpi radialis longus

Origin: from the lateral supracondylar line of the humerus [see Fig. 4.9]. **Insertion**: into the base of the second metacarpal [see Fig. 4.19]. **Nerve supply**: radial nerve. **Actions**: (1) it extends the wrist joint, along with the other extensor; (2) along with the extensor carpi radialis brevis and flexor carpi radialis, it abducts the wrist—this is an important and powerful movement (e.g. in raising a hammer prior to striking with it); and (3) it stabilizes the wrist and fixes it, so that fine movements of the hand can take place on a stable hand.

Extensor carpi radialis brevis

Origin: from the lateral epicondyle of the humerus [see Fig. 4.9]. **Insertion**: into the base of the second and third metacarpals [see Fig. 4.19]. **Nerve supply**: radial and posterior interosseous nerve. **Actions**: it acts with the extensor carpi radialis longus to extend and abduct the wrist. With the other extensors, it stabilizes and fixes the wrist joint.

Extensor digitorum

Origin: from the lateral epicondyle of the humerus [see Fig. 4.9]. **Insertion**: it splits into four tendons which are inserted into the extensor expansion on the medial four digits [Fig. 10.23]. (The extensor expansion is a triangular fascial thickening over the dorsal aspect of the metacarpophalangeal joint and the proximal phalanx, and is described later.) **Nerve supply**: posterior interosseous nerve. **Actions**: extends the wrist and all the joints of the fingers.

Extensor digiti minimi

Origin: from the lateral epicondyle of the humerus [see Fig. 4.9]. **Insertion**: into the extensor expansion of the little finger. **Nerve supply**: posterior interosseous nerve. **Actions**: extends all the joints of the little finger.

Extensor carpi ulnaris

Origin: from the lateral epicondyle of the humerus, and the posterior border of the ulna [see Fig.

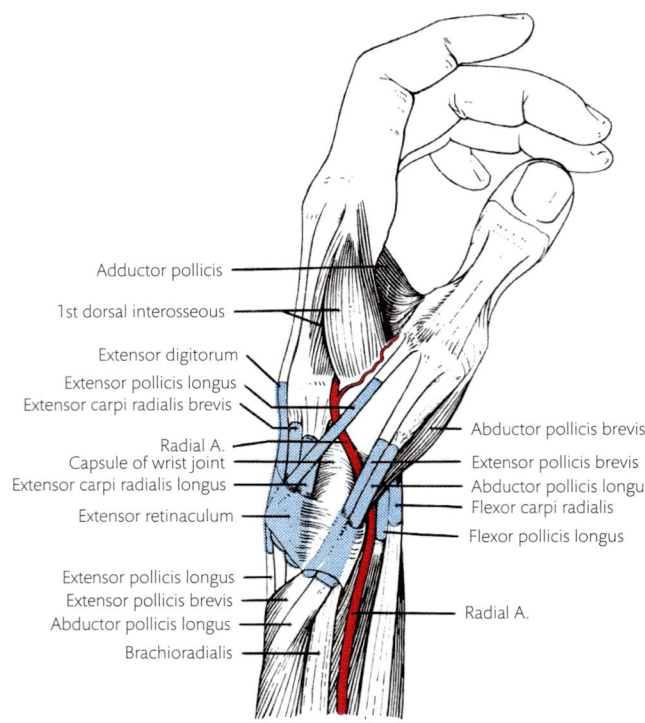

Fig. 10.22 Dissection of the lateral side of the left wrist and hand, showing the tendons in their synovial sheaths.

Labels in Fig. 10.22:
- Adductor pollicis
- 1st dorsal interosseous
- Extensor digitorum
- Extensor pollicis longus
- Extensor carpi radialis brevis
- Radial A.
- Capsule of wrist joint
- Extensor carpi radialis longus
- Extensor retinaculum
- Extensor pollicis longus
- Extensor pollicis brevis
- Abductor pollicis longus
- Brachioradialis
- Abductor pollicis brevis
- Extensor pollicis brevis
- Abductor pollicis longus
- Flexor carpi radialis
- Flexor pollicis longus
- Radial A.

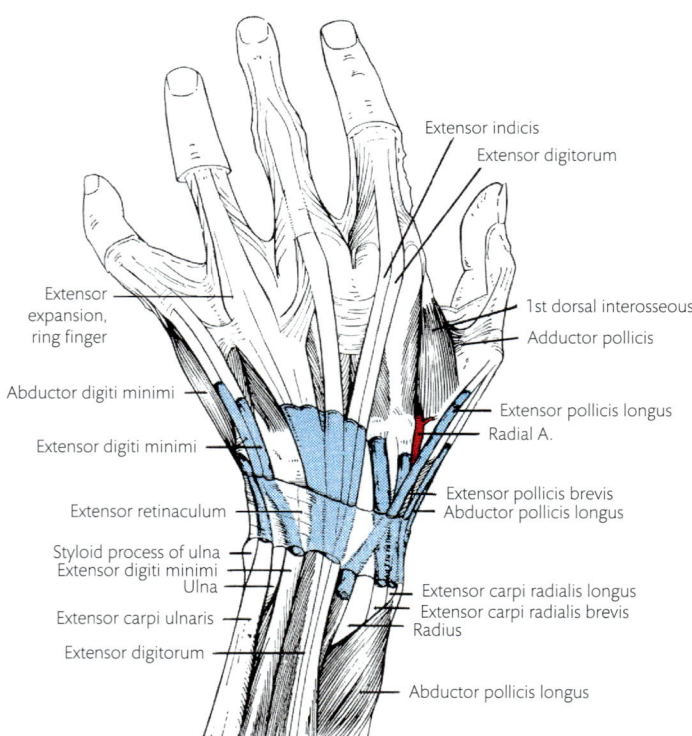

Fig. 10.23 Dissection of the back of the forearm, wrist, and hand, showing the tendons in their synovial sheaths.

Labels in Fig. 10.23:
- Extensor indicis
- Extensor digitorum
- Extensor expansion, ring finger
- 1st dorsal interosseous
- Adductor pollicis
- Abductor digiti minimi
- Extensor digiti minimi
- Extensor pollicis longus
- Radial A.
- Extensor retinaculum
- Extensor pollicis brevis
- Abductor pollicis longus
- Styloid process of ulna
- Extensor digiti minimi
- Ulna
- Extensor carpi ulnaris
- Extensor carpi radialis longus
- Extensor carpi radialis brevis
- Radius
- Extensor digitorum
- Abductor pollicis longus

4.9]. **Insertion**: into the base of the fifth metacarpal [see Fig. 4.19]. **Nerve supply**: posterior interosseous nerve. **Actions**: extension and adduction of the wrist.

Anconeus

Origin: fans out from the lateral epicondyle [see Fig. 4.9]. **Insertion**: to the proximal third of the lateral surface of the ulna. **Nerve supply**: radial nerve in the arm. **Actions**: the **anconeus** holds the ulna firmly against the humerus. It also contracts strongly during pronation, when the distal end of the ulna is displaced laterally by slight medial rotation of the humerus. This occurs in many movements (e.g. in using a screwdriver in the right hand to remove a screw).

Deep muscles of the back of the forearm

There are five muscles in the deep layer: abductor pollicis longus, extensor pollicis brevis, extensor pollicis longus, extensor indices, and supinator. All but the supinator arise from the forearm bones and the interosseous membrane [see Fig. 4.14], and pass to the thumb or index finger.

Abductor pollicis longus

Origin: the abductor pollicis longus arises from the posterior surfaces of the ulna, radius, and interosseous membrane, distal to the supinator [see Fig. 4.14]. **Insertion**: it curves over the posterior surface of the radius and the tendons of the brachioradialis and extensor carpi radialis longus and brevis, and is inserted into the anterolateral surface of the base of the first metacarpal [see Fig. 4.18]. (It forms the anterior/lateral boundary for the anatomical snuffbox.) **Nerve supply**: posterior interosseous nerve. **Actions**: it abducts the thumb at the carpometacarpal joint and produces some extension. It also acts with the supinator, as described below.

Extensor pollicis brevis

Origin: the extensor pollicis brevis arises from the posterior surfaces of the radius and interosseous membrane, immediately distal to the abductor pollicis longus [see Fig. 4.14]. **Insertion**: it curves over the posterior surface of the radius and the tendons of the brachioradialis and extensor carpi radialis longus and brevis, and is inserted on the dorsal surface of the base of the proximal phalanx of the thumb. (It also forms the anterior/lateral boundary

for the anatomical snuffbox.) **Nerve supply**: posterior interosseous nerve. **Actions**: extension of the carpometacarpal and metacarpophalangeal joints of the thumb.

Extensor pollicis longus

Origin: from the middle third of the posterior surface of the ulna and adjacent interosseous membrane [see Fig. 4.14]. **Insertion**: its tendon passes deep to the extensor retinaculum and bends laterally around the dorsal tubercle of the radius. It crosses the tendons of the extensor carpi radialis longus and brevis and runs along the dorsum of the first metacarpal, forming the posterior boundary for the anatomical snuffbox. At the first metacarpophalangeal joint, it is joined by the first palmar interosseous muscle and an extension from the abductor pollicis brevis. It is inserted on the base of the distal phalanx through the extensor expansion of the thumb. **Nerve supply**: posterior interosseous nerve. **Actions**: extension of the carpometacarpal, metacarpophalangeal, and interphalangeal joints of the thumb.

Extensor indices

Origin: from the posterior surface of the ulna below the origin of the extensor pollicis longus and the adjoining interosseous membrane [see Fig. 4.14]. **Insertion**: it is inserted on the extensor expansion of the index finger. **Nerve supply**: posterior interosseous nerve. **Actions**: extension of the metacarpophalangeal and interphalangeal joints of the index finger.

Supinator

Origin: it arises from the lateral surface of the fibrous capsule of the elbow joint and from the supinator crest on the upper end of the ulna, distal to the radial notch. **Insertion**: it curves round the posterior and lateral surfaces of the radius [Fig. 10.8], and is inserted on the radius between the attachments of the flexor digitorum superficialis and pronator teres [see Fig. 4.13]. **Nerve supply**: deep branch of the radial nerve and posterior interosseous nerve. **Actions**: it supinates the forearm. It is a powerful muscle and acts with the biceps brachii. It may be assisted by the abductor pollicis longus which arises, in part, from the ulna and runs parallel to the supinator.

Fig. 10.24, a diagrammatic section through the upper part of the forearm, shows the flexor and extensor muscles.

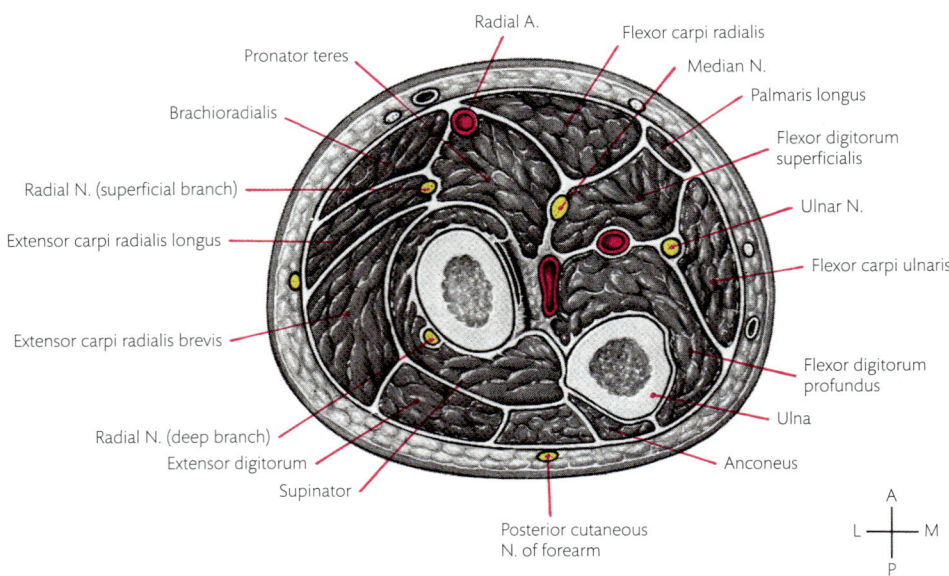

Fig. 10.24 Section through the upper third of the left forearm. A = anterior; P = posterior; M = medial; L = lateral.

Extensor tendons of the fingers

The four tendons of the **extensor digitorum** pass on the dorsum of the hand. They are linked together by oblique strips of tendinous material—the **intertendinous connections**—proximal to the metacarpophalangeal joints [Fig. 10.23]. These connections force the tendons to work together. The index and little fingers each have an additional extensor tendon—**extensor indicis** and **extensor digiti minimi**. These fuse with the corresponding tendon of the extensor digitorum, distal to the connections, and so are able to extend the metacarpophalangeal joints of these fingers alone when the joints of the middle and ring fingers are flexed. The middle and ring fingers cannot be extended individually.

Extensor retinaculum and synovial sheaths of extensor tendons

The extensor retinaculum is a thick strip of deep fascia, 2–3 cm wide. It extends from the triquetrum and styloid process of the ulna to the sharp anterolateral margin of the distal end of the radius. It covers the extensor tendons and holds them in place during extension of the wrist. Distally, the extensor retinaculum is continuous with the thin deep fascia of the dorsum of the hand which fuses, in turn, with the extensor tendons of the fingers.

From the deep surface of the retinaculum, septa extend to the head of the ulna and to each of the bony ridges on the distal end of the radius. The presence of these septa divide the space deep to the retinaculum into six compartments, or tunnels, for separate tendons or groups of tendons and their synovial sheaths [Fig. 10.25]. The tendons and synovial sheaths in the six compartments, from lateral to medial, are: (1) abductor pollicis longus and extensor pollicis brevis; (2) extensor carpi radialis longus and brevis; (3) extensor pollicis longus; (4) extensor digitorum and extensor indices; (5) extensor digiti minimi; and (6) extensor carpi ulnaris [Figs 10.23, 10.25]. The synovial sheaths begin near the proximal edge of the extensor retinaculum and extend for a variable distance on to the corresponding metacarpal. In the case of the extensors of the carpus and the long abductor of the thumb, the sheath ends near the base of the metacarpal to which it is attached. The positions of the extensor tendons on the dorsum of the wrist are shown in Table 10.2.

Deep branch of the radial and posterior interosseous nerves

The deep branch of the radial nerve arises at the level of the lateral epicondyle of the humerus. It descends between the brachialis and the brachioradialis, and gives branches to the extensor carpi radialis brevis and supinator [see Fig. 9.3]. It enters the supinator, winds obliquely round the lateral and posterior surfaces of the radius within that muscle,

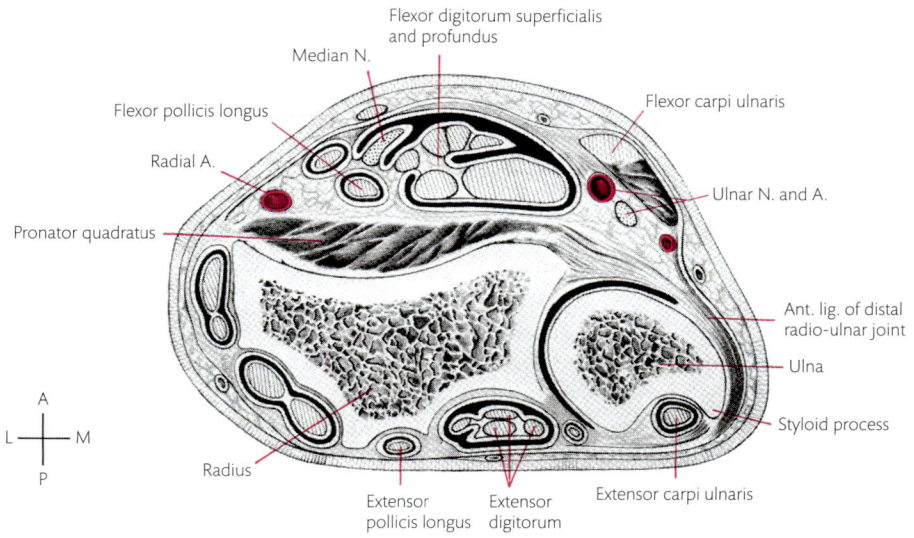

Fig. 10.25 Transverse section through the forearm above the flexor retinaculum, showing the relation of the synovial sheaths to the tendons. A = anterior; P = posterior; M = medial; L = lateral.

and emerges a short distance proximal to the distal border of the supinator, as the posterior interosseous nerve. It gives branches to the surrounding muscles and descends between the extensor digitorum and the abductor pollicis longus on the posterior surface of the interosseous membrane [Fig. 10.8]. In the distal part of the forearm, it accompanies the terminal part of the anterior interosseous artery. It ends on the back of the wrist joint by sending branches to the wrist and the intercarpal joints.

Arteries of the back of the forearm and hand

1. The **posterior interosseous artery** arises from the common interosseous artery in the anterior compartment and passes backwards between the

radius and the ulna, immediately proximal to the interosseous membrane. It appears on the back of the forearm between the supinator and the abductor pollicis longus muscles, close to the posterior interosseous nerve.

Here it gives muscular branches and sends the interosseous recurrent artery, deep to the anconeus, to the elbow joint anastomosis. It ends in the dorsal carpal rete on the dorsal surface of the wrist [Fig. 10.8].

Dissection 10.8 traces the radial artery in the back of the hand.

2. The terminal part of the **anterior interosseous artery** pierces the interosseous membrane, 5–6 cm proximal to the distal end of the radius.

Table 10.2 Position of extensor tendons at the wrist

Tendon palpable	Muscles	Action
Between distal end of radius and base of second and third metacarpals	Extensor carpi radialis longus and brevis	Extension of wrist
Between distal end of ulna and base of fifth metacarpal	Extensor carpi ulnaris	
Centrally over carpus and base of third and fourth metacarpals	Extensor digitorum Extensor indicis	Extension of fingers
On fifth metacarpal	Extensor digiti minimi	
From dorsal tubercle of radius along posterior margin of anatomical snuffbox	Extensor pollicis longus	Extension of thumb
From styloid process of radius along anterior margin of snuffbox	Extensor pollicis brevis	
Between styloid process of radius and base of first metacarpal	Abductor pollicis longus	Abduction of thumb

Objective

I. To identify and trace the radial artery on the lateral side and dorsum of the hand.

Instructions

1. Trace the radial artery distally from the anterior surface of the distal end of the radius. It passes through the anatomical snuffbox, deep to the tendons which form the margins, and enters the proximal end of the first intermetacarpal space.
2. Find its dorsal and palmar carpal branches and the dorsal digital arteries which it gives to the thumb and the lateral side of the index finger.

It descends on the membrane to the dorsal carpal rete with the terminal branch of the posterior interosseous nerve.

3. The **dorsal carpal rete**—a mesh of anastomosing arteries on the dorsal surfaces of the carpal bones—is formed by the **dorsal carpal branches** of the radial and ulnar arteries and the anterior and posterior interosseous arteries. It supplies the dorsal surfaces of the wrist, the carpal and carpometacarpal joints, and the dorsal surface of the hand and fingers. The rete gives rise to a **dorsal digital artery** to the medial side of the little finger and three **dorsal metacarpal arteries** to the medial three intermetacarpal spaces. Each dorsal metacarpal artery receives **perforating branches** from the deep palmar arch and the corresponding palmar metacarpal artery. It ends by dividing into dorsal digital arteries to each side of adjacent fingers.
4. The **radial artery** leaves the anterior surface of the radius and passes deep to the tendons of the abductor pollicis longus and extensor pollicis brevis. In the 'anatomical snuffbox', it lies on the scaphoid and trapezium against which it can be palpated. Here it gives off its **palmar** and **dorsal carpal arteries**. The radial artery then passes deep to the tendon of the extensor pollicis longus and gives off **dorsal digital arteries** to both sides of the thumb and the lateral side of the index finger. The artery then turns medially into the palm through the proximal end of the first intermetacarpal space, between the two heads of the first dorsal interosseous muscle [Figs 10.4, 10.8].

Dissection 10.9 continues the dissection of the back of the hand.

Extensor expansion

The triangular extensor expansion is found on the dorsal aspect of each digit. It forms a hood over the metacarpophalangeal joint and part of the proximal phalanx [Figs 10.17, 10.23]. Just proximal to the metacarpophalangeal joint, the tendon of the extensor digitorum joins the extensor expansion. The base of the triangle extends anteriorly on each side of the metacarpal head to join the deep transverse metacarpal ligament. Over the proximal phalanx, the lateral margins of the expansion are thickened by insertion of part of the tendon of the **interossei muscles** and of the tendon of the **lumbricals** (lateral margin only). These thickened margins pass obliquely backwards to the posterior surface of the proximal interphalangeal joint. As they do so, they send tendinous bundles into the tendon of the long extensor in the midline of the extensor expansion [Fig. 10.17]. The main bulk of the extensor digitorum tendon runs through the middle of the extensor expansion and is held in position by strong transverse tendinous bundles at the metacarpophalangeal joint. Distal to the metacarpophalangeal joint, the tendon splits into three bundles in the expansion. The central part, joined by bundles from the thick margins of the expansion, crosses the proximal interphalangeal joint and is inserted into the base of the middle phalanx. The other two parts fuse with the margins of the extensor expansion at the level of the proximal phalanx, pass over the posterior surface of the proximal interphalangeal and distal interphalangeal joints, and are inserted into the base of the distal phalanx.

At the **metacarpophalangeal** and **interphalangeal joints**, the extensor expansion forms the dorsal part of the **fibrous capsule** of the joint. It is continuous on each side of the joint with the remainder of the fibrous capsule and is held in position by it.

Deep transverse metacarpal ligament and palmar ligaments

The **deep transverse metacarpal ligament** is a strong band which passes between the palmar ligaments of the metacarpophalangeal joints of the medial four digits. These **palmar ligaments** are thick, semi-rigid structures which are firmly attached to the base of the proximal phalanx of

DISSECTION 10.9 Back of the hand-2

Objectives

I. To study the extensor expansion and the insertion of muscles into it.   **II.** To identify and expose the palmar and dorsal interossei and their tendons.

Instructions

1. Follow the extensor tendons into the fingers and thumb. In each finger, the tendon expands into a sheet as it passes towards the metacarpophalangeal joint and forms the **extensor expansion**. Clean the extensor expansion, and note that it is widest at the level of the metacarpophalangeal joint. At this level, the margins of the expansion turn anteriorly on each side of the metacarpal head to be attached to the deep transverse metacarpal ligament.
2. Distal to the metacarpophalangeal joint, the expansion narrows on the dorsal surface of the proximal interphalangeal joint and passes over it to reach the middle and distal phalanges. Over the proximal phalanx, identify the two thickened margins and the central portion of the expansion. Follow the central part to the base of the middle phalanx, and the medial and lateral margins to the base of the distal phalanx.
3. Follow the tendons of the **lumbrical muscles** on the palmar surface of the deep transverse metacarpal ligament into the lateral margins of the expansions.
4. Remove any fat and fascia covering the dorsal surfaces of the intermetacarpal spaces, and expose a **dorsal interosseous muscle** in each space. Trace the muscle distally to its tendon.
5. Divide the dense fibrous tissue which lies between the dorsal parts of the metacarpal heads, and trace part of the tendon into the corresponding margin of the extensor expansion, distal to the metacarpophalangeal joint. (A deeper part of the tendon may be traced into the base of the proximal phalanx.)
6. Identify the deep transverse metacarpal ligament lying between the palmar surfaces of the medial four metacarpal heads.
7. Separate the dorsal interosseous muscle from the two metacarpal bones from which it arises. Turn the muscle distally, and expose the **palmar interosseous muscle** which arises only from the metacarpal of the finger to which it passes. (In the first intermetacarpal space, the first palmar interosseous muscle is only a slender slip.) Trace the tendon of the palmar interosseous to the extensor expansion and sometimes to the base of the proximal phalanx. The first dorsal interosseous muscle is found on the lateral side of the index finger, and the second on the lateral side of the middle finger. The third and fourth dorsal interossei are found on the medial sides of the middle and ring fingers, respectively. The first and second palmar interosseous muscles are found on the medial side of the thumb and index finger, respectively. The third and fourth palmar interossei are found on the lateral side of the ring and little fingers, respectively. The tendons of the interossei in the medial three spaces lie posterior to the deep transverse metacarpal ligament.

each finger. They are only loosely attached to the metacarpal. When the metacarpophalangeal joint is flexed, the palmar ligament slides proximally on the palmar surface of the metacarpal. When the joint is extended, the ligament moves distally on the metacarpal head. In each movement, the attachment of the extensor expansion moves with the ligament.

Muscles inserted into the extensor expansion

In addition to the long extensors of the digits, the lumbrical and interosseous muscles are inserted into the extensor expansion. These muscles **flex the metacarpophalangeal joints and extend the interphalangeal joints of the fingers**—the position of the fingers in writing.

Movements of the fingers

Flexion

The fingers are flexed at all joints by the flexor digitorum profundus. They are flexed at the metacarpophalangeal and proximal interphalangeal joints by the flexor digitorum superficialis, and at the metacarpophalangeal joints by the lumbricals and interossei.
⊃ Inability to flex the distal interphalangeal joints occurs in paralysis of the flexor digitorum profundus—a condition which is evident in the ring and little fingers in damage to the ulnar nerve above the elbow.

Extension

The fingers are extended together at all joints by the extensor digitorum. The index and little fingers are extended independently at all joints by the extensor indicis and extensor digiti minimi. Extension of the interphalangeal joints is assisted by the lumbricals and interossei. When the metacarpophalangeal joints are fully extended, the deep transverse metacarpal ligament is pulled distally on the metacarpal head, making the attachment to the ligament the only effective insertion of the long extensor tendons. The distal parts of the extensor digitorum and extensor indicis in the extensor expansion are slackened and unable to extend the interphalangeal joints. In this situation (when the metacarpophalangeal joints are fully extended), the interphalangeal joints can only be extended by the **lumbricals** and **interossei**.

➲ Flexion of one finger at the metacarpophalangeal and proximal interphalangeal joints, when the other fingers are extended, leaves the distal phalanx of that finger quite lax—it cannot be either extended or flexed. The distal phalanx cannot be flexed because the flexor digitorum profundus cannot act on a single finger (insufficient separation of the parts of the muscle) and it cannot be extended because the extensor expansion can only extend the distal interphalangeal joint when the proximal interphalangeal joint is extended.

Abduction and adduction

Abduction and adduction of the fingers (but not the thumb) occur at the metacarpophalangeal joints. The interossei and abductor digiti minimi produce these movements which have a considerable range when the joints are extended but are limited when they are flexed. Adduction also occurs in flexion of the fingers, and abduction in extension. This is because the axes of the interphalangeal joints do not lie in a straight line, but on an arc of a circle which is concave towards the palm. Confirm this movement in your own hand.

Opposition of the little finger

The little finger has the most mobile metacarpal of all the fingers. It can be rotated laterally to a small degree by the opponens digiti minimi, assisted by the short flexor and abductor. This brings the pulp of the little finger into direct opposition with that of the thumb. Compare this with opposition of the thumb and index finger in your own hand.

Movements of the thumb

Movements of the thumb are intricate and complex. The thumb functions as one half of the hand in grasping objects (the other half being the four fingers). The principal movements of the thumb take place at the **first carpometacarpal joint**. The **first** carpometacarpal joint is a saddle-shaped synovial joint with a great degree of mobility. It allows abduction, adduction, flexion, extension, and medial and lateral rotation.

Movements at the **metacarpophalangeal joint** of the thumb are mainly flexion and extension. A small amount of abduction (produced by the abductor pollicis brevis) and adduction by the adductor pollicis and the first palmar interossei are possible. Only flexion and extension occur at the **interphalangeal joint**—brought about mainly by the flexor pollicis longus and extensor pollicis longus, respectively.

For a complete understanding of the intricate workings of the hand, the following details should be borne in mind: (1) the flexor pollicis longus is used principally when the tip of the thumb is opposed to the tip of a finger or when power is required; (2) the flexor pollicis brevis is used when the main flexion is at the carpometacarpal and metacarpophalangeal joints; (3) the abductor pollicis brevis is used principally when the thumb is opposed to the little finger; (4) the abductor longus is continuously active during opposition and the reverse movement; (5) in straightening the opposed thumb, initial extension is produced by the extensor pollicis brevis, which helps to maintain abduction at the carpometacarpal joint; (6) in the later stages of extending the opposed thumb, the extensor pollicis longus plays a greater part; and (7) the extensor pollicis longus may also act as an adductor of the fully abducted thumb, thus assisting the adductor pollicis or mimicking its activity when paralysed.

[See Tables 11.9 and 11.10 for muscles and movements of the thumb.]

Clinical Applications 10.1, 10.2, and 10.3 explore how the anatomy of the forearm and hand applies to clinical practice.

CLINICAL APPLICATION 10.1 Tenosynovitis

A 12-year-old schoolgirl pricked her thumb with a needle, just proximal to the interphalangeal joint on the palmar aspect. Over the next few days, she developed pain and swelling over the site of injury which soon spread to the lateral side of the hand and palm up to the wrist. Her parents treated her with painkillers and made her rest. The swelling continued to spread and extended to the lower part of the wrist. Movements of the thumb and wrist became excruciatingly painful, and she developed fever on the tenth day. She was taken to a hospital. On examination, her entire thumb, lateral side of the hand, and lower part of the forearm were swollen and tender. Movements of the thumb were restricted due to the swelling and pain. There was no lymphadenopathy.

Study question 1: from your knowledge of the anatomy of this region, which structure, when infected, is most likely to present with these signs and symptoms? (Answer: the synovial sheath of the thumb.)

Study question 2: name the tendon which lies in this sheath. What is the proximal and distal extent of this sheath? (Answer: the tendon of the flexor pollicis longus. The sheath starts from a few centimetres proximal to the flexor retinaculum and ends on the distal phalanx.)

Study question 3: do you think it is possible for the infection to spread outside the sheath? (Answer: Yes.) The radial synovial sheath may communicate with the ulnar (common) synovial sheath for the long flexors, and infection can spread through this communication. The sheath can rupture because of distension, and infection could spread to the surrounding tissue.

CLINICAL APPLICATION 10.2 Carpal tunnel syndrome

A 52-year-old domestic worker developed tingling and burning pain over the palmar aspect of the thumb, and index and middle fingers. On inspection, the doctor noticed that there was flattening of the thenar eminence.

Study question 1: what forms the thenar eminence? (Answer: muscles of the thenar eminence—the abductor pollicis brevis, the flexor pollicis brevis, and the opponens pollicis.) On examination, it was found that she was not able to abduct or oppose her thumb effectively. She also had decreased sensation over the palmar aspect of the thumb, index finger, middle finger, and lateral part of the ring finger.

Study question 2: if this were due to a nerve lesion, which nerve is affected? (Answer: the median nerve.)

Study question 3: how does the affected nerve enter the hand? Name other structures that lie in the same space at the wrist. (Answer: the median nerve passes deep to the flexor retinaculum (in the carpal tunnel) to enter the hand. Other structures in the carpal tunnel are the tendons of the flexor pollicis longus and flexor digitorum superficialis and profundus.) Death of the tendons due to interference with their blood supply can occur in extreme cases.

CLINICAL APPLICATION 10.3 Dupuytren's contracture

Progressive shortening of the palmar aponeurosis results in a condition known as Dupuytren's contracture. The patient's fingers are flexed, because the aponeurosis is attached to the proximal phalanges through the deep transverse metacarpal ligament. The shortening usually affects the medial part of the aponeurosis, and hence the little and ring fingers. Surgical division of the aponeurosis is required to straighten the fingers.

CHAPTER 11
The joints of the upper limb

Elbow joint

Type

Synovial hinge joint.

Articular surfaces

The elbow joint is formed by articulation of the **trochlea** of the humerus with the **trochlear notch** of the ulna, and the **capitulum** of the humerus with the proximal surface of the **head of the radius** [see Figs 4.9, 4.13].

Capsule

Proximally, the anterior part of the fibrous capsule is attached to the medial and lateral epicondyles and to the upper margins of the radial and coronoid fossae of the humerus. The posterior part of the capsule is weak. It stretches from a line joining the epicondyles across the floor of the olecranon fossa to the articular margin of the olecranon.

Distally, it is attached to the coronoid process of the ulna and the annular ligament of the radius. The anterior and posterior parts of the fibrous capsule are weak and contain oblique fibres which permit a full range of movements [Fig. 11.1]. Its fibrous capsule and joint cavity are continuous with those of the proximal radioulnar joint.

Ligaments

The elbow joint is essentially a hinge joint with strong radial and ulnar collateral ligaments. The **radial collateral ligament** is strong. It passes from the distal surface of the lateral epicondyle to the lateral and posterior parts of the annular ligament of the radius [Fig. 11.1].

The **ulnar collateral ligament** radiates from the medial epicondyle of the humerus to

the coronoid process and olecranon of the ulna. The edges of the ligament are thick. The thinner central part is attached to a transverse band that bridges across the interval between these edges [Fig. 11.2]. The ulnar nerve and posterior ulnar recurrent artery lie on the posterior and middle parts of the ligament.

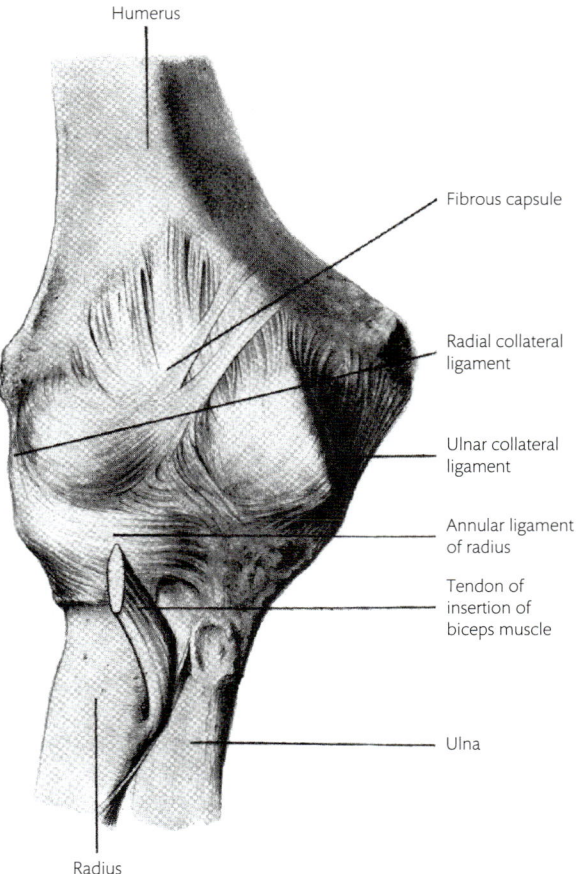

Fig. 11.1 Front of the elbow joint.

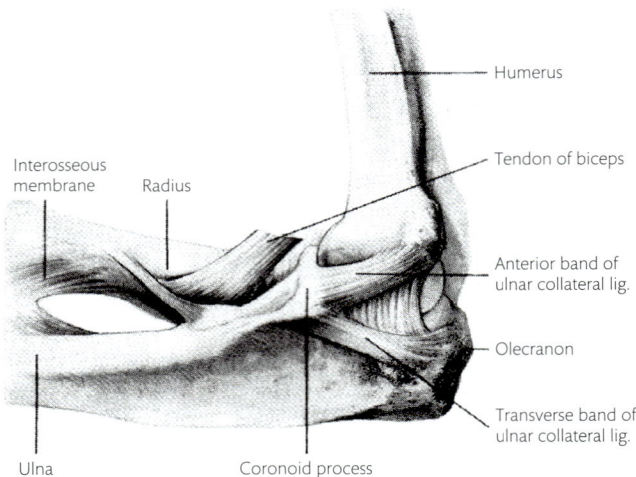

Fig. 11.2 Medial aspect of the elbow joint.

Synovial membrane

As in all synovial joints, the synovial membrane lines the deep surface of the fibrous capsule and the intracapsular non-articular parts of the bones. On the bony fossae—olecranon, coronoid, and radial—the synovial membrane is separated from the fibrous capsule by pads of fat which slide into the fossae when the bony processes are withdrawn during flexion and extension [Fig. 11.3]. The synovial membrane of the elbow joint is continuous with that of the proximal radioulnar joint on the lateral side. Fig. 11.4 shows the bones and joint space of the elbow joint, as seen on a plain X-ray of the elbow in the anteroposterior (AP) and lateral views.

Relations

Anteriorly, the joint is covered by the cubital fossa and its contents: the tendon of the biceps brachii, median nerve, brachial artery, radial artery, pronator teres, brachioradialis, and extensor carpi radialis longus and brevis. Posteriorly, it is covered by the tendon of the triceps brachii.

Dissection 11.1 instructs on cleaning and studying the capsule of the elbow joint.

Dissection 11.2 describes the dissection of the interior of the elbow joint.

Blood and nerve supply

The **nerve supply** of the joint is from all the adjacent nerves—median, ulnar, radial, posterior inter-

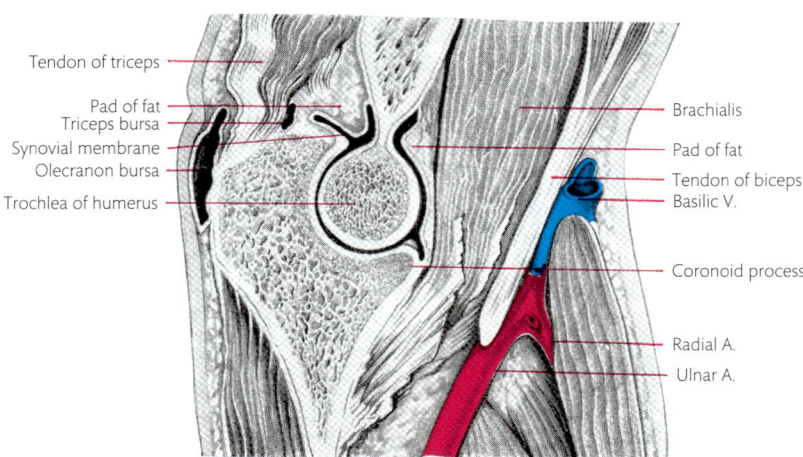

Fig. 11.3 Sagittal section of the right elbow.

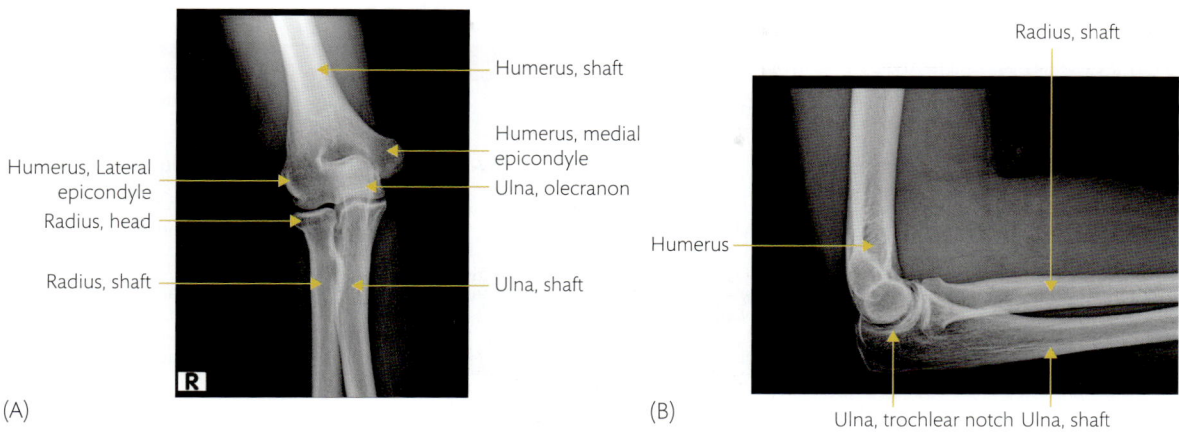

Fig. 11.4 Plain X-ray of the elbow. (A) Anteroposterior view. (B) Lateral view.

Reproduced with kind permission from Aparna Irodi.

osseous, and musculocutaneous. The blood supply is by the numerous collateral and recurrent branches around the elbow—radial, middle, superior and inferior collateral arteries, and the radial, ulnar, and interosseous recurrent arteries.

Movements at the elbow joint

The elbow joint is essentially a hinge joint, the main movements of which are **flexion** and **extension**. (Movements at the elbow joint are distinct from those which take place at the proximal radioulnar joint.) In full flexion, the forearm bones lie parallel to the humerus. In full extension, the ulna is angled laterally, so that the supinated forearm bones make an angle ('**the carrying angle**') of approximately 165 degrees laterally with the humerus. This angulation of the ulna is obscured in pronation when the arm and forearm lie in the same straight line and the radius and ulna cross each other.

In addition to being a hinge joint, the humeroulnar part of the elbow joint is a modified saddle joint which allows some slight abduction and adduction of the ulna. Tables 11.1 and 11.2 summarize the muscles and movements of the elbow joint.

Wrist joint or radiocarpal joint

Dissection 11.3 begins the dissection of the wrist.

Type

Synovial joint, ellipsoid type. The wrist joint has a long radius of curvature transversely, and a short radius of curvature anteroposteriorly.

DISSECTION 11.1 Elbow joint

Objective

I. To study the fibrous capsule of the elbow joint.

Instructions

1. Separate the muscles from the epicondyles, and reflect them distally. Divide the biceps, brachialis, and triceps, a short distance proximal to the elbow, and turn all three distally.
2. Separate all the muscles from the fibrous capsule of the elbow joint, and remove the loose connective tissue from its external surface, so as to define its parts. Retain the median, ulnar, and radial nerves and the brachial artery.

DISSECTION 11.2 Interior of the elbow joint

Objective

I. To study the interior of the elbow joint.

Instructions

1. Make transverse incisions through the anterior and posterior parts of the fibrous capsule, and examine the synovial membrane.

Table 11.1 Muscles acting on the elbow joint

Muscle	Origin	Insertion	Action at elbow	Nerve supply
Brachialis	Humerus, anterior surface, distal half	Ulna, coronoid process	Flexion	Musculocutaneous
Biceps brachii	Scapula, coracoid process, supraglenoid tubercle	Radius, tuberosity	Flexion	Musculocutaneous
Brachioradialis	Humerus, lateral supracondylar line	Radius, base of styloid process	Flexion	Radial
Extensor carpi radialis longus	Humerus, lateral supracondylar line	Second metacarpal	Flexion	Radial
Triceps	Scapula, infraglenoid tubercle—long head. Humerus, posterior surface. Above groove for radial nerve—lateral head, below groove—medial head	Ulna, olecranon	Extension	Radial
Anconeus	Humerus, lateral epicondyle	Ulna, lateral surface, proximal one-third	Holds ulna to humerus Abduction of ulna	Radial
Pronator teres	Humerus, medial supracondylar line, and medial epicondyle	Radius, lateral surface	Flexion—minor role	Median
Flexor carpi radialis	Humerus, medial epicondyle	First metacarpal	Flexion—minor role	Median

Articular surfaces

The radiocarpal or wrist joint consists of the articulation between the convex surface of the **scaphoid**, **lunate**, and **triquetrum**, with the concave distal surfaces of the **radius** and the triangular **articular disc**. The triangular articular disc joins the medial edge of the articular surface of the radius to the styloid process of the ulna [Fig. 11.5]. It separates the ulna from the joint. In the resting position of the hand, only the scaphoid and lateral part of the lunate articulate with the two shallow fossae on the distal surface of the radius. The remainder of the lunate is in contact with the articular disc [Fig. 11.5]. The direction of the radiocarpal joint is oblique—the lateral and dorsal margins of the radius extend further distally than the other margins. This reduces the likelihood of posterior dislocation of any of the carpal bones.

Fibrous capsule and ligaments of the wrist joint

The fibrous capsule passes from the margins of the distal ends of the radius and ulna and the margins of the articular disc to the scaphoid, lunate, and triquetrum. Medially and laterally, slightly thickened portions of the fibrous capsule pass from the styloid processes of the radius and ulna to the scaphoid and triquetrum. These are the **radial** and **ulnar collateral ligaments**. The anterior and posterior parts of the fibrous capsule contain fibres which pass obliquely downwards and medially.

Synovial membrane

The synovial membrane, which lines the fibrous capsule and covers the interosseous ligaments of the carpals, may be continuous with that of the distal radioulnar joint through a defect in the triangular disc.

Relations

Anteriorly, the wrist joint is covered by the long flexor tendons of the front of the arm, the median

Table 11.2 Movements at the elbow joint

Movement	Muscles	Nerve supply
Flexion	Biceps brachii	Musculocutaneous
	Brachialis	Musculocutaneous
	Brachioradialis	Radial
	Extensor carpi radialis longus	Radial
	Flexor carpi radialis	Median
	Pronator teres	Median
Extension	Triceps	Radial
Ulna, abduction	Anconeus	Radial

and ulnar nerves, and the radial and ulnar arteries. Posteriorly lie the extensor retinaculum and the structures under its cover [see Fig. 10.25].

Blood and nerve supply

The **nerve supply** is through the anterior and posterior interosseous nerves and the dorsal branch of the ulnar nerve. The **blood supply** is through the anterior interosseous artery, the anterior and posterior branches of the radial and ulnar arteries, and the branches of the palmar and dorsal metacarpal arteries.

Movements at the wrist joint

The ellipsoid wrist joint has a long radius of curvature transversely which permits abduction and adduction, and a short radius of curvature anteroposteriorly which permits flexion and extension. Because of the two different curvatures of the joint, rotation is not possible.

All these movements are supplemented by similar movements at the **intercarpal joints** [Fig. 11.6]. In adduction, the carpal bones slide laterally, with the lunate passing further on to the radius, while the triquetrum comes into contact with the triangular disc.

The carpal bones move with the radius in pronation and supination. Thus, the movement of adduction when the hand is pronated deviates the hand laterally, but still towards the ulna which is now lateral to the radius in the distal forearm. For this reason, adduction is commonly called 'ulnar deviation', and abduction 'radial deviation'.

Dissection 11.4 continues the dissection of the wrist.

Tables 11.3 and 11.4 summarize the muscles and movements of the wrist joint.

There are three joints between the radius and ulna—proximal, middle, and distal radioulnar joints. The proximal and distal radioulnar joints are synovial joints. The middle radioulnar joint is a fibrous joint.

Radioulnar joints

Proximal radioulnar joint

Type
Synovial, pivot joint.

Articular surfaces
This joint is between the side of the cylindrical **head of the radius** and the **radial notch of the ulna**.

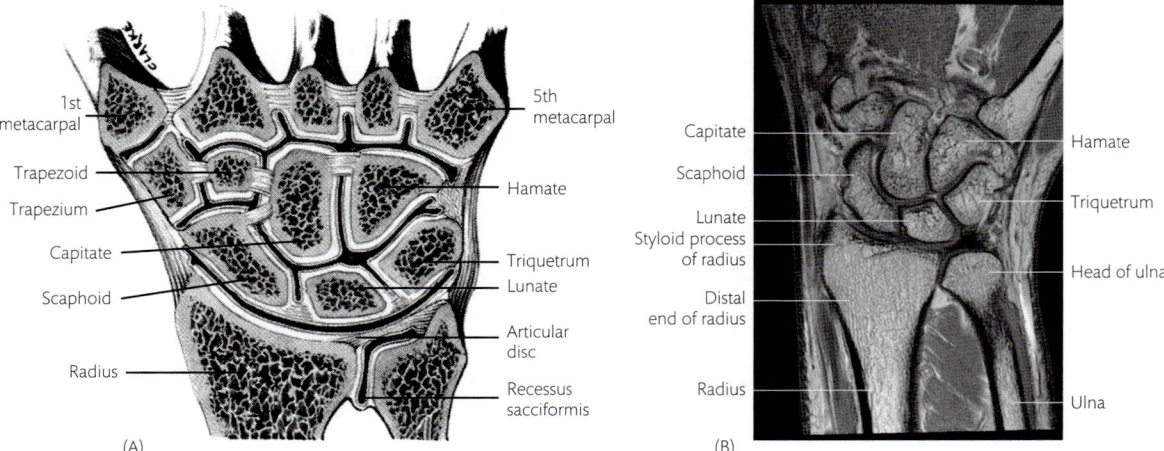

Fig. 11.5 Coronal section through the radiocarpal, intercarpal, carpometacarpal, and intermetacarpal joints to show joint cavities and interosseous ligaments. (A) Schematic drawing. (B) Magnetic resonance imaging.

155

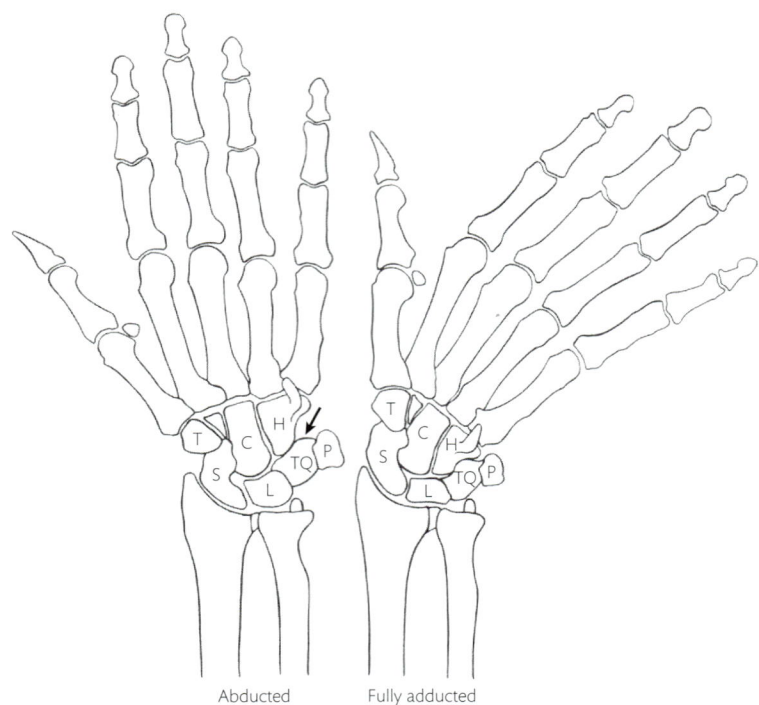

Abducted   Fully adducted

Fig. 11.6 Bones of the hand in partially abducted (left) and fully adducted (right) positions of the wrist. In the partially abducted position, the scaphoid articulates with the radius, and the lunate with the articular disc on the ulna. In adduction, the carpal bones slide laterally; the lunate comes into contact with the radius, and the triquetrum with the articular disc. The distal row of carpal bones slides in the concavity of the proximal row. C = capitate; H = hamate; L = lunate; P = pisiform; S = scaphoid; T = trapezium; TQ = triquetrum.

Fibrous capsule

The fibrous capsule of the joint is continuous with that of the elbow joint [Fig. 11.1].

Annular ligament of the radius

The fibrous capsule is strengthened by the annular and quadrate ligaments. The **annular ligament** is a strong, fibrous collar which encircles the head of the radius and retains it in contact with the radial notch on the ulna [Fig. 11.7]. It is attached to the anterior and posterior margins of the radial notch. It is slightly conical and is loosely attached to the neck of the radius. Thus, the head of the radius is free to turn within the ligament but cannot be pulled down out of it. Proximally, the annular ligament is continuous with, and strengthened by, the lateral and anterior ligaments of the elbow joint [Fig. 11.1]. The weak **quadrate ligament** passes between the neck of the radius and the lower margin of the radial notch of the ulna.

Synovial membrane

The synovial membrane of the proximal radioulnar joint is continuous with that of the elbow joint. It lines the deep surface of the annular ligament and is reflected upwards from its distal margin to surround the intracapsular part of the neck of the radius.

Blood and nerve supply

The **nerve supply** to the joint is by articular branches of the musculocutaneous, median, radial, and ulnar nerves. The **arterial supply** is from the arteries anastomosing around the elbow.

DISSECTION 11.4 Wrist joint-2

Objective

I. To study the interior of the wrist joint.

Instructions

1. Divide the anterior, medial, and lateral ligaments by a transverse incision across the front of the joint.
2. Bend the hand backwards, and expose the articular surfaces.

Table 11.3 Muscles acting on the wrist joint

Muscle	Origin	Insertion	Action at wrist	Nerve supply
Flexor carpi radialis	Humerus, medial epicondyle	Second and third metacarpal bases	Flexion, abduction	Median
Flexor carpi ulnaris	Humerus, medial epicondyle Ulna, subcutaneous border	Pisiform (hamate and fifth metacarpal)	Flexion, adduction	Ulnar
Extensor carpi radialis longus	Humerus, lateral supracondylar line	Second metacarpal base	Extension, abduction	Radial
Extensor carpi radialis brevis	Humerus, lateral epicondyle	Second and third metacarpal bases	Extension, abduction	Radial*
Extensor carpi ulnaris	Humerus, lateral epicondyle Ulna, posterior border	Fifth metacarpal base	Extension, adduction	Radial*
Extensor digitorum	Humerus, lateral epicondyle	Extensor expansions of fingers	Extension, prevented by radial and ulnar flexors of carpus	Radial*
Extensor digiti minimi	Humerus, lateral epicondyle	Extensor expansion, fifth digit		Radial*
Extensor pollicis longus	Ulna, middle third, dorsal surface	Thumb, distal phalanx		Radial*
Extensor indicis	Ulna, distal to extensor pollicis longus	Second digit extensor expansion		Radial*
Flexor digitorum profundus	Ulna, proximal three-fourths, anterior and medial surfaces	Fingers, terminal phalanges	Flexion, prevented by radial and ulnar extensors of carpus	Median** and ulnar
Flexor digitorum superficialis	Humerus, medial epicondyle Ulna, coronoid process Radius, anterior border	Fingers, middle phalanges		Median
Flexor pollicis longus	Radius, anterior surface, middle two-quarters	Thumb, distal phalanx		Median**

* Posterior interosseous branch.
** Anterior interosseous branch.
Note: long flexors and extensors of the digits do not bring about flexion or extension of the wrist. This is because flexion is prevented by radial and ulnar extensors of the wrist, and extension is prevented by radial and ulnar flexors of the wrist.

Movements

The three radioulnar joints move together and the movements are discussed below.

Distal radioulnar joint

Type

Synovial, pivot joint.

Articular surfaces

This joint is between the side of the cylindrical **head of the ulna** and the **ulnar notch of the radius**.

Articular disc at the distal end of the ulna

The triangular, fibrocartilaginous articular disc is the main structure holding the distal ends of the radius and ulna together [Fig. 11.5]. The base of the triangular disc is attached to the distal margin of the ulnar notch of the radius. The apex is attached to a depression at the root of the ulnar styloid process. The disc covers the distal end of the ulna and separates it from the carpal bones. It also separates the distal radioulnar joint from the wrist joint. Occasionally, the disc is perforated and the cavities of the two joints are continuous.

Fibrous capsule of the distal radioulnar joint

The fibrous capsule of the distal radioulnar joint consists of lax fibres passing between the anterior and posterior borders of the disc and the adjacent surfaces of the radius and ulna. Between these bones, the capsule extends proximally to the distal border of the interosseous membrane, enclosing an extension of the joint cavity.

Table 11.4 Movements at the wrist joint

Movement	Muscles	Nerve supply
Flexion	Flexor carpi ulnaris	Ulnar
	Flexor carpi radialis	Median
	Flexor digitorum profundus	Median* and ulnar
	Flexor digitorum superficialis	Median
	Flexor pollicis longus	Median*
	Palmaris longus	Median
Extension	Extensor carpi ulnaris	Radial*
	Extensor carpi radialis longus and brevis	Radial
	Extensor digitorum	Radial*
	Extensor digiti minimi	Radial*
	Extensor pollicis longus	Radial*
	Extensor indicis	Radial*
Abduction	Extensor carpi radialis longus and brevis	Radial
	Flexor carpi radialis	Median
	Abductor pollicis longus	Radial*
Adduction	Flexor carpi ulnaris	Ulnar
	Extensor carpi ulnaris	Radial*

* Anterior or posterior interosseous branch.

Synovial membrane of the distal radioulnar joint

The synovial membrane of the distal radioulnar joint lines the fibrous capsule and the proximal surface of the articular disc.

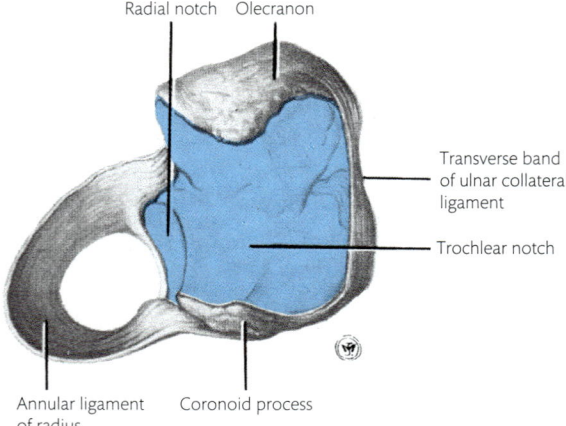

Fig. 11.7 Annular ligament of the radius and proximal articular surfaces of the ulna.

Radial notch   Olecranon

Transverse band of ulnar collateral ligament

Trochlear notch

Annular ligament of radius   Coronoid process

Blood and nerve supply

The **nerve supply** is from the anterior and posterior interosseous nerves; and the **blood supply** is from the palmar and dorsal branches of the anterior interosseous artery.

Dissection 11.5 describes the dissection of the interosseous membrane.

Interosseous membrane of the forearm

This fibrous sheet stretches between the interosseous borders of the radius and ulna and holds these bones together, while allowing movements to take place between them. It begins 2–3 cm distal to the tuberosity of the radius and blends distally with the capsule of the distal radioulnar joint. The fibres run downwards and medially from the radius to the ulna. Forces applied to the radius from the hand (as in a fall on an outstretched arm) are transmitted to the ulna through the interosseous membrane. Such forces are then transferred to the humerus at the elbow joint.

The posterior interosseous vessels pass back between the radius and the ulna, proximal to the interosseous membrane. The anterior interosseous vessels pierce the membrane, 5 cm from its distal end. The membrane increases the area of origin for the deep flexor and extensor muscles of the forearm.

Dissection 11.6 explores the superior and inferior radioulnar joints.

Movements at the radioulnar joints

The movements taking place at the radioulnar joints are **supination** and **pronation**. They occur around an axis that passes through the **centre of the head of the radius** and the **head of the ulna** [Fig. 11.8]. When the limb is supine, the thumb is directed laterally and the radius and ulna lie parallel to each other. When the limb is prone, the thumb and styloid process of the radius

> ### DISSECTION 11.5 Interosseous membrane
>
> **Objective**
>
> I. To study the interosseous membrane.
>
> **Instructions**
>
> 1. Expose the interosseous membrane by removing the muscles from the back or front of the forearm.

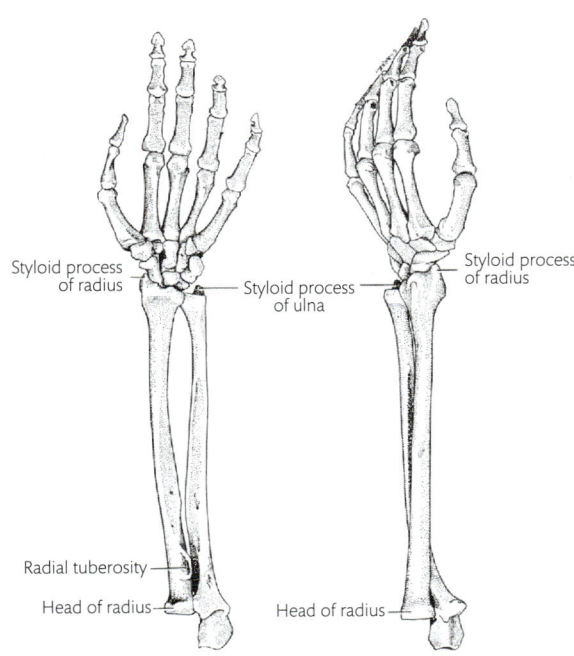

lie on the medial side; the posterior surfaces of the radius and hand face anteriorly, and the body of the radius lies obliquely across the anterior surface of the ulna. In pronation, the head of the radius rotates in the annular ligament, while the distal end turns around the stationary ulna, carrying the hand and articular disc with it. As described above, the ulna remains stationary, and the little finger rotates around its own axis. However, pronation and supination can be carried out around the axis of any one of the fingers.

Pronation

Pronation is produced by muscles on the anterior surface of the forearm which run from medial to lateral—pronator teres, pronator quadratus, and flexor carpi radialis. The pronator teres has the maximum mechanical advantage, because it is inserted into the point of maximum lateral convexity of the radius.

Supination

The biceps brachii is a powerful supinator. Its tendon turns round the medial side of the radius into the posterior aspect of the radial tuberosity. Thus, it rotates the radius laterally on contraction. Muscles which pass from medial to lateral on the posterior aspect of the forearm also produce supination—supinator and abductor pollicis longus. The brachioradialis can help to start supination of the fully pronated forearm.

With the elbow extended, the range of pronation and supination is apparently increased by the added rotation of the humerus. For this reason, clinical tests for the range of movements at the radioulnar joints are always carried out with the elbow flexed at right angles—the position in which the biceps has its maximum supinating power. Tables 11.5 and 11.6 summarize the muscles and movements of the radioulnar joint.

Intercarpal, carpometacarpal, and intermetacarpal joints

There are three separate joint cavities within this complex of articulations between the carpals and metacarpals: (1) the **first carpometacarpal joint** or the carpometacarpal joint of the thumb; (2) the main joint complex; and (3) the pisiform (intercarpal) joint.

The first carpometacarpal joint

Type

Synovial joint, saddle-shaped.

Articular surfaces

The concavoconvex distal aspect of the **trapezium** and the proximal surface of the **base of first metacarpal**.

Styloid process of radius
Styloid process of ulna
Styloid process of radius
Radial tuberosity
Head of radius
Head of radius

Fig. 11.8 Left forearm and hand bones in the supine (left) and semi-prone (right) positions. Note that the ulna remains stationary and the distal end of the radius rotates around it, carrying the hand with it.

Table 11.5 Muscles acting on the radioulnar joint

Muscle	Origin	Insertion	Action on radioulnar joint	Nerve supply
Pronator teres	Humerus, medial epicondyle and medial supracondylar line	Radius, lateral surface, middle of body	Pronation	Median
Pronator quadratus	Ulna, anterior surface, distal one-fourth	Radius, anterior surface, distal one-fourth	Pronation. Holds radius to ulna	Median*
Flexor carpi radialis	Humerus, medial epicondyle	Second and third metacarpal bases	Pronation	Median
Supinator	Humerus, lateral epicondyle. Ulna, supinator crest	Radius, posterior lateral and anterior surfaces of proximal third	Supination	Radial
Biceps brachii	Scapula, coracoid process. Scapula, supraglenoid tubercle	Radius, tuberosity	Supination (with elbow flexion)	Musculocutaneous
Brachioradialis	Humerus, lateral supracondylar line	Radius, distal end, lateral surface	Supination from pronation to mid position	Radial

* Anterior interosseous branch.

Capsule and synovial membrane

It has a loose fibrous capsule lined by synovial membrane.

Blood and nerve supply

This joint receives its **nerve supply** from the posterior interosseous nerve and the superficial branch of the radial nerve. The **blood supply** is from the radial artery.

Movements

The first metacarpophalangeal joint has a wide range of movements: (1) abduction; (2) adduction; (3) flexion; (4) extension; (5) medial; and (6) lateral rotation. When the thumb is adducted, the range of flexion and extension is small; when abducted, the range is greatly increased. The first metacarpal then moves on an arc of a circle, so that it is medially rotated in flexion and laterally rotated in extension.

Table 11.6 Movements at the radioulnar joint

Movement	Muscles	Nerve supply
Pronation	Pronator teres	Median
	Pronator quadratus	Median*
	Flexor carpi radialis	Median
Supination	Supinator	Radial
	Biceps brachii	Musculocutaneous
	Brachioradialis	Radial

* Anterior interosseous branch.

This movement of flexion with medial rotation of the abducted metacarpal of the thumb is known as **opposition**, for it opposes the palmar surface of the thumb to those of the fingers. It is essential in holding or grasping objects and makes the thumb functionally one half of the hand.

The main joint complex

Type

Synovial, sliding joint.

Capsule and synovial membrane

The **intercarpal**, **second to fifth carpometacarpal**, and **intermetacarpal joints** are enclosed in a single joint complex with a common joint cavity and synovial membrane [Fig. 11.5]. The common fibrous capsule unites the exposed surfaces of all these bones. This joint capsule is continuous proximally with the capsule of the wrist joint (though their cavities are separate) and forms three strong intermetacarpal ligaments which limit the cavity distally.

In addition to the intermetacarpal ligaments, **interosseous ligaments** unite each row of carpal bones. The ligaments between the proximal parts of the scaphoid, lunate, and triquetrum bones complete the distal articular surface of the wrist joint and separate its cavity from that of the intercarpal joints. Less regular ligaments unite the distal row of carpal bones and allow continuity of the intercarpal and carpometacarpal joint cavities around them.

Articular surfaces and movements of the intercarpal joints

In each row, the carpal bones articulate with each other by flat surfaces which allow little movement but give some resilience. The joint between the proximal and distal rows is deeply concavoconvex. The capitate and hamate fit into the concavity of the proximal row, and the concave surfaces of the trapezium and trapezoid fit the convex distal surface of the scaphoid [Figs 11.5, 11.9A and B].

The main movements possible at the intercarpal joints are flexion and extension. The distal row also moves on the proximal row around an anteroposterior axis through the centre of the capitate, producing some abduction and adduction [Fig. 11.6]. Thus, the transverse intercarpal joint increases the range of movements at the wrist and the resilience of the region.

Pisiform joint

This specific intercarpal joint is a small, flat area where the pisiform articulates with the palmar surface of the triquetrum. The pisiform is held in position against the pull of the flexor carpi ulnaris by the **pisohamate** and **pisometacarpal ligaments**. The joint allows the pisiform to maintain correct alignment during adduction and abduction of the hand.

Medial four carpometacarpal joints

The metacarpal bone of the index articulates with the trapezoid, and that of the middle finger with the capitate [Fig. 11.5]. Both metacarpals have very limited movements. Confirm this on your own hand. The fourth and fifth metacarpals articulate with the hamate. Note that the metacarpal of the ring finger has considerable movement, while that of the little finger is the most mobile. The carpometacarpal joints of the ring and little fingers are flexed when the fist is clenched or the palm is 'cupped', and the fifth has slight lateral rotation produced by the opponens digiti minimi.

Intermetacarpal joints

These permit slight movement between the bases of the metacarpals.

Nerve supply and blood supply

The **nerve supply** of these joints is from the anterior and posterior interosseous nerves, the dorsal and deep branches of the ulnar nerve, and the superficial branch of the radial nerve. They are supplied to the dorsal and palmar metacarpal arteries.

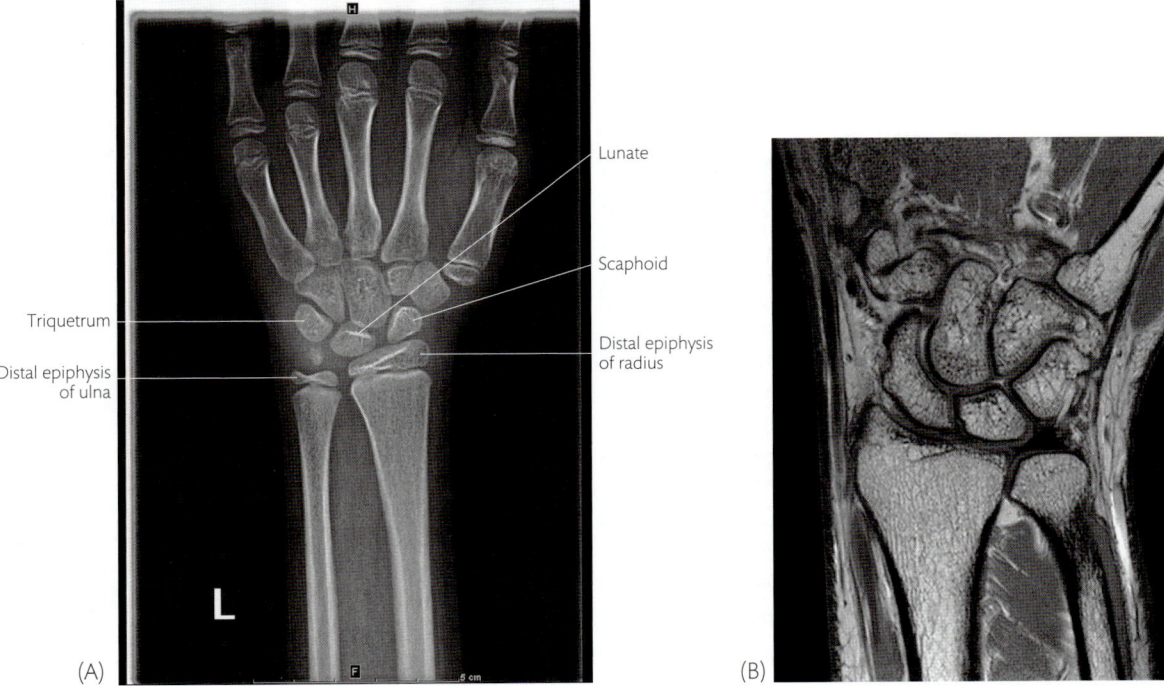

Triquetrum

Distal epiphysis of ulna

Lunate

Scaphoid

Distal epiphysis of radius

(A)

(B)

Fig. 11.9 (A) Anteroposterior radiograph of the wrist of a 12-year-old child. (B) Magnetic resonance imaging of the lower end of the forearm and wrist. See Fig. 11.5B for labels.

The wrist, intercarpal, and intermetacarpal joints are visualized in the living by X-rays and magnetic resonance imaging. Study Figs 11.9A and B, and compare them with Fig. 11.5.

Metacarpophalangeal joints

Type and articulating surfaces

They are condyloid joints in which the convex head of the metacarpal fits into the shallow fossa on the base of the proximal phalanx [see Fig. 4.17]. In the thumb, the metacarpal surface is flatter and less extensive than in the other metacarpals, for the range of movements is smaller in the thumb.

Fibrous capsules of ligaments of the joints

The fibrous capsule is thickened anteriorly to form the palmar ligament, and on each side to form a collateral ligament. On the dorsal surface, it is replaced by the extensor expansion, a mechanism which allows for full range of flexion with continuous support for the dorsal surfaces of the joints.

The **collateral ligaments** are strong, oblique bands which extend from the dorsal surface of the metacarpal heads to the sides of the base of the proximal phalanx anteriorly [see Fig. 10.13]. The **palmar ligament** is a thick, fibrous plate attached firmly to the base of the proximal phalanx, but loosely to the neck of the metacarpal. Its position varies with movements of the metacarpophalangeal joint. In full flexion, the plate lies on the palmar surface of the body of the metacarpus. It moves on to the palmar surface of the head when the joint is straightened and on to its distal surface in full extension. In the medial four digits, the margins of the palmar plate give attachment to the: (1) **deep transverse metacarpal ligaments**; (2) **fibrous flexor sheath**; (3) slips of the **palmar aponeurosis**; (4) **collateral ligaments**; and (5) transverse fibres of the **extensor expansion**. Thus, the palmar aponeurosis and transverse fibres of the extensor expansion are tightened when the plate is drawn distally in full extension. This holds the extensor digitorum tendon in place and prevents it from extending the interphalangeal joints when the metacarpophalangeal joint is extended. When the metacarpophalangeal joints are straight or flexed, the extensor digitorum can extend the interphalangeal joints.

Sesamoid bones

A small, oval sesamoid bone is buried on each side of the palmar ligament of the metacarpophalangeal joint of the thumb where the tendons of the adductor pollicis and flexor pollicis brevis fuse with the ligament. The tendon of the flexor pollicis longus lies in the groove between them. Smaller sesamoid bones may be found in the palmar ligaments of the other joints, particularly in the index and little fingers.

Movements at metacarpophalangeal joints

Based on the shape of the joint surfaces, flexion, extension, abduction, and adduction take place at the metacarpophalangeal joint. (Some amount of passive rotation is possible, but no voluntary rotation.) Abduction and adduction movements of the fingers are possible when the joint is extended and the collateral ligaments are relaxed. These movements are severely restricted in flexion because of the tightening of the collateral ligaments. Because the metacarpals are arranged in an arc convex dorsally, the fingers converge on flexion and diverge on extension.

In **precision movements**, flexion of the metacarpophalangeal joints of the medial four digits is produced mainly by the interossei and lumbricals. Extension of these joints is produced by the extensor digitorum and by the extensor indicis and extensor digiti minimi in the index and little fingers, respectively. The extensors of the index and little fingers permit isolated extension of the index and little fingers, respectively, when the middle and ring fingers are flexed at the metacarpophalangeal joints.

In the **metacarpophalangeal joint of the thumb**, the range of movements is much less in all directions. Flexion is produced by the flexor pollicis brevis and longus, abduction and adduction by the short abductor, adductor, and first palmar interossei, and extension by the long and short extensor muscles.

Interphalangeal joints

Type

Synovial, hinge joint.

Movements of the interphalangeal joints

The articular surfaces of the interphalangeal joints only permit flexion and extension (no abduction and adduction) [see Fig. 4.17].

The flexor digitorum profundus flexes all the joints of the fingers. The flexor digitorum superficialis acts only on the proximal interphalangeal and metacarpophalangeal joints. Extension at the interphalangeal joints is produced by the long extensors of the digits and also by the interosseous and lumbrical muscles acting through the extensor expansion. When the metacarpophalangeal joints of the fingers are fully extended, the long extensors are unable to act on the interphalangeal joints, which can then only be extended by the lumbricals and interossei. The **lumbricals** and **interossei** also flex the metacarpophalangeal joints. ⮞ When the lumbricals and interossei are paralysed, the long extensor muscles produce full extension of the metacarpophalangeal joint, and the long flexor muscles of the fingers act unopposed to flex the interphalangeal joints and produce the '**claw hand**' which is characteristic of this paralysis.

In the thumb, the single interphalangeal joint is acted upon by the flexor and extensor pollicis longus, and sometimes by the extensor pollicis brevis. The abductor pollicis brevis (supplied by the median nerve) may be partly inserted into the extensor expansion and able to produce extension. ⮞ Thus, some extension of the interphalangeal joint of the thumb may still be possible when all the extensor muscles of the thumb are paralysed by damage to the radial nerve.

The muscles acting on, and the movements of, the fingers and thumb are summarized in Tables 11.7, 11.8, 11.9, and 11.10. In these tables, the following abbreviations are used: CM = carpometacarpal joint; DIP = distal interphalangeal joint; IP = interphalangeal joint; MP = metacarpophalangeal joint; PIP = proximal interphalangeal joint.

See Clinical Application 11.1 for a discussion on the claw hand.

Table 11.7 Muscles acting on the fingers

Muscle	Origin	Insertion	Action on fingers	Nerve supply
Flexor digitorum superficialis	Humerus, medial epicondyle. Radius, anterior border	Middle phalanges	Flexion, MP, and PIP	Median
Flexor digitorum profundus	Ulna, proximal two-thirds, anterior and medial surfaces	Distal phalanges	Flexion, all joints	Median* and ulnar
Lumbricals	Tendons of flexor digitorum profundus	Middle and distal phalanges via extensor expansion	Flexion MP, extension IP	Median and ulnar
Extensor digitorum	Humerus, lateral epicondyle	Extensor expansion	Extends all joints of all fingers. If MP is fully extended, IP is not extended	Radial*
Dorsal interossei	Adjacent sides, metacarpals 1–5	Corresponding proximal phalanx, base through extensor expansion	Abduction of MP of index, middle, and ring fingers. Extension IP	Ulnar
Palmar interossei	Metacarpals, medial side 1 and 2, lateral side 4 and 5	Corresponding proximal phalanx, base through extensor expansion	Adduction of MP of thumb, index, ring, and little fingers. Extension IP	Ulnar
Abductor digiti minimi	Pisiform. Flexor retinaculum	Proximal phalanx, base medial side	Abduction MP (and CM) little finger	Ulnar
Flexor digiti minimi	Hamate, hook. Flexor retinaculum	Proximal phalanx, base medial side	Flexion CM and MP	Ulnar
Opponens digiti minimi	Hamate, hook	Fifth metacarpal, medial side	Lateral rotation of metacarpal. Flexion CM	Ulnar
Extensor indicis	Ulna, posterior surface	Extensor expansion	Extension of all joints of index finger	Radial*
Extensor digiti minimi	Humerus, lateral epicondyle	Extensor expansion	Extension of all joints of little finger. Abduction MP	Radial*

* Anterior or posterior interosseous branch.

Table 11.8 Movements of fingers

Movement	Muscles	Nerve supply
Flexion		
All fingers All joints: MP, PIP, DIP	Flexor digitorum profundus	Median* (index and middle) Ulnar (ring and little)
MP and PIP	Flexor digitorum superficialis	Median
MP only	Lumbricals	Median (index and middle)
	Lumbricals	Ulnar (ring and little)
	Interossei	Ulnar
CM and MP, little finger	Flexor digiti minimi	Ulnar
CM only, little finger	Opponens digiti minimi	Ulnar
Extension		
All fingers All joints: MP, PIP, DIP	Extensor digitorum	Radial*
Index: MP, PIP, DIP	Extensor indicis	Radial*
Little finger: MP, PIP, DIP	Extensor digiti minimi	Radial*
IP only when MP fully extended	Lumbricals	Median (index and middle) Ulnar (ring and little)
	Interossei	Ulnar
Abduction at MP		
All fingers, except little finger	Dorsal interossei	Ulnar
Little finger	Abductor digiti minimi	Ulnar
Adduction at MP		
All fingers, except middle	Palmar interossei	Ulnar
Opposition at CM of little finger	Opponens digiti minimi	Ulnar

* Anterior or posterior interosseous branch.

Table 11.9 Muscles acting on the thumb

Muscle	Origin	Insertion	Action on thumb	Nerve supply
Flexor pollicis longus	Radius, anterior surface middle two-quarters	Distal phalanx, base	Flexion all joints	Median*
Flexor pollicis brevis	Trapezium, tubercle. Flexor retinaculum	Proximal phalanx, base	Flexion CM and MP	Median
Abductor pollicis brevis	Scaphoid, tubercle. Flexor retinaculum	Proximal phalanx, anterior aspect	Abduction CM and MP	Median
Opponens pollicis	Trapezium, tubercle. Flexor retinaculum	Metacarpal, anterior surface	Medial rotation and flexion of CM	Median
Abductor pollicis longus	Radius and ulna, dorsal surfaces distal to supinator	Metacarpal base, anterior aspect	Abduction of CM, some extension	Radial*
First palmar interosseous	First metacarpal, base	Proximal phalanx base	Adduction of MP	Ulnar
Adductor pollicis	Metacarpal, base of 2 and 3, body of 3	Proximal phalanx base, posterior aspect	Adduction CM and MP	Ulnar
Extensor pollicis longus	Ulna, posterior surface, middle third	Distal phalanx	Extension of all joints, especially with CM laterally rotated	Radial*
Extensor pollicis brevis	Radius, posterior surface	Proximal (and distal) phalanx base	Extension of CM, MP (and IP), especially when thumb opposed	Radial*

* Anterior or posterior interosseous branch.

Table 11.10 Movements of the thumb

Movement	Muscles	Nerve supply
Flexion		
All joints	Flexor pollicis longus	Median*
CM and MP	Flexor pollicis brevis	Median
CM only	Opponens pollicis	Median
Extension		
All joints	Extensor pollicis longus	Radial*
CM and MP (IP)	Extensor pollicis brevis	Radial*
CM only	Abductor pollicis longus	Radial*
IP only	Abductor pollicis brevis	Median
Abduction		
CM only	Abductor pollicis longus	Radial*
CM and MP	Abductor pollicis brevis	Median
Adduction		
CM and MP	Adductor pollicis	Ulnar
MP only	First palmar interosseous	Ulnar
Opposition		
Medial rotation, CM	Opponens pollicis	Median

* Anterior or posterior interosseous branch.

CLINICAL APPLICATION 11.1 Ulnar claw

The claw hand is an abnormal hand position that develops due to damage to the ulnar and/or median nerves. The affected fingers are hyperextended at the metacarpophalangeal joints, and flexed at the distal and proximal interphalangeal joints. The primary cause of this deformity is paralysis of the lumbricals and interossei which normally flex the metacarpophalangeal joint and extend the interphalangeal joints. When they are paralysed, the extensor action of the long extensors on the metacarpophalangeal joint and the flexor action of the long flexors on the interphalangeal joint are unopposed. Patients with a claw hand will be unable to abduct and adduct their fingers (due to paralysis of the interossei).

An ulnar claw results from a lesion in the ulnar nerve in the hand. The third and fourth lumbricals are paralysed, resulting in clawing of the fourth and fifth fingers. As the ulnar nerve also supplies the interossei, they too are paralysed. The lumbricals of the index and middle fingers are not affected, and clawing of these fingers is not seen (even though the interossei are paralysed).

A paradoxical condition is seen when the ulnar nerve is damaged at the elbow. The effects of the lumbrical paralysis are unchanged. But because, in this condition, the medial half of the flexor digitorum profundus is also denervated, flexion of the interphalangeal joints of the ring and little fingers is weak. The claw-like appearance of the hand is reduced (and not worsened, as one would expect from a higher-level injury). As reinnervation and healing occur along the ulnar nerve after a high lesion, the claw hand deformity will get worse as the patient recovers. The claw hand can be demonstrated in yourself. Fully extend your fingers at all joints, and note the taut extensor tendons on the back of your hand. Keeping the metacarpophalangeal joints fully extended, flex your interphalangeal joints, and note that this can be done without any movement of the extensor tendons.

CHAPTER 12
The nerves and nerve injuries of the upper limb

Effects of nerve injury

An **upper limb neurological examination** is part of the general neurological examination done to assess the integrity of motor and sensory nerves of the upper limb. Most peripheral nerves of the upper limb are mixed nerves, and damage to one would result in both sensory and motor loss. In addition, loss of sympathetic innervation would result in changes in skin texture, absence or decreased sweating, and inability to regulate blood flow in response to changes in temperature.

Sensory loss after nerve injury

Sensory loss is experienced in the skin supplied by the damaged nerve. For example, damage to the axillary nerve results in sensory loss over the lateral part of the upper arm. As cutaneous nerves supplying adjacent areas of skin overlap to a considerable degree, total destruction of one nerve produces an area of complete sensory loss much smaller than the sum of the areas supplied by its individual branches. The areas of skin supplied by cutaneous branches of the medial cord, axillary, median, ulnar, and radial nerves are shown in Fig. 12.1.

Fig. 12.1 shows the cutaneous distribution of the main nerves of the upper limb.

Motor loss after nerve injury

Injury to the motor nerve will result in paralysis of the muscles supplied by it and an inability to move the joint on which the muscles act. For example, injury to the musculocutaneous nerve will paralyse the biceps brachii and brachialis, and make flexion of the elbow difficult or impossible. If only the biceps and brachialis are paralysed, some flexion of the elbow will be possible by other muscles such as the brachioradialis and pronator teres. If the only muscle moving a joint in a certain direction is paralysed, the loss of that movement will be total. For example, paralysis of the flexor digitorum profundus will result in total inability to flex the distal interphalangeal joint of the fingers.

Where a nerve innervates muscles in more than one segment of the limb (shoulder, arm, forearm, hand), the effects of injury to the nerve depend on the level of injury. For example, when the median nerve is damaged at the wrist, muscles supplied by it in the forearm are not paralysed—only those in the hand are.

Injuries to the brachial plexus

Most injuries to the brachial plexus affect either the upper trunk or the lower trunk.

Upper plexus palsies—Erb–Duchenne's palsy

Injury to the upper trunk of the brachial plexus can occur during any forceful stretching of the angle between the shoulder and the neck—as in a difficult vaginal delivery or in a forceful fall on the shoulder. Stretching can damage the C.5 or C.6 root, or the upper trunk (C.5, C.6) of the brachial plexus. (The upper trunk of the brachial plexus is formed by union of the ventral rami of C.5 and C.6. The

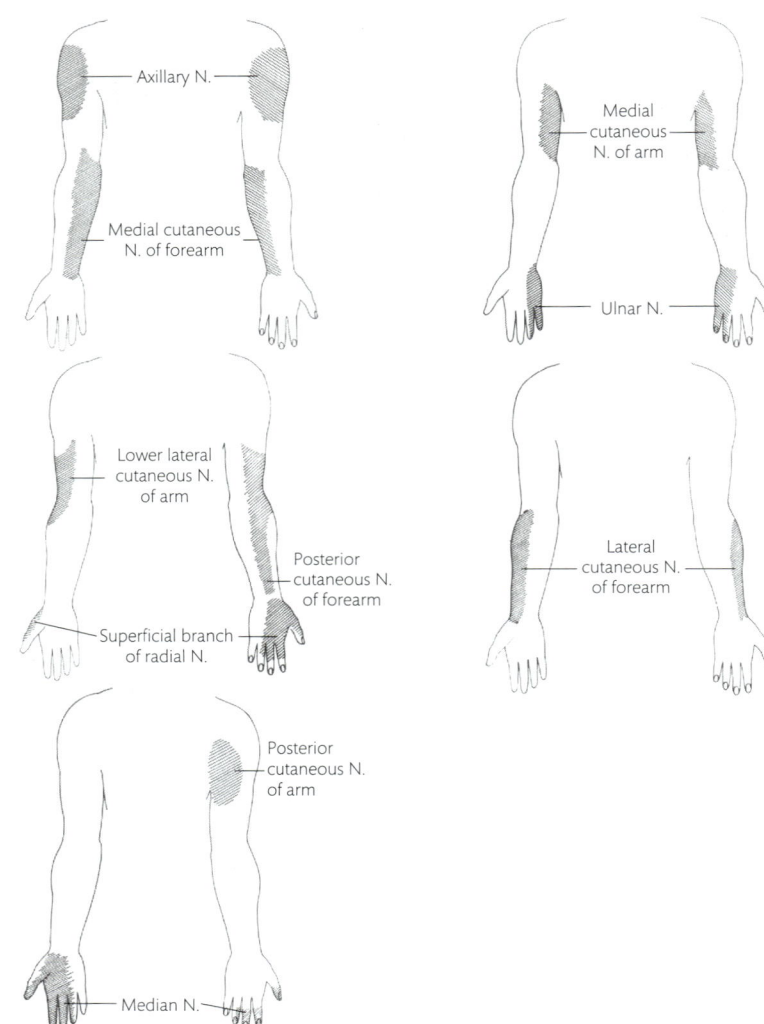

Fig. 12.1 Cutaneous distribution of nerves in the upper limb.

trunk gives off the nerve to the subclavius and the suprascapular nerve, and divides into anterior and posterior divisions). Such an injury will result in the paralysis of muscles supplied by: (1) the axillary nerve—deltoid and teres minor; (2) the suprascapular nerve—supraspinatus and infraspinatus; and (3) the musculocutaneous nerve—biceps brachii and brachialis. As a result, abduction at the shoulder, flexion at the elbow, and supination at the superior radioulnar joint are lost and the arm of the patient is adducted, with the elbow extended and the hand pronated—like that of a waiter hinting for a tip. There is also sensory loss to the skin on the lateral side of the arm. See Table 12.1.

Table 12.1 Effects of injury to the upper trunk

Joint involved	Movement lost or weakened	Explanation for loss/weakness of movement	Deformity produced
Effect on shoulder joint	Abduction Lateral rotation	Deltoid, supraspinatus and infraspinatus are paralysed	Arm is adducted and medially rotated
Effect on elbow joint	Flexion	Biceps brachii and brachialis are paralysed	Forearm is extended
Effect on radio-ulnar joints	Supination	Biceps brachii is paralysed	Forearm is pronated

Lower plexus palsies—Klumpke's palsy

Forced upward traction on the arm may tear the lower root T. 1 of the brachial plexus. The lower roots can also be affected by malignant infiltration from an apical lung tumour or by angulation of the nerve or nerves over a cervical rib. The intrinsic muscles of the hand—lumbricals and interossei—are paralysed and the hand becomes 'clawed' (extended at the metacarpophalangeal joints, flexed at the interphalangeal joints) due to unopposed action of the long flexors and extensors of the fingers. There is loss of sensation along the medial aspect of the forearm [see Fig. 7.4 for area of skin supplied by T. 1]. There can be associated Horner's syndrome (a condition resulting from loss of sympathetic supply) due to traction on the cervical sympathetic chain.

Note: for all tables in this chapter, the fingers of the hand are simply indicated by their name (e.g. 'index' for 'index finger'). CM = carpometacarpal; DIP = distal interphalangeal; IP = interphalangeal; MP = metacarpophalangeal; PIP = proximal interphalangeal.

Median nerve

See Fig. 12.2 for an overview of the median nerve. Table 12.2 shows the effects of injury to the median nerve.

Ulnar nerve

See Fig. 12.3 for an overview of the ulnar nerve. Table 12.3 shows the effects of injury to the ulnar nerve.

Musculocutaneous nerve

See Fig. 12.4 for an overview of the musculocutaneous nerve. Table 12.4 shows the effects of injury to the musculocutaneous nerve.

Axillary nerve

See Fig. 12.5 for an overview of the axillary nerve. Table 12.5 shows the effects of injury to the axillary nerve.

Subscapular nerve

Table 12.6 shows the effects of injury to the subscapular nerve.

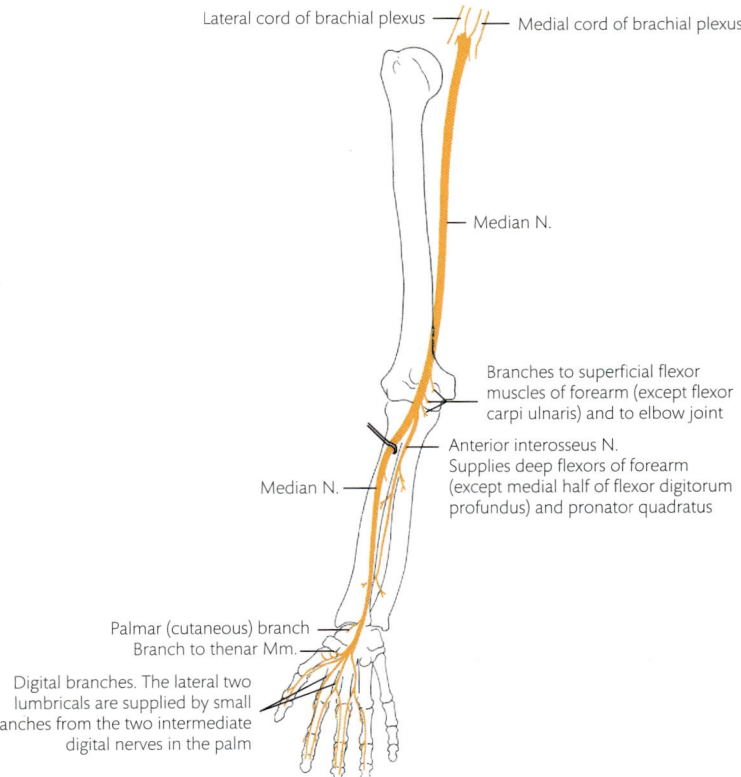

Lateral cord of brachial plexus — Medial cord of brachial plexus

Median N.

Branches to superficial flexor muscles of forearm (except flexor carpi ulnaris) and to elbow joint

Anterior interosseus N. Supplies deep flexors of forearm (except medial half of flexor digitorum profundus) and pronator quadratus

Median N.

Palmar (cutaneous) branch
Branch to thenar Mm.

Digital branches. The lateral two lumbricals are supplied by small branches from the two intermediate digital nerves in the palm

Fig. 12.2 Course and distribution of the median nerve.

Table 12.2 Effects of injury to the median nerve

Joint involved	Movement affected/deformity produced	Explanation for loss/weakness of movement
Effect on shoulder joint	None	None of the muscles that move the shoulder joint are supplied by the median nerve
Effect on elbow joint	Flexion, minimal weakness	Pronator teres and flexor carpi radialis are paralysed (important flexors—the biceps and brachialis are not supplied by the median nerve)
Effect on radioulnar joints	Pronation lost	Pronator teres and pronator quadratus are paralysed
Effect on wrist joint	Flexion weakened	Flexor carpi radialis, palmaris longus, flexor digitorum superficialis, flexor pollicis longus, and part of flexor digitorum profundus are paralysed
	Ulnar deviation	Unopposed action of flexor carpi ulnaris (as flexor carpi radialis is paralysed)
Effect on thumb movements	Flexion of IP joint lost	Flexor pollicis longus (only flexor) is paralysed
	Flexion of CM and MP joints weakened	Flexor pollicis longus is paralysed. Weak movement brought about by adductor pollicis
	Abduction of CM joint weakened	Abductor pollicis brevis is paralysed. Weak movement brought about by abductor pollicis longus
	Opposition lost	Opponens pollicis is paralysed
Effect on MP joints of fingers	Flexion weakened	Flexor digitorum superficialis and flexor digitorum profundus in lateral two fingers are paralysed. Weak flexion brought about by interossei (all fingers), medial two lumbricals (medial two fingers), and flexor digiti minimi (little finger)
Effect on PIP joints of lateral two fingers	Flexion lost	Flexor digitorum superficialis and profundus are paralysed
	Extension weakened	Lumbricals are paralysed. Weak extension brought about by extensor digitorum and interossei
Effect on PIP joints of medial two fingers	Flexion weakened	Flexor digitorum superficialis is paralysed. Weak flexion is brought about by flexor digitorum profundus
Effect on DIP joints of lateral two fingers	Flexion lost	Flexor digitorum profundus is paralysed
Effect on DIP joints of medial two fingers	None	Flexor digitorum profundus is uninvolved

Thoracodorsal nerve

Table 12.7 shows the effects of injury to the thoracodorsal nerve.

Radial nerve

See Fig. 12.5 for an overview of the radial nerve. Table 12.8 shows the effects of injury to the radial nerve.

Suprascapular nerve

Table 12.9 shows the effects of injury to the suprascapular nerve.

Long thoracic nerve

Table 12.10 shows the effects of injury to the long thoracic nerve.

Clinical Applications 12.1 and 12.2 (and Table 12.11) explore the practical application of this knowledge.

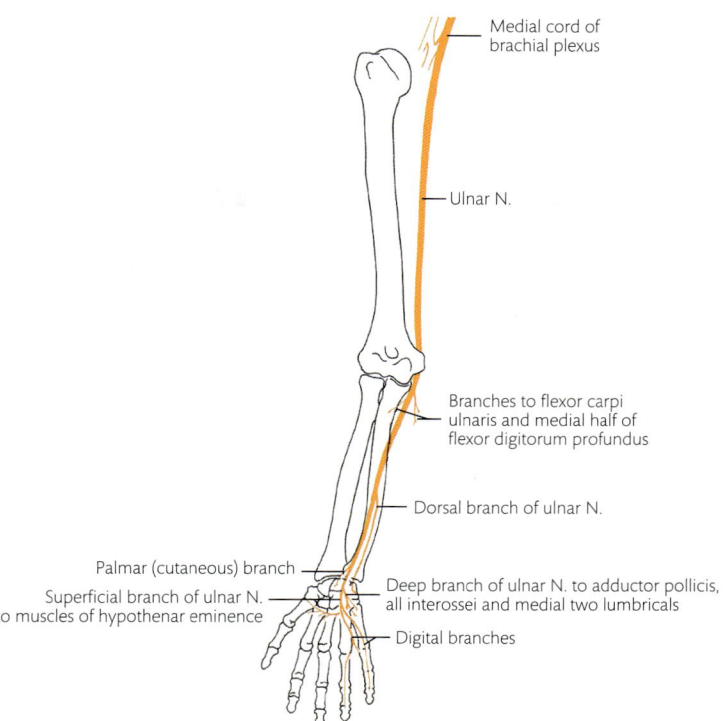

Fig. 12.3 Course and distribution of the ulnar nerve.

Table 12.3 Effects of injury to the ulnar nerve

Joint involved	Movement affected/deformity produced	Explanation for loss/weakness of movement
Effect on shoulder joint	None	None of the muscles that move the shoulder joint are supplied by the ulnar nerve
Effect on elbow joint	None	None of the muscles that move the elbow joint are supplied by the ulnar nerve
Effect on wrist joint	Weakened wrist flexion	Flexor carpi ulnaris and part of flexor digitorum profundus are paralysed
	Radial deviation	Unopposed action of flexor carpi radialis
Effect on radioulnar joints	None	None of the muscles that move the radioulnar joints are supplied by the ulnar nerve
Effect on thumb movement	None	Although adductor pollicis is paralysed, the long flexor and extensor of the thumb together mimic the action of the adductor
Effect on CM joints of little finger	Opposition is lost	Opponens digiti minimi is paralysed
Effect on MP joints of all fingers (medial four digits)	Abduction and adduction lost	All interossei and abductor digiti minimi are paralysed
Effect on MP joints of medial two fingers	Flexion weakened	Flexor digitorum profundus, lumbricals, and flexor digiti minimi are paralysed. Weak flexion is brought about by flexor digitorum superficialis
Effect on PIP joints of medial two fingers	Flexion weakened	Flexor digitorum profundus is paralysed. Weak flexion is brought about by flexor digitorum superficialis
Effect on PIP joints of all fingers	Extension weakened in all fingers	Interossei are paralysed. Lumbricals of medial two fingers are paralysed
	No IP joint extension of medial two fingers if MP joint is fully extended	Extension possible only when extensor digitorum, extensor indicis, and extensor digiti minimi are not extending the MP joint
Effect on DIP joints of medial two fingers	Flexion lost	Flexor digitorum profundus is paralysed

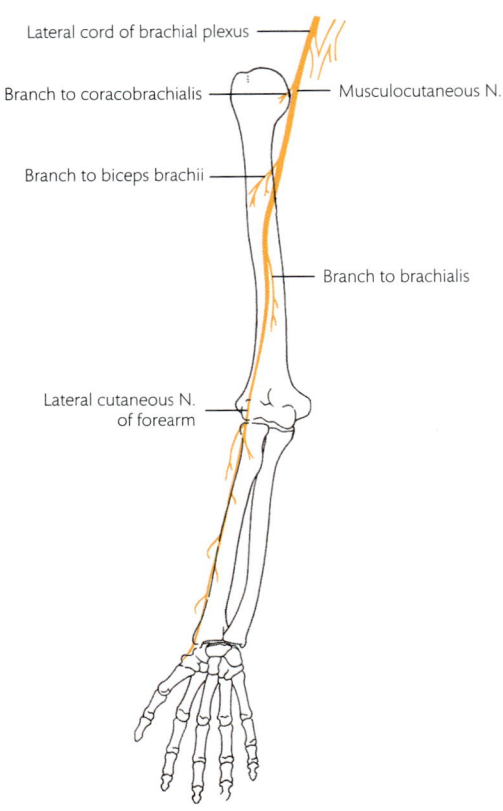

Fig. 12.4 Course and distribution of the musculocutaneous nerve.

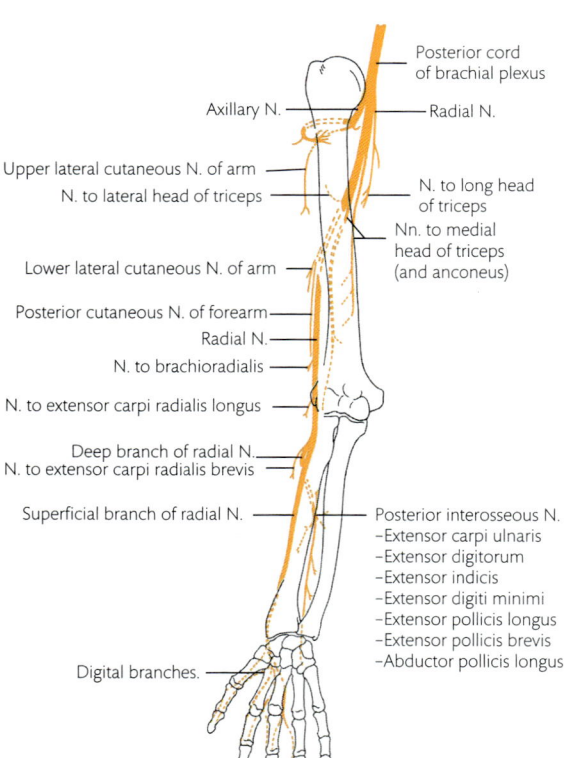

Fig. 12.5 Course and distribution of the axillary and radial nerves.

Table 12.4 Effects of injury to the musculocutaneous nerve

Joint involved	Movement affected	Explanation for loss/weakness of movement
Effect on shoulder joint	Flexion weakened	Coracobrachialis and short head of biceps are paralysed. Weak flexion brought about by deltoid and pectoralis major
	Stability in abduction lost	Long head of biceps is paralysed. Some stability maintained by deltoid, supraspinatus, and subscapularis
Effect on elbow joint	Flexion severely weakened	Biceps brachii and brachialis are paralysed. Some weak flexion is brought about by brachioradialis, extensor carpi radialis longus, pronator teres, and flexor carpi radialis
Effect on radioulnar joints	Supination weakened	Biceps brachii paralysed. Weak supination is brought about by supinator and brachioradialis

Table 12.5 Effects of injury to the axillary nerve

Joint involved	Movement affected	Explanation for loss/weakness of movement
Effect on shoulder joint	Abduction severely weakened	Teres minor and deltoid are paralysed. Weak abduction is brought about by supraspinatus
	Extension severely weakened	Deltoid and teres minor are paralysed. Weak extension is brought about by latissimus dorsi
	Lateral rotation of humerus weakened	Teres minor is paralysed. Weak lateral rotation is brought about by infraspinatus

Table 12.6 Effects of injury to the subscapular nerve

Joint involved	Movement affected	Explanation for loss/weakness of movement
Effect on shoulder joint	Instability and tendency for anterior dislocation	Subscapularis is paralysed
	Medial rotation of humerus weakened	Teres major is paralysed. Weak medial rotation is brought about by pectoralis major and deltoid

Table 12.7 Effects of injury to the thoracodorsal nerve

Joint involved	Movement affected	Explanation for loss/weakness of movement
Effect on shoulder joint	Medial rotation of humerus is weakened	Latissimus dorsi is paralysed
	Inability to pull the body upwards with the upper limb	Latissimus dorsi is paralysed

Table 12.8 Effects of injury to the radial nerve

Joint involved	Movement affected	Explanation for loss/weakness of movement
Effect on shoulder joint	Minor instability of shoulder in abduction, with tendency for downward dislocation in this position	Long head of triceps brachii is paralysed
Effect on elbow joint	Extension lost	Triceps is paralysed
Effect on radioulnar joint	Supination weakened	Supinator is paralysed. Weak supination is brought about by biceps brachii
	Elbow flexion in mid-prone position weakened	Brachioradialis, and extensor carpi radialis longus and brevis are paralysed. Weak movement is brought about by brachialis, biceps brachii, and pronator teres
Effect on wrist joint	Markedly weakened radial deviation	Extensor carpi radialis longus and brevis are paralysed. Weak radial deviation brought about by flexor carpi radialis
	Wrist extension is lost—'wrist drop'	Extensor carpi ulnaris* and extensor digitorum* are paralysed
	Weakened ulnar deviation of wrist	Extensor carpi ulnaris* is paralysed. Weak ulnar deviation is brought about by flexor carpi ulnaris
Effect on MP and IP joints	Extension lost at MP joints Extension weakened at IP joints	Extensor pollicis longus*, extensor pollicis brevis*, and abductor pollicis longus* are paralysed. Weak extension of IP joints brought about by interossei and lumbricals
Effect on MP joints—index	Independent extension lost	Extensor indicis* is paralysed
Effect on MP joints—little	Independent extension lost	Extensor digiti minimi* is paralysed
Effect on CM, MP, and IP joints of thumb	Extension is lost—thumb	Extensor pollicis longus* and extensor pollicis brevis* are paralysed. Some extension is brought about by abductor pollicis brevis
	Thumb abduction is weakened	Abductor pollicis longus* is paralysed. Weak abduction is brought about by abductor pollicis brevis

* Posterior interosseous branch.

Table 12.9 Effects of injury to the suprascapular nerve

Joint involved	Movement affected	Explanation for loss/weakness of movement
Effect on shoulder joint	Difficulty with initiating abduction	Supraspinatus is paralysed. Abduction brought about by deltoid (teres minor and subscapularis assist the deltoid by holding down the humeral head)
	Lateral rotation of humerus weakened	Infraspinatus is paralysed. Some lateral rotation is brought about by posterior fibres of the deltoid and teres minor

Table 12.10 Effects of injury to the long thoracic nerve

Joint involved	Movement affected/deformity produced	Explanation for loss/weakness of movement
Effect on shoulder girdle	Protraction of scapula weakened	Serratus anterior is paralysed. Some protraction is brought about by pectoralis major and minor
	Lateral rotation of scapula weakened	Serratus anterior is paralysed. Some lateral rotation is brought about by trapezius
	Scapula not held against ribs—'winged scapula'	

Table 12.11 Motor assessment of upper limb musculature

Spinal segment (myotome)	Primary movement	Prime muscle causing movement
C.5	Elbow flexion	Biceps brachii and brachialis
C.6	Wrist extension	Extensor carpi radialis longus and extensor carpi radialis brevis
C.7	Elbow extension	Triceps brachii
C.8	Finger flexion*	Flexor digitorum profundus
T.1	Finger abductors (little finger)	Abductor digiti minimi

* Distal interphalangeal joint of the middle finger.
Reference: *Standard neurological classification of spinal cord injury* by American Spinal Injury Association (ASIA): Revised 2019.

CLINICAL APPLICATION 12.1 Cervical rib syndrome

The costal element of the seventh cervical vertebra is normally incorporated into its transverse process soon after birth. Rarely, it may persist and give rise to a condition known as a 'cervical rib'. The presence of a cervical rib can cause thoracic outlet syndrome due to compression of the lower trunk of the brachial plexus. Study question 1: which part/parts of the plexus supplying the upper limb is most likely to be affected in this condition? (Answer: brachial plexus—T.1 root and the lower trunk.)

A patient with a symptomatic cervical rib could experience a dull, aching pain radiating down the medial side of the arm, forearm, and hand, with numbness in the little and ring fingers. Study question 2: name the nerves supplying the skin of these regions. What are they branches of? (Answer: (1) medial cutaneous nerve of the arm—medial cord of the brachial plexus; (2) medial cutaneous nerve of the forearm—medial cord of the brachial plexus; (3) and (4) superficial branch of the ulnar nerve and dorsal branch of the ulnar nerve—ulnar nerve.)

On examination of the patient, there was wasting of all the intrinsic muscles of the hand, except the abductor and flexor pollicis brevis. Study question 3: what is the innervation of these muscles of the hand? (Answer: ulnar nerve.)

Study question 4: would any forearm muscles be affected? If so, which ones? (Answer: yes—medial half of the flexor digitorum profundus and flexor carpi ulnaris.)

CLINICAL APPLICATION 12.2 Motor assessment of upper limb musculature

Motor examination of the upper limb in a patient with spinal cord injury provides a reliable and quick way to localize the level of the lesion. Five muscles of the upper limb, one primarily supplied by each of the five segmental nerves C.5 to T.1, are tested.

The integrity of each spinal segment is evaluated by the ability of a muscle supplied by it to bring about a particular movement of a joint [Table 12.11]. (The strength of the muscle is scored on a 5-point scale, not included here.)

CHAPTER 13
Surface marking of the upper limb

Introduction

Having studied the important structures of the upper limb, it is essential to have the knowledge and skills to mark their position on the surface of the body. This chapter deals with the surface marking of clinically important structures.

Before you begin with the study of surface marking, review the surface anatomy and palpate the bony landmarks of the pectoral region, back, arm, forearm and hand, wrist, palm, and digits, which are given in the preceding chapters.

Cephalic vein

The cephalic vein begins at the base of the first metacarpal, runs upwards on the posterolateral margin of the forearm, curves around the lateral border of the forearm, and ascends in the groove lateral to the belly of the biceps. It continues upwards in the deltopectoral groove and ends by piercing the deep fascia just below the midpoint of the clavicle. To mark it on the surface of the body, first mark the bony and soft tissue landmarks shown in Fig. 13.1A and B:

- Base of the first metacarpal in the anatomical snuffbox (**A**)
- Lateral border of the biceps brachii in the arm (**B**)
- Deltopectoral groove in the pectoral region, just below the inferior third of the clavicle (**C**)
- Mid-clavicular point (**D**).

Join these points by a line that curves around the lateral border of the forearm.

Basilic vein

The basilic vein begins at the medial end of the dorsal venous arch. It curves around the medial border of the forearm to lie on the medial side of the front of the elbow. It then ascends in the groove medial to the belly of the biceps. It ends at the midpoint of the arm by piercing the deep fascia. To mark it on the surface of the body, first mark the bony and soft tissue landmarks shown in Fig. 13.2A and B:

- Base of the fifth metacarpal (**A**)
- Medial border of the biceps brachii in the arm (**B**).

Join these two points by drawing a line that runs along the medial border of the forearm, the anterior and medial aspects of the cubital fossa, and upwards in the middle of the arm.

Axillary artery

The axillary artery runs from the outer border of the first rib to the lower border of the axilla. To mark it on the surface of the body, first mark the bony and soft tissue landmarks shown in the abducted arm in Fig. 13.3:

- Midpoint of the clavicle (**A**)
- Junction of the anterior one-third and posterior two-thirds of a line joining the anterior and posterior axillary folds on the arm (**B**).

Join these two points by a line directed laterally and downwards.

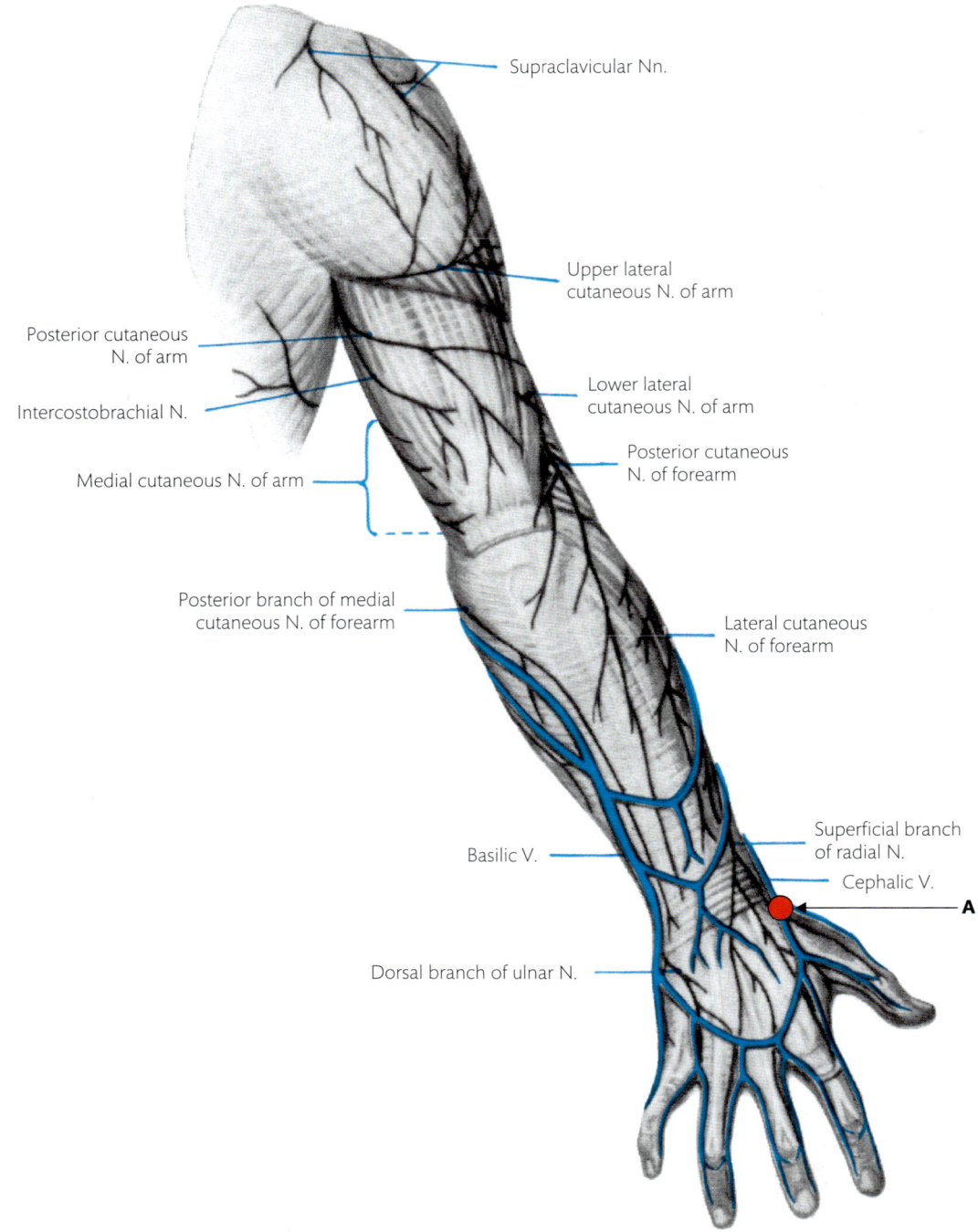

Supraclavicular Nn.

Upper lateral
cutaneous N. of arm

Posterior cutaneous
N. of arm

Intercostobrachial N.

Lower lateral
cutaneous N. of arm

Posterior cutaneous
N. of forearm

Medial cutaneous N. of arm

Posterior branch of medial
cutaneous N. of forearm

Lateral cutaneous
N. of forearm

Superficial branch
of radial N.

Basilic V.

Cephalic V.

A

Dorsal branch of ulnar N.

Fig. 13.1 Bony and soft tissue landmarks for marking the cephalic vein.

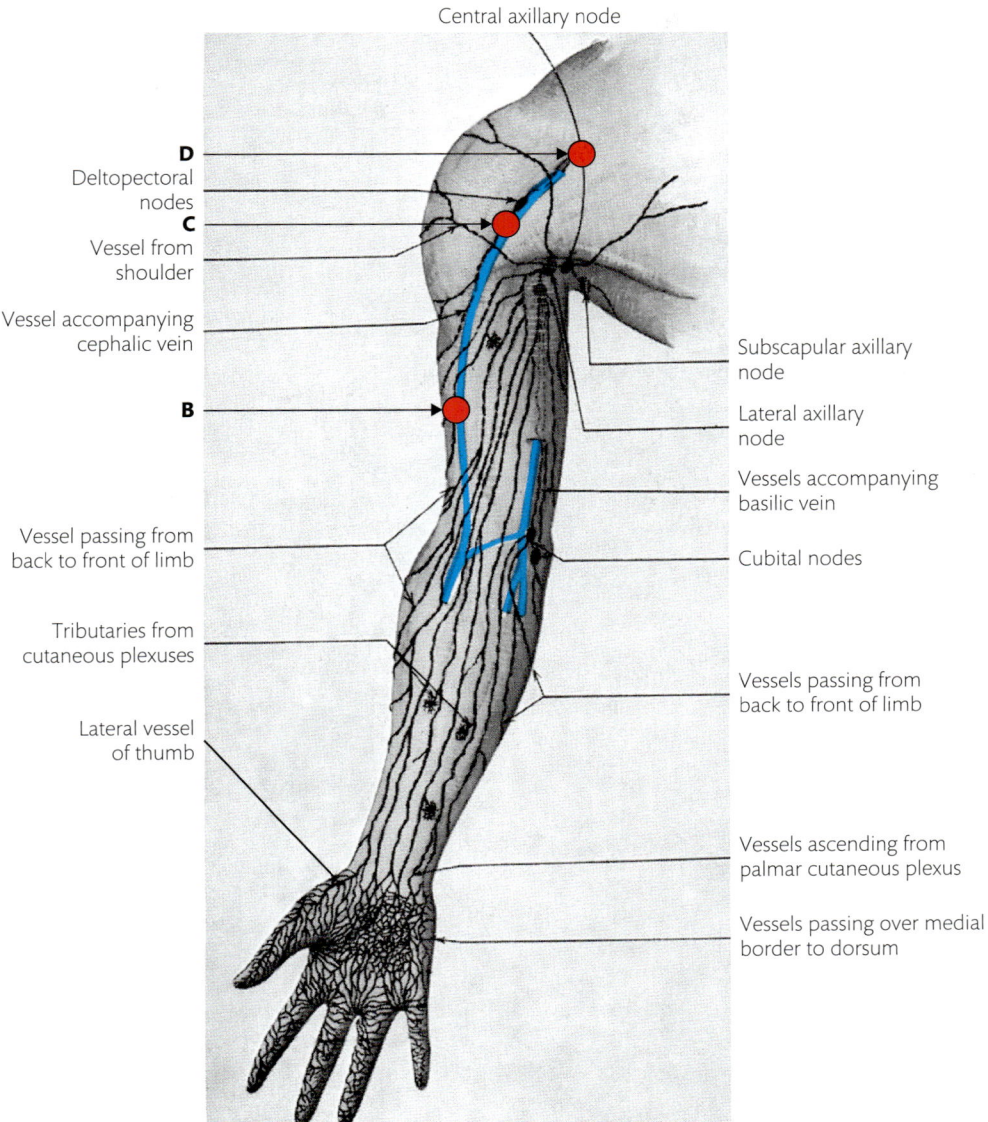

Central axillary node

D
Deltopectoral nodes
C
Vessel from shoulder

Vessel accompanying cephalic vein

B

Vessel passing from back to front of limb

Tributaries from cutaneous plexuses

Lateral vessel of thumb

Subscapular axillary node

Lateral axillary node

Vessels accompanying basilic vein

Cubital nodes

Vessels passing from back to front of limb

Vessels ascending from palmar cutaneous plexus

Vessels passing over medial border to dorsum

Fig. 13.1 *Continued*

Brachial artery

The brachial artery extends from the lower border of the teres major to the neck of the radius.

It can be marked on the surface of the body by a line joining the points shown in Fig. 13.3:

- The junction of the anterior one-third and posterior two-thirds of a line joining the anterior and posterior axillary folds on the arm (**B**)
- At the centre of the cubital fossa on the medial side of the biceps tendon (**C**).

Radial artery

The radial artery extends from the termination of the brachial artery in the cubital fossa to the anterior aspect of the distal end of the radius. It can be marked on the surface of the body by a line joining the points shown in Fig. 13.4:

- The centre of the cubital fossa on the medial side of the biceps tendon (**A**)
- The anterior surface of the radius at the wrist (**B**).

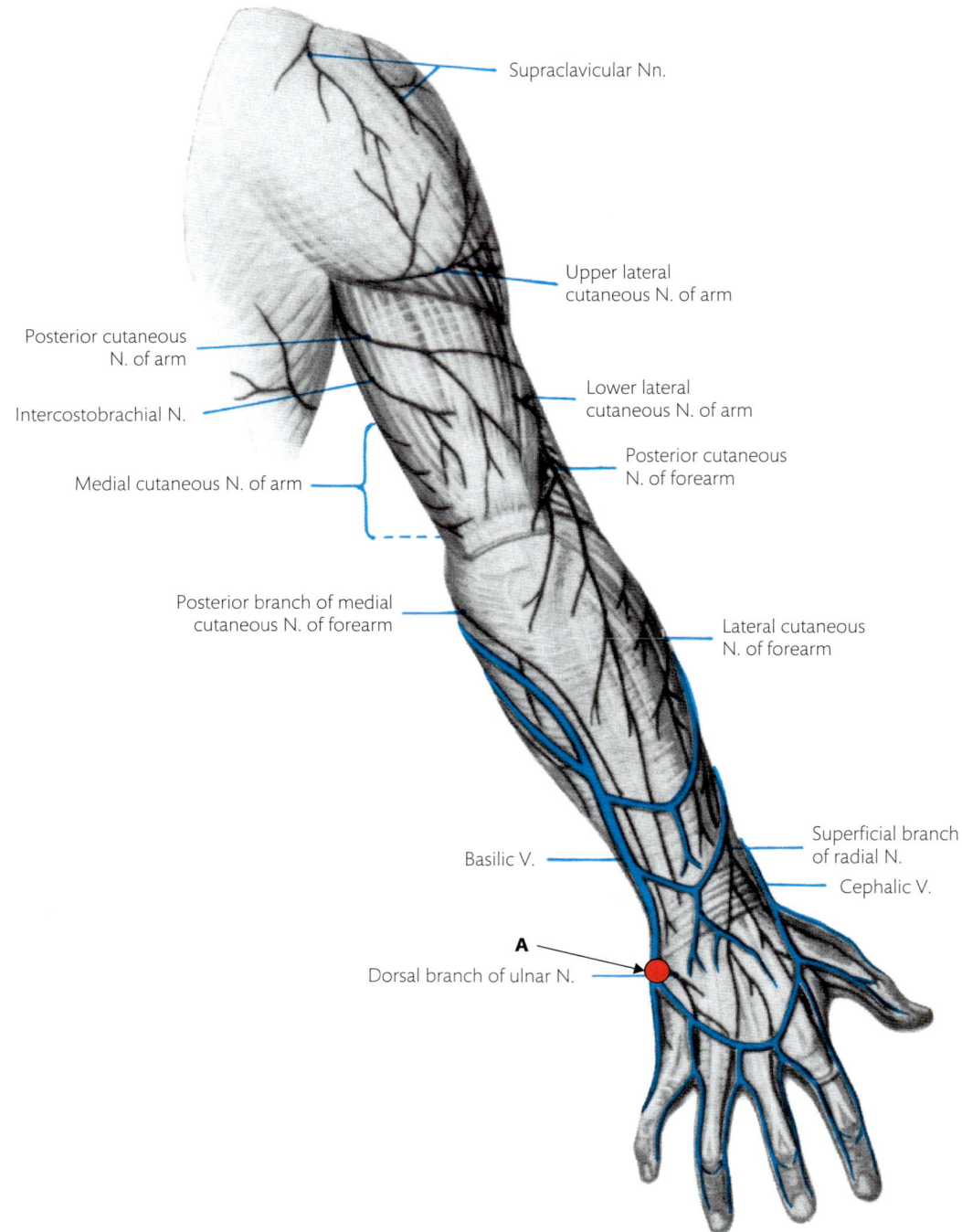

Supraclavicular Nn.

Upper lateral
cutaneous N. of arm

Posterior cutaneous
N. of arm

Intercostobrachial N.

Lower lateral
cutaneous N. of arm

Posterior cutaneous
N. of forearm

Medial cutaneous N. of arm

Posterior branch of medial
cutaneous N. of forearm

Lateral cutaneous
N. of forearm

Superficial branch
of radial N.

Basilic V.

Cephalic V.

A

Dorsal branch of ulnar N.

Fig. 13.2 Bony and soft tissue landmarks for marking the basilic vein.

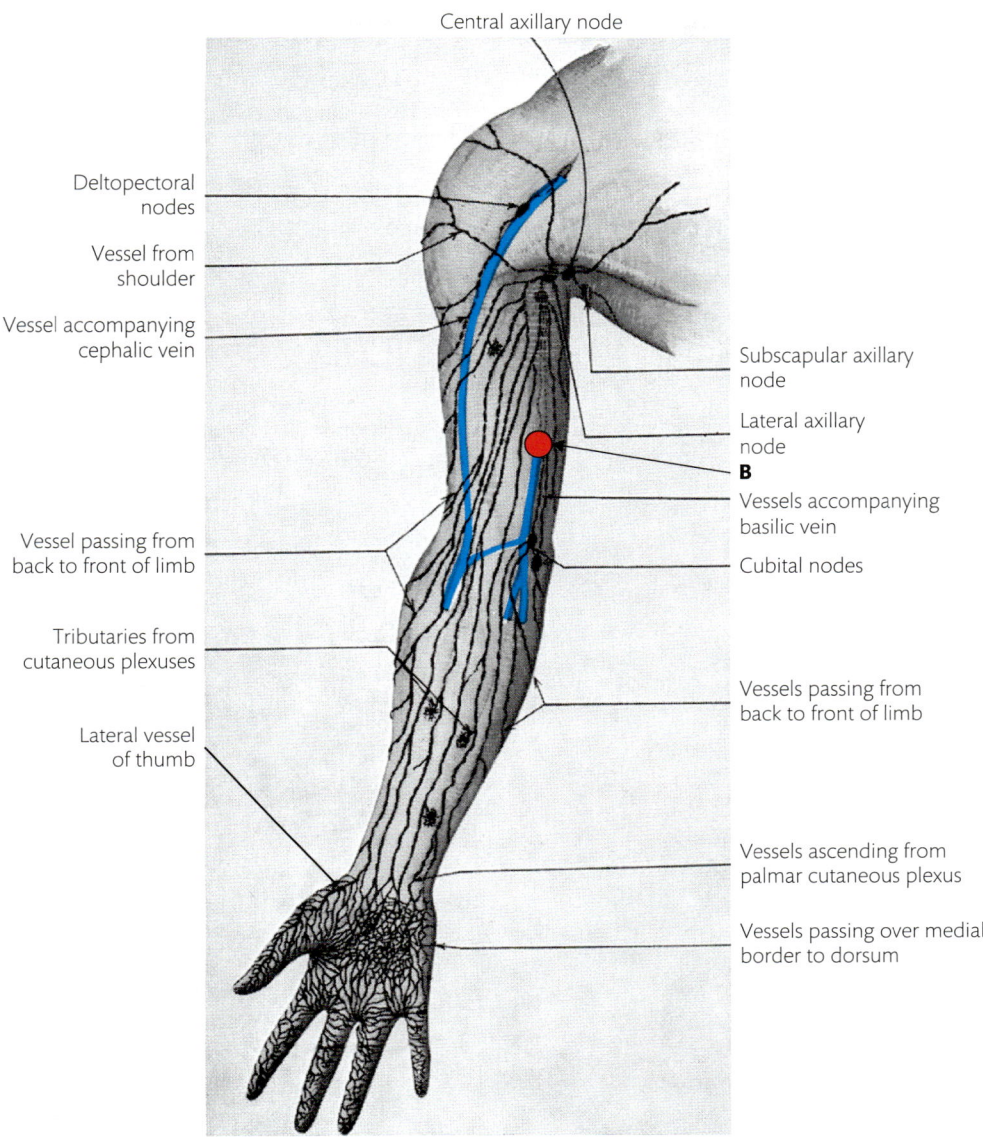

Central axillary node

Deltopectoral nodes

Vessel from shoulder

Vessel accompanying cephalic vein

Subscapular axillary node

Lateral axillary node

B

Vessels accompanying basilic vein

Cubital nodes

Vessel passing from back to front of limb

Tributaries from cutaneous plexuses

Lateral vessel of thumb

Vessels passing from back to front of limb

Vessels ascending from palmar cutaneous plexus

Vessels passing over medial border to dorsum

Fig. 13.2 *Continued*

Ulnar artery

The ulnar artery extends from the termination of the brachial artery in the cubital fossa to the hook of the hamate. It can be marked on the surface of the body by a line joining the points shown in Fig. 13.4:

- The centre of the cubital fossa on the medial side of the biceps tendon (**A**)
- Junction of the upper one-third and lower two-thirds of a line joining the medial epicondyle of the humerus and the styloid process of the ulna (**C**)
- The hook of the hamate (**D**).

Fig. 13.3 Bony and soft tissue landmarks for marking the axillary and brachial arteries.

A

B

C

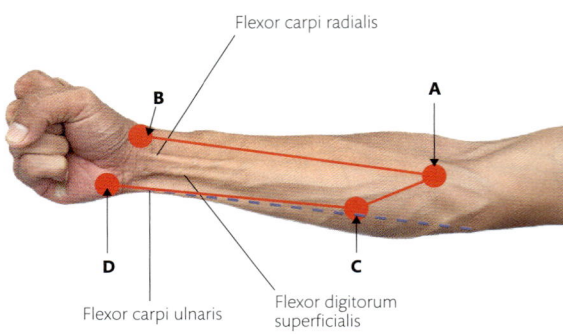

Flexor carpi radialis

Flexor carpi ulnaris

Flexor digitorum superficialis

Fig. 13.4 Bony and soft tissue landmarks for marking the radial and ulnar arteries.

Only background/Shutterstock.com.

Superficial palmar arch

The superficial palmar arch is represented on the surface by a horizontal line, 4 cm long, at the level of the fully extended thumb. See Fig. 8.10.

Deep palmar arch

The deep palmar arch is represented on the surface by a horizontal line, 4 cm long, just distal to the hook of the hamate. See Fig. 8.10.

Axillary nerve

Mark the midpoint of a line joining the acromion to the deltoid tubercle. A horizontal line drawn 2 cm above this point marks the axillary nerve.

Radial nerve

The radial nerve can be marked on the surface by a line joining the following points:

1. The lower limit of the posterior wall of the axilla

2. Junction of the upper one-third and lower two-thirds of a line joining the deltoid tubercle to the lateral epicondyle
3. Anteriorly on the lateral epicondyle of the humerus.

Note: The line joining points 1 and 2 should be drawn on the back of the arm.

Median nerve

The median nerve can be marked on the surface by a line joining the following points:

1. The junction of the anterior one-third and posterior two-thirds of a line joining the anterior and posterior axillary folds on the arm
2. At the centre of the cubital fossa on the medial side of the brachial artery
3. Middle of the front of the wrist just lateral to the tendon of the palmaris longus.

Ulnar nerve

The ulnar nerve can be marked on the surface by a line joining the following points:

1. The junction of the anterior one-third and posterior two-thirds of a line joining the anterior and posterior axillary folds on the arm
2. Back of the medial epicondyle
3. Medial side of the pisiform bone.

Flexor retinaculum

The flexor retinaculum can be marked on the surface of the hand by lines joining the four points:

1. Pisiform
2. Hook of the hamate
3. Tubercle of the scaphoid
4. Tubercle of the trapezium.

CHAPTER 14
MCQs on the upper limb

The following questions have four options. You are required to choose the most correct answer.

1. **The supraclavicular nerves supply the skin down to a horizontal line at the level of the**
   A. clavicle
   B. first costal cartilage
   C. second costal cartilage
   D. third costal cartilage

2. **The anterior axillary wall consists of the following muscles, EXCEPT the**
   A. pectoralis major
   B. pectoralis minor
   C. subclavius
   D. subscapularis

3. **The intercostobrachial nerve communicates with the**
   A. medial cutaneous nerve of the arm
   B. medial cutaneous nerve of the forearm
   C. median nerve
   D. musculocutaneous nerve

4. **The inferior angle of the scapula corresponds approximately to the level of the**
   A. sixth thoracic spine
   B. seventh thoracic spine
   C. eighth thoracic spine
   D. ninth thoracic spine

5. **Retraction of the scapula is caused by the following muscles, EXCEPT the**
   A. rhomboid major
   B. rhomboid minor
   C. upper fibres of the trapezius
   D. middle fibres of the trapezius

6. **The nerve supply of the latissimus dorsi is by the**

    A. long thoracic nerve

    B. dorsal scapular nerve

    C. suprascapular nerve

    D. thoracodorsal nerve

7. **The bones felt in the anatomical snuffbox are all, EXCEPT the**

    A. styloid process of the radius

    B. scaphoid

    C. lunate

    D. trapezium

8. **The upper lateral cutaneous nerve of the arm arises from the**

    A. musculocutaneous nerve

    B. axillary nerve

    C. radial nerve

    D. median nerve

9. **The following arteries are involved in the anastomosis around the scapula, EXCEPT the**

    A. suprascapular artery

    B. subscapular artery

    C. transverse cervical artery

    D. internal thoracic artery

10. **The axis of movement of supination and pronation passes through the centre of the head of the radius proximally and the**

    A. centre of the head of the ulna distally

    B. styloid process of the radius distally

    C. centre of the pisiform distally

    D. hook of the hamate distally

11. **The structures that pass in the carpal tunnel are all, EXCEPT the**

    A. flexor carpi ulnaris

    B. flexor digitorum superficialis

    C. flexor digitorum profundus

    D. flexor pollicis longus

12. **The anterior interosseous nerve is a branch of the**

    A. ulnar nerve

    B. median nerve

    C. radial nerve

    D. musculocutaneous nerve

13. **Carpal tunnel syndrome is caused due to compression of the**

    A. ulnar nerve
    B. median nerve
    C. anterior cutaneous nerve
    D. posterior cutaneous nerve

14. **The muscle producing adduction of the wrist is the**

    A. flexor carpi ulnaris
    B. flexor carpi radialis
    C. extensor carpi radialis longus
    D. extensor carpi radialis brevis

15. **The action of the lumbricals is**

    A. flexion of the metacarpophalangeal joint and extension of the interphalangeal joint
    B. flexion of the metacarpophalangeal joint and flexion of the interphalangeal joint
    C. extension of the metacarpophalangeal joint and flexion of the interphalangeal joint
    D. extension of the metacarpophalangeal joint and extension of the interphalangeal joint

Please see Chapter 27 for the answers.

Find additional MCQs online by searching for *Cunningham's Manual of Practical Anatomy Volume 1 General Anatomy, Upper and Lower Limbs*, 17th edition at https://academic.oup.com/ and go to the online appendix at the end of the book. Use your scratch-off code on the inside cover to access the material. The code will work for 12 months.

PART 3

The lower limb

CHAPTER 15
Introduction to the lower limb

Introduction to the lower limb

The parts of the lower limb are the hip and buttock, the thigh, the leg, and the foot.

The hip and buttock together make up what is called the **gluteal region**. This overlies the side and back of the pelvis. It extends from the waist down to the groove below the buttock—the **gluteal fold**—and to the depression on the lateral side of the hip. The hip and buttock are not clearly distinguished from each other. The **hip** (coxa) is the upper part of the region in a lateral view; the **buttock** (natis) is the rounded bulge behind. The **natal cleft** is the groove between the buttocks.

The lower part of the sacrum and coccyx (the end of the backbone) can be felt in the natal cleft. The perineum lies in front of the buttocks and continues forwards between the thighs.

The skeleton of the hip and buttock is the **hip bone**. It consists of three parts—the **ilium**, **ischium**, and **pubis**. These three bones fuse together at the **acetabulum** [Fig. 15.1] where the head of the femur articulates with the hip bone.

The **thigh** extends from the hip to the knee. The thigh bone—the **femur**—articulates at its upper end with the hip bone to form the hip joint [see Figs 16.6 and 16.7]. At the **knee joint**, the femur articulates with the **tibia** and **patella** (kneecap). The proximal extent of the thigh is the gluteal fold pos-

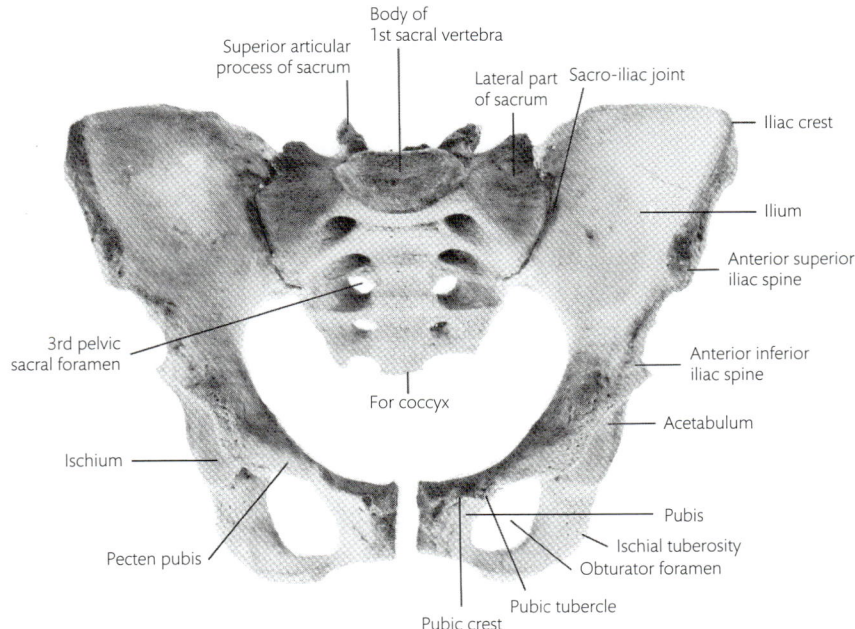

Fig. 15.1 The bony pelvis seen from the front (without the coccyx).

Labels (clockwise):
- Body of 1st sacral vertebra
- Superior articular process of sacrum
- Lateral part of sacrum
- Sacro-iliac joint
- Iliac crest
- Ilium
- Anterior superior iliac spine
- Anterior inferior iliac spine
- Acetabulum
- Pubis
- Ischial tuberosity
- Obturator foramen
- Pubic tubercle
- Pubic crest
- Pecten pubis
- Ischium
- For coccyx
- 3rd pelvic sacral foramen

teriorly, the groove of the groin (**inguinal region**) anteriorly, the perineum medially, and the surface depression on the side of the hip laterally. The ham (poples) is the lower part of the back of the thigh and the back of the knee. The depression on the back of the knee is the **popliteal fossa**.

The **leg** (crus) extends from the knee joint to the ankle joint. The term 'leg' is never used in anatomical descriptions to refer to the entire lower limb, as it frequently is in colloquial speech. The soft, fleshy part of the back of the leg is the **calf** (sura).

The bones of the leg are the **tibia**, or shin bone, and the **fibula**. They lie side by side, with the slender fibula laterally. The tibia and fibula articulate with each other at their upper and lower ends—the superior and inferior tibiofibular joints. Along their length, they are united by the **interosseous membrane**. The lower ends of the tibia and fibula form prominences at the sides of the ankle—the **medial** and **lateral malleoli**, which are readily felt. The medial and lateral malleoli hold the first bone of the foot (the **talus**) between them to form the **ankle joint**. A large part of the tibia is subcutaneous and easily felt. The fibula is mainly covered by muscles which are attached to it, so that only its head and distal quarter are easily felt.

The **foot** extends from the point of the heel to the tips of the toes. Its superior surface is the **dorsum**; its inferior surface is the **sole** (planta) [see Fig. 16.15]. The bones of the foot, from proximal to distal, are the tarsal bones, metatarsals, and phalanges. The five **metatarsal bones** are numbered 1 to 5 from the medial to lateral side. The **toes** (digits) are also numbered from medial to lateral. The first toe is the big toe, or **hallux**; the fifth is the little toe, or **digitus minimus**. The bones of the toes are the **phalanges**. The hallux has two phalanges; each of the other toes has three, though the middle and distal phalanges of the little toe may be fused together.

CHAPTER 16
Osteology of the lower limb

Introduction

The study of the lower limb begins with the study of bones: the hip bone, femur, patella, tibia, fibula, and bones of the foot. By the end of the study, the student should be able to name each bone, determine its position in the body, identify the side to which it belongs, hold it in the anatomical position, describe the important bony features, and demonstrate important muscle attachments on the bone.

Hip bone

General features of the hip bone

The **hip bone** [Figs 16.1, 16.2] is made up of three bones—the **ilium**, **ischium**, and **pubis**. The ilium is large, flat, slightly curved, and directed upwards. The pubis and ischium lie inferiorly, the pubis more anteromedially and the ischium more posterolaterally. The **obturator foramen** is a large aperture in the hip bone between the pubis

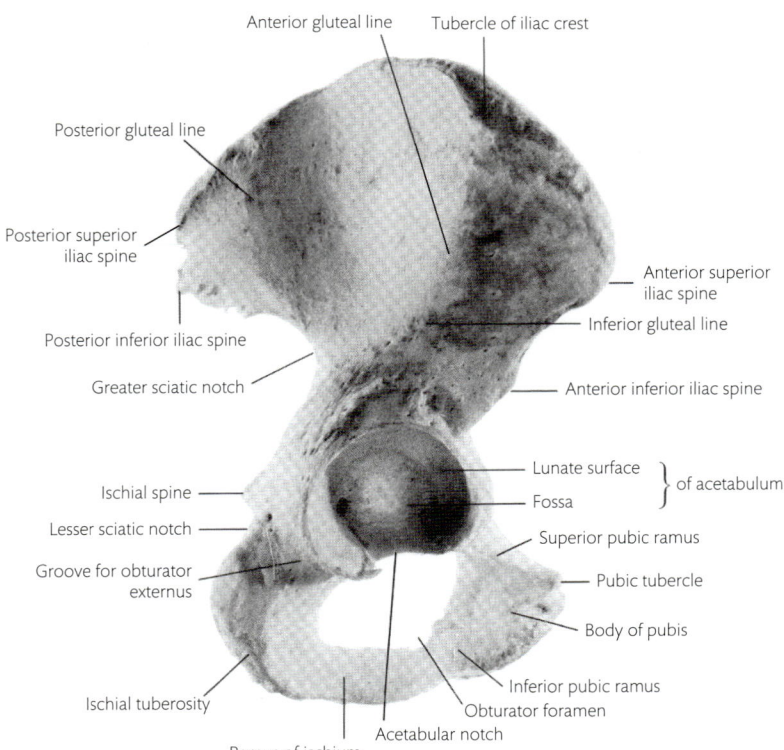

Anterior gluteal line
Tubercle of iliac crest
Posterior gluteal line
Posterior superior iliac spine
Posterior inferior iliac spine
Greater sciatic notch
Anterior superior iliac spine
Inferior gluteal line
Anterior inferior iliac spine
Lunate surface } of acetabulum
Fossa
Ischial spine
Lesser sciatic notch
Groove for obturator externus
Superior pubic ramus
Pubic tubercle
Body of pubis
Ischial tuberosity
Inferior pubic ramus
Ramus of ischium
Acetabular notch
Obturator foramen

Fig. 16.1 Right hip bone seen from the lateral side.

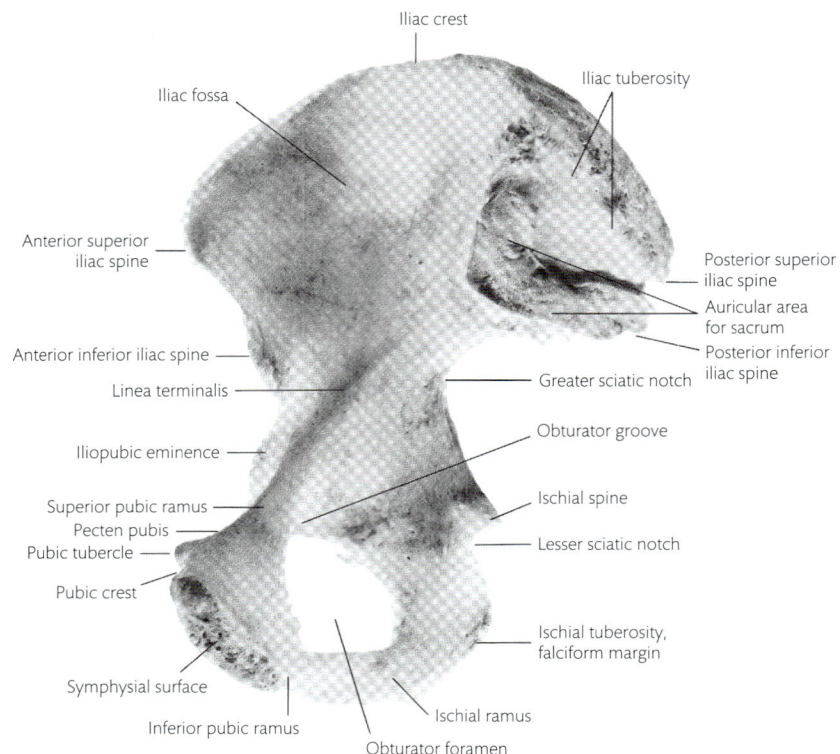

Fig. 16.2 Right hip bone seen from the medial side.

Labels (clockwise): Iliac crest, Iliac tuberosity, Iliac fossa, Anterior superior iliac spine, Posterior superior iliac spine, Auricular area for sacrum, Posterior inferior iliac spine, Anterior inferior iliac spine, Linea terminalis, Greater sciatic notch, Iliopubic eminence, Obturator groove, Superior pubic ramus, Ischial spine, Pecten pubis, Lesser sciatic notch, Pubic tubercle, Pubic crest, Ischial tuberosity, falciform margin, Symphysial surface, Inferior pubic ramus, Ischial ramus, Obturator foramen

and the ischium. The ilium, ischium, and pubis meet at a narrow, thick central part. On the lateral aspect of this central part is the **acetabular fossa** for articulation with the head of the femur. The pubis and ischium are fused together by a bar of bone, inferior to the obturator foramen. This is the **ischiopubic ramus** and is formed by the union of the **inferior ramus of the pubis** and the **ramus of the ischium**. In the region of the acetabulum, the ilium fuses with the superior ramus of the pubis at the **iliopubic eminence** [Fig. 16.2], and with the ischium at the rough ridge on the posterior surface of the acetabulum. The hip bone has a **medial surface**, part of which forms the wall of the pelvis, and a **lateral surface** that has the acetabulum for articulation with the femur [Figs 16.2, 16.3]. The ilium, ischium, and pubis are united in the floor of the acetabulum.

The two hip bones articulate together at the pubic symphysis. Together they form the **pelvic girdle**. Posteriorly, they articulate with the sides of the sacrum at the two **sacroiliac joints**. The right and left hip bones, together with the sacrum and coccyx, make up the skeleton of the pelvis [see Fig. 15.1].

Bony landmarks and muscle attachments

Using Figs 16.1, 16.2, 16.3, 16.4, and 16.5, and dry bones, identify the following bony landmarks and muscle attachments on the hip bone.

Ilium

The upper part of the ilium is flat and wing-like, and provides a large area for attachment of the muscles. It is also known as the **ala** of the ilium. The **body of the ilium** lies inferiorly and joins the ischium and pubis to form the **acetabulum**. The thickened free upper margin of the ilium forms the **iliac crest** which can be felt in the lower margin of the waist. The anterior limit of the iliac crest is the **anterior superior iliac spine**. The iliac crest ends posteriorly at the **posterior superior iliac spine**. The **tubercle of the iliac crest** is a prominence on the external lip of the iliac crest, 5–6 cm posterior to the anterior superior iliac spine. The anterior margin of the ilium has a shallow notch inferior to the anterior superior iliac spine, which separates it from the **anterior inferior iliac spine**. The **posterior inferior iliac spine**

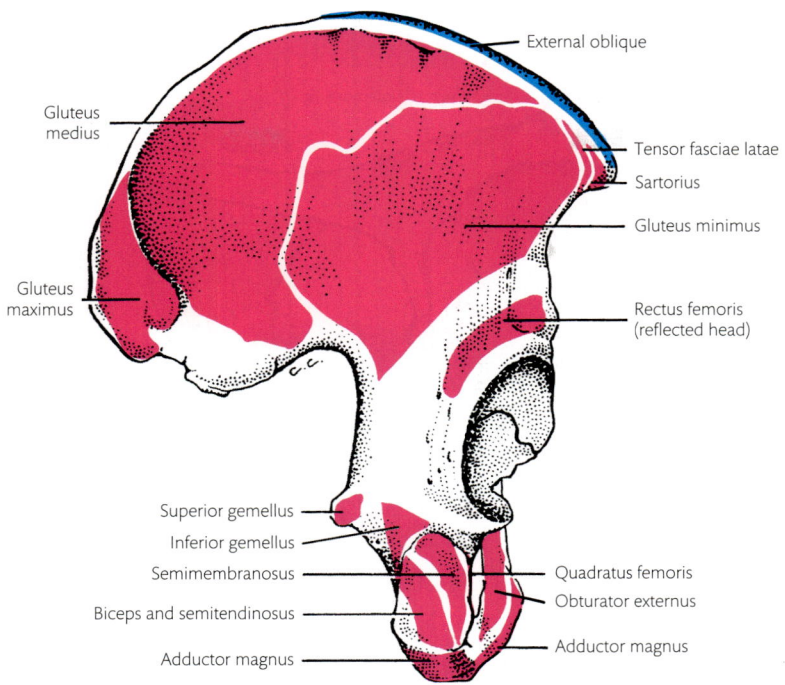

Fig. 16.3 Muscle attachments to the outer surface of the right hip bone.

External oblique

Gluteus medius

Tensor fasciae latae

Sartorius

Gluteus minimus

Gluteus maximus

Rectus femoris (reflected head)

Superior gemellus

Inferior gemellus

Semimembranosus

Biceps and semitendinosus

Adductor magnus

Quadratus femoris

Obturator externus

Adductor magnus

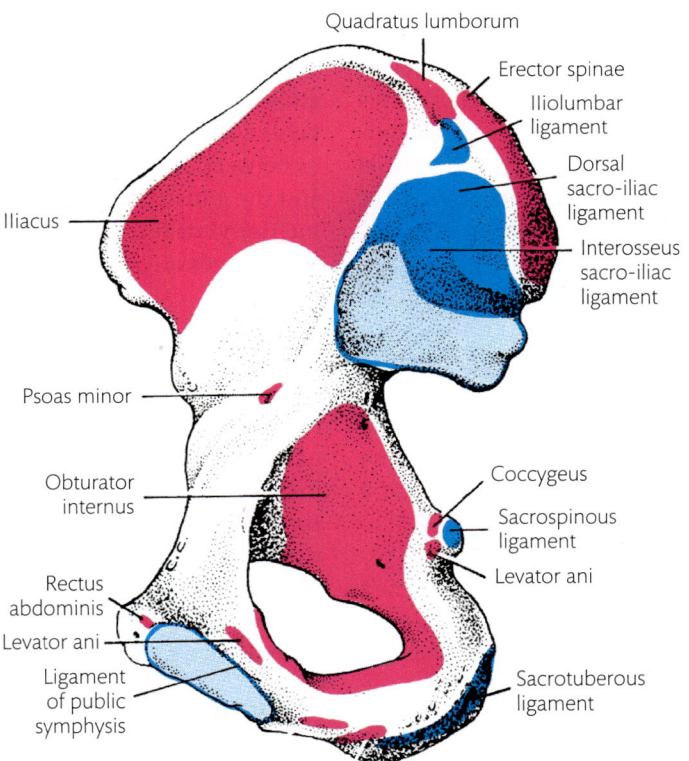

Quadratus lumborum

Erector spinae

Iliolumbar ligament

Dorsal sacro-iliac ligament

Interosseus sacro-iliac ligament

Iliacus

Psoas minor

Obturator internus

Coccygeus

Sacrospinous ligament

Levator ani

Rectus abdominis

Levator ani

Ligament of public symphysis

Sacrotuberous ligament

Fig. 16.4 The medial aspect of the right hip bone. Muscle attachments, red; ligament attachments and cartilage, blue.

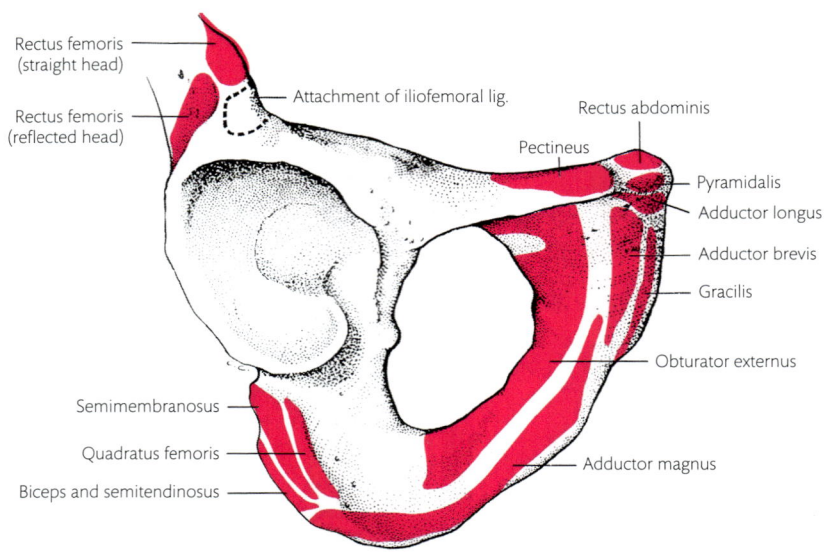

Fig. 16.5 Muscle attachments to the outer surface of the right pubis and ischium.

lies inferior to the posterior superior iliac spine and is separated from it by a shallow notch. The **greater sciatic notch** is a deep, curved depression, or notch, on the posterior margin of the ilium, immediately below the posterior inferior iliac spine, and above the acetabulum.

The lateral or **gluteal surface** of the ilium has three rough, curved lines—the **posterior, anterior**, and **inferior** gluteal lines. Each of these lines begins at the greater sciatic notch and extends outwards. The posterior gluteal line extends upwards and posteriorly to end a short distance in front of the posterior superior iliac spine; the anterior gluteal line ends a short distance posterior to the anterior superior iliac spine, and the inferior gluteal line ends at the anterior margin of the ilium between the anterior superior and anterior inferior iliac spines [Figs 16.1, 16.3].

The **medial** or **sacropelvic surface** of the ilium has a shallow depression anteriorly—the **iliac fossa**. The area behind the iliac fossa has a roughened, ear-shaped **auricular area** for articulation with the sacrum at the sacroiliac joint. Posterior to the auricular area is a further roughened area—the **iliac tuberosity**. A distinct ridge—the **linea terminalis**—extends downwards and forwards on the medial surface of the ilium, from the inferior aspect of the auricular surface. It forms the upper boundary for a region called the true or **lesser pelvis**. The junction of the ilium, ischium, and pubis on the medial side is indistinct and lies at the

level of the acetabulum. (For more details on the pelvic surface of the hip bone, refer to Volume 2.)

The sartorius is attached to the anterior superior iliac spine. The straight head of the rectus femoris is attached to the anterior inferior iliac spine, and the reflected head of the same muscle is attached to the curved area just above the acetabulum. The gluteal surface of the ilium gives origin to the three gluteal muscles. The gluteus maximus is attached behind the posterior gluteal line, the gluteus medius between the posterior and anterior gluteal lines, and the gluteus minimus between the anterior and inferior gluteal lines. The iliac fossa on the medial aspect gives origin to the iliacus.

Ischium

The **ischium** forms the posteroinferior part of the hip bone. The upper part or **body** is fused with the ilium and pubis to form the posteroinferior part of the acetabulum. The posterior border forms the lower part of the **greater sciatic notch**. The **ischial spine** is a prominent triangular projection at the lower end of the greater sciatic notch. Inferior to the ischial spine is the smaller **lesser sciatic notch** on the ischium. The rough bony prominence below the lesser sciatic notch is the **ischial tuberosity**, on which the weight of the body rests while sitting. The **ramus of the ischium** projects upwards and forwards from the ischial tuberosity to meet the inferior ramus of the pubis and form the **ischiopubic rami**. The ischiopubic rami forms the inferomedial boundary for the obturator foramen.

The hamstring muscles, long head of the biceps femoris, semitendinosus, semimembranosus, and hamstring part of the adductor magnus take origin from the ischial tuberosity [Figs 16.3, 16.5]. The superior and inferior gemelli are attached to the ischium on either side of the ischial spine. The obturator externus is attached to the obturator margin of the ischial ramus.

Pubis

The **pubis** forms the anteroinferior part of the hip bone and can be palpated at the lower part of the anterior abdominal wall. The **body of the pubis** lies medially [see Fig. 16.1] and articulates with its fellow of the opposite side through a median fibrous joint—the **pubic symphysis**. The superior border of the body of the pubis forms the pubic crest, which ends laterally in the **pubic tubercle**. The **superior ramus of the pubis** extends laterally and upwards from the body of the pubis. A sharp ridge on the superior pubic ramus—the **pecten pubis**—curves posterolaterally from the pubic tubercle to the **iliopubic eminence**.

Below and behind the pubic symphysis is the **inferior pubic ramus**. The two **inferior pubic rami** diverge to form the **pubic arch**. The inferior pubic ramus unites with the corresponding **ramus of the ischium** to form the **ischiopubic ramus**. Below and behind the pubic symphysis is each inferior pubic ramus, uniting with the corresponding **ramus of the ischium** to form the **ischiopubic ramus**. The ischiopubic ramus forms the boundary between the thigh and the perineum.

The rectus abdominis and pyramidalis (muscles of the abdomen) are attached to the pubic crest, and the pectineus is attached to the superior pubic rami. The anterior surface of the body of the pubis and the inferior pubic rami give origin to the adductor longus, adductor brevis, adductor magnus, and gracilis. The margin of the obturator foramen gives origin to the obturator externus.

Obturator foramen

The obturator foramen is a large triangular to oval foramen in the lower part of the hip bone. It is bounded by the ilium, ischium, and pubis, and covered for the most part by the obturator membrane. The obturator externus takes origin from the outer aspect of the obturator membrane and its anterior and inferior bony margins of the foramen.

Acetabulum

The acetabulum is a cup-shaped depression on the lateral aspect of the hip bone for articulation with the head of the femur at the hip joint. The lunate surface within the acetabulum is a broad strip of articular bone at the periphery of the acetabulum which partially surrounds the central non-articular **acetabular fossa**. This fossa is continuous interiorly with the floor of the **acetabular notch** between the ends of the lunate surface. The acetabular notch is converted into a foramen by the **transverse ligament of the acetabulum** which bridges the notch and completes the acetabular margin.

Exercise 16.1A: to determine the side of the hip bone

Identify the side of the hip bone, using the following instructions. Hold the bone such that:

1. The acetabulum is directed laterally
2. The ilium is directed superiorly
3. The pubis is anterior, and the ischium is posterior.

Exercise 16.1B: to hold the hip bone in the anatomical position

Hold the hip bone in the anatomical position and in the hand of the side to which it belongs.

Femur

General features of the femur

The **femur** is a long bone, and the bone of the thigh. It consists of an **upper end**, a **shaft**, and a **lower end** (Figs 16.6, 16.7). The upper end has three bony prominences—the head, the greater trochanter, and the lesser trochanter. The **head of the femur** is directed medially, upwards, and forwards to articulate with the acetabulum of the hip bone at the hip joint. The **greater trochanter** is directed laterally, and the **lesser trochanter** is directed posteromedially. The **neck of the femur** extends downwards and laterally from the head. It meets the shaft of the femur at the greater and lesser trochanters. The **shaft of the femur** is convex anteriorly, particularly in its proximal half. It has three surfaces and three borders. The middle two-thirds of the posterior border is pronounced

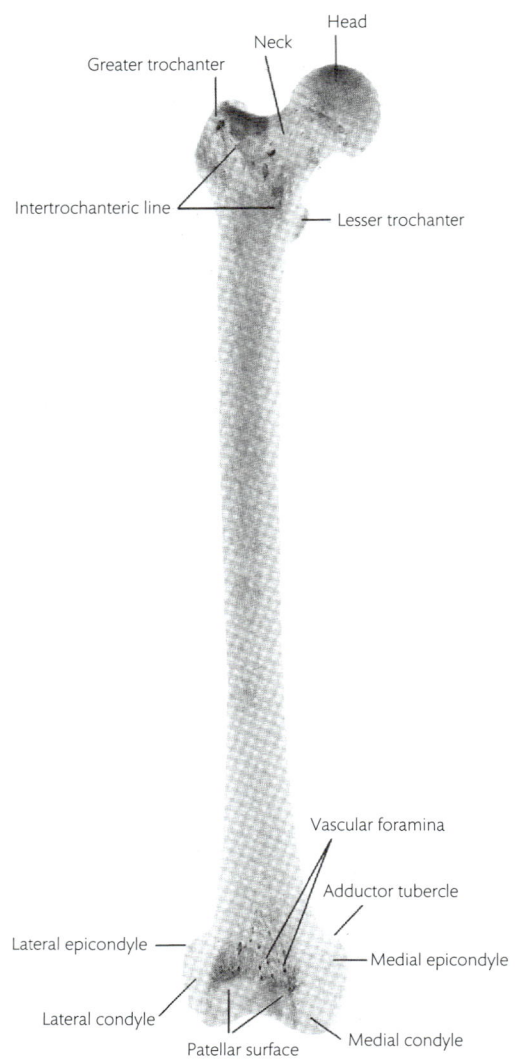

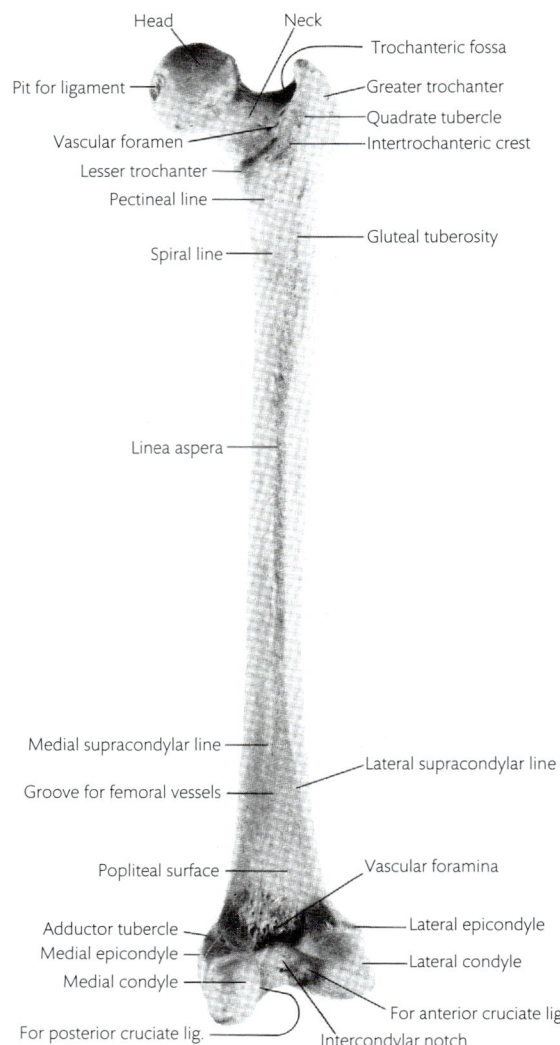

Fig. 16.6 Right femur (anterior aspect).

Fig. 16.7 Right femur (posterior aspect).

and is known as the **linea aspera**. The lower end of the femur consists of the **medial** and **lateral condyles**. Posteriorly, the condyles are separated by a wide **intercondylar fossa**. Anteriorly, the condyles unite in the grooved **patellar surface**.

Bony landmarks and muscle attachments

Using Figs 16.6, 16.7, 16.8, and 16.9, and dry bones, identify the following bony landmarks and muscle attachments on the femur.

Upper end of the femur

The spherical **head of the femur** fits into the acetabulum where it articulates with the C-shaped lunate surface. The head is more than half a sphere and is mostly covered with articular cartilage. It presents a small pit, or fovea, in the centre which lacks cartilage. The ligament of the head of the femur is attached to the fovea. The head of the femur is continuous with the **neck** which joins it to the shaft. Two bony prominences—the **greater** and **lesser trochanters**—mark the junction of the neck with the shaft. The neck meets the shaft posteriorly at a prominent, rounded ridge (the **intertrochanteric crest**) which extends from the **greater trochanter** above to the **lesser trochanter** below. Anteriorly, the neck meets the shaft in a rough **intertrochanteric line** which extends between the two trochanters.

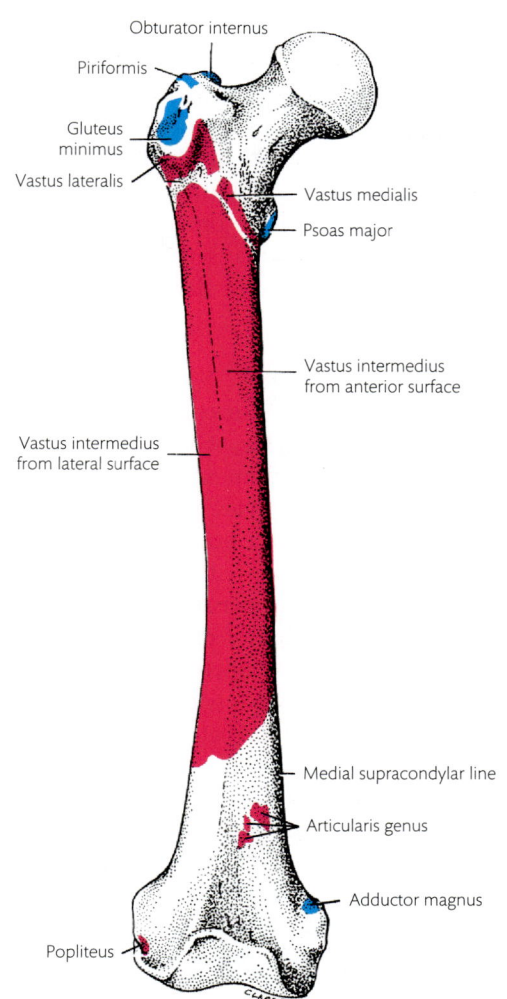

Fig. 16.8 Right femur (anterior aspect) to show muscle attachments.

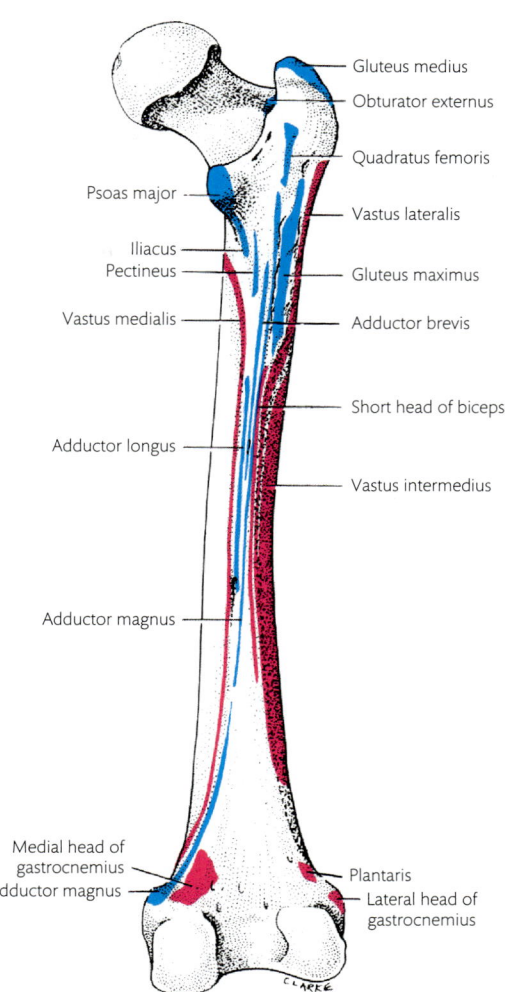

Fig. 16.9 Right femur (posterior aspect) to show muscle attachments.

The **neck of the femur** forms an angle of approximately 125 degrees with the body of the femur. This is the **angle of inclination** and is lesser in the female and greater in the male. A thick bar of bone in the lower part of the neck transmits compressive forces applied by the weight of the body on the head of the femur to the shaft. The surface of the neck is ridged longitudinally by bundles of fibres (**retinaculae**) which are continuous with the fibrous capsule of the hip joint and transmit blood vessels to the neck. Note the foramina on the neck of the femur for these vessels.

The **greater trochanter** is placed laterally and projects upwards and medially over the neck. The bony depression on the medial side of the greater trochanter is the **trochanteric fossa**. The piriformis, gluteus medius, gluteus minimus, and

obturator internus are attached to the greater trochanter. The obturator externus is inserted into the trochanteric fossa.

The lesser trochanter projects posteromedially from the junction of the neck and shaft. It gives insertion to the psoas major. The psoas minor and iliacus are inserted into the bone just below the lesser trochanter.

At the junction of the neck and shaft, the two trochanters are joined on the posterior aspect by a smooth prominent ridge—the **intertrochanteric crest**. The intertrochanteric crest presents a small tubercle—the **quadrate tubercle**—for attachment of the quadratus femoris. The two trochanters are joined on the anterior aspect by the **intertrochanteric line**. This line extends posteriorly, inferior to the lesser trochanter, and becomes continuous with the **spiral line** on the back of the femur. Below the greater trochanter on the posterior surface is a linear rough area—the **gluteal tuberosity**. One-fourth of the fibres of the gluteus maximus are inserted here.

Shaft of the femur

The **shaft of the femur** is distinctly convex anteriorly. It has posterior, medial, and lateral borders and anterior, medial, and lateral surfaces. The medial and lateral borders are indistinct and rounded, which makes the medial, anterior, and lateral surfaces almost continuous. The posterior border has a distinct ridge—the **linea aspera**—in the middle part.

At the upper end, the lateral lip of the linea aspera is continuous with the gluteal tuberosity. The medial lip continues with the spiral line. Superiorly and inferiorly, the medial and lateral lips of the linea aspera separate. Superiorly, the continuation of the medial lip forms the **spiral line** which curves anteriorly, and the continuation of the lateral lip forms the rough **gluteal tuberosity** posteriorly. The spiral line becomes continuous above, with the intertrochanteric line on the anterior surface of the femur. A faint ridge—the pectineal line—extends downwards from the lesser tubercle to the linea aspera. Another faint bony ridge runs along the lower margin of the greater trochanter and joins the intertrochanteric line anteriorly, with the gluteal tuberosity posteriorly.

Inferiorly, the lips of the linea aspera diverge to form the medial and lateral **supracondylar lines** [Fig. 16.7]. These lines form the boundaries of the flattened **popliteal surface of the femur**. The lateral supracondylar line continues down to the lateral epicondyle. The medial line continues to the **adductor tubercle** on the medial epicondyle of the femur but is interrupted where the femoral artery crosses it to become the popliteal artery. The distal end of the shaft of the femur widens into the **medial** and **lateral condyles**.

Most of the shaft of the femur is smooth and gives origin to the vastus intermedius from the anterior and lateral surfaces. Note the area of origin of the articularis genu, a small and deep part of the vastus intermedius. The vastus medialis has a long, linear origin from the posterior aspect of the femur, including the medial lip of the linea aspera. The origin of the vastus lateralis is the root of the greater trochanter on the lateral aspect and the lateral lip of the linea aspera. The three intermuscular septa, the adductor muscles, and the short head of the biceps are also attached to the linea aspera. The order of attachment of the thigh muscles to the linea aspera is that of their position in the thigh—from medial to lateral, the attachments on the linea aspera are the: vastus medialis, medial intermuscular septum, adductor longus, adductor brevis, adductor magnus, posterior intermuscular septum, short head of the biceps, lateral intermuscular septum, and vastus lateralis.

Lower end of the femur

The lower end of the femur consists of the **medial** and **lateral condyles**, which are continuous with each other on the anterior surface. The articular surfaces of the condyles are broad, and cover the anterior, inferior, and posterior aspects of the condyles. They articulate with the femoral aspect of the patella and the upper surface of the tibia at the knee joint. The anterior patellar surface of the femur has a groove—the trochlear groove. The lateral surface of this groove is wider and projects further forwards than the medial surface. The margin of the lateral surface may be felt, proximal to the patella, when the knee is flexed. The tibial surfaces of the medial and lateral condyles are separated by the intercondylar fossa. The tibial surface of the medial condyle is gently curved medially. The tibial surface of the lateral condyle passes straight back. The tibial surface of both condyles extend on to the posterior aspect where they are widely separated by the intercondylar fossa. The intercondylar fossa is separated from the popliteal surface of the femur by the intercondylar line. The medial wall of the fossa is formed by the lateral surface of the medial condyle. It gives attachment to the posterior cruciate ligament. The lateral wall is formed by the medial surface of the lateral condyle and gives attachment to the anterior cruciate ligament.

The **medial** and **lateral epicondyles** are flattened conical projections from the surface of each condyle [Figs 16.6, 16.7]. Each epicondyle shows some additional bony features. The lateral epicondyle gives attachment to the lateral head of the **gastrocnemius**. Below the lateral epicondyle is a fossa with a groove running posteriorly from it. The **tendon of the popliteus** is attached to the fossa and lies in the groove when the knee is flexed. The posterior surface of the medial epicondyle is marked by the attachment of the medial head of the gastrocnemius. The **adductor tubercle** lies superior to the medial epicondyle.

Applied anatomy

The two epiphyses at the knee joint have secondary ossification centres which appear before birth—that of the lower end of the femur and the upper end of tibia (all other secondary centres appear after birth).

The appearance of the secondary ossification centre in the distal end of the femur and proximal tibia is used to determine the gestational age in cases without confirmed gestational age. The secondary ossification centre for the distal end of the femur can be seen by 34 weeks of gestation. The appearance of this secondary ossification centre helps differentiation between intrauterine growth retardation and wrong dating of pregnancy, and prevents unnecessary interventions in a premature fetus.

Exercise 16.2A: to determine the side of the femur

Identify the side of the femur, using the following instructions. Hold the femur such that:

1. The head is directed superiorly and medially
2. The intercondylar fossa (and the lesser trochanter) is directed posteriorly.

Exercise 16.2B: to hold the femur in the anatomical position

Hold the femur in the anatomical position and in the hand of the side to which it belongs.

Exercise 16.3: to articulate the hip bone with the femur

Articulate the femur and hip bone together by bringing the head of the femur and the acetabulum together. (Make sure the two bones belong to the same side.)

Patella

General features and muscle attachments of the patella

The **patella** is a sesamoid bone developed in the tendon of the quadriceps femoris. It has an **anterior** and a **posterior surface**, an **apex** directed downwards, and a **base** directed upwards. The anterior surface is subcutaneous, ridged, and convex. The posterior surface is smooth and covered with articular cartilage for articulation with the femur. This surface has a prominent, vertical ridge which divides it into a slightly larger lateral part and a smaller medial part for articulation with the lateral and medial condyles of the femur. The patella has three borders. The **superior border**, or base, is thick. The **medial** and **lateral borders** are thin and converge towards the apex.

The tendon of the quadriceps femoris is attached to the base of the patella and extends over the anterior surface. The vastus medialis and lateralis have additional attachments to the medial and lateral borders. The ligamentum patellae is attached to the apex of the patella.

Exercise 16.4: to determine the side of the patella

Identify the side of the patella, using the following instructions. Hold the patella such that:

1. The rough convex surface is directed anteriorly
2. The apex is directed downwards
3. The larger of the two facets on the posterior surface is lateral.

Tibia

General features of tibia

The leg bones are the tibia medially and the fibula laterally [Figs 16.10, 16.11]. The tibia has an **upper end**, a **shaft**, and a **lower end**. The upper end is formed by the medial and lateral tibial condyles. The upper surface is flat for articulation with the medial and lateral condyles of the femur. The upper end of the tibia presents a prominent **tibial tuberosity** anteriorly. The shaft of the tibia is roughly triangular in transverse section, with **medial, posterior**, and **lateral surfaces**. The **distal end of the tibia** is slightly expanded and has five surfaces: anterior, posterior, medial, lateral, and inferior surfaces. The tibia articulates with the fibula at the superior and inferior tibiofibular joints, and with the talus at the ankle joint.

Bony landmarks and muscle attachments

Using Figs 16.10, 16.11, 16.12, and 16.13, and dry bones, identify the following bony landmarks and muscle attachments on the tibia.

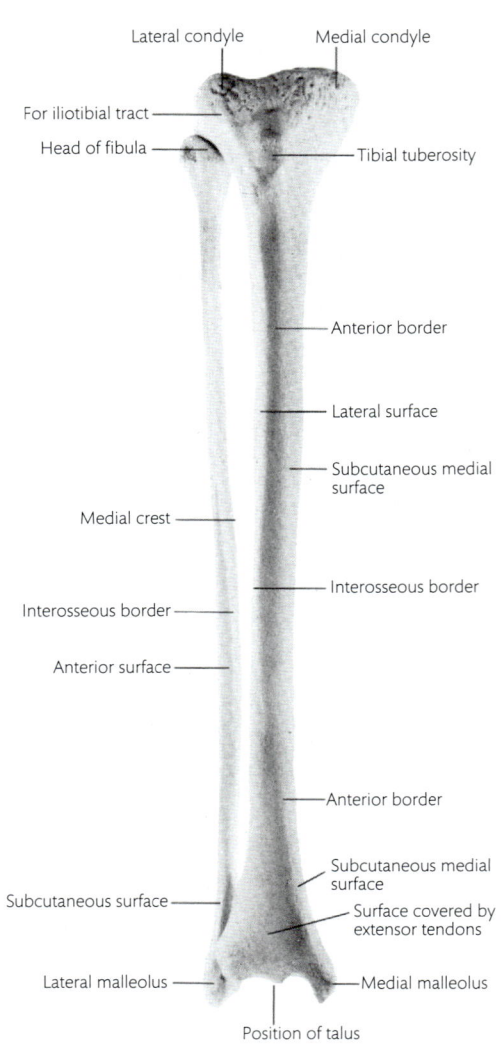

Fig. 16.10 Right tibia and fibula (anterior aspect).

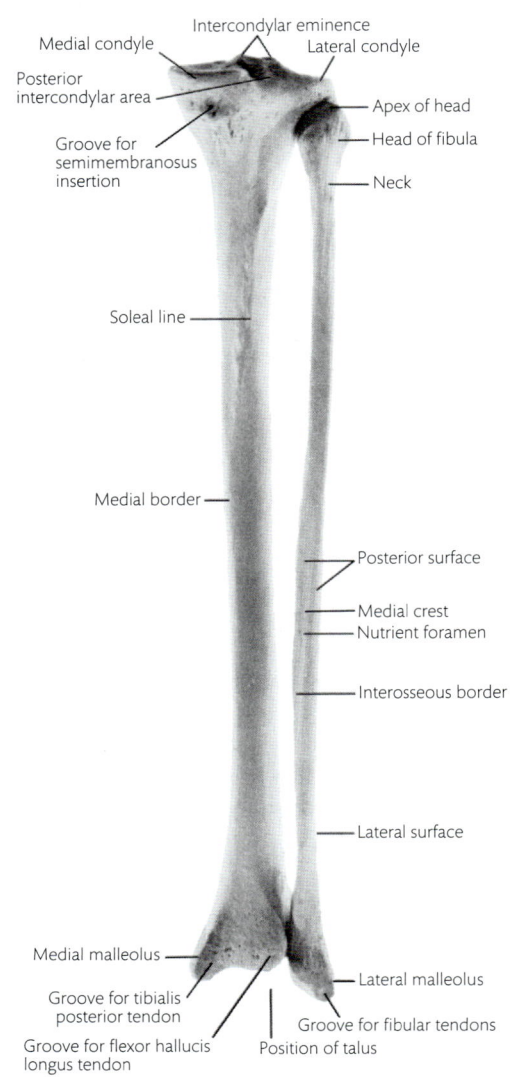

Fig. 16.11 Right tibia and fibula (posterior aspect).

Upper end

The upper end of the tibia is expanded and flattened to form the **medial** and **lateral condyles**. The condyles are limited inferiorly by an almost horizontal line. The upper surfaces of the medial and lateral tibial condyles have smooth, oval, concave areas for articulation with the corresponding condyles of the femur. Between these condyles is the **intercondylar area** which is divided into anterior and posterior parts by an **intercondylar eminence**. On the intercondylar eminence are the **medial** and **lateral intercondylar tubercles**.

The posteromedial surface of the medial condyle has a long, horizontal groove, with a roughened area below it for attachment of the semimembranosus [Fig. 16.13]. In a similar, but slightly lower,

position, the inferior surface of the lateral condyle has a smooth area for articulation with the head of the fibula. The iliotibial tract is attached to the lateral condyle, and the semitendinosus is attached to the medial condyle. The posterior surface is smooth and gives attachment to the popliteus.

The **tibial tuberosity** is a prominent eminence on the anterior aspect. It has a smooth upper part and a rough lower part. The patellar tendon is attached to the smooth upper part of the tibial tuberosity. Inferiorly, the tibial tuberosity is continuous with the sharp **anterior border of the tibia** [Fig. 16.10].

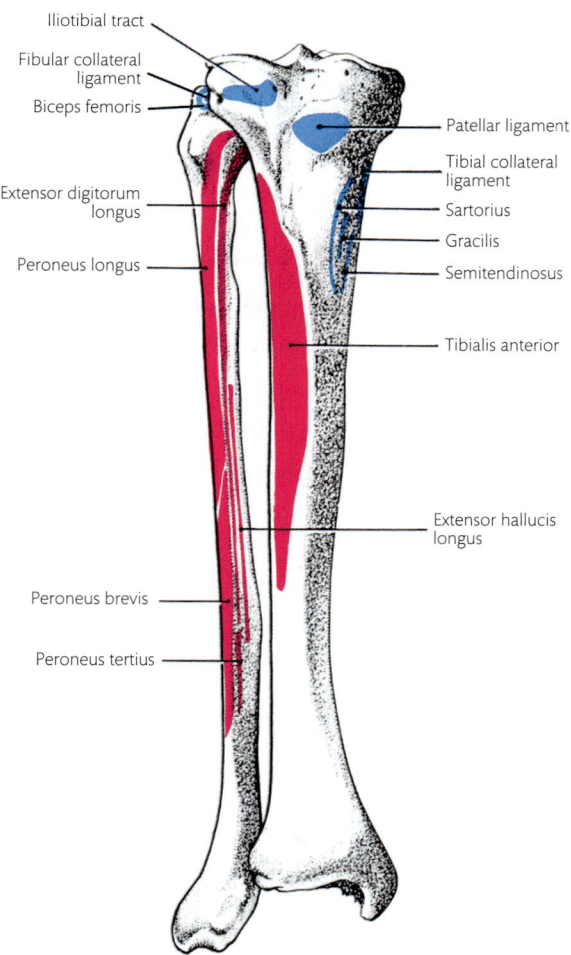

Fig. 16.12 Anterior aspect of bones of the leg to show attachments of muscles.

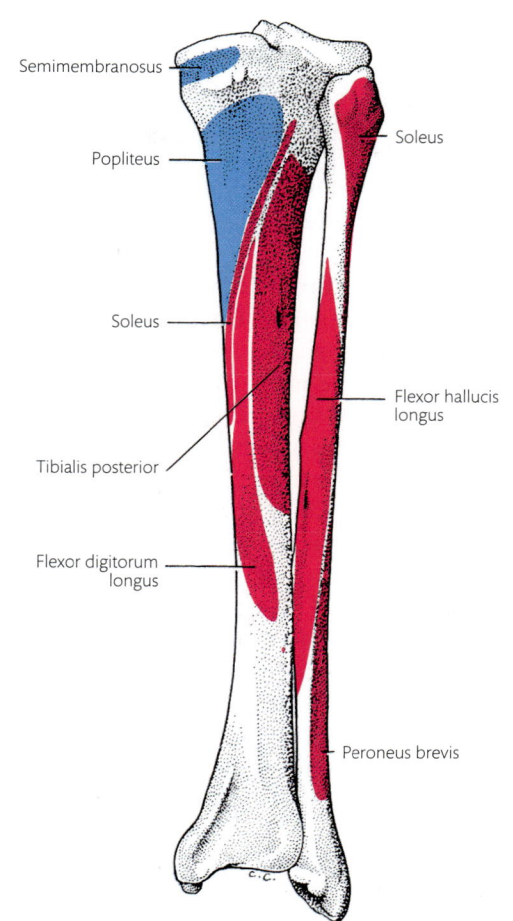

Fig. 16.13 Posterior aspect of bones of the leg to show attachments of muscles.

Shaft

The shaft of the tibia has three borders—**anterior**, **medial**, and **lateral borders**; and three surfaces—**medial**, **lateral**, and **posterior surfaces**. The anterior margin or border is sharp and subcutaneous. It extends from the lower end of the tibial tuberosity and diverges medially at the lower end to form the anterior border of the medial malleolus. This border is palpable throughout its length, though it becomes less sharp in the lower third of the bone.

The **lateral border**, or interosseous border, gives attachment to the interosseous membrane. **The interosseous border of the tibia** begins immediately inferior to the articular area for the fibula. The interosseous membrane between the tibia and the fibula is attached to this border. At the distal end of the tibia, the interosseous border becomes continuous with a rough, triangular, concave area for the thickened part of the interosseous membrane—the interosseous tibiofibular ligament. The interosseous tibiofibular ligament binds the tibia to the fibula above the ankle joint. The interosseous border separates the lateral surface of the shaft of the tibia from the posterior surface, and hence the areas from which the anterior (extensor) muscles and posterior (flexor) muscles of the leg arise.

The **medial border** of the tibia is sharp and continues distally with the medial edge of a groove for the tendon of the tibialis posterior, on the posterior surface of the medial malleolus [Fig. 16.11].

The **medial surface** is subcutaneous, except in its upper quarter. The superficial part of the tibial collateral ligament of the knee is attached to the upper part of the tibia immediately in front of the medial border. The upper quarter of the medial surface gives insertion to three tendons—that of the sartorius, gracilis, and semitendinosus. Inferior to these attachments, the medial surface is smooth and subcutaneous. Inferiorly, it is continuous with the medial surface of the medial malleolus, a thick downwards projection of the distal end of the tibia.

The **lateral surface** gives origin to the tibialis anterior.

On the **posterior surface** of the tibia, an oblique ridge—the **soleal line** [Fig. 16.11]—passes from just below the articular surface for the head of the fibula to the medial border of the tibia, at the junction of the upper and middle thirds of the shaft. This line gives attachment to the tibial head of the soleus muscle and to the popliteal fascia on the posterior surface of the **popliteus muscle**. The popliteus is attached on the tibia to the triangular area, proximal to the soleal line. The nutrient artery to the shaft of the tibia enters the bone midway between the distal part of the soleal line and the interosseous border. Medial to the nutrient foramen is a **vertical ridge** which extends from the middle of the soleal line to the interosseous border distally. The area between the soleal line, interosseous border, and vertical ridge gives attachment to the tibialis posterior [Fig. 16.13].

Lower end

The distal end of the tibia is slightly expanded. The **medial malleolus** is a downward projection on the medial part of the distal end. On the **lateral aspect** of the distal end is a rough triangular area for the attachment of the interosseous tibiofibular ligament, which binds the tibia to the fibula above the ankle joint. The **inferior surface** is concave anteroposteriorly, with a slight, central anteroposterior ridge. It is an articular surface and continuous medially with the articular surface on the lateral aspect of the medial malleolus. Both surfaces articulate with the body of the talus at the ankle joint. The **medial malleolus** projects inferiorly as the apex and has a smooth notch on its posterior aspect. The notch and apex give attachment to the powerful **medial** (deltoid) **ligament of the ankle joint**.

Applied anatomy

The secondary ossification centre for the upper end of the tibia occurs in the ninth month of intrauterine life (i.e. before birth). Visualization of the upper epiphysis indicates that the fetus has reached term.

Exercise 16.5A: to determine the side of the tibia

Identify the side of the tibia, using the following instructions. Hold the tibia such that:

1. The condyles are directed superiorly
2. The medial malleolus is directed medially
3. The tibial tuberosity is directed anteriorly.

Exercise 16.5B: to hold the tibia in the anatomical position

Hold the tibia in the anatomical position and in the hand of the side to which it belongs.

Exercise 16.6: to articulate the tibia with the femur

Articulate the femur and tibia together by bringing the femoral and tibial condyles. (Make sure the two bones belong to the same side.)

Fibula

General features of the fibula

The fibula is the lateral of the two leg bones [Figs 16.10, 16.11]. It has an **upper end**, a **shaft**, and a **lower end**. The upper end consists of the **head** and **neck of the fibula**. The **shaft** is covered with muscles and has three surfaces—**anterior** or **medial surface**, and **lateral** and **posterior surfaces**; and three borders—**interosseous** or **medial border**, and **anterior** and **posterior borders**. The **distal end** of the fibula is the **lateral malleolus**. The head of the fibula articulates with the lateral condyle of the tibia. The lateral malleolus articulates with the distal end of the tibia; and with the talus at the ankle joint.

Bony landmarks and muscle attachments

Using Figs 16.10, 16.11, 16.12, and 16.13, and dry bones, identify the following bony landmarks and muscle attachments on the fibula.

Upper end

The fibula has an expanded head at its proximal end, and a narrow **neck** below the head. On the superomedial surface of the head is the area for articulation with the tibia and the **apex of the**

head lateral to it. The biceps femoris is inserted into the head of the fibula.

Shaft

The **shaft** of the fibula is covered with muscles, and its shape varies considerably with the degree of muscularity. It has three borders—**anterior**, **interosseous**, and **posterior borders**; and three surfaces—**anterior** or **medial surface**, and **lateral** and **posterior surfaces**. The posterior surface is divided by the medial crest into medial and posterior parts. This gives the shaft of the fibula the appearance of having four, not three, surfaces.

The **interosseous border** of the fibula can be traced upwards from the triangular area on the medial side of the lateral malleolus. This border bends forwards as it ascends, so that it comes close to the anterior border.

A narrow **medial** or **anterior surface** lies between the anterior and interosseous borders and gives attachment to the extensor muscles of the leg, extensor digitorum longus, extensor hallucis longus, and peroneus tertius.

Posterior to the interosseous border is the **posterior surface** of the fibula. This **posterior surface** has a curved vertical ridge—the **medial crest**—which separates it into medial and posterior parts. The tibialis posterior is attached to the posterior surface in the area medial to the medial crest. The strong intermuscular septum posterior to the tibialis posterior muscle is attached to the medial crest. Behind the medial crest, the posterior surface of the fibula gives rise to the soleus and flexor hallucis longus [Fig. 16.13].

The **lateral surface** of the fibula lies between the anterior and posterior borders. It gives origin to the peroneal longus and brevis. Inferiorly, these muscles pass behind the subcutaneous triangular area to run over the posterior surface of the lateral malleolus.

Lower end

The distal end of the fibula is expanded to form the **lateral malleolus** which projects downwards. The lateral malleolus projects beyond the medial malleolus and has a large triangular surface on its medial aspect for articulation with the body of the talus. Superior to the articular surface of the lateral malleolus is a rough triangular area for attachment of the **interosseous tibiofibular ligament**. Posterior to the articular area on the lateral malleolus is the **malleollar fossa**. The anterolateral surface of the lateral malleolus is continuous above with a rough, triangular **subcutaneous area**. This is continuous superiorly with the **anterior border** of the fibula.

Exercise 16.7A: to determine the side of the fibula

Identify the side of the fibula, using the following instructions. Hold the fibula such that:

1. The head is directed superiorly
2. The malleolar fossa is directed medially
3. The malleolar fossa lies posteriorly.

Exercise 16.7B: to hold the fibula in the anatomical position

Hold the fibula in the anatomical position and in the hand of the side to which it belongs.

Exercise 16.8: to articulate the tibia with the fibula

Articulate the tibia and fibula together by bringing the facets on the lateral condyle of the femur and the head of the fibula together. (Make sure the two bones belong to the same side.)

Bones of the foot

Bones and muscle attachments

The bones of the foot are shown in Figs 16.14, 16.15, 16.16, and 16.17.

The **tarsal bones** are in two rows. The proximal row consists of two large bones—the **talus** and the **calcaneus**—with the talus resting on the calcaneus.

The calcaneus is the largest bone of the tarsus. It forms the skeleton of the heel. Its long axis is directed forwards and laterally. On the inferior aspect is the **calcaneal tuberosity** which has the medial and lateral tubercles. The **trochlear tubercle** is seen on the lateral surface of the calcaneum. There are three articular facets on the superior surface for articulation with the body and neck of the talus. Between these facets and the equivalents on the talus is the **tarsal sinus**. The medial surface has a horizontal shelf-like projection—the **sustentaculum tali**. This bar of bone is concave above and articulates with the talus. Below it is grooved by the tendon of the flexor hallucis lon-

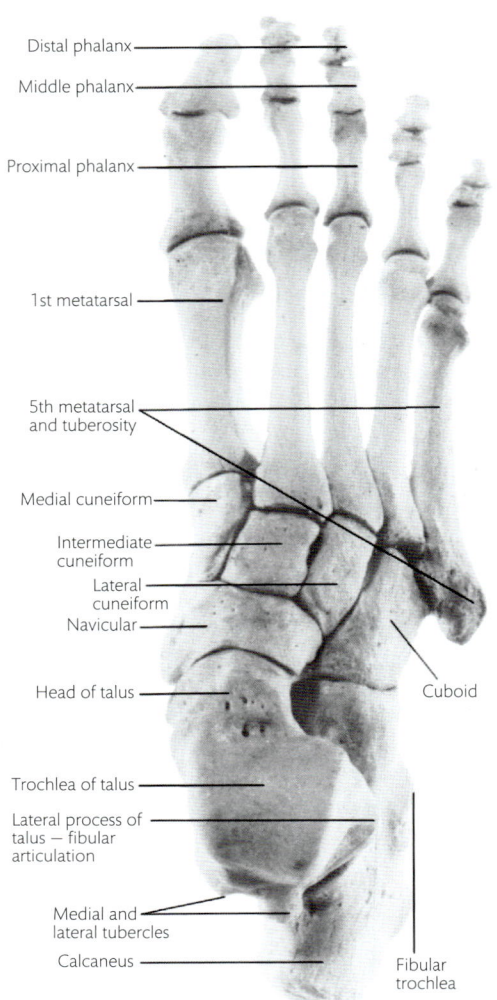

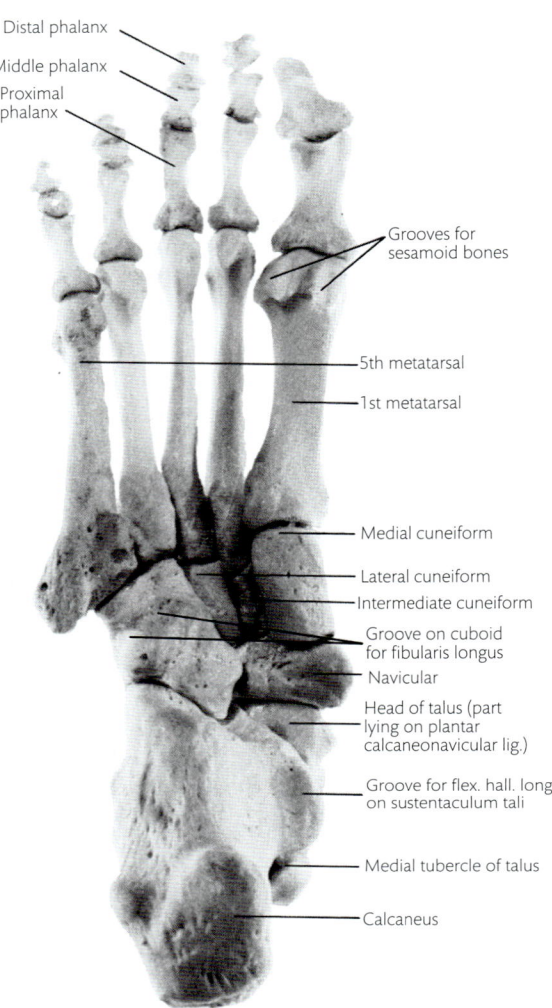

Fig. 16.14 Superior or dorsal surface of bones of the right foot.

Fig. 16.15 Inferior or plantar surface of bones of the right foot.

gus. The sustentaculum tali gives attachment to important ligaments of the ankle and talocalcaneonavicular joints. The calcaneus articulates with the talus which is placed above it, and the cuboid bone which lies in front.

The **talus** consists of a cuboidal **body**, with a **neck** and **head** projecting forwards and slightly medially from it. The **trochlea tali** is the articular surface on the superior aspect of the body. Posterior to the trochlea, the lower part of the talus projects backwards as the **posterior process**. This process is grooved by the tendon of the flexor hallucis longus, resulting in the formation of the lateral and medial **tubercles of the posterior process**. (The posterior process ossifies from a separate centre. Rarely, it fails to fuse with the remainder of the talus and presents in the adult as the **os trigonum**.)

The talus forms numerous joints with adjacent bones. **At the ankle**: (1) the superior surface of the body—the trochlea tali—articulates with the distal end of the tibia; (2) the medial surface of the body articulates with the medial malleolus through a flat, comma-shaped surface; and (3) the lateral surface of the body articulates with the lateral malleolus through a large, triangular area. The talus is firmly held between the tibia and the fibula at the ankle joint.

At the subtalar joint, the inferior surface of the body of the talus articulates with the calcaneus. **At the talocalcaneonavicular joint**, the head and neck of the talus articulate with the calcaneus, navicular, and **plantar calcaneonavicular** (spring) **ligament**.

The navicular lies between the proximal and the distal row of the tarsal bones. It articulates proximally

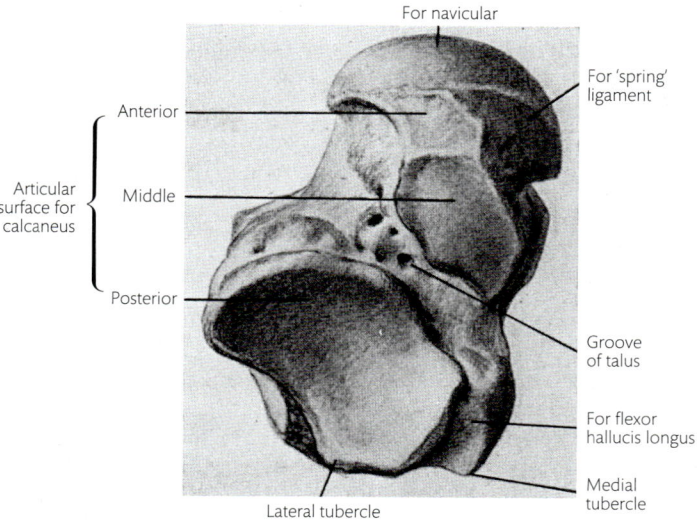

For navicular

For 'spring' ligament

Anterior

Articular surface for calcaneus

Middle

Posterior

Groove of talus

For flexor hallucis longus

Medial tubercle

Lateral tubercle

Fig. 16.16 Right talus (inferior surface). The groove of the talus is the sulcus tali.

with the talus and distally with the three cuneiforms. The distal row of the tarsal bones consists of the **cuboid bone** laterally, and the three wedge-shaped **cuneiform bones** (cuneus = wedge)—**medial**, **intermediate**, and **lateral cuneiforms**—medially. The inferior surface of the cuboid is grooved by the **tendon of the peroneus longus** which passes transversely across the foot. The **cuboid** articulates with the distal end of the calcaneus by a curved, saddle-shaped joint which permits a moderate amount of movement. It also articulates medially with the navicular and lateral cuneiform bone.

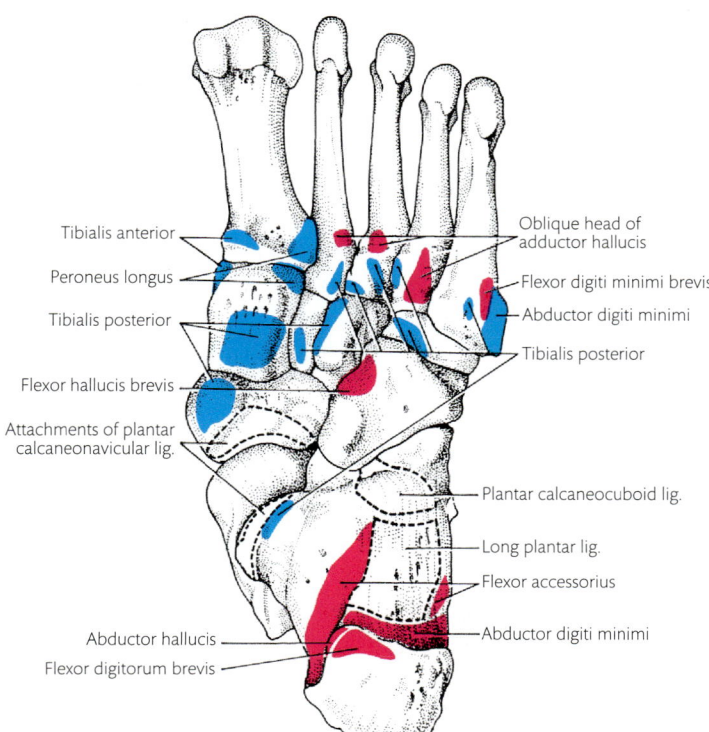

Tibialis anterior

Peroneus longus

Tibialis posterior

Flexor hallucis brevis

Attachments of plantar calcaneonavicular lig.

Abductor hallucis

Flexor digitorum brevis

Oblique head of adductor hallucis

Flexor digiti minimi brevis

Abductor digiti minimi

Tibialis posterior

Plantar calcaneocuboid lig.

Long plantar lig.

Flexor accessorius

Abductor digiti minimi

Fig. 16.17 Muscle attachments to the left tarsus and metatarsus (plantar aspect). Interrupted lines show areas of attachment of ligaments.

The navicular articulates with the three cuneiform bones, and the cuneiforms articulate with each other and the medial three metatarsals. The cuboid articulates with the lateral cuneiform and the lateral two metatarsals. These joint surfaces are flat and permit relatively little movement.

The powerful tendo calcaneus is inserted into the posterior surface of the calcaneus. The plantar surface of this bone gives origin to the abductor hallucis, flexor digitorum brevis, abductor digiti minimi, and flexor accessorius. The extensor digitorum brevis takes origin from the dorsal surface of the calcaneus. The tibialis anterior and peroneus longus are inserted into the navicular and base of the first metatarsal. The peroneus tertius is inserted into the base of the fifth metatarsal, and the tibialis posterior is inserted into all tarsal bones, except the talus, and the base of the middle three metatarsals. The flexor hallucis brevis arises from the cuboid.

The proximal ends—the **base of the metatarsals**—articulate with the tarsal bones at the tarsometatarsal joints, and the base of the medial four metatarsals articulate with each other at the intermetatarsal joints. Each metatarsal has a **head** at the distal end which articulates with the base of the proximal phalanx of the corresponding toe at the metatarsophalangeal joint. The metatarsals give attachment to the oblique head of the adductor hallucis and the interossei.

The bones of the toes are the **phalanges**. The hallux has two phalanges; each of the other toes has three, though the middle and distal phalanges of the little toe may be fused together. The proximal end of the phalanx is its base; the distal end is its head. The phalanges articulate with each other at the **interphalangeal joints**. The extensor expansion and long and short flexors of the toes are inserted on the phalanges.

Exercise 16.9A: to determine the side of the articulated foot

Identify the side of the foot, using the following instructions. Hold the foot such that:

1. The calcaneus is directed posteriorly
2. The big toe is directed medially
3. The talus is placed superiorly.

Exercise 16.9B: to hold the articulated foot in the anatomical position

Hold the articulated foot in the anatomical position and in the hand of the side to which it belongs.

CHAPTER 17
The front and medial side of the thigh

Introduction

Before starting to dissect, study the surface anatomy of the region on yourself, on another living subject, or on the cadaver, and relate this to the appropriate dry bones.

Surface anatomy

The pubic symphysis may be felt at the lower end of the abdominal wall. Draw your finger laterally from the pubic symphysis on the anterosuperior surface of the body of the pubis. This surface is the **pubic crest** which ends in a small, blunt prominence—the **pubic tubercle**—laterally. The tubercle is less easily felt in the male, because it is covered by the spermatic cord. Lateral to the pubic tubercle, a resilient band can be felt in the inguinal groove between the anterior surface of the thigh and the abdomen. This is the **inguinal ligament**. The ischiopubic ramus forms the boundary between the thigh and the perineum, and is palpable through its length [see Fig. 16.2].

Find the **iliac crest** at the lower margin of the waist. Trace it forwards. It slopes downwards and slightly medially to end in a rounded knob—the **anterior superior iliac spine**. This may be grasped between the finger and the thumb in a thin individual. The inguinal ligament stretches from this spine to the pubic tubercle. Trace the outer lip of the iliac crest posteriorly, until you feel a low prominence—the **tubercle of the iliac crest**. This is the widest part of the pelvis. Further posteriorly, the iliac crest turns downwards to end in the **posterior superior iliac spine** at the level of the second sacral vertebra [see Fig. 16.1].

The **greater trochanter of the femur** can be palpated indistinctly, immediately in front of the surface depression on the side of the hip [see Fig. 16.6]. The tip of this trochanter lies at the level of the pubic crest. The **head of the femur** can be felt indistinctly, even though it is deeply buried in muscles. To do this on yourself, place your finger just below the inguinal groove at the **mid-inguinal point** (i.e. midway between the anterior superior iliac spine and the pubic symphysis). Press firmly, and rotate your limb medially and laterally. The head will be felt moving behind the muscles. With lighter pressure, the **femoral artery** can be felt pulsating at the same spot.

Identify the condyles of the femur and their epicondyles on your own knee. The condyles of the tibia and femur can be differentiated by movement of the tibia when the knee is flexed and extended. Grasp the **patella**, and try to move it. The patella is mobile when the knee is extended but becomes rigid when the knee is flexed. Feel the strong **patellar tendon (patellar ligament)** which stretches from the patella to the **tibial tuberosity** (a blunt prominence on the front of the upper end of the tibia) [Fig. 17.1]. This tendon becomes taut when the knee is flexed. During flexion, the patella slides on to the distal end of the femur and the upper part of the patellar surface of the femur is exposed.

With the knee straight, a muscular strip with three tendons can be felt on the medial side of the knee, posterior to the medial epicondyle. When the knee is flexed, these tendons project back. The muscles and tendons on the posterior medial side of the knee are the **sartorius**, **gracilis**, and **semitendinosus**. Another tendon, more deeply placed and less read-

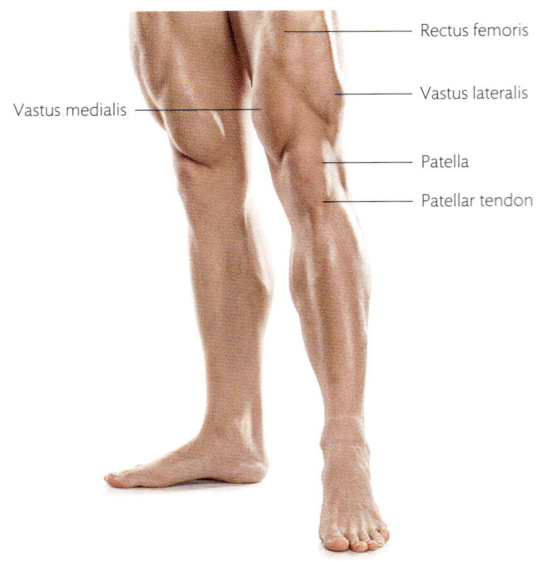

Rectus femoris

Vastus lateralis

Vastus medialis

Patella

Patellar tendon

Fig. 17.1 Front of the knee, lower part of the thigh, and upper leg, illustrating the visible bony elevations and muscle masses.

I T A L O/Shutterstock.com.

ily felt, is that of the **semimembranosus**. On the lateral side, a single stout tendon can be felt, posterior to the lateral epicondyle, when the knee is bent. This is the tendon of the **biceps femoris** [Fig. 17.2].

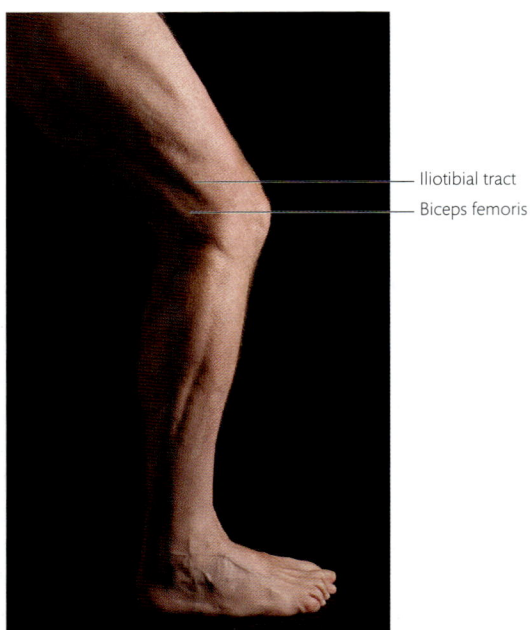

Iliotibial tract

Biceps femoris

Fig. 17.2 Lateral side of the knee, showing surface projection of the iliotibial tract and biceps tendon.

threerocksimages/Shutterstock.com.

Trace it down to the **head of the fibula**. Anterior to this tendon and separated from it by a depression is a broad, tendon-like structure which is best felt when standing with the knee slightly bent. This is the **iliotibial tract**, a strip of thickened deep fascia of the thigh [Fig. 17.2]. Through the iliotibial tract, two muscles—the gluteus maximus and tensor fasciae latae—are inserted into the lateral condyle of the tibia.

Proximal to the medial epicondyle of the femur is a fleshy swelling. This is the lowest part of the **vastus medialis** muscle [Fig. 17.1]. When the knee is bent, a shallow groove appears, posterior to this part of the muscle. Press your finger into the groove, and feel the tendon of the **adductor magnus** muscle. Slide your finger distally on the tendon to the adductor tubercle where the tendon is attached. The fleshy swelling proximal to the lateral epicondyle is the lowest part of the vastus lateralis [Fig. 17.1].

Front of the thigh

Dissection 17.1 (and Fig. 17.3) instructs how to reflect the skin on the front of the thigh.

Superficial fascia

The superficial fascia of the thigh close to the inguinal ligament is in two layers that are continuous with the same two layers in the abdomen. There is a thick, superficial fatty layer and a deeper membranous layer. The membranous layer of the superficial abdominal fascia descends and is attached to the deep fascia of the thigh along a line parallel and approximately 1 cm inferior to the inguinal ligament [Fig. 17.4]. At the pubic tubercle, the line of fusion extends downwards across the front of the

---

DISSECTION 17.1 Skin reflection

Objective

I. To reflect the skin on the front of the thigh.

Instructions

1. Make incisions 9 and 10 [Fig. 17.3] through the skin.
2. Reflect the skin from the superficial fascia, and turn it laterally.

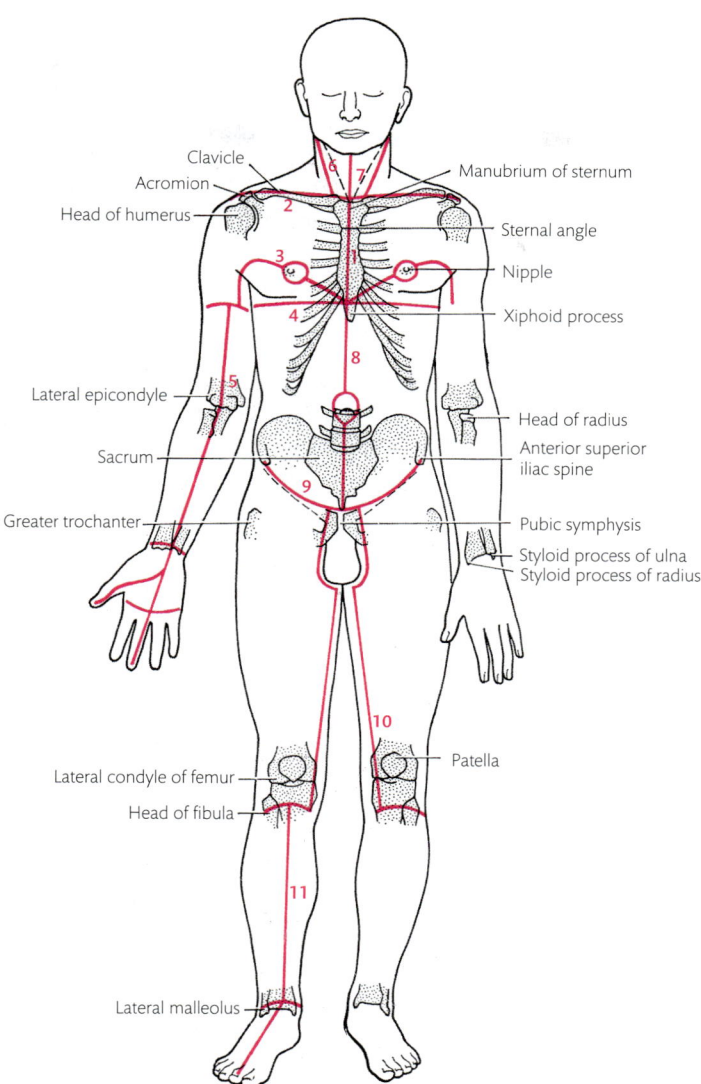

Fig. 17.3 Landmarks and incisions.

body of the pubis and the margin of the inferior pubic ramus to the ischial tuberosity. This arrangement permits communication between the perineum and the tissue deep to the membranous layer in the anterior abdominal wall. The fusion of the abdominal fascia to the fascia of the thigh separates the tissue of the anterior abdominal wall and perineum from the thigh.

Dissection 17.2 (and Fig. 17.5) explores the superficial structures of the inguinal region.

Saphenous opening

The saphenous opening is a defect in the superficial fascia. It overlies the upper part of the femoral vein. The saphenous opening lies approximately 3–4 cm

inferolateral to the pubic tubercle and is about 3 cm long and 1.5 cm wide. It is limited in the upper, lateral, and posterior sides by the sharp **falciform margin**, formed by the thicker deep fascia which surrounds it [Fig. 17.5]. The fatty superficial fascia over the opening is thin and perforated, and gets the name **cribriform fascia** (*cribrum* = sieve). The cribriform fascia and saphenous opening transmit: (1) the long saphenous vein; (2) one or more of the superficial inguinal arteries; and (3) efferent lymph vessels from the superficial inguinal lymph nodes.

Superficial inguinal lymph nodes

The superficial inguinal lymph nodes lie in the superficial fascia and are arranged in the shape of

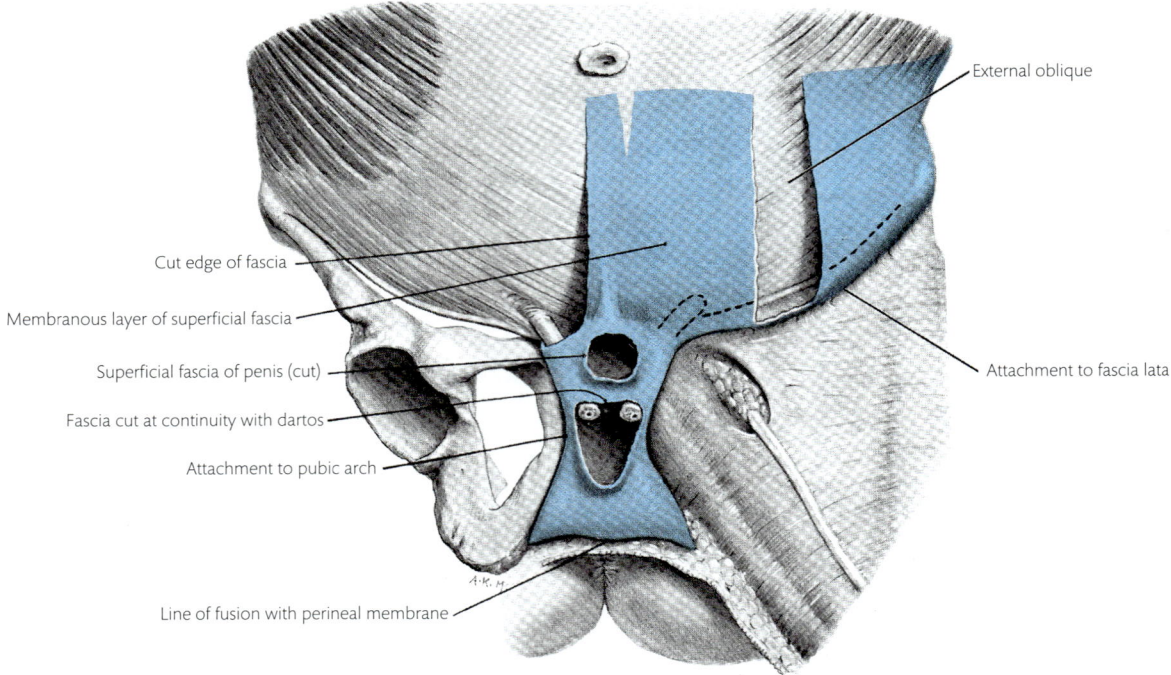

Cut edge of fascia

Membranous layer of superficial fascia

Superficial fascia of penis (cut)

Fascia cut at continuity with dartos

Attachment to pubic arch

External oblique

Attachment to fascia lata

Line of fusion with perineal membrane

Fig. 17.4 Diagram showing continuity of the membranous layer of the superficial fascia of the abdominal wall, perineum, and thigh.

DISSECTION 17.2 Superficial veins

Objectives

I. To explore the continuity of fascial planes and spaces from the abdomen to the perineum and from the abdomen to the lower limb.   **II.** To expose the upper part of the long saphenous vein and the tributaries in this area. **III.** To identify the superficial arteries. IV. To demonstrate the saphenous opening and cribriform fascia.

Instructions

1. Make a horizontal incision through the entire thickness of the superficial fascia of the anterior abdominal wall from the anterior superior iliac spine to the midline.
2. Raise the superficial fascia inferior to the cut, and pass the fingers downwards between the membranous layer of the fascia and the underlying aponeurosis of the external oblique muscle [Fig. 17.4].
3. Appreciate that little resistance is felt to the passage of the fingers, till the line of fusion of the membranous layer with the deep fascia of the thigh is reached at the fold of the groin.
4. Note that the fingers cannot be carried into the thigh because of this line of fusion.

5. Pass the fingers medially along this line, and find the opening into the perineum, just medial to the pubic tubercle. Note that a finger can easily be passed into the perineum.
6. In the male, the finger passes beside the spermatic cord towards the scrotum; in the female, it passes into the base of the labium majus.
7. Revert back to the dissection of the lower limb. Find the **long saphenous vein** in the superficial fascia of the medial part of the anterior surface of the thigh. Trace the vein downwards to the knee and upwards to the point where it turns sharply backwards through the deep fascia to enter the femoral vein.
8. As the upper part is exposed, note the lower group of **superficial inguinal lymph nodes** scattered along the vein and the delicate, thread-like lymph vessels which enter them.
9. Three small veins enter the long saphenous vein at its upper end. Follow these and the small superficial inguinal branches of the femoral artery. They pierce the deep fascia and supply the adjacent skin and lymph nodes. The **superficial external pudendal vessels** pass medially to the external

genital organs; the **superficial epigastric** runs superiorly to the anterior abdominal wall, and the **superficial circumflex iliac** runs towards the lateral part of the groin [Fig. 17.5]. (When tracing these vessels, note the upper group of superficial inguinal lymph nodes which lie scattered along the lower border of the inguinal ligament. They vary greatly in number and size.)

10. Find the ilioinguinal nerve just below the pubic tubercle. Trace its branches to the skin of the upper medial part of the thigh. It also sends branches to the external genital organs.

11. Lift the upper end of the long saphenous vein, and note that it turns backwards over a sharp edge of the deep fascia.

12. Follow this edge round the lateral side of the vein and upwards towards the inguinal ligament. This is the falciform margin of the **saphenous opening** [Fig. 17.5]. From this margin, the thin **cribriform fascia** passes in front of the opening and the femoral vessels in the femoral sheath.

13. Remove the cribriform fascia to expose the femoral sheath. Take care not to damage the structures which pierce the cribriform fascia or lie posterior to it.

a T. The upper nodes are below, and roughly parallel to, the inguinal ligament. The lower nodes are placed vertically along the upper part of the long saphenous vein [Fig. 17.5].

Afferents: the superficial inguinal lymph nodes receive almost all the lymph from the skin and superficial fascia below the level of the umbilicus. This includes: (1) the skin and superficial fascia of the trunk below the level of the umbilicus, including the perineum (anal canal, lower vagina, and urethra); and (2) the skin and superficial fascia of the lower limb, with a few exceptions. In addition, (3) a few lymph vessels from the fundus and body of the uterus reach the superficial inguinal lymph nodes along the round ligament of the uterus.

In exception to the above, the testis drains to the lumbar lymph nodes; the glans penis or glans clitoris drains to the deep inguinal lymph nodes; and the skin and superficial fascia of the heel and lateral part of the foot drain into the popliteal nodes in the popliteal fossa.

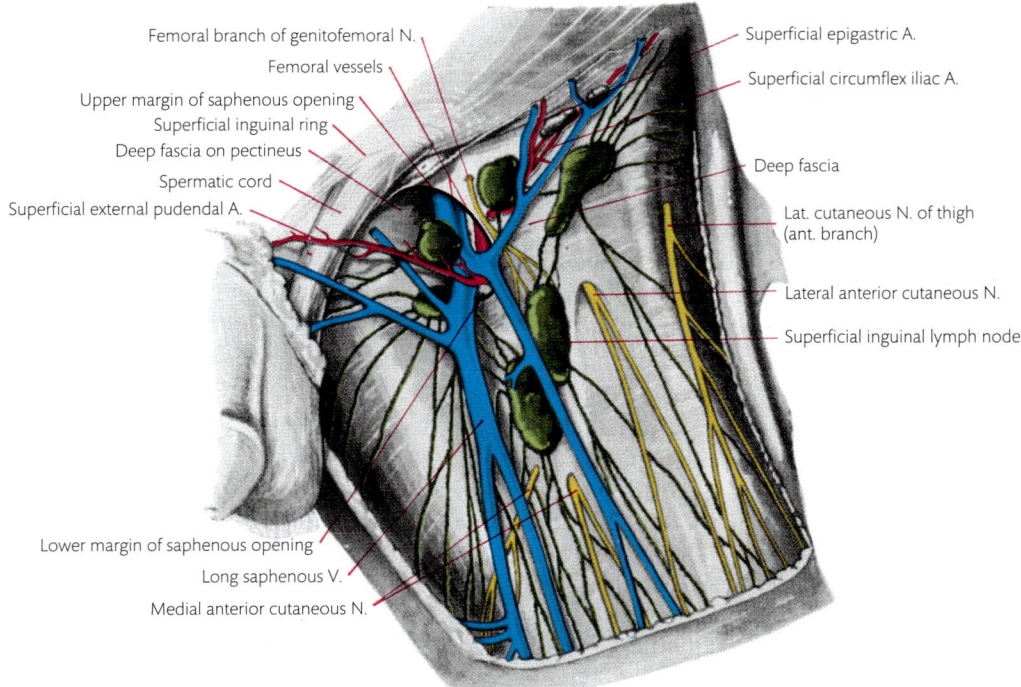

Femoral branch of genitofemoral N.
Femoral vessels
Upper margin of saphenous opening
Superficial inguinal ring
Deep fascia on pectineus
Spermatic cord
Superficial external pudendal A.

Superficial epigastric A.
Superficial circumflex iliac A.
Deep fascia
Lat. cutaneous N. of thigh (ant. branch)
Lateral anterior cutaneous N.
Superficial inguinal lymph node

Lower margin of saphenous opening
Long saphenous V.
Medial anterior cutaneous N.

Fig. 17.5 Superficial dissection of the proximal part of the front of the thigh. The saphenous opening and the superficial lymph nodes and lymph vessels of the groin are displayed.

The **efferents** from the superficial inguinal nodes pass through the cribriform fascia to the deep inguinal lymph nodes on the femoral vessels and the external iliac nodes on the external iliac vessels in the abdomen.

Long saphenous vein

This is the longest vein of the lower limb and lies in the superficial fascia. It **begins** on the medial side of the dorsum of the foot as the continuation of the medial end of the dorsal venous arch, and **ends** in the femoral vein by piercing the cribriform fascia in the thigh. In the leg, it lies anterior to the medial malleolus, on the medial surface and medial border of the tibia, and on the posteromedial surface of the knee. In the thigh, it ascends to enter the femoral vein through the saphenous opening [Fig. 17.6].

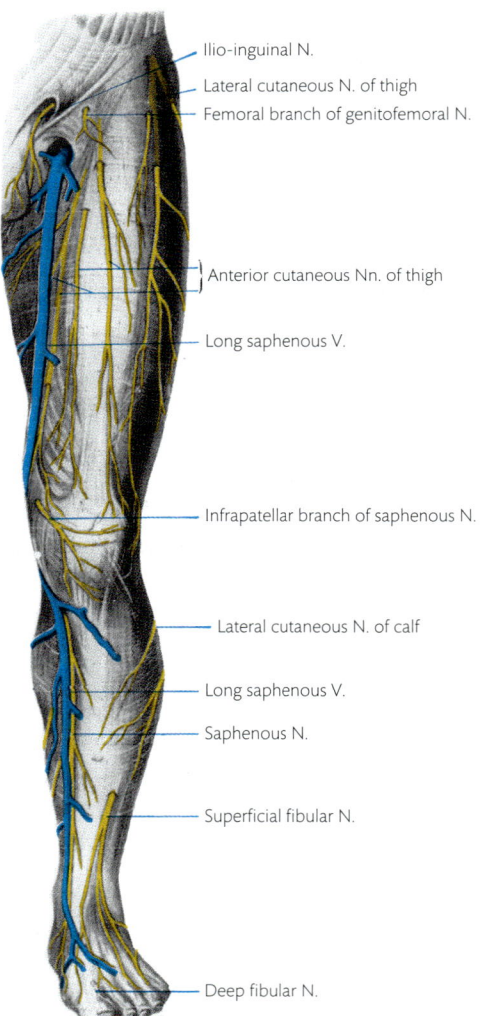

- Ilio-inguinal N.
- Lateral cutaneous N. of thigh
- Femoral branch of genitofemoral N.
- Anterior cutaneous Nn. of thigh
- Long saphenous V.
- Infrapatellar branch of saphenous N.
- Lateral cutaneous N. of calf
- Long saphenous V.
- Saphenous N.
- Superficial fibular N.
- Deep fibular N.

Fig. 17.6 Cutaneous nerves on the front of the lower limb.

DISSECTION 17.3 Cutaneous nerves on the front of the thigh

Objective

I. To clean and trace the cutaneous nerves on the front of the thigh.

Instructions

1. Strip the superficial fascia down from the front and lateral side of the thigh by blunt dissection. Leave the deep fascia in place.
2. With the assistance of Figs 17.5 and 17.6, find the cutaneous nerves as they pierce the deep fascia, and follow them distally.
3. Note how most of these nerves terminate in the patellar plexus, anterior to the patella.
4. Check for the presence of a prepatellar bursa between the skin and the lower part of the patella.

The long and short saphenous veins form parallel channels with the deep veins (plantar and tibial veins) of the lower limb. Venous blood in the lower limb has to flow against gravity, and several mechanisms exist in both superficial and deep veins to aid this. These mechanisms and the clinical importance of these veins are discussed later in Chapter 22 on the leg and foot.

Dissection 17.3 traces the cutaneous nerves.

Cutaneous nerves

The cutaneous nerves for the front and medial sides of the thigh are from three sources: (1) directly from the lumbar plexus; (2) branches of the femoral nerve; and (3) branches of the obturator nerve.

1. The **ilioinguinal nerve** (L. 1) is a branch of the lumbar plexus. It emerges just lateral to the pubic tubercle (through the superficial inguinal ring) and is distributed to the scrotum or labium majus and the medial side of the thigh.
2. The **femoral branch of the genitofemoral nerve** (L. 1, 2) (a branch of the lumbar plexus) is small and difficult to find. It enters the thigh, posterior to the inguinal ligament, and pierces the deep fascia, lateral to the saphenous opening. This nerve supplies an area of skin immediately below the inguinal ligament.
3. The **lateral cutaneous nerve of the thigh** (L. 2, 3) is a branch of the lumbar plexus. It enters the thigh posterior to the lateral part of the inguinal ligament. It gives a posterior branch that pierc-

es the deep fascia and supplies an area of skin over the greater trochanter. The remainder of the nerve pierces the deep fascia lower down. It descends on the lateral side of the thigh to the patella, sending branches to the skin of the lateral and anterior surfaces of the thigh.

4. Three **anterior cutaneous branches** arise from the femoral nerve (L. 2, 3) in the front of the thigh. They supply the skin of the anterior and medial surfaces of the thigh and the upper part of the medial surface of the leg. The more medial branches pierce the deep fascia more distally [Fig. 17.6].

5. The **saphenous nerve** (L. 3, 4) arises from the femoral nerve and descends with the femoral artery, deep to the sartorius muscle. It sends an **infrapatellar branch** through that muscle to supply the skin medial to the knee and distal to the patella. The main nerve pierces the deep fascia, posterior to the sartorius, at the knee. It supplies the skin of the medial surface of the leg and the medial surface of the foot up to the proximal end of the big toe. It descends on the leg with the long saphenous vein [Fig. 17.6].

6. From the obturator nerve, an occasional branch supplies the medial side of thigh.

Peripatellar nerve plexus

The peripatellar plexus is a fine subcutaneous nerve plexus formed by branches of the lateral cutaneous nerve of the thigh, the medial and intermediate anterior cutaneous branches of the femoral nerve, and the infrapatellar branch of the saphenous nerve. The plexus supplies the skin over the patella, the patellar ligament, and the proximal part of the tibia anteriorly.

Dissection 17.4 is the dissection of the deep fascia of the thigh.

Fascia lata

The deep fascia of the thigh is called the **fascia lata**. Like the deep fascia elsewhere in the body, it is continuous with the periosteum of the underlying bones, either directly where the bone is subcutaneous or indirectly through intermuscular septa. The upper part of the fascia lata is attached around the root of the lower limb to: (1) the iliac crest laterally; (2) the inguinal ligament anteriorly; (3) the body of the pubis, ischiopubic rami, and ischial tuberosity medially; and (4) the sacrotuberous ligament and sacrum posteriorly. Below, at the knee, the fascia lata fuses with the patella, the femoral and tibial condyles, and the head of the fibula. Posteriorly, it is continuous with the dense fascia covering the popliteal fossa.

The fascia lata is thin medially. Laterally, it forms the **iliotibial tract**, a thickened band stretching from the iliac crest to the lateral tibial condyle. Two muscles—**gluteus maximus** and **tensor fasciae latae**—are inserted into the tract. Through the insertion into the iliotibial tract, they help to stabilize the pelvis on the thigh and maintain extension of the knee while standing.

From the deep surface of the fascia lata, three intermuscular septa pass to the linea aspera of the femur. These **medial**, **posterior**, and **lateral intermuscular septa** separate the thigh muscles into three compartments [Fig. 17.7]: (1) the **anterior compartment** lies anteriorly and laterally, and contains the extensor muscles of the knee and the femoral nerve; (2) the **medial compartment** contains the adductor muscles of the hip and the obturator nerve; and (3) the **posterior compartment** contains the flexor muscles (hamstrings) of the knee and the sciatic nerve.

Superficial patellar bursae

A number of superficial and deep fluid-filled bursae are present around the knee joint and enable free movement of the skin on the underlying tissues, as in kneeling, and movement of the deep tissues on one another. There are two or three subcutaneous bursae between the skin and the front of the patella, the lower part of the patellar ligament, and the tibial tuberosity.

Inguinal ligament

The free lower border of the aponeurosis of the external oblique muscle of the abdomen forms the **inguinal ligament**. The inguinal ligament

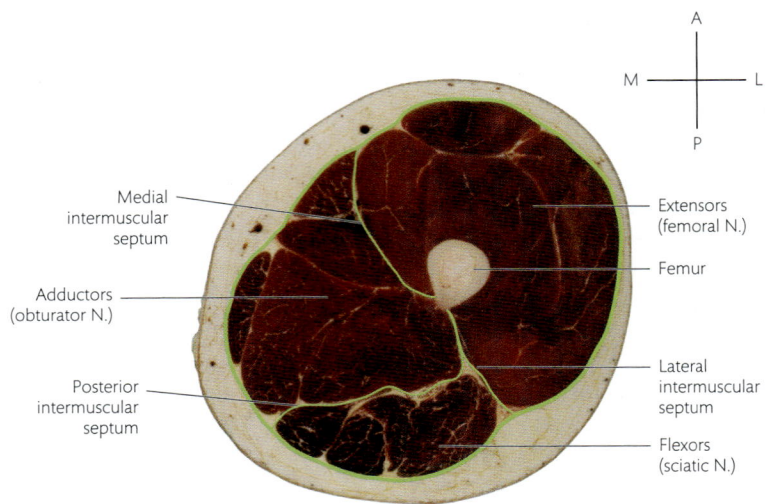

Fig. 17.7 Section of the thigh to show the arrangement of muscles and intermuscular septa forming the osteofascial compartments of the right thigh. A = anterior; P = posterior; M = medial; L = lateral.

Image courtesy of the Visible Human Project of the US National Library of Medicine.

is a major landmark and separates the anterior abdominal wall from the front of the thigh. **Attachments**: it extends from the anterior superior iliac spine laterally to the pubic tubercle medially. The free edge of the aponeurosis curves back on itself to form a groove on the abdominal aspect. The fascia lata is attached to the length of the ligament. It exerts traction on it and makes the inguinal ligament convex inferiorly. Later-

al to the pubic tubercle, an extension from the deep surface of the inguinal ligament forms the **lacunar ligament**. It extends posteriorly to the pecten pubis. This triangular lacunar ligament [Fig. 17.8A] has an apex attached to the pubic tubercle, and a sharp and curved base directed laterally. The free base of the lacunar ligament lies medial to the aperture through which the femoral vessels enter the thigh.

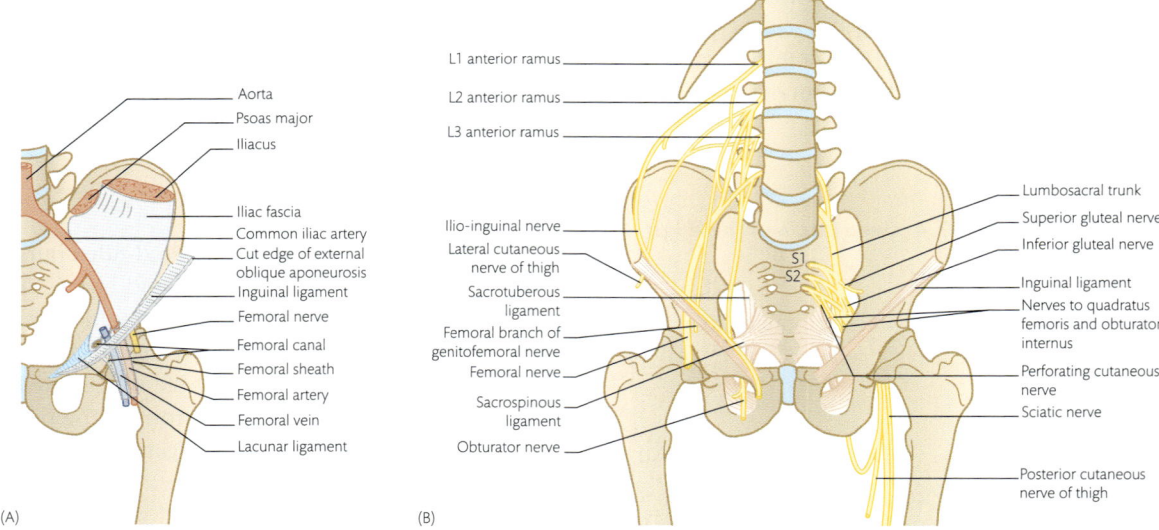

Fig. 17.8 (A) Diagram to show the routes of entry of femoral nerves and blood vessels into the lower limb. A portion of the aponeurosis of the external oblique muscle of the abdomen and the inguinal and lacunar ligaments are shown. (B) Diagram to show the course of sciatic, femoral, and obturator nerves as they enter the lower limb.

Junction of the anterior abdominal wall and the thigh

At the inguinal ligament, the anterior and posterior abdominal walls come together, as seen in Fig. 17.9]. The **transversalis fascia** lining the deep surface of the anterior abdominal wall and the **iliac fascia** covering the lower part of the posterior abdominal wall meet at the inguinal ligament. Between the inguinal ligament and the hip bone is a gap through which muscles, vessels, and nerves pass from the abdomen into the thigh [Fig. 17.10]. Deep to the lateral part of the inguinal ligament are the iliacus and psoas covered by fascia, and the femoral and lateral cutaneous nerves of the thigh. Deep to the medial part are the external iliac artery and vein, continuing in the thigh as the femoral artery and vein.

Femoral sheath

The femoral artery and vein are covered by a funnel-shaped extension of the fascial lining of the abdomen—the **femoral sheath**. The femoral sheath is formed by the transversalis fascia anteriorly and the fascia iliaca posteriorly [Fig. 17.11]. It lies immediately lateral to the lacunar ligament [Figs 17.8A, 17.10]. Two anteroposterior septa within the femoral sheath divide the space into three compartments. The lateral compartment contains the **femoral artery** and the **femoral branch of the genitofemoral nerve**. The middle compartment contains the **femoral vein**. The medial compartment is the **femoral canal**.

Femoral canal

The femoral canal is the most medial compartment within the femoral sheath. It narrows inferiorly and ends where the sheath fuses with the adventitia of the vessels at the lower margin of the saphenous opening. The wide upper end of the femoral canal is the **femoral ring**. It is separated from the abdominal cavity only by the smooth innermost lining of the abdominal wall—the peritoneum. The **boundaries** of the femoral ring are: **anteriorly** the inguinal ligament; **medially** the sharp edge of the lacunar ligament; **posteriorly** the pecten pubis; and **laterally** the femoral vein. Inferiorly, the canal lies posterior to the saphenous opening and cribriform fascia, and anterior to the fascia covering the pectineus muscle. The canal contains **loose fatty tissue** (the femoral septum), a small **deep inguinal lymph node** (the node of Cloquet), and some lymph vessels.

Dissection 17.5 explores these features.

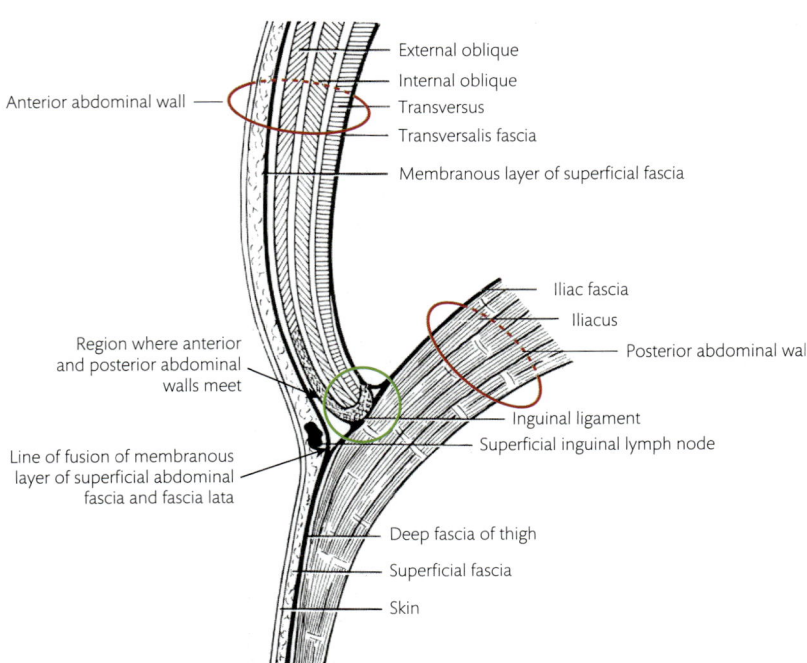

Fig. 17.9 Diagram of fasciae and muscles of the inguinal and subinguinal regions lateral to the femoral sheath.

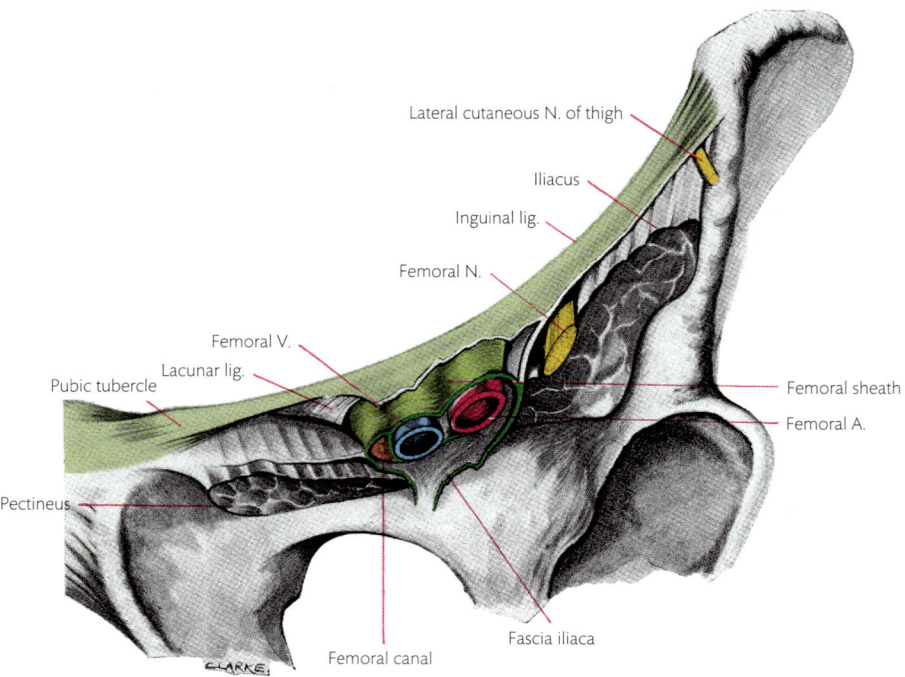

Fig. 17.10 Dissection to show the femoral sheath and structures which pass between the inguinal ligament and the hip bone—seen from below and in front.

Femoral triangle

The femoral triangle is an intermuscular space on the front of the thigh. **Base**: the base lies superiorly and is formed by the inguinal ligament. **Lateral border**: is formed by the medial border of the

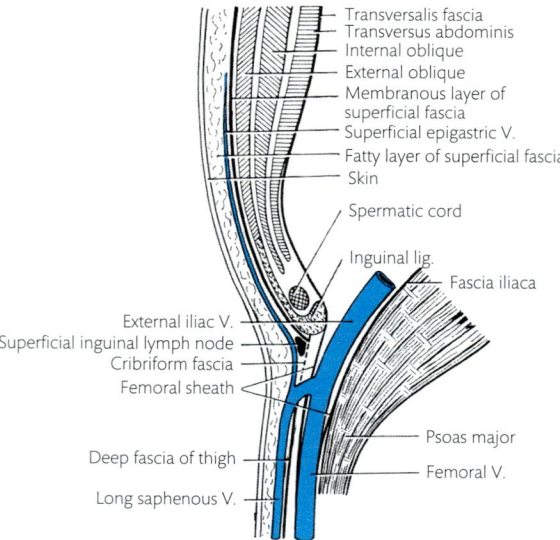

Fig. 17.11 Diagram of fasciae and muscles of the inguinal and subinguinal regions in the line of the femoral vein.

sartorius. **Medial border**: is formed by the medial border of the adductor longus. Inferiorly is the **apex** of the triangle where the medial border of the sartorius and the lateral border of the adductor longus meet. It is continuous with a narrow intermuscular space—the **adductor canal** [Fig. 17.12].

The **roof**, or **anterior wall**, of the triangle is made up of the deep fascia. The superficial inguinal lymph nodes and vessels, the upper part of the long saphenous vein, the femoral branch of the genitofemoral nerve, branches of the ilioinguinal nerve, and superficial branches and tributaries of the femoral vessels lie over the deep fascia covering the femoral triangle.

The **floor**, or **posterior wall**, is formed by the adductor longus, pectineus, psoas major, and iliacus, from the medial to lateral sides. The floor slopes posteriorly, and the femoral vessels lie in the central hollow.

Dissection 17.6 studies the femoral triangle.

Contents of the femoral triangle

Important blood vessels, nerves, lymph nodes, and lymphatics lie in the femoral triangle.

1. The **femoral vessels** traverse the triangle from the base to the apex. The vein is medial to the artery at the base, but behind it at the apex.

DISSECTION 17.5 Femoral sheath, canal, and ring

Objective

I. To clean and study the femoral sheath and its contents.   **II.** To study the margins of the femoral ring.

Instructions

1. Follow the long saphenous vein through the anterior wall of the femoral sheath to the femoral vein, and expose the femoral vein.
2. Split the femoral sheath, lateral and medial to the vein, to expose the femoral artery and femoral canal, respectively. Note the septa of the sheath which separate the compartments in which the artery, vein, and canal lie.
3. Note that the canal is shorter than the spaces which contain the vessels. Introduce your little finger into the canal, and push it upwards. It is possible to enter the abdomen through the canal.
4. At the abdominal opening of the canal (the **femoral ring**), feel the edge of the lacunar ligament medially, the inguinal ligament anteriorly, and the pecten pubis posteriorly.

2. The **profunda femoris artery** is a branch of the femoral artery. It arises from the posterolateral side of the femoral artery, curves down behind it, and goes posterior to the adductor longus. The profunda vein is anterior to its artery and ends in the femoral vein.
3. The **lateral** and **medial circumflex femoral arteries** arise from the profunda femoris artery near its origin. The lateral circumflex femoral artery runs laterally among the branches of the femoral nerve and passes posterior to the sartorius. The medial circumflex femoral artery passes backwards between the psoas and pectineus muscles. The circumflex veins end in the femoral vein.
4. The **deep external pudendal** artery arises from the femoral artery near the base of the triangle. It runs medially to the scrotum in the male and to the labium majus in the female.
5. Three or four **deep inguinal lymph nodes** lie along the medial side of the femoral vein. They receive lymph vessels from the superficial inguinal and popliteal lymph nodes, and from the deep structures of the limb. Efferent lymph vessels pass from the deep inguinal nodes to the

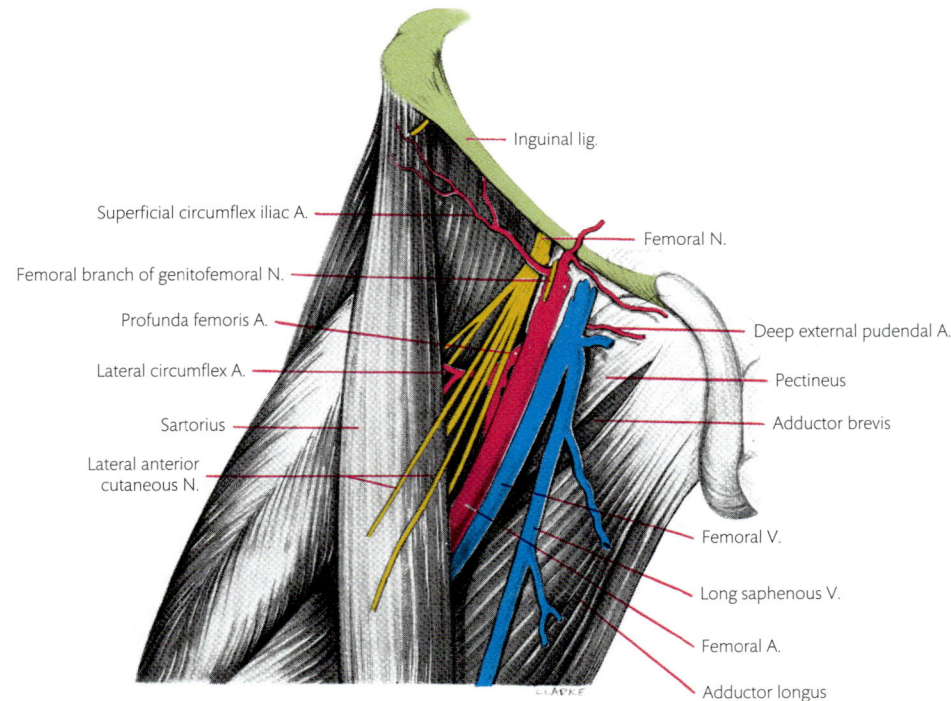

Fig. 17.12 Dissection of the right femoral triangle.

Labels: Inguinal lig. — Superficial circumflex iliac A. — Femoral N. — Femoral branch of genitofemoral N. — Profunda femoris A. — Deep external pudendal A. — Lateral circumflex A. — Pectineus — Sartorius — Adductor brevis — Lateral anterior cutaneous N. — Femoral V. — Long saphenous V. — Femoral A. — Adductor longus

DISSECTION 17.6 Femoral triangle

Objective

I. To identify the muscles, vessels, and nerves of the femoral triangle.

Instructions

1. Expose the sartorius and adductor longus muscles down to the apex of the triangle where they meet. Preserve the nerves close to the sartorius [Fig. 17.12].
2. Place a block under the knee to flex the hip joint, and relax the structures in the triangle. Find the femoral nerve lateral to the artery in the groove between the psoas and iliacus muscles.
3. Note that the femoral nerve divides almost immediately into a number of cutaneous and muscular branches.
4. Find the nerve to the pectineus passing medially behind the femoral artery.
5. Follow the other branches of the femoral nerve, till they leave the triangle. Avoid injury to the lateral circumflex artery which passes laterally among these nerves near their origin.
6. Remove the venae comitantes of the smaller arteries to get a clear picture of the arrangement of the vessels.
7. Clean the upper part of the femoral artery. Find the deep external pudendal artery which arises from the upper part of the femoral artery and runs medially.
8. Identify the root of the large profunda femoris artery which arises from the posterolateral surface of the femoral artery, about 5 cm below the inguinal ligament. Follow it downwards with the profunda vein behind the femoral vessels, until it leaves the triangle.
9. Find the lateral and medial circumflex femoral arteries which arise from the profunda near its origin or from the adjacent femoral artery. Trace the lateral artery as far as the sartorius, and the medial artery backwards as far as possible behind the femoral vessels. Preserve the proximal parts of the circumflex veins which enter the femoral vein.
10. Trace the nerve to the pectineus behind the femoral vein.
11. Remove the fascia from the pectineus, and find the anterior branch of the obturator nerve in the interval between it and the adductor longus. The nerve descends behind both muscles in front of the adductor brevis.
12. Strip the fascia from the surface of the iliacus and psoas major. Place your finger on the anterior surface of the tendon of the psoas, and push it downwards and backwards, following the line of the tendon. It is usually possible to reach the lesser trochanter of the femur to which the tendon is attached.
13. The medial circumflex artery passes backwards between the psoas and pectineus muscles, parallel to your finger. Expose it as far as possible.

external iliac nodes on the external iliac vessels in the abdomen.
6. The **femoral branch of the genitofemoral nerve** supplies the skin over the femoral triangle [Fig. 17.6].
7. The **lateral cutaneous nerve of the thigh** crosses the lateral angle of the triangle.
8. The **femoral nerve** ends in the femoral triangle.

Dissection 17.7 (and Figs 17.13 and 17.14) continues the dissection of the front of the thigh.

Sartorius

Origin: this long, strap-like muscle arises from the anterior superior iliac spine and runs across the front of the thigh. **Insertion**: it descends posterior to the medial side of the knee and forms a thin tendinous sheet in the leg, and is inserted into the upper part of the medial surface of the tibia [see Fig. 16.12]. This tendon is separated by a bursa (**bursa anserina**) from the tendons of the gracilis and semitendinosus which are inserted on the tibia posterior to it.

The sartorius forms the lateral boundary of the femoral triangle, the roof of the adductor canal, and produces a vertical, fleshy ridge far back on the medial side of the extended knee. When the knee is flexed, the muscle slips backwards into the medial boundary of the popliteal fossa. **Nerve supply**: femoral nerve. **Actions**: it flexes the hip joint and the knee joint, and rotates the thigh laterally to bring the limb into the position adopted when sitting cross-legged (**sartorius** comes from the Latin word *sartor*, meaning tailor. The name bears reference to the cross-legged position in which tailors once sat). When the knee is flexed, the sartorius medially rotates the tibia on the femur.

Objective

I. To identify the tensor fasciae latae and the parts of the quadriceps femoris. **II.** To trace the medial and lateral circumflex femoral arteries and their branches. **III.** To demonstrate the boundaries and contents of the adductor canal.

Instructions

1. Expose the sartorius down to its insertion onto the tibia.
2. Make a vertical incision through the fascia lata from the tubercle of the iliac crest to the lateral margin of the patella. Remove the fascia lata between the incision and the sartorius. This uncovers the tensor fasciae latae and parts of the four elements of the quadriceps muscle but leaves the greater part of the iliotibial tract in position.
3. Find: (1) the rectus femoris in the middle of the front of the thigh; (2) part of the vastus lateralis lateral to the rectus femoris; (3) the vastus intermedius deep to the rectus femoris; and (4) part of the vastus medialis between the lower parts of the rectus femoris and sartorius [Fig. 17.13].
4. Lift the rectus femoris, and follow it to its origin on the hip bone.
5. Trace the lateral circumflex femoral artery behind the sartorius and the rectus femoris. Follow its three branches: (1) the descending branch along the anterior border of the vastus lateralis; (2) the ascending branch that runs between the sartorius and the tensor fasciae latae on a deeper plane; and (3) a small transverse branch that enters the vastus lateralis.
6. Pull the middle third of the sartorius laterally. This exposes a narrow strip of the fascia, the roof of the adductor canal, between the vastus medialis and the adductor muscles. Divide the fascia longitudinally, and find the femoral vessels, the saphenous nerve, and the nerve to the vastus medialis in the canal [Fig. 17.14].

Adductor canal

The adductor canal is a deep furrow on the medial side of the thigh. **Boundaries**: it is bounded *anteriorly* by the vastus medialis; *posteriorly* by the adductor longus and magnus muscles; and the *roof*, or *anteromedial* boundary, is formed by the strong fascia that stretches from the vastus medialis to the adductors and has the sartorius lying on it. **Extent**: from the apex of the femoral triangle to the adductor hiatus—an opening in the attachment of the adductor magnus. **Contents**: (1) the femoral vessels; (2) the saphenous nerve; and (3) the nerve to the vastus medialis lie in this canal. Inferiorly, the saphenous nerve leaves the canal by piercing the roof. The femoral vessels pass into the popliteal fossa through the adductor hiatus, and the nerve to the vastus medialis descends in the canal and enters that muscle.

Femoral artery

The femoral artery is the main artery of the lower limb. It **begins** as the continuation of the external iliac artery of the abdomen. As this vessel passes behind the inguinal ligament at the mid-inguinal point [Figs 17.8A, 17.11], its name changes to the femoral artery. **Relations**: at the mid-inguinal point, the femoral artery is medial to the femoral nerve, lateral to the femoral vein, and anterior to the psoas major and hip bone. ➲ By compressing the artery against the bone, it is possible to control bleeding from a more distal point when a distal branch of the artery is cut. The artery enters the femoral triangle, anterior to the head of the femur, and is covered only by skin and the fascia in the triangle. It leaves the triangle at its apex and runs through the adductor canal with the femoral vein, the saphenous nerve, and the nerve to the vastus medialis. Here it lies close to the shaft of the femur and receives a branch from the obturator nerve. It **ends** by continuing as the popliteal artery as it passes through the tendinous opening in the adductor magnus [Fig. 17.12].

Branches

In the femoral triangle, the femoral artery gives off the profunda femoris, three small superficial arteries of the groin—the superficial circumflex iliac artery, the superficial epigastric artery, and the superficial external pudendal artery—and the deep external pudendal artery. In the adductor canal, it gives muscular branches and the descending genicular artery to the knee. The descending genicular artery supplies adjacent muscles and the knee joint, and sends a branch with the saphenous nerve to the medial side of the knee and leg [see Fig. 16.1].

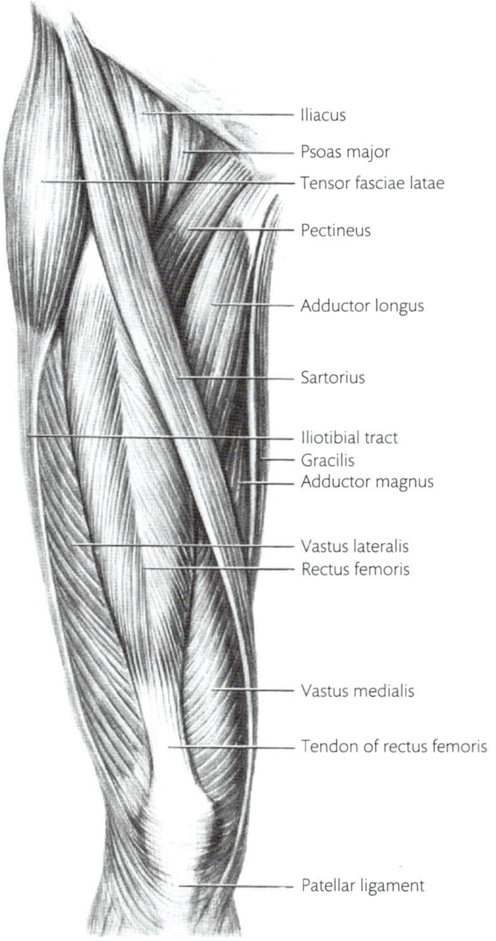
Iliacus
Psoas major
Tensor fasciae latae
Pectineus
Adductor longus
Sartorius
Iliotibial tract
Gracilis
Adductor magnus
Vastus lateralis
Rectus femoris
Vastus medialis
Tendon of rectus femoris
Patellar ligament

Fig. 17.13 Muscles of the front of the right thigh.

Femoral vein

The femoral vein **begins** as the continuation of the popliteal vein, at the opening in the adductor magnus. It **ends** as the external iliac veins, behind the inguinal ligament. The relationship between the femoral vein and artery changes—the vein is posterior to the artery in the lower part of the femoral triangle, and medial in the upper part [Fig. 17.12]. The femoral vein contains several **valves**. One is constantly present, proximal to the entry of the profunda vein. Open the vein, and examine the valve.

Tributaries

The tributaries of the femoral vein in the thigh correspond to the branches of the arteries, with some exceptions. The superficial veins of the groin end in the long saphenous vein (the corresponding arteries are branches of the femoral artery), and the medial and lateral circumflex veins enter the femoral vein (the corresponding arteries are usually branches of the profunda femoris artery).

Femoral nerve (L. 2, 3, 4)

The femoral nerve arises from the **lumbar plexus** in the abdomen [Figs 17.8B, 17.15, 17.16]. It descends between the iliacus and the psoas major muscles behind the iliac fascia and enters the thigh, posterior to the inguinal ligament [Fig. 17.10]. It ends by dividing into branches 2cm below the inguinal ligament. The branches are muscular, articular, or

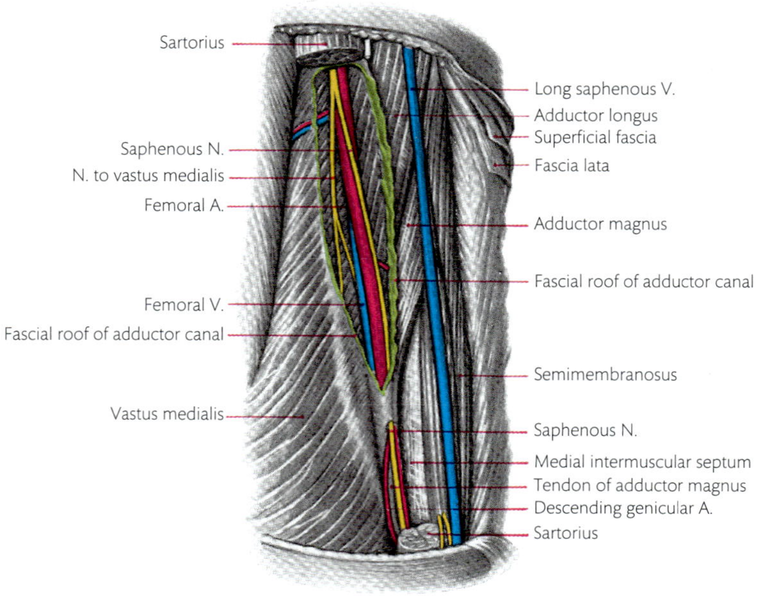
Sartorius
Saphenous N.
N. to vastus medialis
Femoral A.
Femoral V.
Fascial roof of adductor canal
Vastus medialis
Long saphenous V.
Adductor longus
Superficial fascia
Fascia lata
Adductor magnus
Fascial roof of adductor canal
Semimembranosus
Saphenous N.
Medial intermuscular septum
Tendon of adductor magnus
Descending genicular A.
Sartorius

Fig. 17.14 Dissection of the adductor canal in the right thigh. A portion of the sartorius has been removed.

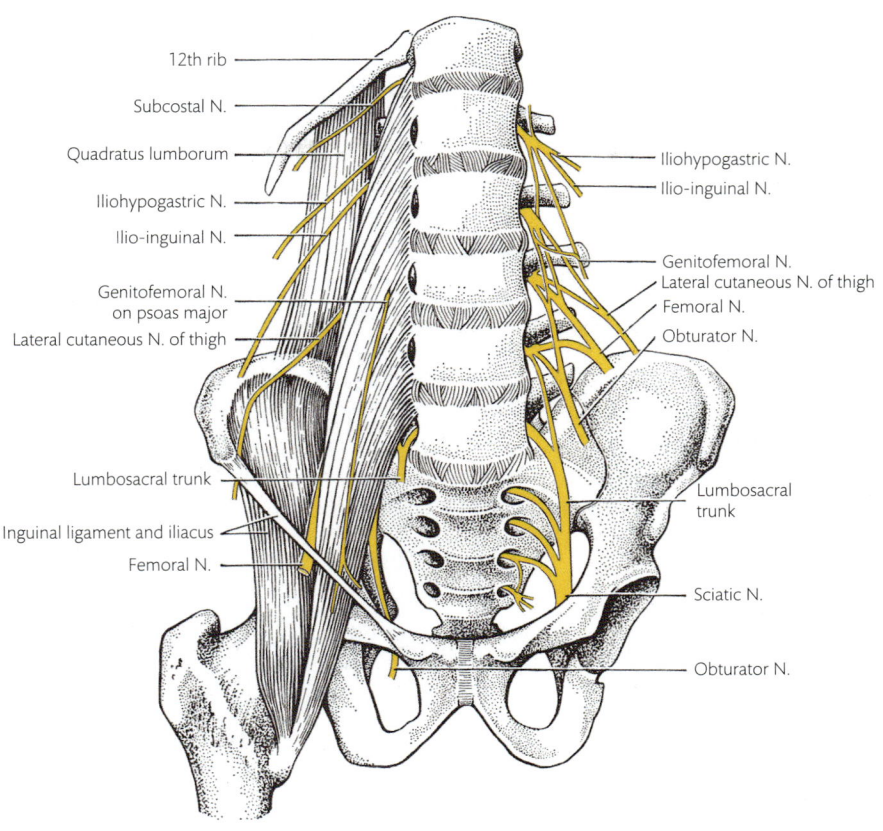

Labels on Fig. 17.15 (clockwise/by region):

12th rib
Subcostal N.
Quadratus lumborum
Iliohypogastric N.
Ilio-inguinal N.
Genitofemoral N. on psoas major
Lateral cutaneous N. of thigh
Lumbosacral trunk
Inguinal ligament and iliacus
Femoral N.

Iliohypogastric N.
Ilio-inguinal N.
Genitofemoral N.
Lateral cutaneous N. of thigh
Femoral N.
Obturator N.
Lumbosacral trunk
Sciatic N.
Obturator N.

Fig. 17.15 Lumbar plexus (semi-diagrammatic) shown in relation to the iliopsoas muscles and other muscles on the posterior abdominal wall.

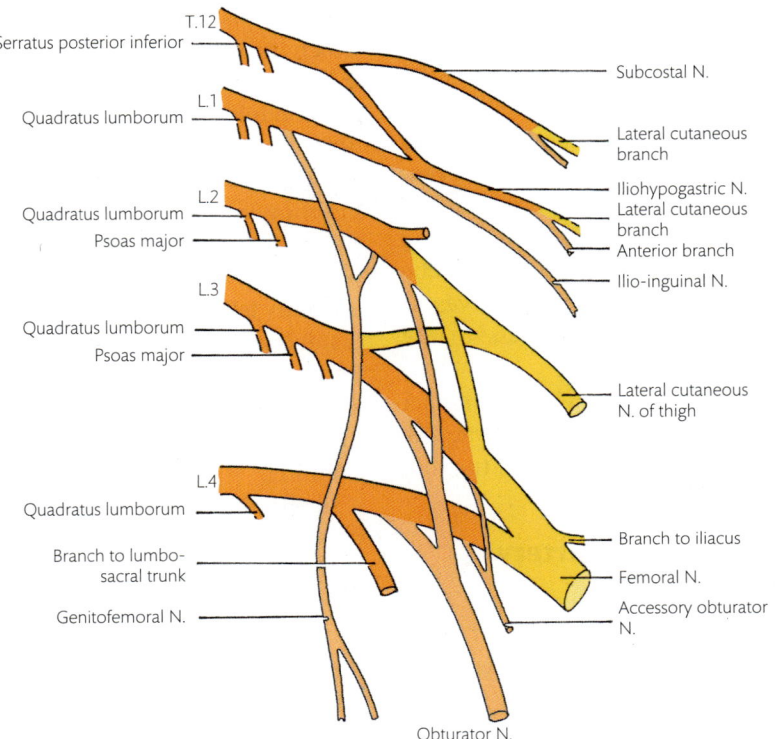

Labels on Fig. 17.16:

T.12
Serratus posterior inferior
L.1
Quadratus lumborum
L.2
Quadratus lumborum
Psoas major
L.3
Quadratus lumborum
Psoas major
L.4
Quadratus lumborum
Branch to lumbo-sacral trunk
Genitofemoral N.

Subcostal N.
Lateral cutaneous branch
Iliohypogastric N.
Lateral cutaneous branch
Anterior branch
Ilio-inguinal N.
Lateral cutaneous N. of thigh
Branch to iliacus
Femoral N.
Accessory obturator N.
Obturator N.

Fig. 17.16 Diagram of the lumbar plexus. Ventral divisions, light orange; dorsal divisions, yellow.

cutaneous. Muscular branches are to the pectineus, sartorius, and quadriceps femoris. Articular branches are to the hip and knee joints. Cutaneous branches include the anterior cutaneous nerves of the thigh (medial and lateral) and the saphenous nerve.

Cutaneous branches

The **lateral** and **medial anterior cutaneous nerves** (L. 2, 3) run along the medial margin of the sartorius and pierce the deep fascia to supply the skin and subcutaneous tissue.

The **saphenous nerve** (L. 3, 4) is the longest branch of the femoral nerve and the only one that has its main distribution in the leg and foot. It accompanies the femoral vessels in the adductor canal and pierces the fibrous roof of the canal and the deep fascia at the posterior border of the sartorius, medial to the knee [Fig. 17.6].

Muscular and articular branches

The branch to the **pectineus** runs medially and downwards behind the femoral vessels to the pectineus. Two or three nerves to the **sartorius** usually arise in common with the lateral anterior cutaneous nerve. The nerves to the **rectus femoris** (usually two) enter the deep surface of the muscle, and the upper one supplies the hip joint. The nerve to the **vastus medialis** enters the adductor canal and supplies the muscle at different levels. It sends a branch to the knee joint. The nerve to the **vastus lateralis** passes deep to the rectus femoris and accompanies the descending branch of the lateral circumflex artery to the anterior border of the muscle. It usually gives a branch to the knee joint. Two or three nerves to the **vastus intermedius** enter its anterior surface. The most medial nerve is a long, slender branch which runs along the medial edge of the vastus intermedius to the **articularis genus** muscle (a part of the vastus intermedius). Its terminal filaments pass to the knee joint.

The nerves supplying the vasti, which only act on the knee, send branches to the knee joint. The nerve to the rectus femoris, which acts on the hip, sends branches to the hip joint.

Lateral circumflex femoral artery

The lateral circumflex femoral artery is the largest branch of the profunda femoris artery. It arises from the profunda femoris artery near its origin and runs laterally among the branches of the femoral nerve, and then deep to the rectus femoris. It ends by dividing into: (1) ascending; (2) transverse; and (3) descending branches. It supplies the structures on the lateral side of the hip and thigh.

The **ascending branch** passes along the intertrochanteric line of the femur to the gluteal surface of the ilium. It supplies the surrounding muscles and the hip joint, and anastomoses with the **superior gluteal artery**. The small **transverse branch** passes backwards through the vastus lateralis. It anastomoses with other arteries, posterior to the femur. The **descending branch** runs along the anterior border of the vastus lateralis. It supplies a large part of the quadriceps and sends a long branch through the vastus lateralis to the anastomosis at the knee joint.

Tensor fasciae latae

This muscle lies between the gluteal region and the front of the thigh [Fig. 17.13]. **Origin**: it arises from the anterior part of the iliac crest. **Insertion**: it is inserted into the iliotibial tract, 3–5 cm below the level of the greater trochanter. It is enclosed between two layers of the iliotibial tract. **Nerve supply**: superior gluteal nerve. **Actions**: (1) flexion and medial rotation of the hip joint; (2) extension of the knee through the iliotibial tract.

Iliotibial tract

The iliotibial tract is a thick band of fascia lata which runs vertically on the lateral side of the thigh. **Attachments**: from the iliac crest to the lateral condyle of the tibia [Fig. 17.2]. The greater part of the gluteus maximus and the tensor fasciae latae are inserted into it. These muscles, through their insertion into the tract, help to steady the pelvis on the thigh and keep the knee extended in the erect position. While standing, the tensed iliotibial tract is readily felt on the lateral side of the thigh, immediately proximal to the lateral condyle of the femur. By comparison, the palpably relaxed quadriceps and mobile patella indicate that the quadriceps is not responsible for maintaining knee extension on standing.

Superiorly, most of the posterior part of the tract passes deep to the gluteus maximus. Its anterior part splits to enclose the tensor fasciae latae, and the intermediate part passes directly to the iliac crest. Inferiorly, the tract is continuous with the rest of the fascia lata and the lateral intermuscular septum.

Dissection 17.8 exposes the lateral intermuscular septum.

Intermuscular septa

There are three intermuscular septa in the thigh—lateral, medial, and posterior intermuscular septa [Fig. 17.7]. The lateral intermuscular septum is strong and fibrous. It passes from the deep surface of the iliotibial tract to the lateral supracondylar line and the linea aspera. The medial and posterior septa are thin fascial sheets. All three septa pass to the linea aspera and the corresponding supracondylar line. The septa separate the muscles of the three fascial compartments. The **anterior compartment** lies between the medial and lateral intermuscular septa, and contains the extensors of the knee and the femoral nerve. The **medial compartment** lies between the medial and posterior septa, and contains the adductors of the hip and the obturator nerve. The **posterior compartment** lies between the posterior and lateral intermuscular septa and contains the flexors of the knee and the sciatic nerve [Fig. 17.7].

Quadriceps femoris

The extensor group consists principally of four large muscles—the **quadriceps femoris**—which are inserted into the patella. Their tendinous fibres continue over the anterior surface of the patella as the patellar tendon which attaches the patella to the tibial tuberosity [Fig. 17.13].

Rectus femoris

Origin: this muscle arises by two heads. The **straight head** from the anterior inferior iliac spine, and the **reflected head** from a groove immediately above the acetabulum [see Fig. 16.5]. In varying degrees of flexion of the hip joint, one or other of these heads takes the major part of the strain. **Insertion**: the muscle runs vertically down the front of the thigh in a groove between the iliopsoas and tensor fasciae latae superiorly, and between the vastus lateralis and vastus medialis inferiorly. It overlies the anterior part of the vastus intermedius. It joins the common tendon to be inserted into the proximal part of the patella. **Nerve supply**: femoral nerve. **Actions**: extension of the knee and flexion of the hip joint.

Vastus lateralis

Origin: it has a long linear origin from the upper part of the intertrochanteric line, the root of the greater trochanter, the gluteal tuberosity, and the upper half of the lateral lip of the linea aspera [see Fig. 16.9]. The lowest fibres lie 3–4 cm proximal to the patella. Together with the vastus intermedius with which it is partly fused, the vastus lateralis muscle covers the lateral aspect of the femur. **Insertion**: the muscle fibres run downwards and forwards to the base of the patella, the anterolateral part of the fibrous capsule of the knee joint, and the lateral condyle of the tibia. **Nerve supply**: femoral nerve. **Actions**: extension of the knee joint.

Vastus medialis

Origin: the vastus medialis muscle has a long linear origin from the intertrochanteric and spiral lines, the linea aspera, the medial supracondylar line [see Figs 16.7, 16.9], the medial intermuscular septum, and the tendons of the adductor longus and magnus. **Insertion**: most of the muscle bundles are directed downwards and forwards onto the proximal surface of the patella. The lowest fibres run horizontally into the medial aspect of the upper half of the patella. These lowest fibres help to hold the patella medially and prevent lateral displacement of the patella. They form a prominent bulge, just proximal to the medial condyle of the femur [Fig. 17.1]. Some of the fibres of the vastus medialis are inserted into the anteromedial part of the fibrous capsule of the knee joint and the medial condyle of the tibia. **Nerve supply**: femoral nerve. **Actions**: extension of the knee.

Vastus intermedius

Origin: the vastus intermedius has an extensive origin from the lateral and anterior surfaces of the shaft of the femur, and the lower part of the lateral intermuscular septum [see Fig. 16.8]. **Insertion**: it passes to the common tendon of the quadriceps which is inserted into the lateral border of the patella, and the lateral condyle of the tibia. **Nerve supply**: femoral nerve. **Actions**: extension of the knee joint.

Articularis genu

Some of the lowest fibres of the vastus interme- dius constitute the **articularis genus** muscle. **Origin**: they arise from the front of the femur. **Insertion**: into the suprapatellar bursa of the knee joint. **Nerve supply**: femoral nerve. **Actions**: these fibres retract the bursa during extension of the knee joint.

All four parts of the quadriceps femoris are insert- ed into the patella and through it and the patellar tendon on the tibial tuberosity. The three vasti act solely on the knee joint. The rectus femoris is also a flexor of the hip joint. It is more separate than the vasti and only fuses with them, as it approaches the insertion into the patella. The tendon of the quadri- ceps, the patella, and the patellar tendon together form the anterior part of the fibrous capsule of the knee joint.

Deep patellar bursae

There are two deep patellar bursae: (1) a large **supra- patellar bursa** which separates the tendon of the quadriceps femoris from the front of the femur—it extends approximately a hand breadth above the patella and is usually continuous with the cavity of the knee joint; and (2) a **deep infrapatellar bursa** which lies between the tibial tuberosity and the patellar ligament.

Psoas major and iliacus

Origin: the psoas major and iliacus arise within the abdomen—the iliacus from the iliac fossa [see Fig. 16.4] and the psoas major from the vertebral column. (Details of the bony attachments in the abdomen are discussed in Volume 2.) They fuse with each other as they enter the thigh, posterior to the inguinal ligament [Fig. 17.15]. They are sep- arated posteriorly from the capsule of the hip joint by a **bursa** which may communicate with the joint cavity. **Insertion**: the muscles pass inferior to the neck of the femur. The psoas major is inserted into the lesser trochanter. The iliacus is inserted into the surface of the femur below it [see Fig. 16.9]. **Nerve supply**: the ventral rami of L.2 and L.3. **Actions**: the iliopsoas are the chief flexors of the hip joint. If the limb is fixed, it flexes the trunk on the thigh. It also produces medial rotation of the

thigh, because its insertion is lateral to the axis of rotation of the femur.

➲ Its action is important clinically, because spasm of the psoas produces flexion and medial rotation of the hip joint—a position taken up by the right lower limb in appendicitis when the inflamed appendix causes spasm on the underlying right psoas.

When the neck of the femur is fractured, the ili- opsoas produces marked lateral rotation of the dis- tal segment of the femur (and of the distal part of the limb). As a result, the toes of the affected limb point laterally in the supine patient.

➲ The psoas major is covered by the psoas fas- cia, a covering which extends from the lumbar ver- tebrae to the insertion of the psoas on the lesser tubercle. An abscess formed in the lumbar vertebrae can track within the psoas fascia and present in the front of the thigh—a **psoas abscess**.

Medial side of the thigh

The muscles on the medial side of the thigh pro- duce adduction at the hip joint. These are the **adductor muscles**, pectineus, adductor longus, adductor brevis, adductor magnus, and gracilis. They are arranged in three layers. The anterior layer has the pectineus and adductor longus. The middle layer has the adductor brevis. The posterior layer has the adductor magnus. These muscles are attached proximally to the hip bone [see Fig. 16.5] and distally to the back of the femur [see Fig. 16.9].

The two branches of the **obturator nerve**—the anterior and posterior divisions—descend between the muscles and are separated from each other by the adductor brevis. The nerve supplies these mus- cles and the obturator externus. The **profunda femoris artery** descends posterior to the adduc- tor longus, close to the femur.

Adductor longus

Origin: this triangular muscle takes origin by a narrow tendon from the front of the body of the pubis, immediately below the pubic crest [see Fig. 16.5]. **Insertion**: it widens as it passes inferolat- erally, and is inserted into the linea aspera of the femur between the vastus medialis and the other adductors [see Fig. 16.9]. **Nerve supply**: anterior branch of the obturator nerve. **Actions**: adduction of the thigh.

DISSECTION 17.9 Medial compartment of the thigh-1

Objectives

I. To study the adductor longus, gracilis, and obturator externus. II. To identify and trace the anterior and posterior divisions of the obturator nerve.

Instructions

1. Remove the fascia from the adductor longus, gracilis, and obturator externus and the nerves that supply them.
2. Divide the adductor longus transversely, 2–3 cm below its origin. Turn the distal part towards the femur.
3. Find the nerve supplying it, and trace it to the anterior branch of the obturator nerve.
4. Follow the anterior branch of the obturator nerve inferiorly to the gracilis, and find a small branch entering the adductor canal.
5. Trace the gracilis to its attachments.
6. Define the attachments of the pectineus [see Figs 16.5, 16.9]. Avoid injury to the branches of the obturator nerve behind it and the medial circumflex artery superolateral to it. Detach the pectineus from its origin, and turn it laterally.
7. Trace the anterior branch of the obturator nerve and the medial circumflex artery as far as possible.
8. Identify the obturator externus. It lies superior to the medial circumflex artery and has the anterior branch of the obturator nerve passing anterosuperior to it [Fig. 17.17].

Dissection 17.9 (and Fig. 17.17) begins the dissection of the medial compartment of the thigh.

Pectineus

Origin: the pectineus arises from the pectineal surface of the pubis [see Fig. 16.5]. **Insertion**: it is inserted into the upper half of a line joining the lesser trochanter of the femur to the linea aspera [see Fig. 16.9]. In the base of the femoral triangle, the pectineus lies between the adductor longus and the iliopsoas [Fig. 17.13]. **Nerve supply**: femoral nerve and accessory obturator nerve, when present. The pectineus often receives an additional nerve supply from the anterior division of the obturator nerve. **Actions**: adduction and flexion of the thigh.

Adductor brevis

Origin: the adductor brevis arises from the pubis, inferior to the origin of the adductor longus. It lies behind the pectineus and adductor longus. **Insertion**: it is inserted into the linea aspera behind the pectineus and adductor longus [see Figs 16.5, 16.9]. **Nerve supply**: obturator nerve. **Actions**: adduction of the thigh.

Dissection 17.10 continues the dissection of the medial compartment of the thigh.

Gracilis

Origin: the gracilis arises from the lower half of the body of the pubis close to the symphysis and from the anterior part of the inferior pubic ramus [see Fig. 16.5]. It lies on the medial side of the thigh.

Insertion: it is inserted into the upper part of the medial surface of the tibia, posterior to the sartorius. At its insertion, it is separated from the sartorius and the tibial collateral ligament of the knee by a complex bursa—the bursa anserina. **Nerve supply**: anterior branch of the obturator nerve. **Actions**: adduction of the thigh, and flexion and medial rotation of the knee joint.

Adductor magnus

Origin: this muscle takes origin from the ischiopubic ramus and the lower part of the ischial tuber-

DISSECTION 17.10 Medial compartment of the thigh-2

Objective

I. To expose and clean the adductor brevis and magnus, and the nerve which supplies them.

Instructions

1. Divide the adductor brevis close to its origin. Turn it laterally, preserving the anterior branch of the obturator nerve.
2. Find and trace the posterior branch of the obturator nerve behind the muscle.
3. Remove the fascia from the surface of the obturator externus and adductor magnus, without damaging the branches of the obturator nerve. Define the attachments of the adductor magnus.

osity [see Fig. 16.5]. It lies posterior to the other adductor muscles. **Insertion**: it is inserted into the back of the femur, from the gluteal tuberosity to the adductor tubercle [see Fig. 16.9]. At intervals, the insertion to bone is interrupted and the muscle fibres are inserted into tendinous slips which arch over the **perforating arteries** on the surface of the femur. The opening through which the femoral vessels pass—the **adductor hiatus**—is the largest of these arches. It lies on the medial supracondylar line approximately at the junction of the middle and lower thirds of the thigh. The adductor magnus is fan-shaped [see Fig. 16.1]. It has horizontal anterior fibres, oblique middle fibres, and nearly vertical posterior fibres. The vertical fibres pass from the ischial tuberosity to the adductor tubercle. At the adductor tubercle, the tendon of the adductor magnus is continuous with the medial intermuscular septum and gives attachment to the lower fibres of the vastus medialis. **Nerve supply**: the adductor magnus is a hybrid muscle. The part originating from the ischiopubic ramus is supplied by: (1) the posterior branch of the obturator nerve; and (2) the part originating from the ischial tuberosity with the hamstring muscles is supplied by the tibial part of the sciatic nerve. **Actions**: the ischial part of the adductor magnus acts with the hamstring muscles to extend the hip joint. The horizontal and oblique fibres adduct the hip.

Stabilizing the hip bone

An important action of the adductor muscles is to stabilize the hip bone on the femur. They prevent the hip bone from tilting laterally when standing on one leg. During walking, they are continuously active in the supporting limb.

Obturator externus

Origin: this fan-shaped muscle arises from the anterior aspect of the obturator membrane and from the anterior and inferior margins of the obturator foramen [see Fig. 16.5]. **Insertion**: it passes back, curving upwards on the inferior and posterior surfaces of the neck of the femur, to be inserted into the trochanteric fossa [see Fig. 16.7]. **Nerve supply**: posterior branch of the obturator nerve. **Actions**: flexion and lateral rotation of the thigh; it also supports the hip joint and functions as an extensile ligament of the hip joint.

Obturator nerve (L. 2, 3, 4)

The obturator nerve arises from the lumbar plexus in the abdomen [Fig. 17.15]. It descends medial to the psoas muscle, on to the lateral wall of the lesser pelvis where it lies lateral to the ovary. Here it joins the obturator vessels and enters the **obturator canal**, a passage between the obturator membrane and the superior pubic rami. In the obturator canal, it divides into anterior and posterior divisions. The **anterior division** descends in the thigh, anterior to the obturator externus and adductor brevis. It supplies the adductor longus, adductor brevis, and gracilis, and the hip joint [Fig. 17.17]. Distal to the adductor longus, it enters the adductor canal and forms a plexus with branches from the medial anterior cutaneous nerve of the thigh and the saphenous nerve. Through this plexus, it may supply parts of the medial side of the thigh. The **posterior division** supplies and pierces the obturator externus and descends between the adductor brevis and magnus, supplying both. An articular branch passes through the lower part of the adductor magnus to the back of the knee joint.

Accessory obturator nerve

The accessory obturator nerve, when present, is a branch of either the lumbar plexus or the obturator nerve. It descends along the medial side of the psoas major and crosses the superior ramus of the pubis into the thigh (it does not pass through the obturator canal). It may end in the hip joint or in the pectineus, or it may pass between the psoas and the pectineus to replace part of the obturator nerve [Fig. 17.16].

Medial circumflex femoral artery

The medial circumflex femoral artery is a branch of the profunda femoris. It passes back superior to the pectineus and adductor muscles, and inferior to the psoas, obturator externus, and quadratus femoris muscles [Fig. 17.17]. It gives branches to adjacent muscles and supplies the hip joint through the acetabular notch. The terminal part of this artery is seen in the gluteal region where it takes part in the **cruciate anastomosis**. (Refer to Chapter 18.)

Obturator artery

The obturator artery is a branch of the internal iliac artery. It accompanies the obturator nerve in the

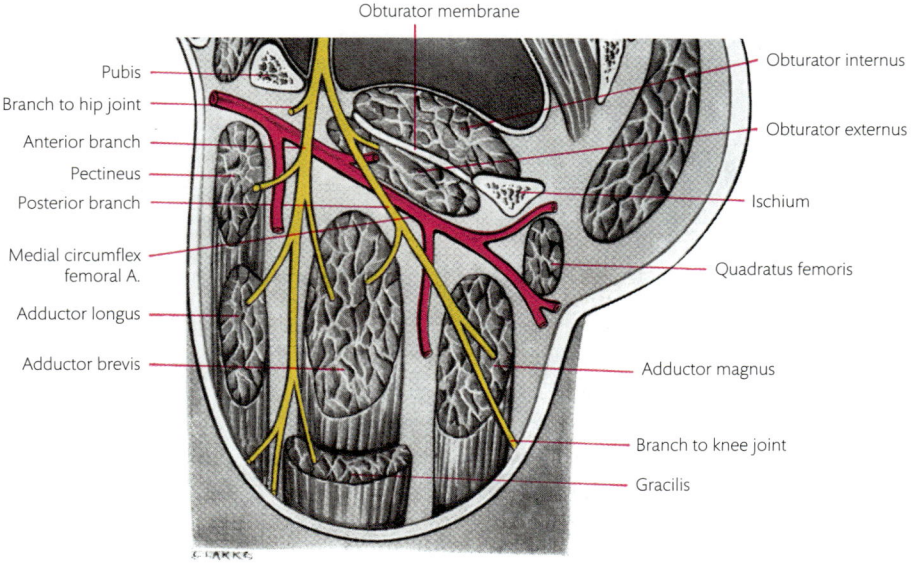

Pubis

Branch to hip joint

Anterior branch

Pectineus

Posterior branch

Medial circumflex femoral A.

Adductor longus

Adductor brevis

Obturator membrane

Obturator internus

Obturator externus

Ischium

Quadratus femoris

Adductor magnus

Branch to knee joint

Gracilis

Fig. 17.17 Schematic diagram of the adductor muscles and obturator nerve.

obturator canal. It divides into branches which form an arterial circle on the obturator membrane, deep to the obturator externus. It supplies the adjacent muscles, bone, and hip joint. Its articular branch runs through the acetabular notch and enters the ligament of the head of the femur, sometimes playing a minor role in the supply of the femoral head.

See Clinical Applications 17.1 and 17.2.

CLINICAL APPLICATION 17.1 Femoral hernia

In the erect position, the weight of the abdominal contents presses down on the inguinal region. The femoral ring forms a point of weakness and may allow the entry of a loop of the intestine or other abdominal contents into the femoral canal. Such protrusion of abdominal contents into the thigh constitutes a femoral hernia. As the femoral ring is limited anteriorly by the inguinal ligament, any event which stretches the inguinal ligament enlarges the femoral ring. This could happen as a result of repeated pregnancies that weaken the abdominal muscles. Any other condition which chronically raises the intra-abdominal pressure (e.g. repeated coughing or straining) will also predispose to the development of such a hernia. Femoral hernias are more common in women.

When a loop of the intestine enters the femoral ring, it carries the peritoneum covering of the abdominal opening of the canal in front of it. The peritoneum forms a hernial sac which descends in the femoral canal and bulges forwards through the cribriform fascia into the superficial fascia of the thigh. If the sac continues to enlarge, it expands superolaterally in the superficial fascia, so that the entire hernia becomes U-shaped. This course of the hernia should be kept in mind when external pressure is applied in an attempt to return the hernial sac and its contents to the abdomen. The sac should first be pushed down and medially towards the saphenous opening, then through the cribriform fascia, and only then should an attempt be made to return it through the distended femoral canal.

As the hernial sac expands in the subcutaneous tissue, the margins of the femoral ring may constrict the neck of the sac. This tends to obstruct the passage of intestinal contents in the loop of gut and occlude the blood vessels to it. This could lead to strangulation of the hernia, possibly resulting in gangrene and rupture. Surgical reduction of an obstructed or a strangulated hernia commonly requires division of the lacunar ligament. Care should be taken in dividing the lacunar ligament, as an abnormal obturator artery may lie on it. When present, this abnormal artery arises from the inferior epigastric artery, instead of from the internal iliac artery, and commonly crosses the abdominal aspect of the lacunar ligament.

CLINICAL APPLICATION 17.2 Deep tendon reflexes

The patellar tendon reflex is a deep tendon reflex routinely done to test L.3 and L.4 segments of the spinal cord. The patient sits on the edge of the examination table, with his legs hanging freely. The physician strikes the patellar tendon sharply with a reflex hammer. This causes the leg to extend at the knee. Mostly, the response is evaluated visually by watching for the extension of the knee. The contraction of the quadriceps muscle can be evaluated by palpation as well.

The impact of the reflex hammer stretches the patellar tendon. This triggers sensory nerves that innervate the quadriceps to send information from the tendon to the spinal cord—segments L.3 and L.4. In the spinal cord, small internuncial neurons are activated which, in turn, stimulate the motor neurons supplying the quadriceps. This leads to contraction of the quadriceps and extension of the knee

Some important points about the deep tendon reflexes are:

1. Sensory fibres relaying a stimulus to the spinal cord form the afferent limb of the reflex arc. They travel through the ventral ramus and dorsal root to the spinal cord [see Fig. 1.22].
2. Motor fibres supplying the quadriceps form the efferent limb of the arc. They travel through the ventral root and ventral rami to reach the muscle [see Fig. 1.22].
3. Deep tendon reflexes are withdrawal reflexes involving the spinal cord (no involvement from higher centres).
4. Both afferent and efferent nerves have to be intact for a reflex action to occur.
5. Abnormal reflexes include reflexes that are lost, diminished, or heightened (of increased power and/or speed).
6. Responses are graded by using standard criteria.

CHAPTER 18
The gluteal region

Surface anatomy

The gluteal region is bound by the **iliac crest** superiorly, the **gluteal fold** of the buttock inferiorly, a **line** joining the anterior superior iliac spine to the front of the greater trochanter laterally, and the **natal cleft** between the buttocks medially [Fig. 18.1]. The horizontal gluteal fold is due to adherence of the skin to the deep fascia over the gluteus maximus, the large buttock muscle. Deep to the lower part of this muscle is the **ischial tuberosity** [Fig. 18.1]. This can be felt by pressing the fingers upwards into the medial part of the gluteal fold but is most easily identified as the rounded bony mass on which one sits.

The natal cleft begins near the third sacral spine. The lower part of the **sacrum** and the **coccyx** are in its floor. Palpate the sacrum and coccyx. The coccyx can be identified by its relative mobility. Between the lower part of the sacrum and the ischial tuberosity, a deep resistance can be felt through the posterior part of the gluteus maximus. This is the **sacrotuberous ligament**. It holds the lower part of the sacrum and prevents the upper part from being pushed down by the weight of the body.

Trace the iliac crest forwards to the anterior superior iliac spine and backwards to the posterior superior iliac spine. The posterior superior iliac spine lies in a skin dimple at the level of the second sacral spine. The posterior surface of the sacrum lies between the right and left posterior superior iliac spines.

Dissection 18.1 (and Fig. 18.2) looks at the cutaneous nerves in the gluteal region.

Superficial fascia

This is dense and contains a lot of fat, especially at the upper and lower margins of the gluteus maximus.

Cutaneous nerves

The cutaneous nerves reach the gluteal region from all directions [see Fig. 18.2].

1. From above: the **lateral cutaneous branches** of the **subcostal** (T. 12) and **iliohypogastric** (L. 1) nerves pass downwards, anterior and posterior to the tubercle of the iliac crest. They supply the skin down to the level of the greater trochanter.
2. From below: branches of the **posterior cutaneous nerve of the thigh** curve over the lower border of the gluteus maximus to the posteroinferior part of the gluteal region.
3. From the lateral side: the posterior branch of the **lateral cutaneous nerve of the thigh** (L. 2, 3) supplies the anteroinferior part.
4. From the medial side: cutaneous branches of the **dorsal rami** of L. 1–3 and S. 1–3, and the **perforating cutaneous nerve** (S. 2, 3—ventral rami) supply the medial and intermediate parts. The lumbar nerves are long and descend obliquely across the region, almost to the gluteal fold. The sacral branches are short. The perforating cutaneous nerve pierces the sacrotuberous ligament and the gluteus maximus midway between the coccyx and the ischial tuberosity [Fig. 18.2].

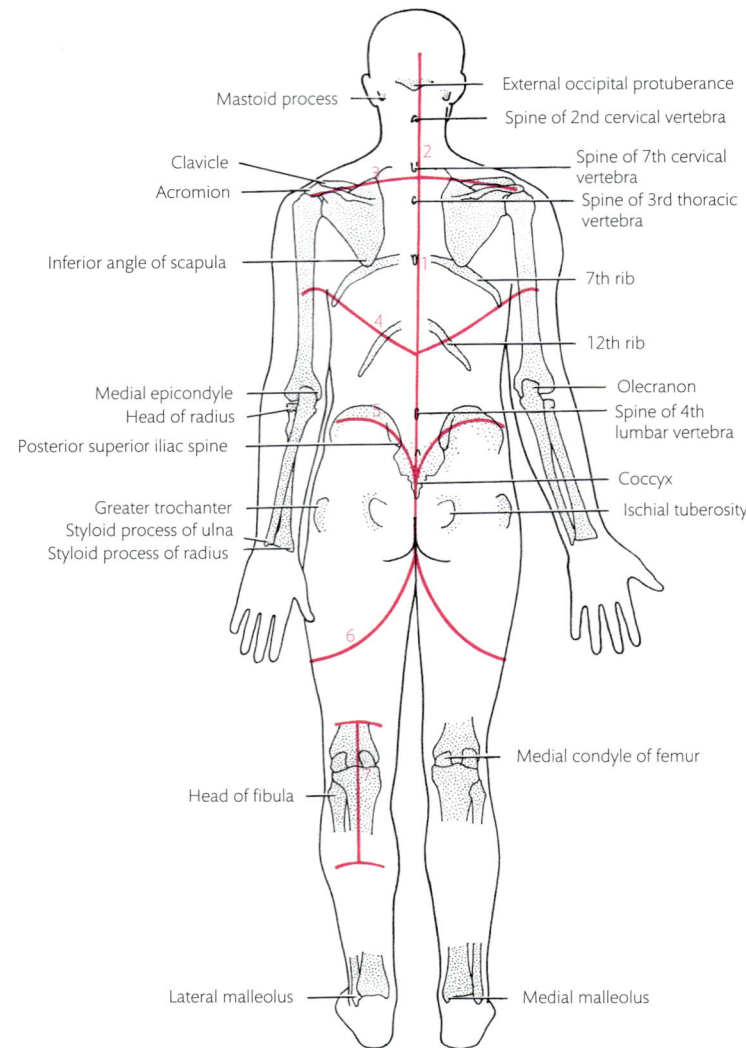

Fig. 18.1 Landmarks and incisions.

DISSECTION 18.1 Skin reflection and cutaneous nerves-1

Objective

I. To reflect the skin and identify the cutaneous nerves.

Instructions

1. Make skin incisions 5 and 6 [Fig. 18.1]. Reflect the flap of skin and superficial fascia laterally.
2. Attempt to find the cutaneous nerves of the gluteal region. They are difficult to find because of the dense superficial fascia, but it is usually possible to identify the branches of the lumbar nerves [Fig. 18.2].

Deep fascia

The deep fascia is thick over the anterior border of the gluteus maximus where the iliotibial tract splits to enclose the muscle. Everywhere else, the fascia is thin over the muscle, and thick deep to it.

Dissection 18.2 looks at the gluteus maximus.

Gluteus maximus

Origin: this powerful muscle takes origin from: (1) the external surface of the ilium behind the posterior gluteal line [see Fig. 16.3]; (2) the back of

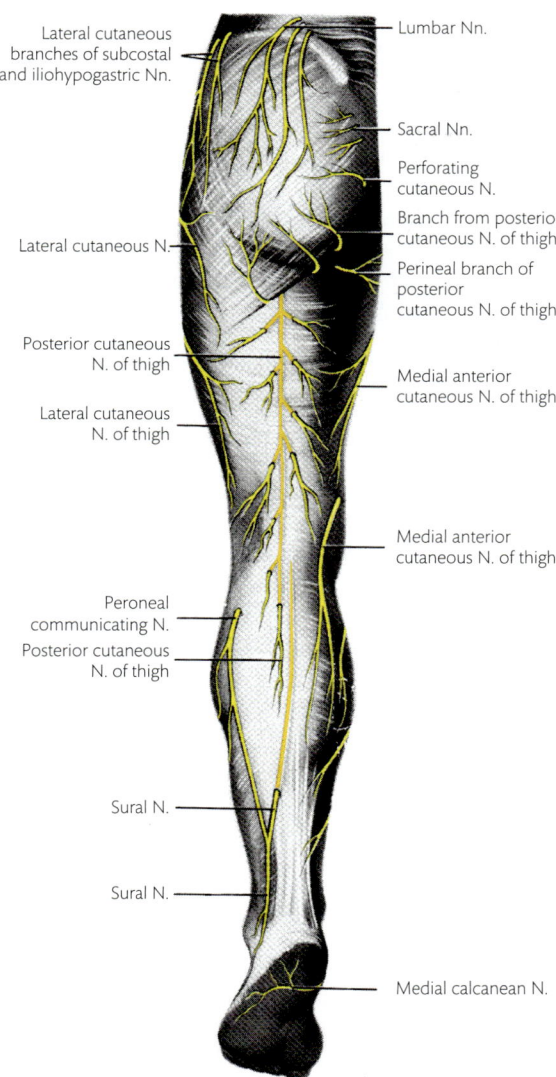

Lateral cutaneous branches of subcostal and iliohypogastric Nn.

Lumbar Nn.

Sacral Nn.

Perforating cutaneous N.

Branch from posterior cutaneous N. of thigh

Lateral cutaneous N.

Perineal branch of posterior cutaneous N. of thigh

Posterior cutaneous N. of thigh

Medial anterior cutaneous N. of thigh

Lateral cutaneous N. of thigh

Medial anterior cutaneous N. of thigh

Peroneal communicating N.

Posterior cutaneous N. of thigh

Sural N.

Sural N.

Medial calcanean N.

Fig. 18.2 Cutaneous nerves on the back of the lower limb.

the sacrum and coccyx; and (3) the sacrotuberous ligament. Its fibres run downwards and laterally, and abruptly become aponeurotic. This abrupt thinning produces the hollow of the hip, posterior to the greater trochanter of the femur. **Insertion**: fibres of the deeper one-fourth of the muscle are inserted into the gluteal tuberosity of the femur [see Fig. 16.9]. The remainder are inserted into the iliotibial tract. Aponeurotic fibres passing to the iliotibial tract run superficial to the greater trochanter and the upper part of the vastus lateralis, whereas the lower part of the muscle crosses the ischial tuberosity. The gluteus maximus is separated from all three deeper structures (greater trochanter, vastus lateralis, and ischial tuberosity) by large **bursae**. **Nerve supply**: inferior gluteal nerve. **Actions**: it is (1) a powerful extensor of the hip joint, used especially when strength is required (e.g. when the erect position has to be regained while lifting heavy weights from the floor). It is also used in running and climbing, more especially in achieving full extension of the hip joint. It acts jointly with the tensor fasciae latae (2) to stabilize the pelvis on the thigh (supporting the trunk) in an anteroposterior plane and (3) to extend the knee through the iliotibial tract.

Dissection 18.3 (and Fig. 18.3) looks at structures deep to the gluteus maximus.

Structures deep to the gluteus maximus

Begin by studying an articulated pelvis, preferably one with the sacrotuberous and sacrospinous ligaments attached [Fig. 18.4]. The **sacrotuberous ligament** passes from the medial side of the ischial tuberosity to the posterior iliac spine. The **sacrospinous ligament** runs from the ischial spine to the side of the lower part of the sacrum and coccyx, deep to the sacrotuberous ligament. These two ligaments convert the two sciatic notches into foraminae—an upper **greater sciatic foramen** and a lower **lesser sciatic foramen**. The sacrospinous ligament lies edge to edge with the levator ani muscle. Together with the muscle of the opposite side, the levator ani forms the muscular floor of the pelvis and separates the pelvis above from the perineum below. The greater sciatic foramen (which lies superior to the ischial spine) leads from the gluteal region into the pelvis. The lesser sciatic

---

DISSECTION 18.2 Gluteus maximus

Objective

I. To define the extent of the gluteus maximus.

Instructions

1. If any branches of the posterior cutaneous nerve of the thigh have been found, follow them back to the trunk of the nerve.
2. Remove the thin deep fascia from the gluteus maximus, and define the attachments of the muscle. (Leave the insertion of the muscle into the iliotibial tract intact.)

DISSECTION 18.3 Structures deep to gluteus maximus-1

Objectives

I. To expose the piriformis.   **II.** To identify and trace the superior and inferior gluteal vessels and nerves, and the posterior cutaneous nerve of the thigh.

Instructions

1. Cut across the gluteus maximus from its inferior margin upwards, 2–3 cm medial to its femoral insertion, and reflect it. This is difficult because the vessels (superior and inferior gluteal) and the inferior gluteal nerve enter its deep surface and are easily destroyed before they are seen. Avoid this by passing two fingers deep to the lower edge of the muscle and cutting upwards between the fingers to the upper border at a point directly superior to the greater trochanter.

2. As you reflect the lateral part of the muscle to its insertion, identify the bursae which separate it from the greater trochanter and the upper part of the vastus lateralis.

3. Reflect the medial part of the muscle. Keep close to the deep surface of the muscle to avoid injury to the posterior cutaneous nerve of the thigh [Fig. 18.2].

4. Find the inferior gluteal vessels and nerve entering the lower part of the muscle [Fig. 18.3].

5. As the ischial tuberosity is uncovered, look for the bursa superficial to the origin of the hamstring muscles.

6. Identify the piriformis muscle.

7. Trace the branch of the superior gluteal artery to where it emerges between the gluteus medius superiorly and the piriformis inferiorly.

8. Remove the fascia from the piriformis muscle, and trace it to its attachment to the greater trochanter.

9. Find and follow the posterior cutaneous nerve of the thigh upwards to the point where it emerges at the lower border of the piriformis. A perineal branch of this nerve curves medially, below the ischial tuberosity, towards the perineum.

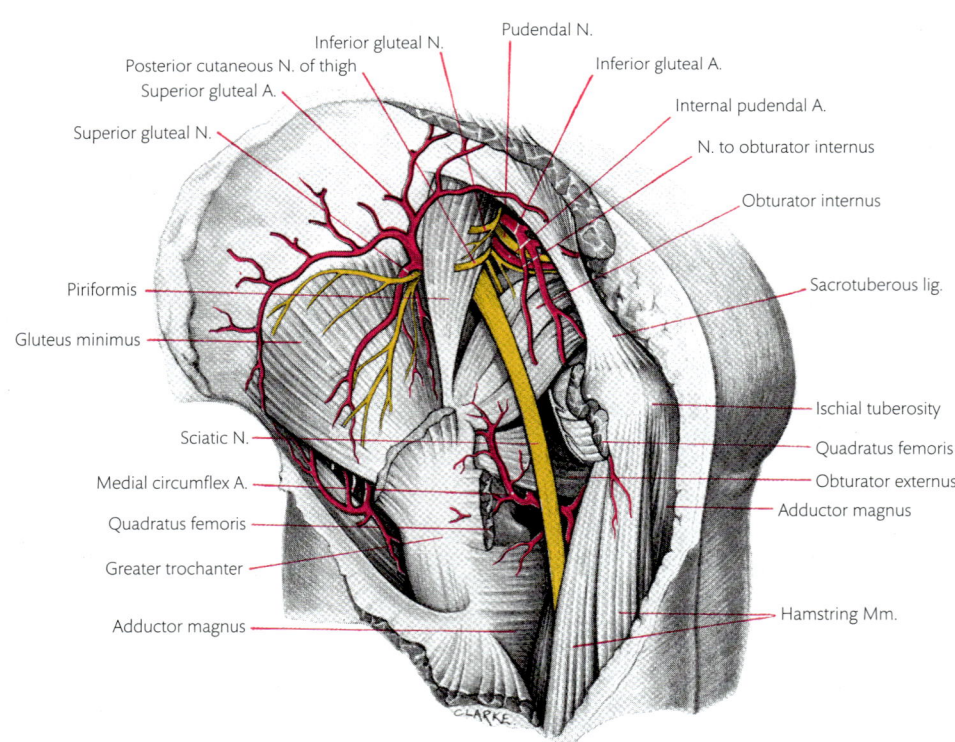

Fig. 18.3 Dissection of the left gluteal region. The gluteus maximus and gluteus medius have been removed, and the quadratus femoris has been reflected. In the specimen, the inferior gluteal artery is medial to the internal pudendal, instead of lateral to it.

Structures deep to the gluteus maximus

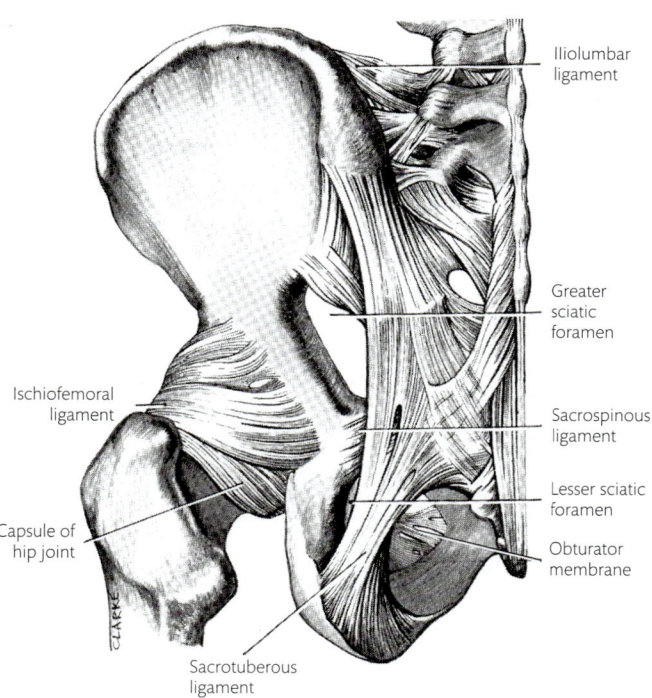

Iliolumbar ligament

Greater sciatic foramen

Sacrospinous ligament

Lesser sciatic foramen

Obturator membrane

Ischiofemoral ligament

Capsule of hip joint

Sacrotuberous ligament

Fig. 18.4 Dorsal view of the pelvic ligaments and the hip joint.

foramen (which lies inferior to the ischial spine) leads from the gluteal region into the perineum. This arrangement allows for structures to pass between the gluteal region and the pelvis through the greater sciatic foramen, and between the gluteal region and the perineum through the lesser sciatic foramen.

Vessels and nerves which enter the gluteal region from the pelvis may remain in the gluteal region, descend into the back of the thigh, or enter the perineum by turning forwards through the lesser sciatic foramen. **Structures confined to the gluteal region** include the gluteal vessels and nerves. **Structures descending to the back of the thigh** are the sciatic nerve, the posterior cutaneous nerve of the thigh, and branches of the inferior gluteal vessels. **Structures entering the perineum** are the internal pudendal vessels, the pudendal nerve, and the nerve to the obturator internus. The lesser sciatic foramen also allows passage of the obturator internus from the lateral wall of the perineum into the gluteal region.

Sacrotuberous ligament

The sacrotuberous ligament passes upwards from the medial side of the **ischial tuberosity** to the margins of the **sacrum and coccyx**, and to the

posterior superior and inferior iliac spines of the hip bone. The lateral edge of the ligament forms the posteromedial border of the greater and lesser sciatic foramina. The medial edge forms the posterior boundary of the perineum. The ligament holds down the posterior part of the sacrum and prevents the weight of the body from depressing the anterior part [Fig. 18.5]. When loading on the anterior part of the sacrum is severe (e.g. in landing

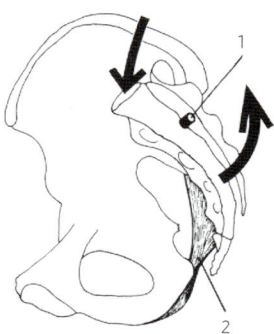

Fig. 18.5 Diagram to show the forces applied to the sacrum due to the weight of the body acting through the vertebral column (thick, straight arrow). Note how the sacrotuberous ligament (2) acts as a shock absorber, permitting only slight movement of the sacrum at the sacroiliac joint around its axis of movement (1). This causes the lower part of the sacrum to swing upwards (thick, curved arrow).

on the feet when jumping from a height), the ligament gives resilience by allowing slight movement at the sacroiliac joint.

Sacrospinous ligament

This thick, triangular band is the aponeurotic posterior surface of the coccygeus muscle. It passes from the **spine of the ischium** to the margin of the **coccyx** and of the last piece of the **sacrum**, deep to the sacrotuberous ligament.

Dissection 18.4 (and Fig. 18.6) continues to explore the gluteal region.

Inferior gluteal nerve (L. 5; S. 1, 2)

The inferior gluteal nerve is a branch of the sacral plexus. It enters the gluteal region with the posterior cutaneous nerve of the thigh, inferior to the piriformis. It breaks into a number of nerves which

enter the deep surface of the gluteus maximus—the only structure it supplies [Figs 18.3, 18.7].

Inferior gluteal artery

The inferior gluteal artery is a branch of the internal iliac artery. It enters the gluteal region below the piriformis. It sends large branches into the deep surface of the gluteus maximus and cutaneous branches to the buttock and to the back of the thigh with the posterior cutaneous nerve of the thigh. The artery also gives the slender **companion artery of the sciatic nerve** and anastomoses with the circumflex femoral arteries.

Sciatic nerve (L. 4, 5; S. 1, 2, 3) in the gluteal region

The sciatic nerve is the thickest nerve in the body. It arises from the sacral plexus and passes through

DISSECTION 18.4 Structures deep to gluteus maximus-2

Objectives

I. To expose the sciatic nerve and its branches. **II.** To identify the nerve to the obturator internus, the internal pudendal artery, the pudendal nerve, the nerve to the quadratus femoris and gemelli, and the medial circumflex femoral artery. **III.** To identify the tendon of the obturator internus, superior and inferior gemelli, quadratus femoris, and adductor magnus.

Instructions

1. Find the large sciatic nerve, as it emerges from the pelvis at the lower border of the piriformis. Carefully split the fascia surrounding the nerve. Trace the nerve downwards to where it gives branches to the hamstring muscles near the ischial tuberosity. The vessels running with these branches arise from the medial circumflex femoral artery [Figs 18.3, 18.6].
2. Push the upper part of the sciatic nerve laterally to expose the posterior surface of the acetabulum.
3. Find the slender nerve to the quadratus femoris.
4. Medial to the upper part of the sciatic nerve, identify the ischial spine and the sacrospinous ligament. The ligament can be felt as a tough resistance medial to the spine. On the surface of the spine and ligament, find the nerve to the obturator internus, the internal pudendal vessels, and the pudendal nerve.

5. Remove the fascia from the muscles deep to the sciatic nerve. From above downwards, these are [Figs 18.3, 18.6]: (1) the tendon of the obturator internus overlapped by the superior and inferior gemelli—separate the gemelli, expose the tendon, and follow the tendon to the greater trochanter; (2) the quadratus femoris passing from the ischial tuberosity to the back of the femur; and (3) the posterior surface of the adductor magnus.
6. Find and trace the branches of the medial circumflex femoral artery which appear both above and below the quadratus femoris [see Figs 17.17, 18.3].
7. Inferior to this, the first perforating artery (a branch of the profunda femoris) may be found piercing the adductor magnus, close to the gluteal tuberosity of the femur [Fig. 18.6].
8. Separate the gemellus inferior from the quadratus femoris.
9. Lift the gemelli and obturator internus, and cut across them, lateral to the nerve to the quadratus femoris.
10. Follow the nerve to the quadratus femoris and its branch to the inferior gemellus.
11. Separate the quadratus femoris from the adductor magnus, and remove the quadratus femoris to expose the lesser trochanter of the femur, the medial circumflex femoral artery, the posterior part of the capsule of the hip joint, and the tendon of the obturator externus.

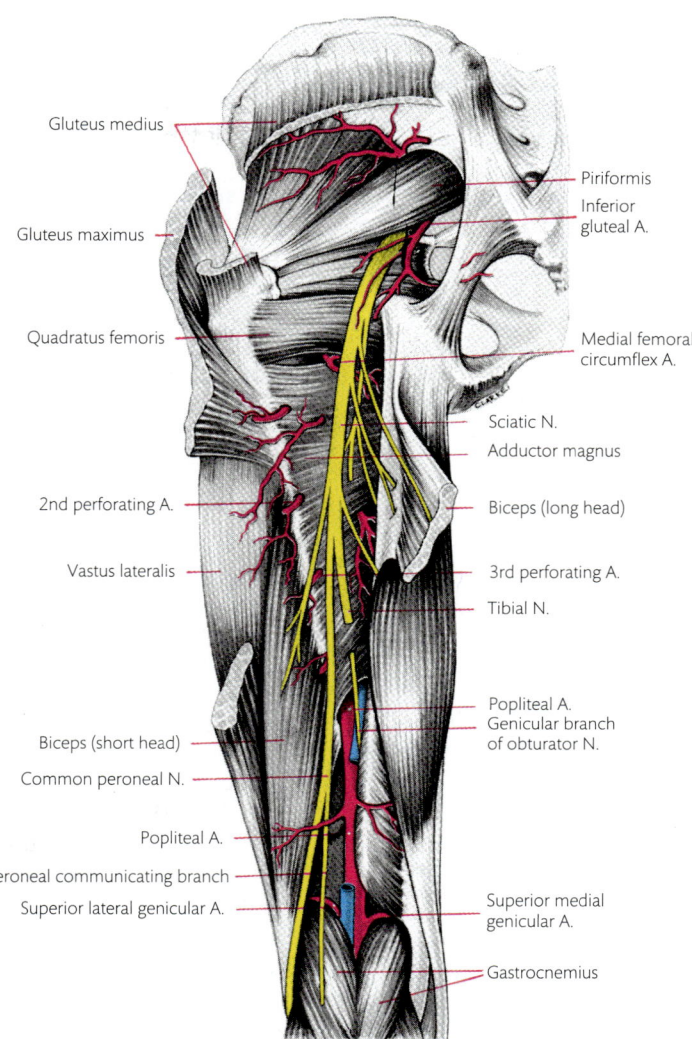

Gluteus medius

Gluteus maximus

Quadratus femoris

2nd perforating A.

Vastus lateralis

Biceps (short head)

Common peroneal N.

Popliteal A.

Peroneal communicating branch

Superior lateral genicular A.

Piriformis

Inferior gluteal A.

Medial femoral circumflex A.

Sciatic N.

Adductor magnus

Biceps (long head)

3rd perforating A.

Tibial N.

Popliteal A.
Genicular branch of obturator N.

Superior medial genicular A.

Gastrocnemius

Fig. 18.6 Dissection of the gluteal region and the back of the thigh.

the lower part of the greater sciatic foramen into the gluteal region. It lies deep to the gluteus maximus. From above downwards, it lies on the: (1) ischial wall of the acetabulum and the nerve to the quadratus femoris; (2) obturator internus tendon, with the two gemelli on either side; and (3) quadratus femoris [Fig. 18.3]. At this level, one or more nerves leave its medial side to supply the hamstring muscles. The sciatic nerve then enters the thigh on the posterior surface of the adductor magnus [Fig. 18.6]. It usually ends halfway down the back of the thigh by dividing into the **common peroneal** and **tibial nerves**. The point of division of the sciatic nerve is variable. If it occurs before the nerve leaves the pelvis, the tibial nerve emerges below the piriformis and the common peroneal nerve pierces that muscle.

Pudendal nerve, internal pudendal artery, and nerve to the obturator internus

These three structures enter the gluteal region through the lowest part of the greater sciatic foramen. They lie on the posterior surface of the junction of the ischial spine and the sacrospinous ligament, with the pudendal nerve most medial, and the artery and nerve to the obturator internus lateral to it. They turn forwards immediately and enter the perineum through the lesser sciatic foramen [Fig. 18.3].

Small muscles on the back of the hip joint

A series of small muscles lie deep to the gluteus maximus on the back of the hip joint.

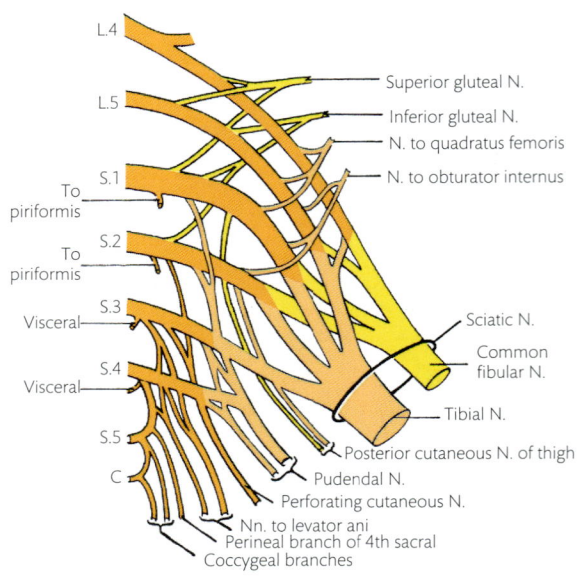

Fig. 18.7 Diagram of the sacral plexus. Ventral divisions, light orange; dorsal divisions, yellow.

Labels on figure:
L.4
L.5
To piriformis
S.1
To piriformis
S.2
Visceral
S.3
Visceral
S.4
S.5
C
Superior gluteal N.
Inferior gluteal N.
N. to quadratus femoris
N. to obturator internus
Sciatic N.
Common fibular N.
Tibial N.
Posterior cutaneous N. of thigh
Pudendal N.
Perforating cutaneous N.
Nn. to levator ani
Perineal branch of 4th sacral
Coccygeal branches

Piriformis

Origin: the piriformis takes origin from the pelvic surface of the middle three pieces of the sacrum and passes through the greater sciatic foramen. **Insertion**: it is inserted into the upper border of the greater trochanter of the femur [see Fig. 16.8], immediately lateral to the tendon of the obturator internus. **Nerve supply**: branches of the first and second sacral nerves in the pelvis. **Actions**: lateral rotation of the hip joint.

Obturator internus

Origin: the obturator internus is a large, fan-shaped muscle which arises from the pelvic surface of the obturator membrane and most of the bone surrounding the foramen [see Fig. 16.4]. **Insertion**: the muscle fibres converge posteriorly to the lesser sciatic foramen, turn sharply over the lesser sciatic notch, and run laterally to be inserted into the upper medial part of the greater trochanter [see Fig. 16.8]. The tendon is separated from the notch by a bursa. In the lesser pelvis, the **levator ani muscle**—which separates the pelvis from the perineum—arises from the fascia covering the pelvic surface of the obturator internus. As such, the obturator internus is in the lateral wall of both the pelvis and the perineum. **Nerve supply**: nerve to the obturator internus (L. 5; S. 1, 2). **Actions**: lateral rotation of the hip joint.

Superior and inferior gemelli

Origin: the superior and inferior gemelli arise from the superior and inferior margins of the lesser sciatic notch. **Insertion**: they are inserted into the posterior surface of the tendon of the obturator internus. **Nerve supply**: the superior gemellus is supplied by a branch of the nerve to the obturator internus; the inferior gemellus is supplied by a branch of the nerve to the quadratus femoris. **Actions**: they are lateral rotators of the hip joint.

Quadratus femoris

The **quadratus femoris** originates from the lateral margin of the ischial tuberosity and is inserted into the back of the greater trochanter of the femur in the region of the quadrate tubercle [see Fig. 16.9]. The muscle lies between the inferior gemellus and the superior margin of the adductor magnus. **Nerve supply**: nerve to the quadratus femoris (L. 4, 5; S. 1). **Actions**: it is a lateral rotator of the hip joint.

Medial circumflex femoral artery

The **medial circumflex artery** arises from the profunda femoris (or femoral) artery in the femoral triangle. (The initial part of this artery was seen on the medial side of the thigh.) It passes backwards between the pectineus and the iliopsoas, and then inferior to the obturator externus and the hip joint, above the adductor muscles [see Fig. 17.17]. It sends muscular branches down between the adductor muscles and a branch to the acetabulum, and anastomoses with the obturator artery. Posteriorly, the artery divides into an **ascending branch** which runs to the trochanteric fossa and a **transverse branch** to the hamstring muscles. The ascending branch anastomoses with branches of both gluteal arteries and sends branches to the neck of the femur. These branches supply a large part of the **femoral head**.

The cruciate anastomosis

The **cruciate anastomosis** is formed at the back of the upper part of the femur by four arteries: (1) the transverse branch of the medial circumflex femoral artery; (2) the transverse branch of the lateral femoral circumflex artery; (3) the inferior gluteal artery; and (3) the first perforating artery.

Dissection 18.5 continues with dissection of the gluteal region.

DISSECTION 18.5 Structures deep to gluteus maximus-3

Objectives

I. To expose the gluteus medius and minimus muscles. **II.** To identify and trace the superior gluteal nerves and vessels.

Instructions

1. Remove the fascia from the superficial surface of the gluteus medius, and define its attachments.
2. Lift the inferior border of the gluteus medius away from the piriformis, and define the plane of separation between it and the gluteus minimus beneath. Separate the muscles from behind forwards by pushing your fingers between them.
3. Cut across the gluteus medius, 5 cm above the greater trochanter, and reflect its parts. Confirm its insertion. Find the bursa between the tendon of insertion and the greater trochanter.
4. Find the branches of the superior gluteal vessels and nerve, deep to the upper part of the muscle. Trace these between the gluteus medius and minimus to the deep surface of the tensor fasciae latae. Here the artery anastomoses with the ascending branch of the lateral femoral circumflex artery, and the nerve enters and supplies the tensor fascia latae.

Gluteus medius

Origin: this powerful muscle arises from the ilium between the anterior and posterior gluteal lines [see Fig. 16.3]. It is overlapped posteriorly by the gluteus maximus and anteriorly by the tensor fasciae latae. Between these muscles, it is felt in the living through the skin and iliotibial tract. To check this on yourself, feel it contracting when the opposite foot is raised from the ground. **Insertion**: the flattened tendon is inserted into the posterosuperior angle of the greater trochanter and the oblique ridge on its lateral surface [see Fig. 16.9]. A small bursa lies between the tendon and the trochanter in front of the insertion. **Nerve supply**: superior gluteal nerve. **Actions**: it abducts the thigh when the limb is free to move. When the limb is supporting the weight of the body, the action of the gluteus medius is reversed and it tilts the pelvis on the hip joint, so that the opposite foot can be raised from the ground. The anterior fibres are medial rotators of the thigh.

Gluteus minimus

Origin: this thick muscle arises from the ilium between the anterior and inferior gluteal lines [see Fig. 16.3]. **Insertion**: it is inserted into the front of the greater trochanter [see Fig. 16.8] and fuses with the fibrous capsule of the hip joint. A bursa separates the tendon from the upper anterior part of the trochanter. **Nerve supply**: superior gluteal nerve.

Actions: like the gluteus medius, the gluteus minimus abducts the thigh. When the limb is supporting the weight of the body, it tilts the pelvis to the same side, enabling the opposite foot to clear the ground while walking. The anterior fibres are medial rotators of the thigh.

Gluteus medius gait

Stabilizing the hip, when one foot is off the ground, requires abduction of the opposite (supporting) hip. When the left foot is raised off the ground, the right abductors act to: (1) prevent the pelvis from sagging to the left side; and (2) create space to swing the left foot in taking a forward step. If the hip abductors—gluteus medius and minimus—are weak or paralysed, the patient is unable to stabilize the hip when standing on one leg or in walking.

Trendelenburg's sign: is elicited by asking the patient to stand on one leg. When the sign is positive, the pelvis sags to the unsupported side. It indicates damage to the superior gluteal nerve or to the gluteus medius and minimus.

Trendelenburg's gait: is an abnormal gait adopted by a patient who is unable to abduct the hip. When one foot is off the ground, the patient compensates for the tendency to fall on the unsupported side (due to lack of stability in the supporting hip) by leaning laterally to the affected side. A person with this gait will be seen lurching towards the weakened side every time the opposite foot is raised. If both hips are affected, the person will have a 'waddling gait'.

Superior gluteal nerve (L. 4, 5; S. 1)

The superior gluteal nerve is a branch of the **sacral plexus**. It enters the gluteal region through the greater sciatic foramen above the piriformis [Figs 18.3, 18.7]. It runs between the gluteus medius and minimus, and divides into a number of branches. The upper branches enter the gluteus medius. The lowest branch crosses the minimus, gives small branches to the gluteus medius and minimus, and supplies the tensor fasciae latae.

Superior gluteal artery

The superior gluteal artery arises from the posterior division of the internal iliac artery and enters the gluteal region with the corresponding nerve. Here it divides into a superficial and a deep branch. The **superficial branch** passes between the gluteus medius and the piriformis to supply the gluteus maximus. The **deep branch** runs with the branches of the superior gluteal nerve to supply the gluteus medius and minimus, the hip joint, and the tensor fasciae latae. It ends by anastomosing with the ascending branch of the lateral circumflex femoral artery. Together, they give branches to the neck and greater trochanter of the femur.

Anastomosis between branches of the internal and external iliac arteries

There are many anastomoses between branches of the internal iliac and external iliac (or femoral) arteries: (1) an important cruciate anastomosis occurs between the medial femoral circumflex artery, the obturator artery, the superior and inferior gluteal arteries, the lateral circumflex femoral artery, and the perforating branches of the profunda femoris; (2) the superior gluteal artery anastomosis with the lateral femoral circumflex artery and the superficial circumflex iliac artery; (3) in the perineum, the internal pudendal artery (a branch of the internal iliac artery) anastomoses with the deep and superficial external pudendal branches of the femoral artery; (4) in the abdomen, the external iliac artery may anastomose with, or form, the obturator artery; (5) in addition, the external iliac artery may communicate with the subclavian artery through an anastomosis between the superior and inferior epigastric arteries in the anterior abdominal wall.

Dissection 18.6 provides instruction on exposure of the capsule of the hip joint.

See Clinical Application 18.1 for the practical implications of the anatomy discussed in this chapter.

CHAPTER 19
The popliteal fossa

Surface anatomy

The popliteal fossa is an intermuscular space behind the knee, posterior to the lower third of the femur, the knee joint, and the upper part of the tibia. It forms a hollow when the knee is flexed, as the tendons which form its boundaries stand out from the femur. The fossa bulges slightly when the knee is straight.

Review the major palpable structures around the knee joint by examining your own knee. Find the **tendon of the biceps femoris** behind the lateral condyle of the femur [see Fig. 17.2]. With the knee extended, press your finger against the posterior surface of the lateral condyle, immediately medial to the biceps tendon, and move the finger from side to side. The **common peroneal nerve** can be felt in this location as a rounded cord. Follow the tendon of the biceps to the head of the fibula, and repeat the process on the back of the head. The same nerve can be felt again. Slide the finger down to the posterolateral side of the neck of the fibula. Move the finger up and down, maintaining some pressure to feel the nerve on the fibular neck. In all these positions, the common peroneal nerve may be damaged by an injury to the bone.

With the knee flexed and the lateral side of the foot pressed against the floor, feel the **fibular collateral ligament** of the knee joint—a firm resistance between the head of the fibula and the lateral femoral condyle.

Place a finger in the middle of the popliteal fossa, with the knee bent. Press firmly, and feel the pulsations of the popliteal artery. In the lower part of the fossa, the two heads of the gastrocnemius form rounded swellings which merge inferiorly in the calf.

Immediately above the popliteal fossa, the back of the thigh is smooth and rounded. But the distal part of the belly of the semimembranosus may be seen to bulge near the midline, as it contracts in walking [Fig. 19.1].

Dissection 19.1 explores the popliteal region.

Fascia of the popliteal region

There is relatively little fat in the superficial fascia. The deep fascia, though thin, is strong and firmly bound to the tendons which form the boundaries of the popliteal fossa.

The popliteal fossa

The **popliteal fossa** is a narrow space between the diverging hamstring muscles superiorly and the converging heads of the gastrocnemius inferiorly. When dissected, the popliteal fossa is a diamond-shaped space.

Boundaries of the popliteal fossa

The fossa has **four boundaries**: upper lateral, upper medial, lower lateral, and lower medial. The **upper lateral** boundary is formed by the biceps femoris. The **upper medial** boundary is formed by the semimembranosus and semitendinosus. The gracilis, sartorius, and tendon of the adductor magnus lie close to the semimembranosus and semitendinosus as they approach the knee. The **lower**

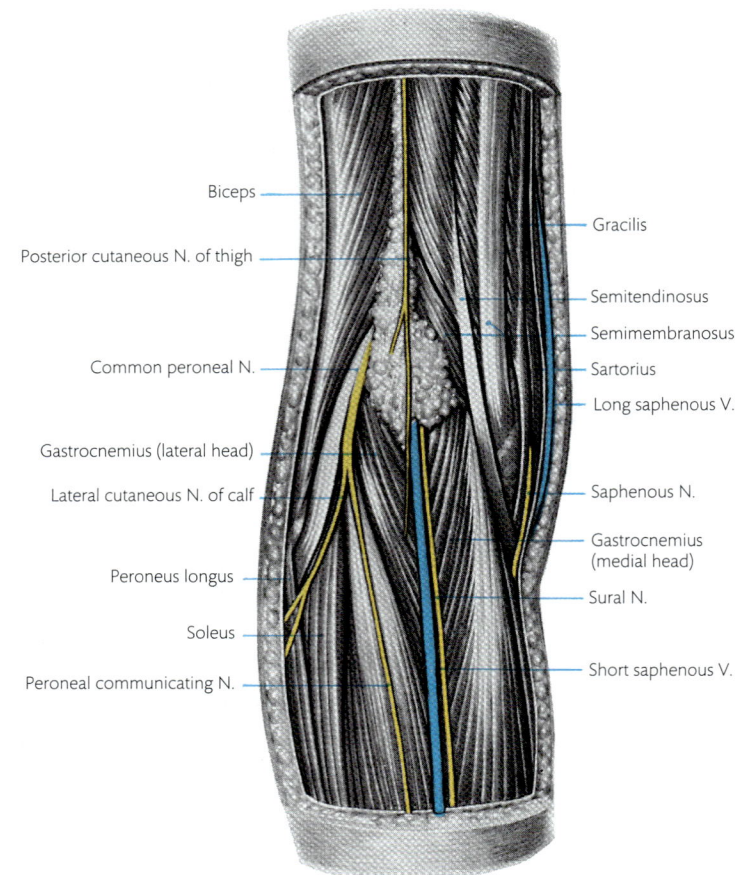

Biceps

Posterior cutaneous N. of thigh

Common peroneal N.

Gastrocnemius (lateral head)

Lateral cutaneous N. of calf

Peroneus longus

Soleus

Peroneal communicating N.

Gracilis

Semitendinosus

Semimembranosus

Sartorius

Long saphenous V.

Saphenous N.

Gastrocnemius (medial head)

Sural N.

Short saphenous V.

Fig. 19.1 Left popliteal region after removal of the deep fascia.

lateral and **lower medial** boundaries are formed by the two heads of the gastrocnemius. The space between the boundaries is narrow, and only a small part of the popliteal vessels just above the knee joint is not covered by muscles. The **posterior wall** of the fossa is the deep fascia. Its **anterior wall** is the popliteal surface of the femur, the capsule of the knee joint, and the fascia covering the popliteus.

DISSECTION 19.1 Skin reflection, and cutaneous nerves and vessels of the popliteal region

Objective

I. To identify and trace the posterior cutaneous nerve of the thigh, the short saphenous vein, and the sural nerve in the roof of the popliteal fossa.

Instructions

1. Make skin incision 7 [see Fig. 18.1] and reflect the skin flaps, leaving the superficial fascia intact. Strip the superficial fascia from the deep fascia, starting proximally.
2. Look for branches of the posterior cutaneous nerve of the thigh piercing the deep fascia in the proximal part of the fossa.
3. Look for the short saphenous vein, with the terminal part of the posterior cutaneous nerve in the distal part of the fossa.
4. Follow the medial anterior cutaneous nerve of the thigh downwards in the posteromedial part of the calf [see Fig. 18.2].
5. The peroneal communicating nerve (a branch of the common peroneal nerve) may be found in the lower lateral part of the popliteal area. It may pierce the deep fascia at a much lower level, in which case it will be found later.

DISSECTION 19.2 Popliteal fossa

Objectives

I. To identify and define the muscles and tendons forming the boundaries of the popliteal fossa.   **II.** To identify the plantaris.   **III.** To identify and trace the tibial nerve, the common peroneal nerve, the posterior cutaneous nerve of the thigh, and the popliteal artery and its branches. **IV.** To trace the short saphenous vein to the popliteal vein.

Instructions

1. Cut through the deep fascia along the biceps femoris. Expose the muscle and its tendon to the insertion.
2. Make a similar incision over the semitendinosus and semimembranosus. Follow the tendon of the semitendinosus to the medial surface of the tibia, then lift the semimembranosus and follow it distally to the tibia. Find the bursa between the semimembranosus and the medial head of the gastrocnemius.
3. Find and follow the gracilis and its tendon. As you free the gracilis from the posterior surface of the sartorius, look for the saphenous nerve which emerges between them. Follow the nerve down with the long saphenous vein.
4. Follow the posterior cutaneous nerve of the thigh upwards to the upper angle of the popliteal fossa. Strip the deep fascia from the posterior surface of the popliteal fossa, and remove the fat from its upper angle to expose the tibial nerve. Follow this large nerve down.
5. Find the branches of the tibial nerve [Figs 19.2, 19.3]. The cutaneous branch from it—the sural nerve—lies between the two heads of the gastrocnemius. There are three articular branches to the knee joint—superior medial, inferior medial, and middle genicular nerves. The superior medial genicular nerve arises near the upper angle of the fossa, and the other two below this. Trace them as far as possible.

The muscular branches to the gastrocnemius, plantaris, soleus, and popliteus arise near the middle of the fossa. Separate the heads of the gastrocnemius, and follow these branches as far as possible.
6. Find the common peroneal nerve, medial to the tendon of the biceps femoris. Trace the nerve proximally to the upper angle of the fossa and distally to the back of the head of the fibula. Find the superior lateral and inferior lateral genicular branches. They arise near the upper limit of the fossa. The superior lateral genicular leaves the fossa above the lateral femoral condyle. The inferior lateral genicular accompanies the trunk of the nerve. The peroneal communicating nerve and the lateral cutaneous nerve of the calf arise near the lateral angle of the fossa. Follow these branches downwards.
7. Remove the fascia from the heads of the gastrocnemius. Identify and separate the plantaris from the posteromedial surface of the lateral head, but avoid injury to the nerve to the lateral head of the gastrocnemius which passes between them.
8. Lift the upper part of the tibial nerve from the popliteal vessels. The vein is posterolateral to the artery, with the genicular branch of the obturator nerve in the groove between them. If this nerve is found, trace it proximally and distally.
9. Remove the fascia from the popliteal vein, the popliteal artery, and the venous channels which run on it. Remove these venous channels to expose the artery, but retain the short saphenous vein.
10. Find the large muscular branches of the popliteal artery, and divide these. They pass with the nerves to the muscles. Gently scrape the fat from the popliteal surface of the femur, and find the genicular branches of the artery [Fig. 19.4]. Follow the lateral and medial superior and inferior genicular arteries and the middle genicular artery piercing the posterior capsule of the knee joint.

Dissection 19.2 (and Figs 19.2, 19.3, 19.4) describes the dissection of the popliteal fossa.

Contents of the popliteal fossa

The popliteal fossa has: (1) the **tibial nerve**; (2) the **common peroneal nerve** (both terminal branches of the sciatic nerve); (3) the **popliteal artery**; and (4) the **popliteal vein** running through it. It also contains: (5) the **popliteal lymph nodes**; and (6) the **posterior cutaneous nerve of the thigh**. The popliteal vessels lie close to the posterior surface of the knee joint and are minimally affected by its movements. The tibial nerve runs with the vessels in the centre of the fossa. The common peroneal nerve deviates laterally and runs along the medial side of the biceps femoris.

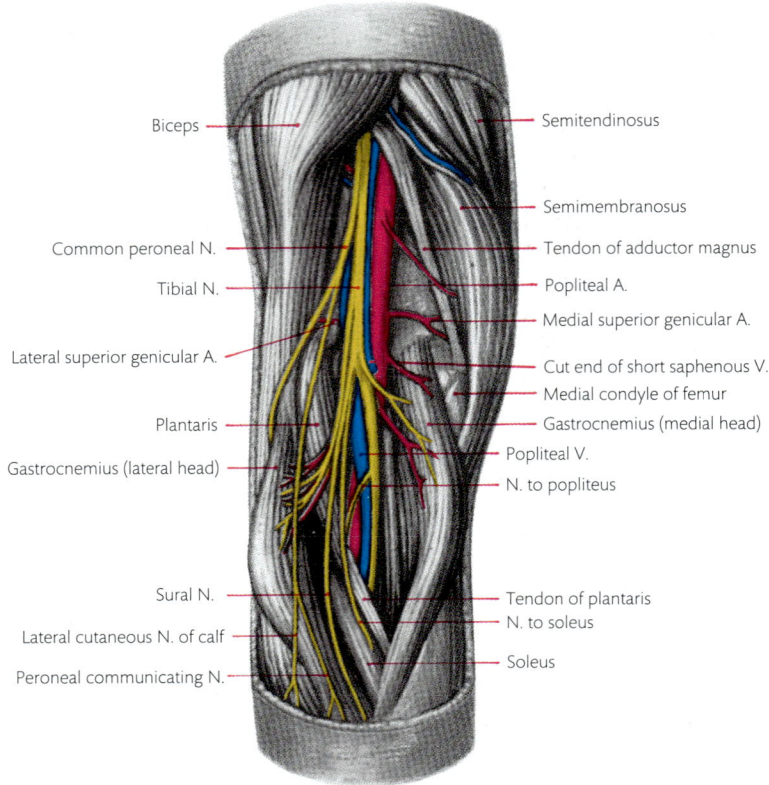

Biceps

Common peroneal N.

Tibial N.

Lateral superior genicular A.

Plantaris

Gastrocnemius (lateral head)

Sural N.

Lateral cutaneous N. of calf

Peroneal communicating N.

Semitendinosus

Semimembranosus

Tendon of adductor magnus

Popliteal A.

Medial superior genicular A.

Cut end of short saphenous V.

Medial condyle of femur

Gastrocnemius (medial head)

Popliteal V.

N. to popliteus

Tendon of plantaris

N. to soleus

Soleus

Fig. 19.2 Dissection of the left popliteal fossa. The boundaries have been pulled apart and separated.

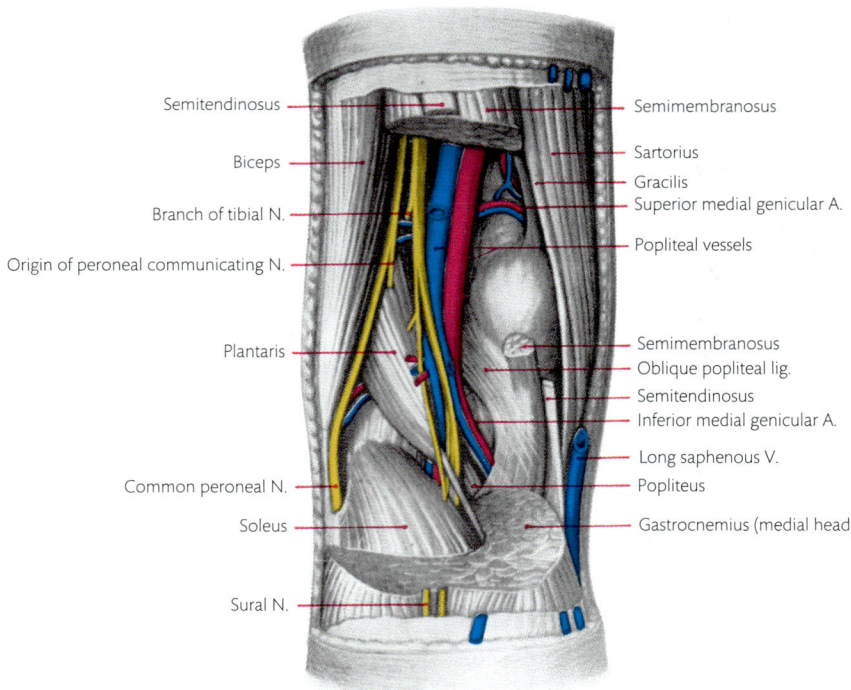

Semitendinosus

Biceps

Branch of tibial N.

Origin of peroneal communicating N.

Plantaris

Common peroneal N.

Soleus

Sural N.

Semimembranosus

Sartorius

Gracilis

Superior medial genicular A.

Popliteal vessels

Semimembranosus

Oblique popliteal lig.

Semitendinosus

Inferior medial genicular A.

Long saphenous V.

Popliteus

Gastrocnemius (medial head)

Fig. 19.3 Dissection of the left popliteal fossa. The two heads of the gastrocnemius and portions of the semimembranosus and semitendinosus have been removed.

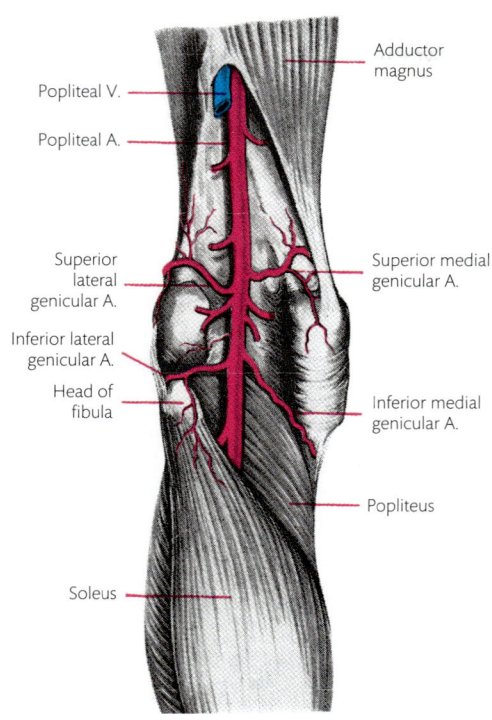

Fig. 19.4 Left popliteal artery and its branches.

Labels in figure:
Popliteal V.
Popliteal A.
Superior lateral genicular A.
Inferior lateral genicular A.
Head of fibula
Soleus
Adductor magnus
Superior medial genicular A.
Inferior medial genicular A.
Popliteus

between the lateral head of the gastrocnemius and the plantaris to enter the superficial surface of the soleus. The nerve to the popliteus descends over the muscle and curves around its inferior border to reach its anterior surface.

Articular branches

Three articular branches arise in the upper part of the fossa and supply the knee joint. They descend to join the corresponding arteries [Fig. 19.4]. (1) The superior medial genicular nerve runs above the medial condyle of the femur, deep to the muscles; (2) the middle genicular nerve pierces the posterior capsule of the knee joint and supplies the structures in the intercondylar notch of the femur; (3) the inferior medial genicular nerve runs inferomedially on the upper border of the popliteus and forwards, deep to the superficial part of the tibial collateral ligament of the knee. The nerve to the popliteus supplies the superior tibiofibular joint and sends branches to the interosseous membrane.

Tibial nerve (L. 4, 5; S. 1, 2, 3)

The tibial nerve is the larger of the two terminal branches of the sciatic nerve and arises in the middle of the back of the thigh. It runs vertically through the popliteal fossa, posterior to the popliteal vessels [Fig. 19.2]. This nerve supplies the muscles of the back of the leg and the sole of the foot, and the skin of the lower half of the back of the leg and the lateral side and sole of the foot.

Branches in the popliteal fossa

Cutaneous branches

The **sural nerve** arises in the middle of the fossa. It descends between the heads of the gastrocnemius, pierces the deep fascia in around the middle of the back of the leg, and supplies the skin on the back of the leg and along the lateral side of the dorsum of the foot up to the little toe.

Muscular branches

Muscular branches arise in the distal part of the fossa and pass to the gastrocnemius, plantaris, soleus, and popliteus. The nerve to the soleus passes

Common peroneal nerve (L. 4, 5; S. 1, 2)

The common peroneal nerve is the smaller of the two terminal branches of the sciatic nerve [Figs 19.2, 19.3]. It supplies the muscles on the lateral and anterior compartments of the leg and the dorsum of the foot, and the skin on the lateral side of the leg and the greater part of the dorsum of the foot. The common peroneal nerve runs along the medial border of the biceps femoris to the back of the head of the fibula and curves around the neck of the fibula, deep to the upper fibres of the peroneus longus muscle. At the neck of the fibula, it ends by dividing into the **superficial** and **deep peroneal nerves**.

Branches in the popliteal fossa

Cutaneous branches

The **peroneal communicating nerve** arises in the upper part of the popliteal fossa. It joins the sural nerve at a variable level [see Fig. 18.2]. The **lateral cutaneous nerve of the calf** arises on the lateral head of the gastrocnemius. It supplies the skin on the lateral side of the upper half of the leg.

Articular branches

The **superior** and **inferior lateral genicular nerves** are small. They accompany the corresponding arteries to the knee joint [Fig. 19.4]. The **recurrent genicular nerve** also supplies the knee joint.

Genicular branch of the obturator nerve

This slender continuation of the posterior branch of the obturator nerve pierces the distal part of the adductor magnus, descends on the popliteal artery, and pierces the posterior surface of the fibrous capsule of the knee joint to supply it.

Posterior cutaneous nerve of the thigh

The terminal part of the **posterior cutaneous nerve of the thigh** passes through the popliteal fossa, immediately deep to the deep fascia. It gives branches to the overlying skin and enters the superficial fascia with the short saphenous vein.

Popliteal artery

The popliteal artery **begins** at the tendinous opening in the adductor magnus as a continuation of the femoral artery. It **ends** at the lower border of the popliteus muscle by dividing into the **anterior** and **posterior tibial arteries**. The artery lies on the anterior wall of the popliteal fossa. From above downwards, it is anterior to the semimembranosus, popliteal vein, tibial nerve, gastrocnemius, and plantaris [Figs 19.2, 19.3, 19.4].

Branches

Muscular branches

The muscular branches pass to the lower parts of the hamstring muscles and to the upper parts of the muscles of the calf. These are large and give rise to cutaneous branches, one of which accompanies the sural nerve. The branches to the hamstrings anastomose superiorly with the perforating branches of the profunda femoris artery.

Articular branches

Five named genicular arteries arise from the popliteal artery and lie on the anterior wall of the popliteal fossa. They are the **superior medial**, **superior lateral**, **middle**, **inferior medial**, and **inferior lateral genicular arteries** [Fig. 19.4].

The popliteal artery is the only significant route through which blood can reach the leg and foot from the thigh. It may be compressed when sitting on a hard-edged seat or with the legs crossed. ➲ When obliterative arterial disease affects the popliteal artery, blood supply to the large muscles of the leg is seriously compromised. This results in ischaemic pain in muscles on exercise, which is relieved by rest—a condition known as **intermittent claudication**.

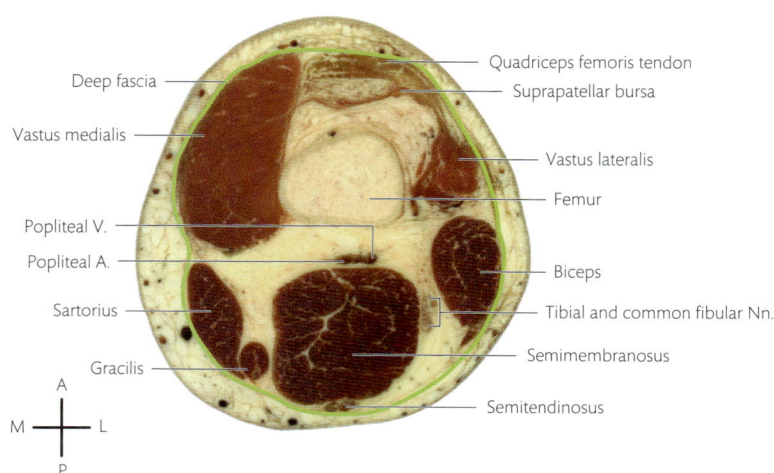

Fig. 19.5 Transverse section through the proximal part of the popliteal region of the thigh. A = anterior; P = posterior; M = medial; L = lateral. Image courtesy of the Visible Human Project of the US National Library of Medicine.

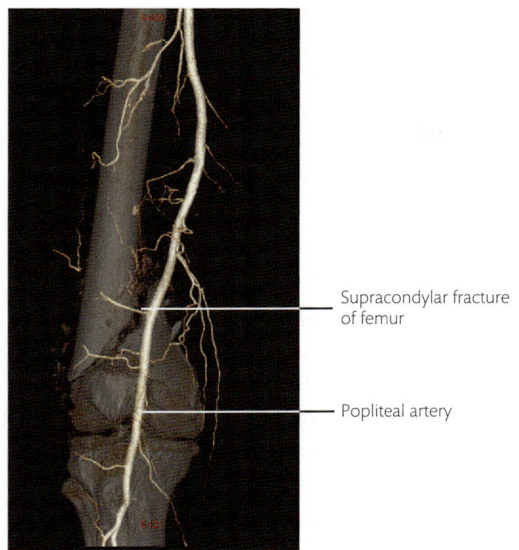

Supracondylar fracture
of femur

Popliteal artery

Fig. 19.6 Computed tomography angiogram of the popliteal artery showing the undamaged artery close to the fracture.

of the popliteal artery. The tributaries of the popliteal vein correspond to the branches of the artery. It also receives the short saphenous vein. Open the popliteal vein, and note its numerous valves. The vein **ends** at the adductor hiatus by continuing as the femoral vein.

Fig. 19.5 is a cross-section through the lower part of the thigh, showing the upper boundaries and contents of the popliteal fossa.

Popliteal lymph nodes

The popliteal nodes are deep nodes which lie in the popliteal fossa. **Afferents**: they drain the deep tissues of the leg, foot, and knee. They also receive superficial lymph vessels from the lateral side of the foot, heel, and back of the calf, and lymph vessels that run along the short saphenous vein. These afferents pierce the deep fascia over the lower part of the popliteal fossa. **Efferents**: the popliteal nodes drain to the deep inguinal lymph nodes.

Clinical Applications 19.1 and 19.2 (and Fig. 19.6) demonstrate the practical implications of the anatomy described in this chapter.

Popliteal vein

The popliteal vein **begins** by the union of the anterior and posterior tibial veins at the lower border of the popliteus. It ascends on the posterior surface

CLINICAL APPLICATION 19.1 Closed supracondylar femoral fracture

A 25-year-old man suffered a closed supracondylar fracture of the femur. Because the popliteal artery is closely applied to the popliteal surface of the femur and the joint capsule, fractures of the distal femur may rupture the artery, resulting in haemorrhage. A computed tomography angiography was done to rule out damage to the popliteal artery and its branches [Fig. 19.6].

CLINICAL APPLICATION 19.2 Popliteal pulse and popliteal aneurysm

The rhythmical throbbing of arteries produced by the regular contractions of the heart can be felt at a number of peripheral locations. The popliteal pulse is palpable in the popliteal fossa by pressing the artery against the anterior surface of the fossa. It is especially important, as a strong popliteal pulse is an indicator of good blood flow to the distal parts of the lower limb. Because the popliteal artery is deep, it may be difficult to feel the pulsations. Palpation is commonly performed with the person in the prone position, with the knee flexed to relax the popliteal fascia and hamstrings.

A popliteal aneurysm is an abnormal thinning of the wall of the popliteal artery and dilatation of the vessel. It presents as a pulsatile swelling and should be differentiated from a popliteal cyst or a Baker's cyst. A popliteal cyst is an outpouching of the synovial membrane from the bursa between the medial head of the gastrocnemius and the semimembranosus tendon.

CHAPTER 20
The back of the thigh

Introduction

The main structures in the back of the thigh are the flexors of the knee (hamstrings) and the sciatic nerve.

Posterior cutaneous nerve of the thigh (S. 1, 2, 3)

The posterior cutaneous nerve of the thigh is a branch of the **sacral plexus**. It arises in the pelvis and enters the gluteal region through the lower part of the greater sciatic foramen, close to the inferior gluteal vessels and nerve. In the gluteal region, it is deep to the gluteus maximus and lies on the sciatic nerve. It then runs down the middle of the back of the thigh, immediately deep to the deep fascia, and is separated from the sciatic nerve by the long head

of the biceps. It pierces the deep fascia at the back of the knee and supplies the skin as far as the middle of the back of the calf [see Fig. 18.2].

Branches

It is a purely cutaneous nerve which supplies a long strip of skin at the back. **In the gluteal region**, it gives off: (1) the gluteal branches which curve around the lower border of the gluteus maximus and supply a small area of skin in the lower part of the buttock; and (2) the perineal branch which turns medially across the back of the hamstring muscles to the perineum. The perineal branch supplies the skin on the upper medial side of the thigh and the external genital organs. **In the thigh and leg**, the posterior cutaneous nerve of the thigh gives numerous small branches to the medial side and back of the thigh, and the upper half of the posterior surface of the leg [see Fig. 18.2].

Dissection 20.1 describes the dissection of the hamstring muscles and adjacent vessels and nerves.

DISSECTION 20.1 Back of the thigh-1

Objectives

I. To reflect the skin and deep fascia on the back of the thigh. II. To expose the sciatic nerve and trace its branches. III. To identify and study the semimembranosus, semitendinosus, and biceps femoris.

Instructions

1. Make a vertical incision through the skin remaining on the back of the thigh. Strip the skin and superficial fascia from the deep fascia by blunt dissection. Look for the branches of the posterior cutaneous nerve of the thigh as you do so.
2. Find and follow the branches of the medial anterior and lateral cutaneous nerves of the thigh into this region [see Fig. 18.2].

3. Divide the deep fascia vertically. Find the posterior cutaneous nerve of the thigh, and trace it to the gluteal region.
4. Remove the fascia from the posterior surfaces of the hamstring muscles.
5. Follow the sciatic nerve down from the gluteal region. Trace its branches into the hamstring muscles, including the short head of the biceps and adductor magnus. Find the branches of the perforating arteries which run with the nerves.
6. Separate the hamstring muscles from one another, and trace them to their attachments.

Flexors of the knee

The flexors of the knee form a mass on the back of the thigh and have their tendons in the back of the knee. The tendons of these muscles in the popliteal region, or back of the thigh, give them the name **hamstrings** (ham=Old English word for the depression behind the knee). The muscles—**semitendinosus**, **semimembranosus**, and **long head of the biceps femoris**—arise from the ischial tuberosity and are inserted into the bones of the leg. They extend the hip and flex the knee.

Biceps femoris

Origin: the biceps femoris has two heads. The **long head** arises with the semitendinosus from the medial part of the ischial tuberosity [see Fig. 16.3]. It is partly continuous with the sacrotuberous ligament. The **short head** arises from the linea aspera and the upper half of the lateral supracondylar line. **Insertion**: the two heads unite and are inserted into the head of the fibula [see Figs 19.1, 16.12]. Near its insertion, the tendon is pierced by the **fibular collateral ligament** of the knee joint. **Nerve supply**: the long head is supplied by the tibial part of the sciatic nerve, and the short head is supplied by the common peroneal part of the sciatic nerve. Each head may be separately paralysed as a result of a wound. **Actions**: it extends the hip joint, flexes the knee joint, and rotates the leg laterally when the knee is flexed. The short head acts only on the knee joint.

Semitendinosus

Origin: the semitendinosus originates with the long head of the biceps from the medial part of the ischial tuberosity [see Fig. 16.3]. In the distal third of the thigh, the muscle forms a cylindrical tendon. **Insertion**: on the medial side of the knee, the tendon turns forwards and spreads out to be inserted into the upper part of the medial surface of the tibia, posterior to the tendons of the gracilis and sartorius. A complex **bursa** separates these tendons from one another and communicates with a bursa between the semitendinosus and the tibial collateral ligament of the knee. **Nerve supply**: branches from the tibial part of the sciatic nerve. **Actions**: it extends the hip joint, flexes the

knee, and medially rotates the leg when the knee is flexed.

Semimembranosus

Origin: the semimembranosus arises from the lateral part of the ischial tuberosity [see Fig. 16.3]. The broad tendon of origin passes down and medially deep to the biceps and semitendinosus. **Insertion**: at the back of the knee, it forms a thick, flattened tendon which is inserted chiefly into the groove on the posteromedial surface of the medial condyle of the tibia [see Fig. 16.13]. Fibres from the tendon extend to form: (1) the **oblique popliteal ligament** of the knee; and (2) the **fascia covering the popliteus**. The semimembranosus bursa lies between the tendon and the medial head of the gastrocnemius. **Nerve supply**: tibial part of the sciatic nerve. **Actions**: like the semitendinosus, the semimembranosus extends the hip, flexes the knee, and medially rotates the leg when the knee is flexed.

Sciatic nerve (L. 4, 5; S. 1, 2, 3) in the back of thigh

The sciatic nerve arises from the **sacral plexus**. It leaves the pelvis through the greater sciatic foramen. In the back of the thigh, the nerve lies deep to the long head of the biceps on the posterior surface of the adductor magnus. It ends by dividing into the **tibial** and **common peroneal** nerves midway down the thigh.

Branches from the trunk of the nerve

The tibial part of the sciatic nerve supplies the **ischial part of the adductor magnus, semitendinosus, semimembranosus**, and **long head of the biceps**. The common peroneal part of the nerve supplies the **short head of the biceps femoris**.

Dissection 20.2 continues to look at the back of the thigh.

Profunda femoris artery

This is the chief artery to the muscles of the thigh. It arises from the lateral side of the femoral artery in the femoral triangle and curves medially behind the femoral vessels, giving off the lateral and medial circumflex femoral arteries. It then passes posteriorly between the pectineus and adductor longus, and descends close to the femur, posterior to the adductor longus. Here it gives off the first three per-

forating arteries. A little below the middle of the thigh, the profunda femoris continues as the fourth perforating artery [Fig. 20.1].

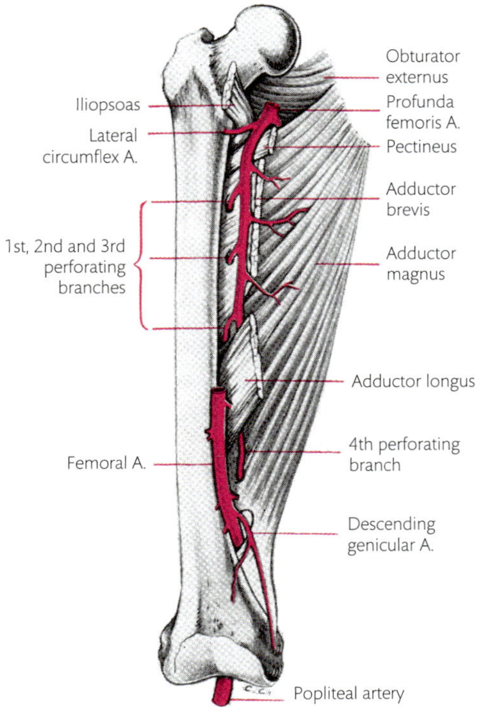

Branches

The medial and lateral circumflex femoral arteries have been described earlier. The muscular branches of the profunda supply the adductor muscles. The four perforating arteries arise in series from the profunda femoris [Fig. 20.1]. They wind round the back of the femur to the vastus lateralis [see Fig. 16.9]. They give branches to the adductor and hamstring muscles, and the second or third forms the nutrient artery to the femur.

Arterial anastomoses

A series of communications between neighbouring arteries in the back of the thigh establishes a longitudinal anastomosis between the internal iliac, femoral, and popliteal arteries. From proximal to distal, this anastomosis is formed by communications between: (1) the inferior gluteal artery; (2) terminal branches of the circumflex femoral arteries; (3) perforating arteries; and (4) branches of the popliteal artery. An alternative supply to the distal part of the limb is established by this route.

Clinical Application 20.1 looks at the condition of sciatic neuritis.

Iliopsoas
Lateral circumflex A.
1st, 2nd and 3rd perforating branches
Femoral A.

Obturator externus
Profunda femoris A.
Pectineus
Adductor brevis
Adductor magnus
Adductor longus
4th perforating branch
Descending genicular A.
Popliteal artery

Fig. 20.1 Profunda femoris artery and its branches. The upper part of the adductor longus has been removed to expose the artery.

CLINICAL APPLICATION 20.1 Sciatic neuritis

A 45-year-old woman consulted her neurologist for low back pain which radiated to the hip and back of the thigh. The neurologist investigated her complaints and diagnosed her to have a 'lumbar radiculopathy' leading to 'sciatic neuritis'.

Study question 1: which lumbar nerves contribute to the formation of the sciatic nerve? (Answer: L. 4 and L. 5.)

Study question 2: using your knowledge of the parts of the typical spinal nerve, deduce which part/parts are involved in a 'radiculopathy'. (Answer: the root of the nerves. *radicle* = roots.)

Using Fig. 17.15, review the proximity of the intervertebral discs to the spinal nerve roots. The sciatic neuritis seen in this patient could well be the result of a spinal disc herniation pressing on one of the lumbar roots of the sciatic nerve.

The doctor tests to see if the pain is worse on stretching the sciatic nerve. Study question 3: what movement of the hip would stretch the sciatic nerve? (Answer: flexion of the hip.) This forms the basis for the 'straight leg raising' test. The patient lies supine, and the doctor lifts the patient's leg while the knee is straight. The patient experiences sciatic pain when the leg is raised to 30–40 degrees.

CHAPTER 21
The hip joint

Type and articular surfaces

The hip joint is a ball-and-socket synovial joint, in which the head of the femur articulates with the acetabulum of the hip bone. Compared to the shoulder joint, the range of movement is less, but strength and stability of the hip joint are much greater.

Articular capsule

The fibrous capsule surrounds the joint on all sides. It is strong anteriorly, and thin posteriorly. Proximally, it is attached to the margin of the acetabulum and to the transverse ligament of the acetabulum. Distally, it is attached to the intertrochanteric line of the femur and to the root of the greater trochanter in front, and to the neck of the femur about 1.5 cm medial to the intertrochanteric crest at the back.

The fibres which comprise the capsule run in two different directions. The majority run obliquely downwards and laterally from the acetabulum to the femur. The oblique fibres are best seen on the anterior surface. Other bundles encircle the capsule approximately parallel to the margin of the acetabulum. These circular bundles form the **zona orbicularis** and are best seen on the posterior and inferior parts of the fibrous capsule.

From the attachment of the fibrous capsule to the femur, a number of fibre bundles turn back on the neck of the femur. These are called **retinacula**. Small arteries and veins pass along the retinacula to supply the neck and head of the femur (note the multiple foramina in the bone, especially on the posterosuperior surface of the neck). ➲ When the neck of the femur is fractured within the capsule, the retinacula help to hold the fragments together. If the retinacula are torn when the neck is fractured, the blood supply of the head of the femur may be destroyed and affect repair of the fracture.

Ligaments

The fibrous membrane has three thickenings that form **extracapsular ligaments** of the hip joint. They are the iliofemoral, pubofemoral, and ischiofemoral ligaments. Within the capsule of the hip joint are the **intracapsular ligaments**: the transverse acetabular ligament; the ligament of the head of the femur; and the labrum acetabulare.

Iliofemoral ligament

This iliofemoral ligament lies on the front of the joint. It forms the thickest and most powerful part of the articular capsule. **Attachments**: proximally, it is attached to the inferior part of the anterior inferior iliac spine and to the surface of the ilium, immediately lateral to the spine. Distally, it widens to be attached to the intertrochanteric line of the femur. It is thicker at the sides than in the middle and has the appearance of an inverted Y [Fig. 21.1].

The iliofemoral ligament is more than 0.5 cm thick. It is one of the strongest ligaments in the body—its only rival being the interosseous sacroiliac ligament. It is rarely torn. ➲ In hip dislocation, the surgeon may use the iliofemoral ligament as a support to lever the head of the femur back into the acetabulum.

In the erect posture, the centre of gravity of the body falls slightly behind a line joining the centres of

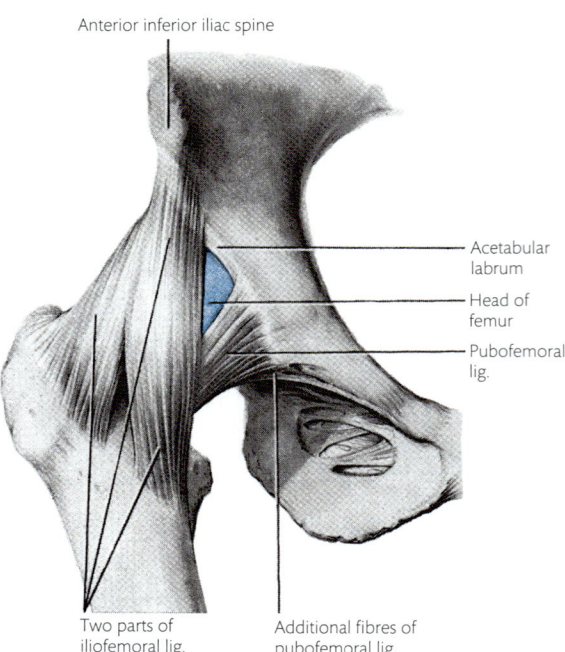

Fig. 21.1 Dissection of the hip joint from the front.

the two hip joints. As such, there is a natural tendency for the body to fall backwards on the hip joints. This is resisted by the iliofemoral ligaments, which maintain the erect posture without muscular activity at these joints. Dissection 21.1 exposes this ligament.

Pubofemoral ligament

This ligament arises from the pubic bone and the obturator membrane [Fig. 21.1]. It lies in the lower and anterior parts of the fibrous capsule.

DISSECTION 21.1 Hip joint

Objectives

I. To expose the iliofemoral ligament.   II. To open the hip joint.

Instructions

1. Cut through the femoral vessels and nerve, immediately inferior to the inguinal ligament. Divide the sartorius and rectus femoris about 5 cm from their origins, and turn them downwards.
2. Cut through the iliopsoas near its insertion, and reflect the two parts to expose the psoas bursa and the capsule of the hip joint. Remove the bursa, and identify the margins of the thick iliofemoral ligament.

Ischiofemoral ligament

This ligament is a weak band which arises from the ischium below the acetabulum. It passes upwards and laterally into the fibrous capsule.

Transverse ligament of the acetabulum

The transverse ligament of the acetabulum is a strong band of fibres. It bridges across the acetabular notch. It completes the rim of the acetabulum and converts the notch into a foramen, through which vessels and nerves enter the acetabular fossa. It gives attachment to the ligament of the head of the femur [Fig. 21.2].

Ligament of the head of the femur

The ligament of the head of the femur extends from the transverse ligament of the acetabulum to the pit on the head of the femur [Fig. 21.2]. Its narrow, cylindrical end is implanted into the **pit on the head of the femur**, and its broad, flattened end is attached to the transverse ligament of the acetabulum and the adjacent margins of the acetabular fossa. It is a weak band of connective tissue surrounded by a synovial membrane. The ligament is too weak to play a part in strengthening the hip joint. Sometimes it transmits a small blood vessel to the head of the femur.

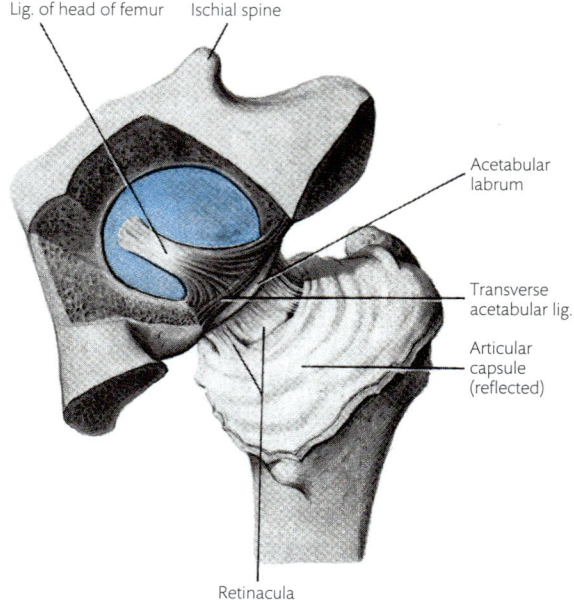

Fig. 21.2 Dissection of the right hip joint from the pelvic side. The floor of the acetabulum has been removed, and the articular capsule of the joint drawn laterally towards the trochanters.

Acetabular labrum

The acetabular labrum is a fibrocartilaginous ring attached to the rim of the acetabulum and the transverse ligament. It deepens the cavity of the acetabulum and narrows its mouth by sloping inwards. The labrum fits tightly on the head of the femur and plays an important function in maintaining the head of the femur in the acetabulum. Both surfaces of the labrum are covered by the synovial membrane. The free margin is relatively thin, and the attached margin is much thicker.

Synovial membrane

The synovial membrane lines the inner surface of the fibrous capsule and covers the neck of the femur as far as the margin of the articular cartilage of the head. At the acetabular attachment of the articular capsule, the synovial membrane is reflected on to the acetabular labrum, the transverse ligament, and the ligament of the head of the femur.

A mass of fat occupies the non-articular fossa of the acetabulum. It is covered by the synovial membrane which extends from the inner margin of the lunate surface of the acetabulum and is reflected on to the ligament of the head of the femur at the acetabular notch. Acetabular branches from the obturator and medial circumflex femoral arteries, and the obturator nerve enter the ligament through the acetabular notch.

Relations

Anterior

The femoral artery and vein lie in front of the hip joint. They are separated from the joint capsule by the iliopsoas and the bursa deep to the psoas tendon.

Inferior

Inferiorly, the obturator externus lies in intimate contact with the joint capsule. The adductor muscles lie a little further away.

Posterior relations

The posterior surface of the hip joint is covered by the piriformis, the tendon of the obturator internus with the two gemelli, and the sciatic nerve.

Superior relations

The gluteus minimus, gluteus medius, and part of the gluteus maximus lie on the superior aspect of the hip joint.

Blood vessels and nerves

The hip joint is supplied by arterial anastomosis formed around the neck of the femur between the following pairs of vessels: (1) ascending branches of the medial and lateral circumflex femoral arteries; (2) acetabular branches of the obturator and medial circumflex femoral arteries; and (3) branches of the superior and inferior gluteal arteries. Arteries which run in the ligament of the head of the femur sometimes enter the head of the femur, but they do not form a significant part of its blood supply.

Nerves enter the joint from: (1) the nerve to the quadratus femoris; (2) the femoral nerve through the nerve to the rectus femoris; and (3) the anterior division of the obturator nerve, and occasionally the accessory obturator nerve.

Identify the hip joint and the important bony landmarks on the radiograph of the pelvis [Fig. 21.3]. The important muscles surrounding the hip joint and acting on it are seen in sectional images [Figs 21.4, 21.5].

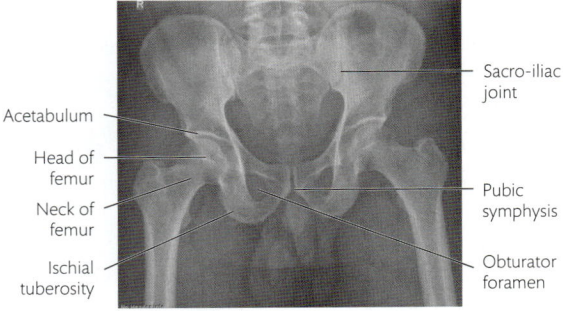

Fig. 21.3 Anteroposterior radiograph of the adult pelvis.

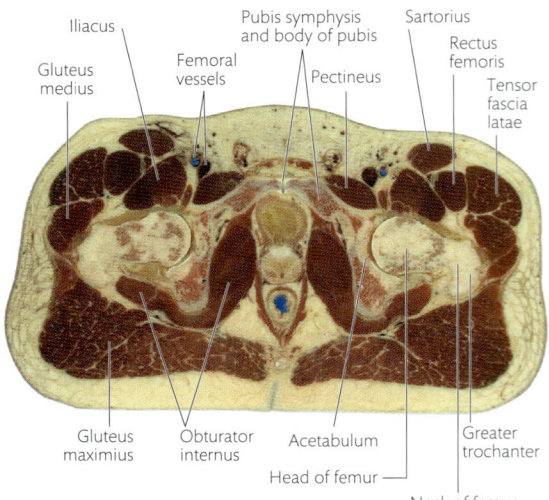

Fig. 21.4 Section through the pubic symphysis and the two hip joints.

Image courtesy of the Visible Human Project of the US National Library of Medicine.

Movements of the hip joint

Movements possible at the hip joint are **flexion**, **extension**, **abduction**, **adduction**, and **medial** and **lateral rotation**. Test the range of movements of your own joint. **Flexion** is very free, and limited only when the thigh comes into contact with the anterior abdominal wall. **Extension** from the anatomical position is restricted by the iliofemoral ligament. Extending the limb posteriorly to the horizontal plane is only possible by tilting the pelvis forwards on the opposite limb. **Abduction** is restricted by the pubofemoral ligament. **Adduction** (as in crossing one thigh over the other) is limited by the lateral part of the iliofemoral ligament and the upper part of the fibrous capsule. **Medial rotation** tightens the ischiofemoral ligament and is limited by it. **Lateral rotation** is limited by the

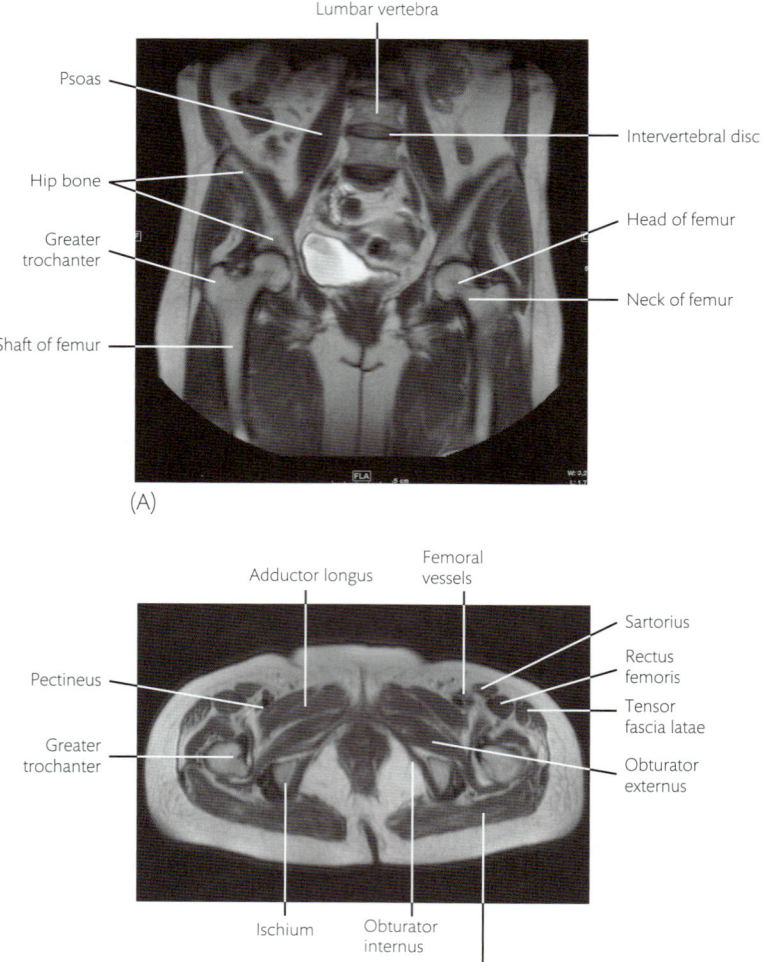

Fig. 21.5 Magnetic resonance imaging of the pelvis. (A) Coronal view. (B) Axial view.

pubofemoral ligament and the lateral parts of the iliofemoral ligament.

Movements of the hip joint are brought about by muscles which cross the joint. Table 21.1 lists the muscles which act on the hip joint, together with the origin, insertion, relation to the hip joint, nerve supply, and action of each muscle. In Table 21.2, the muscles are grouped according to the actions which they produce on the hip joint, and the nerve supply of each muscle is repeated. This allows for easy assessment of the degree of impairment of a particular movement following destruction of a particular nerve.

Table 21.1 Muscles acting on the hip joint

Muscle	Origin	Insertion	Relation to hip joint	Action	Nerve supply
Adductor longus	Pubis, body	Femur, linea aspera	Medial	Adduction	Obturator nerve
Adductor brevis	Pubis, body, and inferior ramus	Femur, linea aspera	Medial	Adduction	Obturator nerve
Adductor magnus	Pubis, inferior ramus Ischium, ramus, and tuberosity	Femur, linea aspera, medial supracondylar line, adductor tubercle	Medial and posterior	Adduction, extension (medial rotation)	Obturator nerve
Psoas major	Lumbar vertebrae	Femur, lesser trochanter	Anterior	Flexion (medial rotation)	Lumbar plexus
Iliacus	Iliac fossa	Femur, line inferior to lesser trochanter	Anterior	Flexion	Lumbar plexus
Pectineus	Pubis, body, and superior ramus	Femur, back of lesser trochanter to linea aspera	Anterior and medial	Adduction, flexion	Femoral nerve
Gluteus minimus	Ilium between anterior and inferior gluteal lines	Femur, greater trochanter, anterior surface	Lateral and anterolateral	Abduction, medial rotation, fixes pelvis on thigh*	Superior gluteal nerve
Gluteus medius	Ilium between anterior and posterior gluteal lines	Femur, greater trochanter, lateral surface	Lateral and anterolateral	Abduction, medial rotation, fixes pelvis on thigh*	Superior gluteal nerve
Gluteus maximus, deep one-fourth	Sacrum, posterior surface, sacrotuberous ligament	Femur, gluteal tuberosity	Posterior	Extension, lateral rotation	Inferior gluteal nerve
Gluteus maximus, superficial three-fourths	Ilium, posterior to posterior gluteal line Sacrum, posterior aspect Sacrotuberous ligament	Tibia lateral condyle, via iliotibial tract	Posterior	Extension, lateral rotation	Inferior gluteal nerve
Piriformis	Sacrum pelvic surface, middle three pieces	Femur, greater trochanter	Posterior and lateral	Abduction, lateral rotation	Sacral nerves
Obturator externus	Hip bone, obturator membrane external surface, margin of obturator foramen	Femur, trochanteric fossa	Inferior	Lateral rotation (flexion)	Obturator nerve
Obturator internus	Internal surfaces of ilium, pubis, ischium, and obturator membrane	Femur, greater trochanter, medial surface	Inferior	Lateral rotation	Nerve to obturator internus
Gemelli	Ischium, each side of lesser sciatic notch	With obturator internus	Posterior	Lateral rotation	Nerve to obturator internus Nerve to quadratus femoris

(continued)

Table 21.1 *Continued*

Muscle	Origin	Insertion	Relation to hip joint	Action	Nerve supply
Biceps femoris, long head	Ischial tuberosity	Head of fibula	Posterior	Extension	Tibial branch of sciatic nerve
Semimembranosus	Ischial tuberosity	Tibia, medial condyle, posteromedial surface	Posterior	Extension	Tibial branch of sciatic nerve
Semitendinosus	Ischial tuberosity	Tibia, medial surface, upper one-fourth	Posterior	Extension	Tibial branch of sciatic nerve
Gracilis	Pubis body and inferior ramus	Tibia, medial surface, upper one-fourth	Posterior and medial	Adduction	Obturator nerve
Sartorius	Ilium, anterior superior iliac spine	Tibia, medial surface, upper one-fourth	Anterior	Flexion, lateral rotation	Femoral nerve
Tensor fasciae latae	Iliac crest, anterior one-fourth	Tibia lateral condyle, via iliotibial tract	Anterior	Flexion, abduction Stabilizes pelvis on thigh	Superior gluteal nerve
Rectus femoris	Ilium, anterior inferior iliac spine, and area above acetabulum	Patella through quadriceps tendon	Anterior	Flexion	Femoral nerve

* This action prevents the pelvis from sagging towards the opposite side when the opposite lower limb is raised from the ground.

Table 21.2 Movements at the hip joint

Movement	Muscles	Nerve supply
Flexion	Iliacus and psoas major	Lumbar ventral rami
	Rectus femoris	Femoral
	Sartorius	Femoral
	Tensor fasciae latae	Superior gluteal
	Pectineus	Femoral
	Adductors—longus and brevis	Obturator
Extension	Gluteus maximus	Inferior gluteal
	Semimembranosus	Sciatic (tibial part)
	Semitendinosus	Sciatic (tibial part)
	Biceps femoris, long head	Sciatic (tibial part)
	Adductor magnus, ischial part	Sciatic (tibial part)
Adduction	Adductors—longus, brevis, and magnus	Obturator
	Gracilis	Obturator
	Pectineus	Femoral
	Quadratus femoris	L. 4, 5 and S. 1 ventral rami
Abduction	Gluteus medius and minimus	Superior gluteal
	Tensor fasciae latae	Superior gluteal
	Piriformis, in flexion	L. 5, S. 1 and 2 ventral rami
	Obturator internus in flexion	L. 5, S. 1 and 2 ventral rami

Table 21.2 *Continued*

Movement	Muscles	Nerve supply
Medial rotation	Tensor fasciae latae	Superior gluteal
	Gluteus minimus	Superior gluteal
	Gluteus medius, anterior fibres	Superior gluteal
	Adductors	Obturator
	Iliopsoas	L. 1, 2 ventral rami
Lateral rotation	Sartorius	Femoral
	Gluteus maximus	Inferior gluteal
	Obturator internus and gemelli	L. 5, S. 1 and 2 ventral rami
	Obturator externus	Obturator
	Quadratus femoris	L. 4 and 5, S. 1 ventral rami
	Piriformis, in extension	L. 5, S. 1 and 2 ventral rami

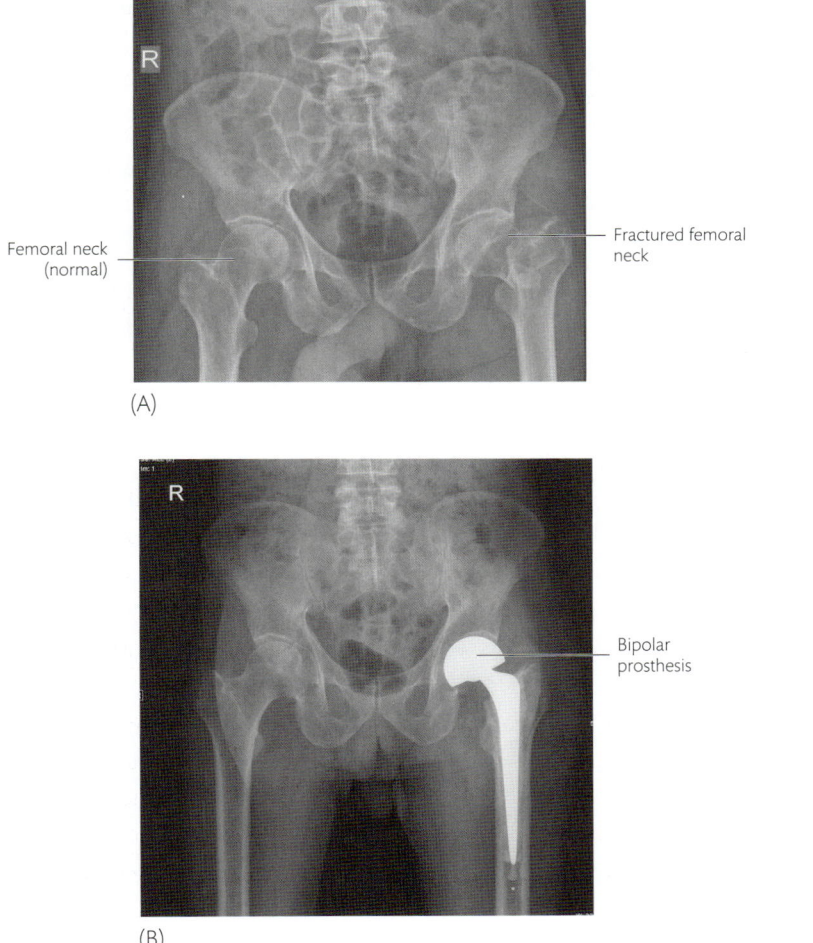

Femoral neck (normal)

Fractured femoral neck

(A)

Bipolar prosthesis

(B)

Fig. 21.6 (A) Plain X-ray of the hip showing fracture of the neck of the femur. (B) Hip X-ray after hemiarthroplasty (different patient).

In movements of the hip, as in all movements in the joints of the lower limb, it should be remembered that when the foot is on the ground, the muscles of that limb are being used in a reversed manner (i.e. the trunk and proximal parts of the limb are moving while the distal part is stationary).

➔ From Table 21.2, it is obvious that medial rotation and abduction are seriously disturbed by injuries to the superior gluteal nerve. Adduction is severely impaired when the obturator nerve is damaged. Extension is most usually produced by the hamstrings, with the gluteus maximus used mainly when extra power is required or in extremes of extension when the hamstrings may be actively insufficient to produce the movement.

Bursae around the hip joint

The tendon of the psoas major muscle passes close to the hip joint, and there is constant movement of it against the joint as the hip flexes and extends. This bursa frequently communicates with the cavity of the hip joint.

Other bursae are present between the greater trochanter and the insertion of the gluteus medius and minimus, and between the ischial tuberosity and the gluteus maximus.

See Clinical Application 21.1 (and Fig. 21.6) for a practical application of the anatomy discussed in this chapter.

CLINICAL APPLICATION 21.1 Fractured neck of the femur with partial hip replacement

A 72-year-old man slipped and fell in the bathroom. He had severe pain in his left hip and was unable to stand up. On examination, his left leg was laterally rotated and he was unable to lift his left heel off the bed. The greater trochanter on the left side appeared to be higher than on the right. On palpation, there was tenderness in the femoral triangle in front of the hip joint. A diagnosis of a fractured neck of the femur was made. An X-ray confirmed the diagnosis and showed a fracture just below the head of the femur [Fig. 21.6A].

Study question 1: what are the attachments of the fibrous capsule? Are all neck fractures intracapsular? (Answer: the fibrous capsule is attached to the margin of the acetabulum. On the femur, it is attached anteriorly to the intertrochanteric line and posteriorly 1 cm medial to the intertrochanteric crest. Fractures of the lateral-most part of the neck are extracapsular.)

One cause for concern in a fractured neck of femur is the possibility of inadequate blood supply. Study question 2: what is the most important blood vessel supplying the head and neck of the femur? (Answer: the most important source of blood supply to the femoral head and neck is the medial circumflex femoral artery. Vessels which arise from it pierce the capsule and run medially on the posterior and superior aspects of the neck.)

The patient was operated upon, with removal of the femoral head and replacement with an artificial prosthesis. He had an uneventful recovery. An X-ray after hip replacement is shown in Fig. 21.6B.

CHAPTER 22
The leg and foot

Introduction

Surface anatomy of the leg

On the leg, identify the head of the fibula, and the condyles and tuberosity of the tibia. Feel the anterior and medial borders of the tibia and the subcutaneous medial surface between them. Follow this surface to the **medial malleolus**. The **long saphenous vein** and the **saphenous nerve** run along the upper two-thirds of the medial border of the tibia.

Find the **neck of the fibula**, and roll the common peroneal nerve on its posterolateral surface. Trace the fibula downwards, feeling it through the muscles. Press on the middle of the bone, and note that it can be pushed inwards to a slight extent, like a firm spring. Find the **lateral malleolus**, and confirm that it projects further distally than the medial malleolus.

On the back of the leg, feel the fleshy mass of the **gastrocnemius** and **soleus**. These two muscles can be made to contract by standing on your toes. When this is done, the two heads of the more superficial gastrocnemius stand out. Deep to the gastrocnemius, the soleus bulges laterally, medially, and inferiorly. Trace these muscles to the **tendo calcaneus**—the thick tendon through which they are inserted into the tip of the heel. Note that the tendo calcaneus lies more than 2 cm posterior to the bones of the leg. Place your fingers in the depressions anterior to the tendo calcaneus, and press forwards. The malleoli can be felt indistinctly through the tendons which cover their posterior surfaces. On the back of the medial malleolus, feel the pulsations of the **posterior tibial artery**.

Surface anatomy of the foot

On your own foot, identify the following structures.

Grip the posterior part of the **calcaneus**. Feel its lateral and medial **processes**. These are blunt prominences on each side of the plantar surface of the calcaneus. Feel the subcutaneous lateral surface of the calcaneus and the resistance of the peroneal tendons, immediately posteroinferior to the lateral malleolus. Follow the tendons inferiorly.

Find the **tuberosity on the base of the fifth metatarsal**. It lies midway between the point of the heel and the little toe. The **head of the fifth metatarsal** is felt as a rounded bulge at the root of the little toe.

Extend the toes firmly. A bulge appears on the lateral side of the dorsum of the foot, a thumb-breadth anterior to the lateral malleolus. This is the **extensor digitorum brevis** muscle which arises from the dorsal surface of the distal part of the calcaneus.

Repeatedly plantar flex and dorsiflex the foot. The anterior part of the **trochlea tali** can be felt under the anterior part of the distal end of the tibia in plantar flexion. Invert the foot. The **tendon of the tibialis anterior** becomes prominent, as it passes to the junction of the medial cuneiform and the base of the first metatarsal. Appreciate the bony prominence on the dorsum of the foot, distal to the lateral malleolus. This is the lateral part of the **head of the talus**. Confirm that it disappears on eversion.

On the medial side of the ankle, feel the **sustentaculum tali** below the medial malleolus. Note that this moves with the posterior part of the calcaneus, as the foot is inverted and everted. Evert the foot as far as possible, and feel the bony prominence anteroinferior to the medial malleolus. This is the medial surface of the head of the talus. Keep your finger on it, and invert the foot. Note that it disappears and that the proximal edge of the **navicular** becomes prominent, distal to it. Follow the navicular to its **tuberosity** on the plantar aspect. The tendon which passes forwards to this tuberosity is the tendon of the tibialis posterior.

Note the parts of the foot which are normally in contact with the ground. Note also that it is the head of the first metatarsal which takes the pressure when you push off to take a step.

The arrangement of the phalanges with each other and with the metatarsals (metatarsophalangeal and interphalangeal joints) is similar to that in the fingers. The main difference lies in the massive phalanges in the big toe and the relatively small phalanges in the others. The appearance of these bones indicates that the major stresses to the foot are carried by the big toe.

Intermuscular septa

These intermuscular septa are present in the leg. They separate the groups of muscles, give attachments to muscle fibres proximally, and hold the muscle groups tightly within a fascial space. Due to this arrangement, when the muscles contract, they are compressed by the fascia, and the blood within them is pumped upwards in the veins. This is essential for proper part venous return from the lower limb. ➔ Extensive division of the fascia or wasting of the muscles within the fascial compartments impairs venous return and lead to swelling of the limb (oedema).

There are three named intermuscular partitions in the leg: (1) the interosseous membrane between muscle groups B and E [Fig. 22.1]; (2) the anterior septum between groups A and B; and (3) the posterior septum between groups A and H. In addition, the muscles of the back of the leg are divided into three layers by two coronal sheets of fascia between groups E and G, and between groups G and H.

Front of the leg and dorsum of the foot

See Dissection 22.1.

Superficial veins of the front of the leg and dorsum of the foot

There are two **dorsal digital veins** in each toe. Each joins the corresponding vein from the adjacent toe to form a **dorsal metatarsal vein**. The dorsal metatarsal vein drains into the **dorsal venous arch**. The dorsal digital veins on the medial side of the big toe and the lateral side of the little toe join the ends of the arch to form the long and short saphenous veins, respectively. The dorsal venous arch lies on the distal parts of the shafts of the metatarsals. It drains the dorsum of the foot and toes.

A. Peroneal muscles
B. Extensors
C. Tibia
D. Fibula
E. Tibialis posterior
G. Long flexors of toes
H. Superficial muscles of calf
A. Anterior.
P. Posterior.
M. Medial.
L. Lateral.

Superficial peroneal N.
Deep peroneal N.

Tibial N.

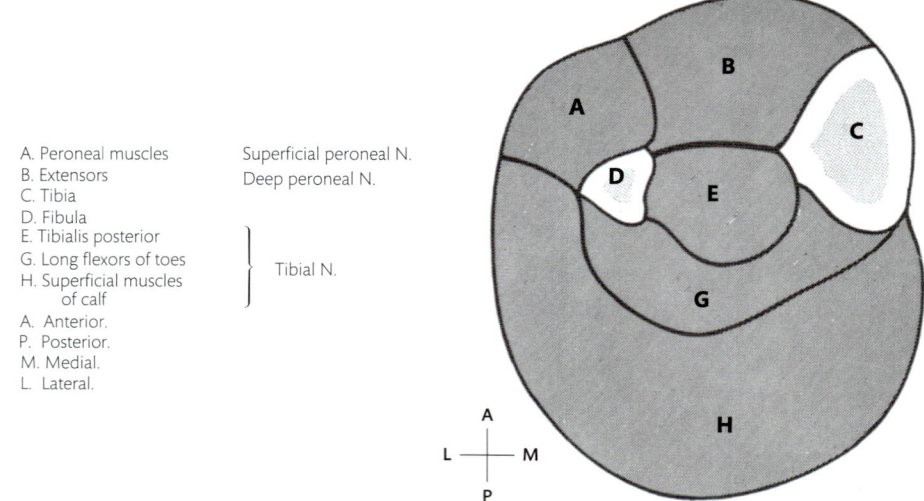

Fig. 22.1 Diagram of osteofascial compartments of the leg. The interosseus membrane lies between B and E.

DISSECTION 22.1 Skin reflection, and cutaneous nerves and vessels

Objectives

I. To reflect the skin.   II. To identify and trace the lateral cutaneous nerve of the calf, sural nerve, superficial peroneal nerve, dorsal digital nerves, and long and short saphenous veins in the leg.

Instructions

1. Place the limb in a convenient position (e.g. with a block under the knee and the foot bent down—plantar flexed). Make incision 11 [see Fig. 17.3]. Reflect the skin from the front of the leg and the dorsum of the foot, but retain the superficial fascia so as not to destroy the veins.
2. Find the lateral cutaneous nerve of the calf and the long saphenous vein, with the saphenous nerve beside it. Follow all three to the foot [see Fig. 17.6]. Note the branches of the nerves and the tributaries of the vein.
3. Establish the continuity of the long saphenous vein with the medial extremity of the dorsal venous arch

of the foot. This arch lies transversely on the anterior parts of the metatarsals. Follow the arch to the lateral side of the foot, finding its dorsal metatarsal tributaries. Laterally, the arch is continuous with the short saphenous vein.

4. Trace the beginning of the short saphenous vein to a point below the lateral malleolus, and find the sural nerve beside it. Follow the sural nerve along the lateral side of the foot to the little toe. It gives a communicating branch to the superficial peroneal nerve.
5. Find the superficial peroneal nerve, as it pierces the deep fascia at the junction of the middle and distal thirds of the leg. Trace the nerve and its branches into the dorsum of the foot. Note that the most medial branch is to the medial side of the big toe and that a branch does not pass to the first interdigital cleft.
6. Find the dorsal digital nerves of the adjacent sides of the big and second toes. Follow them proximally. They arise from a branch of the deep peroneal nerve, proximal to the cleft.

The **short saphenous vein** runs back inferior and posterior to the lateral malleolus. It ascends to the popliteal fossa in the back of the leg [Fig. 22.2].

The **long saphenous vein** passes posteriorly on the medial side of the foot. It ascends anterior to the medial malleolus, obliquely across the distal third of the medial surface of the tibia [Fig. 22.3]. The further course and termination are described in Chapter 17.

Cutaneous nerves of the front of the leg and dorsum of the foot

The upper two-thirds of the front of the leg are supplied by two nerves—the **saphenous nerve** (L. 3, 4) medially, and the **lateral cutaneous nerve of the calf** laterally. The lower third is supplied by the **superficial peroneal** and **saphenous nerves**.

The dorsum of the foot is supplied mainly by the medial and intermediate cutaneous branches of the **superficial peroneal nerve**. The lateral margin is supplied by the **sural nerve,** and the medial margin by the **saphenous nerve** proximally and the **superficial peroneal nerve** distally. The first interdigital cleft and the skin immediately proximal to it are supplied by the **deep peroneal nerve**. The **dorsum of the toes** is supplied by the

digital branches of these nerves [see Figs 17.6, 24.1]. On the terminal phalanges, the supply is from the plantar nerves.

Superficial peroneal nerve (L. 4, 5; S. 1)

The superficial peroneal nerve arises from the common peroneal nerve on the lateral side of the neck of the fibula. It descends between the peroneus longus and brevis supplying them. It enters the superficial fascia at the junction of the middle and distal thirds of the leg and branches into the **medial** and **intermediate dorsal cutaneous nerves**. Each of these supplies the skin in the lower part of the front of the leg and the dorsum of the foot, and divides into two **dorsal digital nerves** of the foot. Those from the medial nerve pass to the medial side of the big toe and the adjacent sides of the second and third toes. Those from the intermediate nerve pass to the adjacent sides of the third and fourth, and fourth and fifth toes [see Fig. 24.1]. All dorsal digital nerves lie deep to the dorsal venous arch.

Sural nerve (S. 1, 2)

The sural nerve arises from the tibial nerve in the popliteal fossa and descends in the back of the leg to the posterior surface of the lateral malleolus. It lies with the short saphenous vein in the

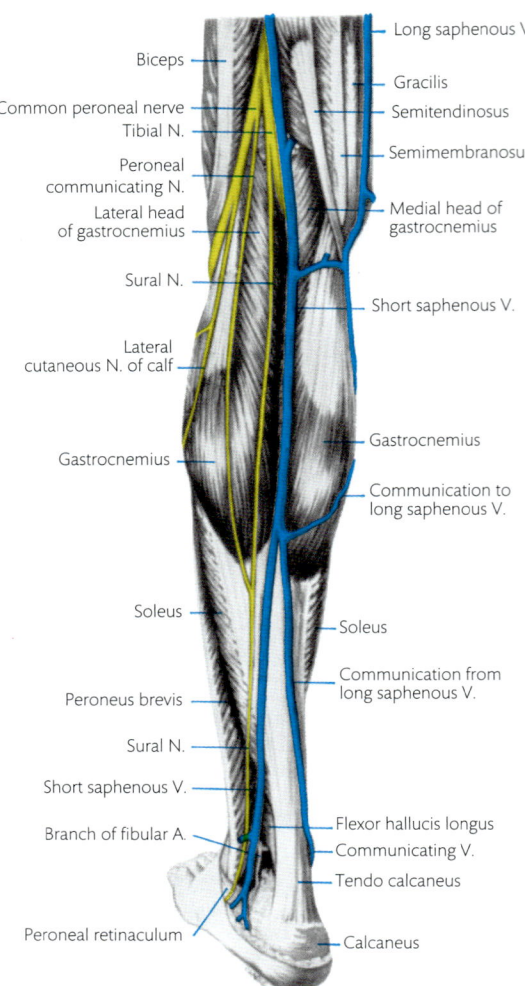

Fig. 22.2 Superficial dissection of the leg, viewed from the posterolateral side, showing veins and nerves. (Usually the short saphenous vein ends in the popliteal fossa.)

Labels for Fig. 22.2:
Biceps
Common peroneal nerve
Tibial N.
Peroneal communicating N.
Lateral head of gastrocnemius
Sural N.
Lateral cutaneous N. of calf
Gastrocnemius
Soleus
Peroneus brevis
Sural N.
Short saphenous V.
Branch of fibular A.
Peroneal retinaculum

Long saphenous V.
Gracilis
Semitendinosus
Semimembranosus
Medial head of gastrocnemius
Short saphenous V.
Communication to long saphenous V.
Soleus
Communication from long saphenous V.
Flexor hallucis longus
Communicating V.
Tendo calcaneus
Calcaneus

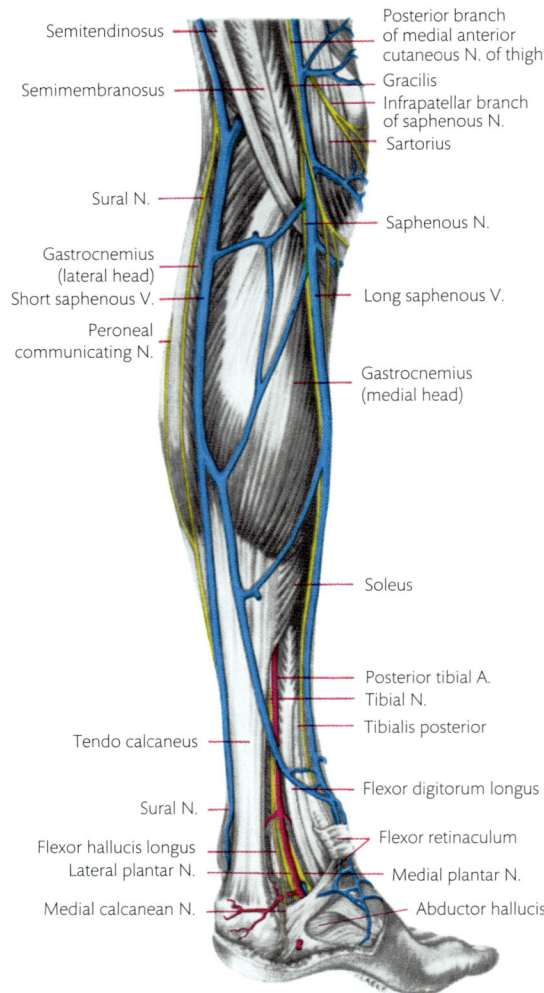

Fig. 22.3 Superficial dissection of the leg, viewed from the posteromedial side, showing veins and nerves.

Labels for Fig. 22.3:
Semitendinosus
Semimembranosus
Sural N.
Gastrocnemius (lateral head)
Short saphenous V.
Peroneal communicating N.
Tendo calcaneus
Sural N.
Flexor hallucis longus
Lateral plantar N.
Medial calcanean N.

Posterior branch of medial anterior cutaneous N. of thigh
Gracilis
Infrapatellar branch of saphenous N.
Sartorius
Saphenous N.
Long saphenous V.
Gastrocnemius (medial head)
Soleus
Posterior tibial A.
Tibial N.
Tibialis posterior
Flexor digitorum longus
Flexor retinaculum
Medial plantar N.
Abductor hallucis

superficial fascia of the lower leg and supplies the overlying skin. It is joined by the **peroneal communicating** branch of the common peroneal nerve [see Fig. 18.2]. It turns forwards along the lateral border of the foot and little toe, supplying the skin of both.

Dissection 22.2 (and Fig. 22.4) looks at the deep fascia of the leg.

Deep fascia of the front of the leg

The deep fascia of the leg is very strong. It is fused with the periosteum where the bones are subcutaneous. At the ankle, it forms thickened bands (retinacula) which hold the tendons close to the joint

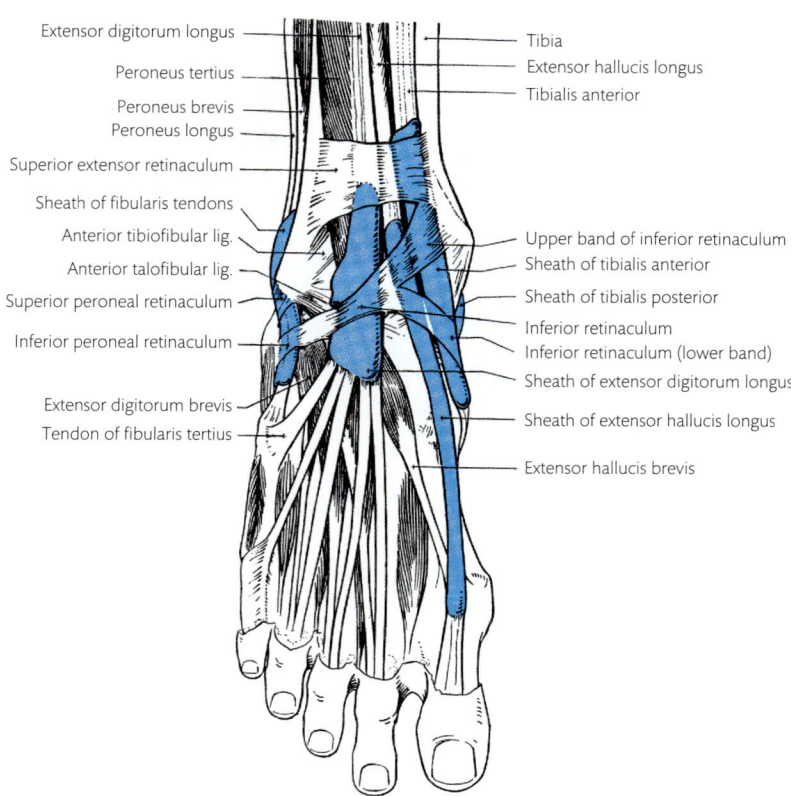

Extensor digitorum longus
Peroneus tertius
Peroneus brevis
Peroneus longus
Superior extensor retinaculum
Sheath of fibularis tendons
Anterior tibiofibular lig.
Anterior talofibular lig.
Superior peroneal retinaculum
Inferior peroneal retinaculum
Extensor digitorum brevis
Tendon of fibularis tertius

Tibia
Extensor hallucis longus
Tibialis anterior

Upper band of inferior retinaculum
Sheath of tibialis anterior
Sheath of tibialis posterior
Inferior retinaculum
Inferior retinaculum (lower band)
Sheath of extensor digitorum longus
Sheath of extensor hallucis longus
Extensor hallucis brevis

Fig. 22.4 Synovial sheaths of the dorsum of the foot.

and prevent them from springing forwards in dorsiflexion of the ankle.

Retinacula

The **superior extensor retinaculum** is a broad band between the triangular subcutaneous area of the fibula and the anterior border of the tibia.

The **inferior extensor retinaculum** is Y-shaped. The stem of the Y is attached to the upper surface of the anterior part of the calcaneus [Fig. 22.4]. Medially, the limbs of the Y separate. The upper limb is attached to the medial malleolus; the lower fuses with the fascia of the sole on the medial side of the foot. The deep fibres of the retinaculum form two loops around the tendons—one around the extensor digitorum longus and peroneus tertius, and another around the extensor hallucis longus. The retinacula prevent the tendons from slipping medially when the foot is inverted and there is a marked medial angulation of the tendon of the extensor hallucis at the retinaculum.

The **superior** and **inferior peroneal retinacula** hold the tendons of the peroneus longus and brevis in position, as they pass posteroinferior to the lateral malleolus and on the lateral surface of the calcaneus [see Fig. 16.15].

A similar **flexor retinaculum**, posteroinferior to the medial malleolus, holds the tendons of the deep muscles of the back of the leg in position, as they pass into the foot [Fig. 22.5].

See Dissection 22.3 and Figs 22.6 and 22.7.

Synovial sheaths of extensor tendons

There are three synovial sheaths [Fig. 22.4] in the front of the ankle: (1) the synovial sheath surrounding the tendon of the **tibialis anterior**—it extends from the upper border of the superior extensor retinaculum almost to the insertion of the tendon; (2) this sheath on the **extensor hallucis longus** begins between the retinacula and reaches to the proximal phalanx of the big toe; and (3) the third around the tendons of the

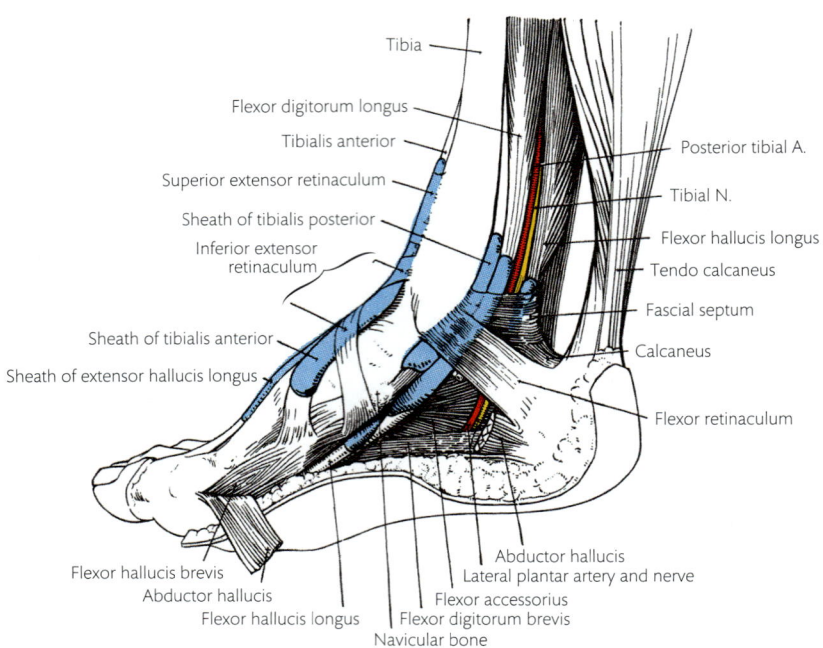

Tibia

Flexor digitorum longus

Tibialis anterior

Superior extensor retinaculum

Sheath of tibialis posterior

Inferior extensor retinaculum

Sheath of tibialis anterior

Sheath of extensor hallucis longus

Posterior tibial A.

Tibial N.

Flexor hallucis longus

Tendo calcaneus

Fascial septum

Calcaneus

Flexor retinaculum

Flexor hallucis brevis

Abductor hallucis

Flexor hallucis longus

Abductor hallucis

Lateral plantar artery and nerve

Flexor accessorius

Flexor digitorum brevis

Navicular bone

Fig. 22.5 Dissection of the leg and foot, showing the synovial sheaths.

DISSECTION 22.3 Front of the leg-1

Objectives

I. To study the tibialis anterior, extensor hallucis longus, extensor digitorum longus, peroneus tertius, and extensor digitorum brevis. **II.** To identify and trace the deep peroneal nerve and anterior tibial vessels in the leg and foot. **III.** To dissect the extensor expansion.

Instructions

1. Divide the deep fascia of the front of the leg longitudinally between the tibia and fibula. Extend the incision onto the dorsum of the foot. Leave the superior and inferior extensor retinacula intact. Near the superior retinaculum, pass a blunt seeker deep to these retinacula. This will allow you to define their margins more easily. Avoid injury to the synovial sheaths of the tendons which lie deep to the retinacula [Fig. 22.4].
2. Turn the deep fascia medially and laterally, and confirm its attachments to the bones. Open the synovial sheaths of the tendons, proximal to the superior extensor retinacula, and pass a blunt probe along them to define their extent.

3. Follow the tendons upwards and downwards.
4. Define the individual muscles and their attachments to the bones. Separate the tibialis anterior from the other three muscles down to the interosseous membrane.
5. Find the anterior tibial vessels and the deep peroneal nerve on the membrane. Trace these upwards and downwards.
6. Pull the peroneus tertius muscle aside, and find the perforating branch of the fibular artery as it pierces the lower part of the interosseous membrane. Follow it into the foot [Fig. 22.6].
7. Find the extensor digitorum brevis muscle on the dorsum of the foot. Trace its tendons to the medial four toes [Fig. 22.7].
8. Follow the anterior tibial artery and the deep peroneal nerve into the foot. Here the artery is known as the dorsalis pedis artery [Fig. 22.7].
9. On the second or third toe, clean the extensor expansion formed by the extensor tendons. Trace the expansion (which is similar to that in fingers) to the distal phalanx.

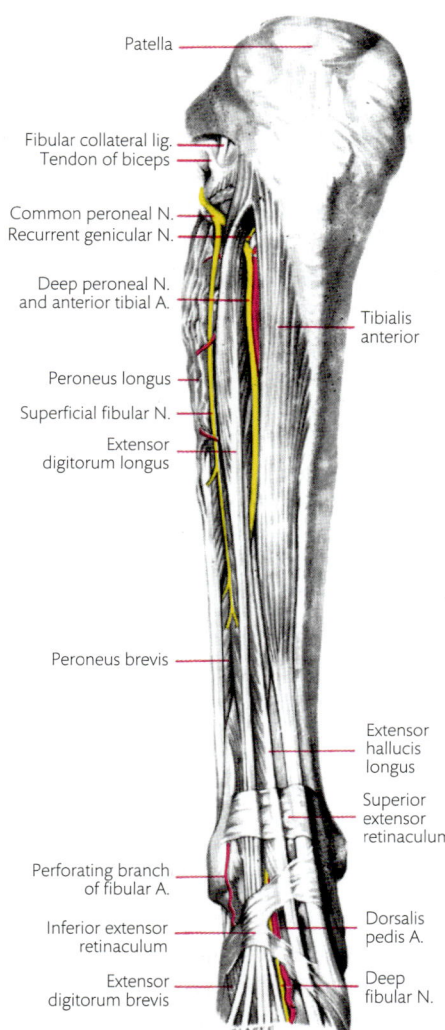

Patella

Fibular collateral lig.
Tendon of biceps

Common peroneal N.
Recurrent genicular N.

Deep peroneal N.
and anterior tibial A.

Tibialis anterior

Peroneus longus

Superficial fibular N.

Extensor digitorum longus

Peroneus brevis

Extensor hallucis longus

Superior extensor retinaculum

Perforating branch of fibular A.

Inferior extensor retinaculum

Dorsalis pedis A.

Extensor digitorum brevis

Deep fibular N.

Fig. 22.6 Dissection of the front and lateral sides of the leg.

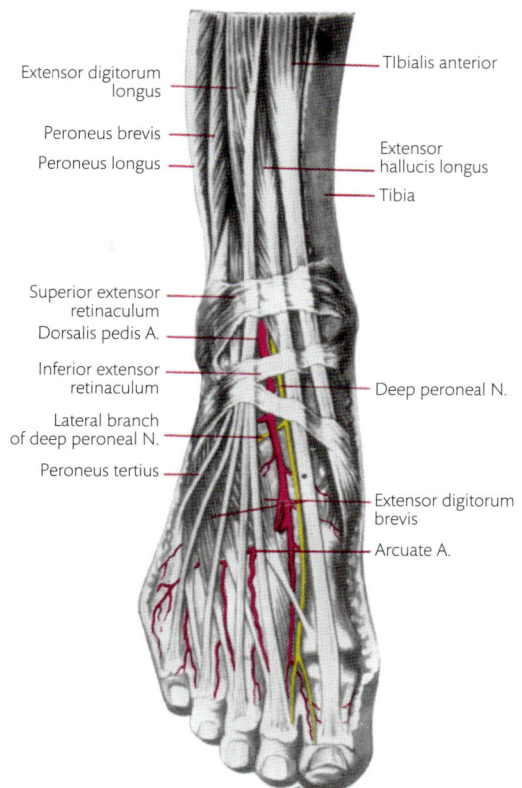

Extensor digitorum longus

Peroneus brevis

Peroneus longus

TIbialis anterior

Extensor hallucis longus

Tibia

Superior extensor retinaculum

Dorsalis pedis A.

Inferior extensor retinaculum

Lateral branch of deep peroneal N.

Peroneus tertius

Deep peroneal N.

Extensor digitorum brevis

Arcuate A.

Fig. 22.7 Dissection of the dorsum of the foot.

extensor **digitorum longus** and **peroneus tertius** extending from the lower part of the superior extensor retinaculum to the middle of the dorsum of the foot.

Contents of the anterior compartment of the leg

The anterior compartment of the leg contains: (1) the tibialis anterior; (2) the extensor digitorum longus; (3) the extensor hallucis longus; and (4) the peroneus tertius. The compartment also contains the anterior tibial vessels and the deep peroneal nerve which supply these muscles and continue into the dorsum of the foot [Figs 22.6, 22.7].

Tibialis anterior

Origin: the tibialis anterior takes origin from the upper half of the lateral surface of the tibia and from the interosseous membrane [see Fig. 16.12]. **Insertion**: the tendon passes deep to the retinacula, bends medially, and is inserted into the medial surface of the medial cuneiform and adjacent part of the first metatarsal close to their plantar aspects. This insertion is almost continuous with that of the peroneus longus (see later). **Nerve supply**: deep peroneal nerve and recurrent genicular nerves. **Actions**: it is a dorsiflexor and powerful invertor of the foot when the foot is raised from the ground. When the foot is on the ground, the muscle helps to balance the leg and talus on the other tarsal bones, so that the leg is kept vertical, even when walking on uneven ground.

Extensor digitorum longus

Origin: the extensor digitorum longus is a long, thin muscle that arises mainly from the upper part of the medial surface of the fibula [see Fig. 16.12]. The tendon passes deep to the retinacula. anterior to the ankle joint. **Insertion**: it divides into four parts which pass to the lateral four toes. As each tendon approaches the metatarsophalangeal joint, it forms an **extensor expansion**, similar to that in the fingers. Each extensor expansion fuses with the fibrous capsule on the dorsal surface of the metatarsophalangeal joint and extends on each side of the joint to the deep transverse metatarsal ligament. The thick central part of the expansion continues on the dorsal surface of the proximal phalanx and is inserted into the base of the middle phalanx. On the second to fourth toes, it is joined by the greater part of the tendon of the extensor digitorum brevis. The lateral and medial parts of the expansion continue distally, fused to the median portion. They cross the dorsolateral surfaces of the distal interphalangeal joint and are inserted into the base of the distal phalanx. The tendons of the lumbricals may join the medial side of the expansion in each toe. The interossei may send delicate extensions into it, although, in the foot, interossei are principally inserted into the proximal phalanx. The little toe has only the long extensor tendon, but the expansion is otherwise the same as in the second to fourth toes. **Nerve supply**: deep peroneal nerve. **Actions**: it extends the interphalangeal and metatarsophalangeal joints of the lateral four toes. When the metatarsophalangeal joints of the foot are extended, extension of the interphalangeal joints depends solely on the lumbricals. This is because the interossei are not inserted to a significant degree into the extensor expansion. The extensor digitorum longus also dorsiflexes the ankle joint and acts with the peroneal muscles to evert the foot.

Peroneus tertius (fibularis tertius)

This small muscle is continuous at its origin with the extensor digitorum longus and appears to be a part of it [see Fig. 16.12]. **Origin**: it arises from the distal part of the medial surface of the fibula and the adjacent interosseous membrane. **Insertion**: it is inserted into the base of the fifth metatarsal bone. Frequently, it is partly fused with the tendon of the extensor digitorum longus to the fifth toe. **Nerve supply**: deep peroneal nerve. **Actions**: it everts the foot and dorsiflexes the ankle.

Extensor hallucis longus

Origin: it arises from the middle of the medial surface of the fibula, medial to the extensor digitorum

longus [see Fig. 16.12]. The tendon passes deep to the extensor retinacula and crosses in front of the anterior tibial artery. The inferior extensor retinaculum holds the tendon laterally, so that it inclines forwards and medially to the big toe. **Insertion**: on the head of the metatarsal, the tendon forms an **extensor expansion**. The central part of the expansion passes to the distal phalanx. The lateral part of the expansion passes to the deep transverse metatarsal ligament, and the medial part joins the tendon of the abductor hallucis (see later). It is this part of the expansion that is stretched over the head of the first metatarsal when the big toe deviates laterally in 'hallux valgus'. **Nerve supply**: deep peroneal nerve. **Actions**: it extends the phalanges of the big toe and dorsiflexes the ankle joint. It plays a part in inversion when the big toe is extended.

Deep peroneal nerve (deep fibular nerve)

The deep peroneal nerve is a branch of the common peroneal nerve. It arises between the neck of the fibula and the peroneus longus muscle. It pierces the anterior intermuscular septum to enter the anterior compartment of the leg and descends with the anterior tibial vessels [Fig. 22.6]. It lies on the interosseous membrane between the tibialis anterior and the long extensors of the toe. Near the ankle joint, it is crossed superficially by the extensor hallucis longus tendon [Fig. 22.7] and enters the dorsum of the foot midway between the medial and lateral malleoli. It lies between the tendons of the extensor digitorum longus and extensor hallucis longus and the dorsalis pedis artery. It almost immediately divides into two branches: (1) a **medial branch** that continues towards the first interdigital space—it supplies the adjacent joints and the first dorsal interosseous muscle, and ends by forming the **dorsal digital nerves** for the adjacent sides of the big and second toes; (2) the **lateral branch** supplies the extensor digitorum brevis and the surrounding joints. The deep peroneal nerve supplies all the muscles of the anterior compartment of the leg and the extensor digitorum brevis on the foot. ⊙ If the nerve is destroyed, dorsiflexion of the ankle and extension of the metatarsophalangeal joints are lost, and inversion is weakened. This condition is known as 'foot drop'.

Anterior tibial artery

This artery **begins** as a branch of the popliteal artery at the lower border of the popliteus. It enters the anterior compartment of the leg above the interosseous membrane, and runs on the

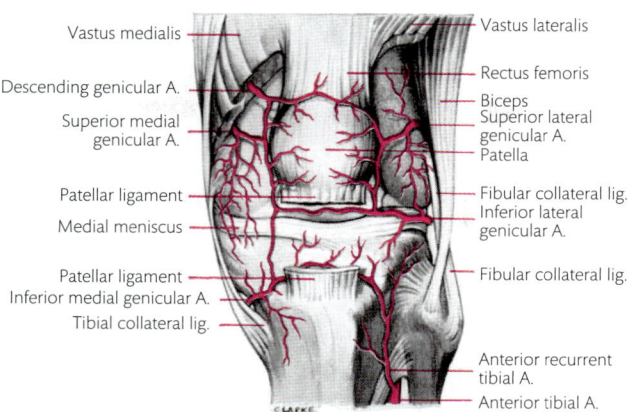

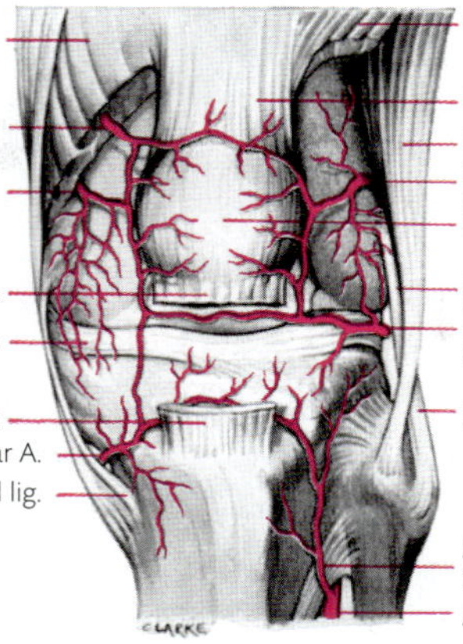
Vastus medialis
Descending genicular A.
Superior medial genicular A.
Patellar ligament
Medial meniscus
Patellar ligament
Inferior medial genicular A.
Tibial collateral lig.
Vastus lateralis
Rectus femoris
Biceps
Superior lateral genicular A.
Patella
Fibular collateral lig.
Inferior lateral genicular A.
Fibular collateral lig.
Anterior recurrent tibial A.
Anterior tibial A.

Fig. 22.8 Arterial anastomosis on the front and sides of the left knee joint.

anterior surface of that membrane with the deep peroneal nerve. It becomes progressively more superficial, as it descends and **ends** as the dorsalis pedis artery, midway between the medial and lateral malleoli [Fig. 22.7]. The anterior tibial veins are closely applied to the artery.

Branches

The anterior tibial artery supplies the muscles of the anterior compartment of the leg and sends an anterior tibial recurrent artery upwards to the knee joint [Fig. 22.8]. Other branches of the anterior tibial artery are the **medial** and **lateral anterior malleolar arteries** which join the plexus on each malleolus. The lateral one anastomoses with the perforating branch of the fibular artery (a branch of the posterior tibial artery).

Dorsalis pedis artery

The dorsalis pedis artery is the continuation of the anterior tibial artery [Fig. 22.7]. It begins on the anterior surface of the ankle joint and runs with the deep peroneal nerve, deep to the inferior extensor retinaculum, to the proximal end of the first intermetatarsal space. Here it divides into the **arcuate artery** and the **first dorsal metatarsal artery**. On the dorsum of the foot, it lies on the tarsal bones and is readily palpated against them between the tendons of the extensor hallucis longus and extensor digitorum longus.

Branches

It gives medial and lateral **tarsal branches** to the tarsal bones and extensor digitorum brevis. The **arcuate artery** runs laterally across the base of the metatarsals, deep to the extensor tendons. It gives **dorsal metatarsal arteries** to the second, third, and fourth intermetatarsal spaces. Each of these communicates with the plantar arch in the sole of the foot through perforating branches at the proximal end of the intermetatarsal space. The perforating artery in the first space—the **deep plantar branch** of the arcuate artery—is larger than the others and carries blood to the medial end of the plantar arch [Figs 22.9, 22.10].

Extensor digitorum brevis

Origin: the extensor digitorum brevis originates from the anterior part of the dorsal surface of

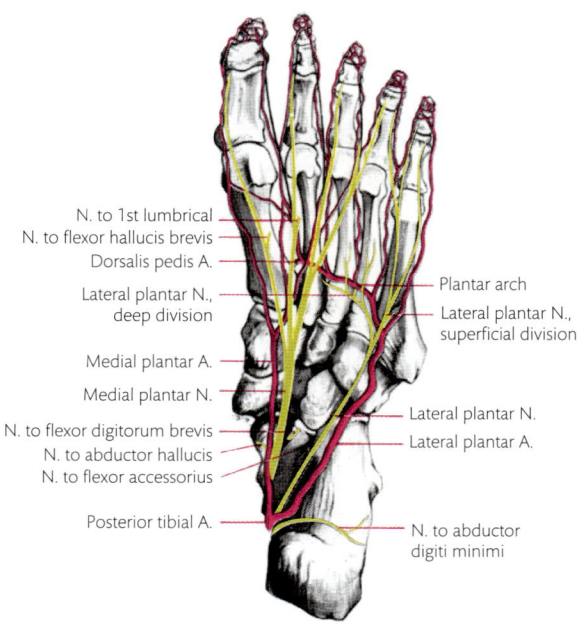

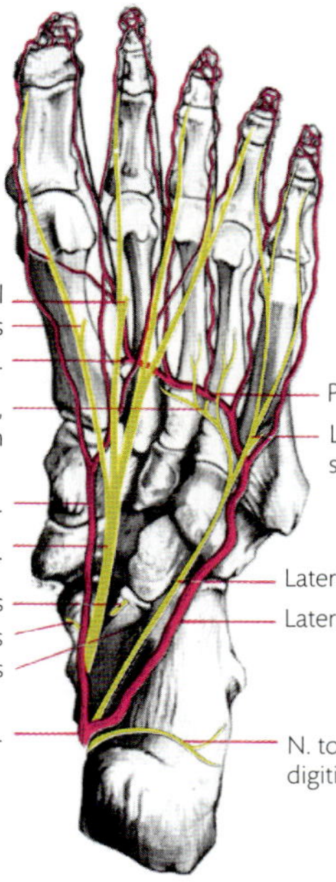
N. to 1st lumbrical
N. to flexor hallucis brevis
Dorsalis pedis A.
Lateral plantar N., deep division
Medial plantar A.
Medial plantar N.
N. to flexor digitorum brevis
N. to abductor hallucis
N. to flexor accessorius
Posterior tibial A.
Plantar arch
Lateral plantar N., superficial division
Lateral plantar N.
Lateral plantar A.
N. to abductor digiti minimi

Fig. 22.9 Arteries and nerves of the sole of the foot.

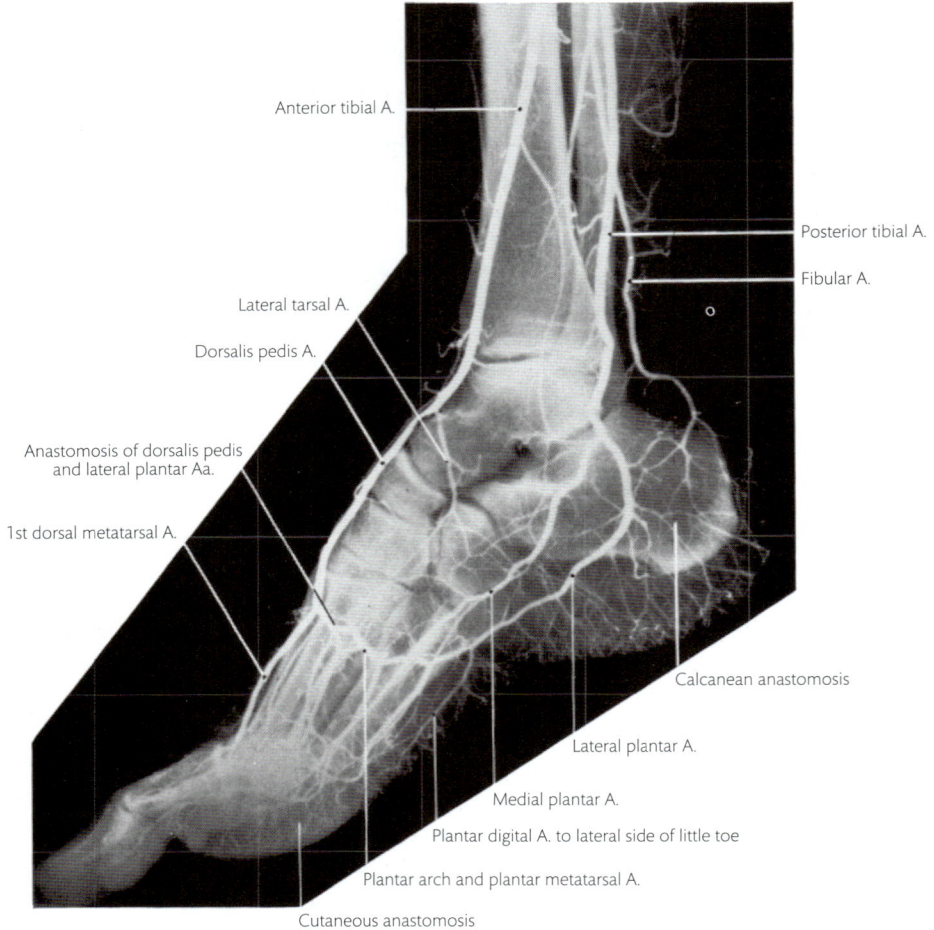

Fig. 22.10 Lateral radiograph of the foot after injection of the arteries with X-ray-opaque material.

the calcaneus and from the stem of the inferior extensor retinaculum. It divides into four tendons; the most medial is the **extensor hallucis brevis**. **Insertion**: the tendon of the extensor hallucis brevis runs obliquely across the dorsum of the foot to the base of the proximal phalanx of the big toe. The remaining three tendons join the long extensor tendons of the second to fourth toes. They are inserted into the middle and terminal phalanges of these toes through the extensor expansions. **Nerve supply**: deep peroneal nerve. **Actions**: the extensor hallucis brevis extends the metatarsophalangeal joint of the big toe. The other tendons extend the metatarsophalangeal and interphalangeal joints of the second, third, and fourth toes.

Dissection 22.4 continues with the dissection of the front of the leg.

Lateral side of the leg

The peroneus longus and brevis muscles lie on the lateral surface of the fibula in the lateral compart-

DISSECTION 22.4 Front of the leg-2

Objective

I. To study the extensor retinacula.

Instructions

1. Cut through the extensor retinacula, and reflect their attachments.
2. Note the extensions from the deep surface that hold the tendons in place.

ment of the leg. They lie between the anterior and posterior intermuscular septa and take partial origin from them [Fig. 22.1].

Peroneus longus (fibularis longus)

Origin: the peroneus longus takes origin from the upper two-thirds of the lateral surface of the fibula [see Fig. 16.12]. It overlaps the peroneus brevis as they descend on the lateral surface of the fibula. The tendon passes behind the triangular subcutaneous area of the fibula to reach the posterior border of the lateral malleolus. It lies in a **common synovial sheath** with the tendon of the peroneus brevis and is held in position by the **superior peroneal retinaculum**—a thickened part of the deep fascia between the fibula and the calcaneus [Fig. 22.11]. Below the lateral malleolus, the tendon runs on the lateral surfaces of the calcaneus and cuboid, inferior to the tendon of the peroneus brevis, and the peroneal trochlear. At the lateral border of the foot, it turns over the lateral margin of the cuboid, obliquely across the sole of the foot in a groove on the plantar surface of the cuboid. **Insertion**: it is inserted into the medial cuneiform and the base of the first metatarsal [see Fig. 16.16]. **Nerve supply**: superficial peroneal nerve. **Actions**: the peroneus longus is

an evertor of the foot and supports the transverse arch of the foot.

Peroneus brevis (fibularis brevis)

Origin: the peroneus brevis arises from the lower two-thirds of the lateral surface of the fibula [see Fig. 16.12]. It descends on the lateral surface of the fibula deep to the peroneus longus, passes behind the triangular subcutaneous area of the fibula and the lateral malleolus, and passes deep to the superior peroneal retinaculum. It runs on the lateral side of the calcaneus superior to the tendon of peroneus longus and the peroneal trochlear. **Insertion**: the peroneus brevis is attached to the base of the fifth metatarsal. **Nerve supply**: superficial peroneal nerve. **Actions**: the peroneus brevis is an evertor of the foot and a plantar flexor of the ankle joint. Fig. 22.12A and B shows eversion and inversion of the foot.

Dissection 22.5 looks at the peroneal compartment of the leg.

Terminal branches of the common peroneal nerve

The common peroneal nerve turns round the lateral surface of the neck of the fibula. It gives off a small recurrent genicular branch to the knee joint

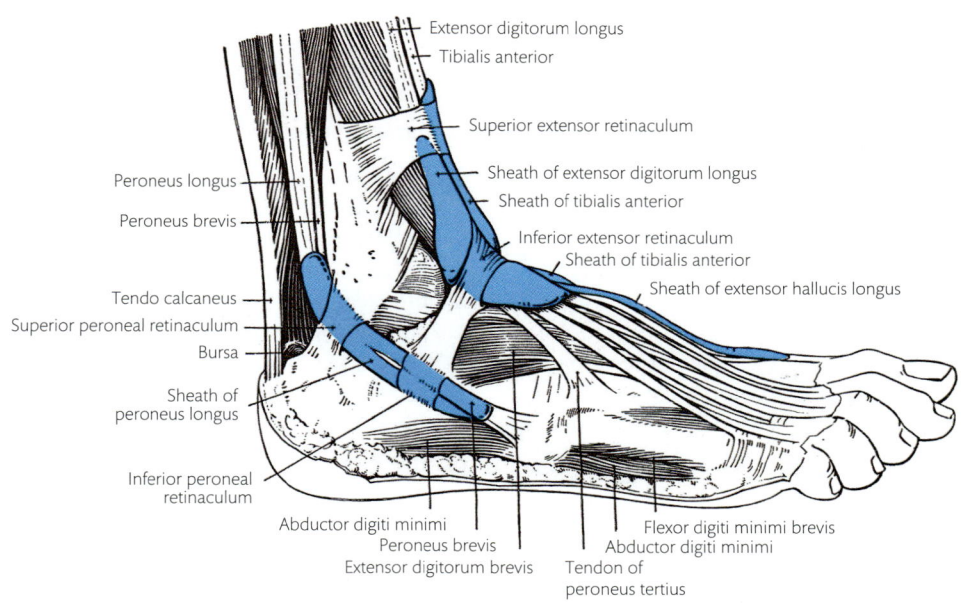

Extensor digitorum longus
Tibialis anterior
Superior extensor retinaculum
Peroneus longus
Sheath of extensor digitorum longus
Peroneus brevis
Sheath of tibialis anterior
Inferior extensor retinaculum
Sheath of tibialis anterior
Sheath of extensor hallucis longus
Tendo calcaneus
Superior peroneal retinaculum
Bursa
Sheath of peroneus longus
Inferior peroneal retinaculum
Abductor digiti minimi
Peroneus brevis
Extensor digitorum brevis
Flexor digiti minimi brevis
Abductor digiti minimi
Tendon of peroneus tertius

Fig. 22.11 Dissection showing the synovial sheaths of tendons of the lateral aspect of the foot.

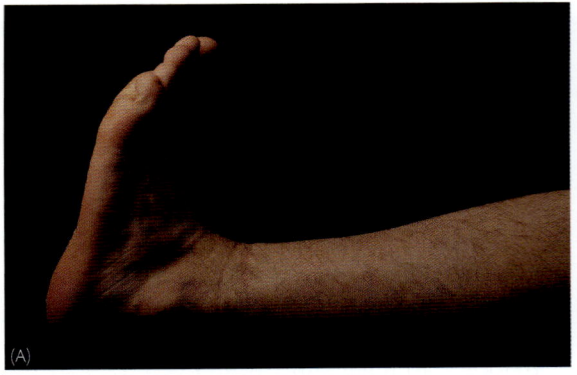
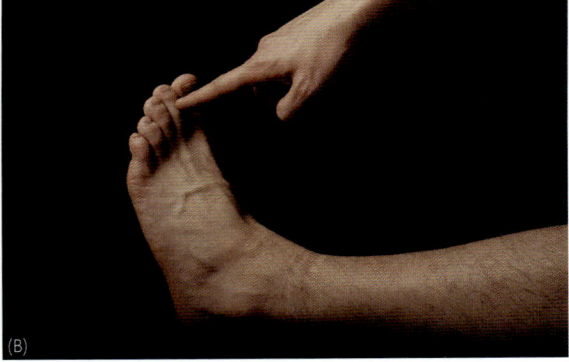

Fig. 22.12 (A) Eversion of foot. (B) Inversion of foot.

DreamBig/Shutterstock.com.

and the superior tibiofibular joint. The common peroneal nerve then divides into the superficial and deep peroneal nerves. The deep peroneal nerve pierces the extensor digitorum to enter the anterior compartment of the leg.

Superficial peroneal nerve (L. 4, 5; S. 1)

The superficial peroneal nerve descends in the lateral compartment. It pierces the deep fascia in the distal third of the leg and divides into **medial** and **intermediate dorsal cutaneous nerves**. It supplies the peroneus longus and brevis, the skin of the lower part of the front of the leg, the large part of the dorsum of the foot, and most of the dorsal surfaces of the toe.

Medial side of the leg

This consists of the medial surface of the tibia. It is subcutaneous, except for a small part at the upper end which is covered by the tibial collateral ligament of the knee joint and the tendons of the sartorius, gracilis, and semitendinosus [see Fig. 16.12].

Long saphenous vein

This important vein begins at the medial border of the foot by the union of the **dorsal venous arch** and the **medial dorsal digital vein** of the big toe. It ascends in front of the medial malleolus, and

---

DISSECTION 22.5 Peroneal compartment of the leg

Objectives

I. To expose and study the peroneus longus and brevis.
II. To identify and trace the superficial peroneal nerve.

Instructions

1. Divide the deep fascia over the peroneal muscles by a longitudinal incision. Turn the flaps aside, and demonstrate their continuity with the anterior and posterior intermuscular septa. Retain the peroneal retinacula.
2. Separate the muscles from each other. Determine their attachments, and find the nerves supplying them.
3. Trace the common peroneal nerve to the peroneus longus, and divide the muscle to follow the

nerve between the muscle and the fibula. Trace the superficial peroneal nerve downwards. Follow the deep peroneal nerve into continuity with the part already found in the anterior compartment of the leg.
4. Find and open the common synovial sheath of the peroneal muscles. It begins 3–5 cm above the superior peroneal retinaculum. Pass a blunt seeker into it, and try to define its extent. Distally, it divides to surround each tendon separately and continues into the sole of the foot around the tendon of the peroneal longus.
5. Do not follow the latter tendon into the sole, but follow the tendon of the peroneus brevis to its insertion, and define the inferior peroneal retinaculum.

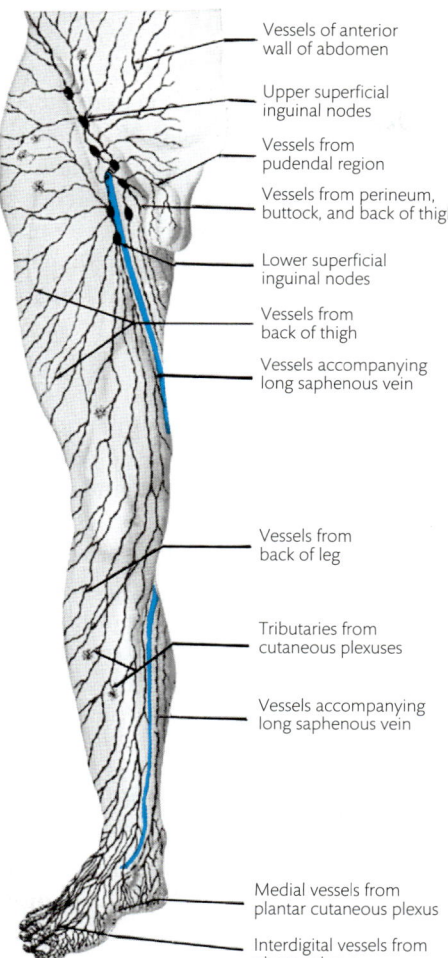

Vessels of anterior
wall of abdomen

Upper superficial
inguinal nodes

Vessels from
pudendal region

Vessels from perineum,
buttock, and back of thigh

Lower superficial
inguinal nodes

Vessels from
back of thigh

Vessels accompanying
long saphenous vein

Vessels from
back of leg

Tributaries from
cutaneous plexuses

Vessels accompanying
long saphenous vein

Medial vessels from
plantar cutaneous plexus

Interdigital vessels from
plantar plexus

Fig. 22.13 Superficial lymph vessels of the anterior surface of the lower limb.

passes obliquely backwards across the distal third of the medial surface of the tibia and along the medial border of the tibia to the posteromedial side of the knee [Fig. 22.13]. The further course and termination of this vein are described in Chapter 17. In the leg, the long saphenous vein lies between two layers of the membranous fascia and is crossed by a more superficial plexus of veins which tend to join it near the knee. The long saphenous vein receives many tributaries throughout its length and communicates through them with the short saphenous vein. The long saphenous vein communicates through valved perforating veins with the deep veins of the leg.

Saphenous nerve (L. 3, 4)

The saphenous nerve arises from the femoral nerve in the femoral triangle and accompanies the

femoral artery into the adductor canal. It leaves the canal through the roof and lies deep to the sartorius. Here it gives off the **infrapatellar branch**, which pierces the sartorius to reach the peripatellar plexus. The nerve then emerges between the sartorius and the tendon of the gracilis a little above the knee. The saphenous nerve lies posteromedial to the knee and pierces the deep fascia inferior to it. In the leg and foot, it accompanies the long saphenous vein and ends in the skin on the medial side of the foot. It supplies the skin on the medial side of the knee and leg, and the proximal part of the dorsum of the foot [see Fig. 17.6].

Dissection 22.6 looks at the medial side of the leg.

Back of the leg

The muscles of the back of the leg are arranged in three layers separated by two sheets of fascia [Fig. 22.1]. The **superficial layer of muscles** is inserted into the heel by the large **tendo calcaneus**. This layer consists of the gastrocnemius, soleus, and plantaris—muscles that are powerful plantar flexors of the ankle joint (H in Fig. 22.1). These muscles move independently of the middle layer of muscles.

> **DISSECTION 22.6 Medial side of the leg**
>
> **Objectives**
>
> **I.** To expose and clean the tendons of the sartorius, gracilis, and semitendinosus.   **II.** To study the medial (tibial) collateral ligament.
>
> **Instructions**
>
> 1. Trace the tendons of the sartorius, gracilis, and semitendinosus to their attachments.
> 2. Turn them forwards, and note the complex bursa which lies between them and the tibial collateral ligament of the knee joint.
> 3. Turn the tendons forwards, and clean the surface of the ligament. Note that the tendon of the semimembranosus and the inferior medial genicular vessels and nerves pass deep to the superficial part of the ligament.
> 4. The deep part of the ligament is attached to the margin of the tibial condyle, superior to the insertion of the semimembranosus.

The **middle layer of muscles** in the back of the leg are the long flexors of the toes—flexor hallucis longus and flexor digitorum longus (G in Fig. 22.1). They are separated from the superficial muscles by a well-defined fascial layer which stretches from the medial border of the tibia to the posterior border of the fibula. This fascia encloses the muscles, together with their vessels and nerve, and is attached above to the soleal line on the tibia. At the ankle, this fascia is thickened to form part of the flexor retinaculum [Fig. 22.5].

The **deepest layer of muscles** consists of the tibialis posterior which lies on the interosseous membrane between the tibia and fibula (E in Fig. 22.1). The fascia covering the posterior surface of the tibialis posterior is attached above to the soleal line, laterally to the medial crest of the fibula and medially to the vertical ridge on the tibia. A muscle not shown in this diagram, but included in the back of the leg, is the popliteus. It is limited to the proximal part of the leg and is inserted above the soleus.

Dissection 22.7 looks at the cutaneous vessels and nerves of the back of the leg.

DISSECTION 22.7 Cutaneous vessels and nerves—back of the leg

Objectives

I. To reflect the skin.   II. To identify and trace the cutaneous nerves and veins on the back of the leg.

Instructions

1. Make a transverse incision through the skin on the distal part of the heel. Carry the incision along the borders of the foot to join the previous incisions.
2. Find the sural nerve and short saphenous vein below the lateral malleolus. Then strip the skin and superficial fascia upwards from the back of the leg, retaining the nerve and vein. The small medial calcanean nerves may be found, as the skin is removed from the medial side of the heel [Fig. 22.3].
3. Find the meeting point of the sural and peroneal communicating nerves. Follow both upwards through the deep fascia to the tibial and common peroneal nerves.
4. Trace the short saphenous vein into the popliteal vein.

Short saphenous vein

The short saphenous vein is formed by the union of the lateral end of the dorsal venous arch of the foot and the dorsal digital vein of the lateral side of the little toe. It passes backwards, inferior to the lateral malleolus, and ascends in the back of the leg. It pierces the deep fascia at the lower border of the popliteal fossa and enters the popliteal vein. It drains the lateral side of the foot, ankle, and back of the leg.

Sural nerve (S. 1, 2)

The sural nerve arises from the tibial nerve in the popliteal fossa. It descends on the posterior surface of the gastrocnemius and enters the superficial fascia at the middle of the back of the leg. It is joined by the **peroneal communicating nerve** [see Fig. 18.2] and accompanies the short saphenous vein to the lateral side of the foot. The sural nerve supplies the skin on the lower lateral side of the back of the leg, the lateral border and lateral part of the dorsum of the foot, and the lateral side of the little toe.

Peroneal communicating nerve (L. 5; S. 1, 2)

The peroneal communicating nerve arises from the common peroneal nerve in the popliteal fossa. It pierces the deep fascia over the lateral head of the gastrocnemius and descends to join the sural nerve [see Fig. 18.2]. It supplies the skin on the proximal two-thirds of the posterolateral surface of the leg and the territory of the sural nerve.

The posterior branch of the **medial anterior cutaneous nerve of the thigh** (L. 2, 3) descends into the calf and supplies the skin on the upper posteromedial part of the leg.

Lymph vessels and lymph nodes of the lower limb

Very little of this system can be demonstrated by dissection as the lymph nodes are poorly seen, unless enlarged by disease. However, knowledge of the course of the lymph vessels, and the location of the nodes to which they drain, is essential in determining the possible site of disease when an enlarged lymph node is found.

Study Figs 22.13 and 22.14, and note the following points.

1. **Superficial lymph vessels and nodes** drain the skin and subcutaneous tissues. They lie superficial to the deep fascia, which separates them from the **deep lymph vessels** and **nodes**. The superficial and deep systems communicate with each other only at certain points. In the lower limb, such communications occur through the cribriform fascia in the thigh and the popliteal fascia. (In the upper limb, it occurs in the axilla and where the basilic vein pierces the deep fascia of the arm.)

2. The superficial vessels are much more numerous than the deep vessels. There are few lymph vessels in muscle, but very many in skin, synovial membranes, synovial sheaths, and bursae.

3. Superficial lymph vessels take a direct course to the superficial lymph nodes. Deep lymph vessels run with the deep blood vessels and enter the deep nodes.

4. **Superficial nodes** in the lower limb are virtually restricted to the inguinal region. These drain almost all the superficial tissues of the lower limb, except the superficial tissue of the lateral side of the foot and the back of the leg, which drains to the popliteal nodes.

5. The **superficial inguinal lymph nodes** are arranged in the shape of a T. The nodes forming the stem of the T drain the lower limb. The lateral nodes of the horizontal part of the T drain the upper lateral gluteal region and the posterior and lateral parts of the trunk. The medial nodes in the horizontal part of the T drain the upper medial part of the thigh, perineum, medial gluteal region, and anteromedial part of the abdominal wall below the umbilicus.

6. There is a 'lymphshed' along the back of the lower limb. (The term lymphshed refers to the imaginary line which separates areas draining medially and laterally.) Vessels from the medial half of the limb pass round the medial surface of the limb and those from the lateral half pass round the lateral surface of the limb, to converge on the inguinal lymph nodes [Fig. 22.14]. This 'lymphshed' is split distally by the vessels flowing to the popliteal nodes.

7. **Deep lymph vessels** of the leg and foot enter the **popliteal nodes**. The lymph from the popliteal nodes drains into the deep inguinal lymph nodes through lymphatics that run along the femoral vessels. The deep inguinal lymph nodes also receive the efferent vessels from the superficial inguinal lymph nodes. They drain lymph to the **external iliac nodes** by passing behind the inguinal ligament.

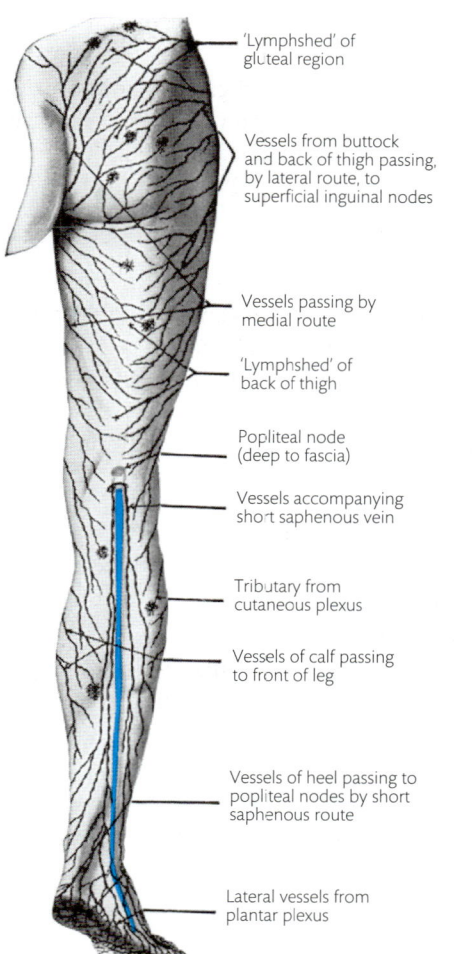

'Lymphshed' of gluteal region

Vessels from buttock and back of thigh passing, by lateral route, to superficial inguinal nodes

Vessels passing by medial route

'Lymphshed' of back of thigh

Popliteal node (deep to fascia)

Vessels accompanying short saphenous vein

Tributary from cutaneous plexus

Vessels of calf passing to front of leg

Vessels of heel passing to popliteal nodes by short saphenous route

Lateral vessels from plantar plexus

Fig. 22.14 Superficial lymph vessels of the posterior surface of the lower limb.

8. Deep lymph vessels of the perineum and gluteal region drain with the corresponding blood vessels (gluteal and internal pudendal) through the greater sciatic foramen into the pelvis (**internal iliac nodes**).

Deep fascia

The deep fascia of the leg is dense. At the ankle, it is thickened to form the peroneal and flexor retinacula [Figs 22.5, 22.11] on the lateral and medial sides, respectively. The **flexor retinaculum** stretches from the calcaneus to the medial malleolus. It covers the tendons of the deep flexor muscles of the back of the leg, the tibial nerve, and the posterior tibial vessels as they pass into the foot, posterior and inferior to the medial malleolus. Distally, the flexor retinaculum gives partial attachment to the abductor hallucis muscle of the foot.

DISSECTION 22.8 Back of the leg-1

Objectives

I. To dissect the flexor retinaculum. II. To study the gastrocnemius, soleus, plantaris, and tendo calcaneus. III. To identify and trace the tibial nerve and posterior tibial vessels.

Instructions

1. Define the flexor retinaculum posteroinferior to the medial malleolus. The medial calcanean nerves and vessels may be found passing through it to the skin.
2. Extend the division of the deep fascia (which was made when the sural nerve was followed) down to the calcaneus. Reflect the fascia.
3. Identify and follow the bellies of the gastrocnemius to their attachments. Lift the muscle from the underlying soleus and cut across the medial head, close to its attachment to the femur. Turn the medial head laterally to expose the lower part of the popliteal vessels and tibial nerve in the popliteal fossa. Find the large muscular branches to the gastrocnemius.
4. Lift the tendon of the semimembranosus from the proximal part of the medial head of the gastrocnemius, and find the bursa which separates them. Lift this part of the medial head of the gastrocnemius,

and find the bursa which separates it from the fibrous capsule of the knee joint. This bursa may be continuous with the bursa under the semimembranosus and with the joint cavity through the articular capsule.
5. Trace the nerve to the soleus from the tibial nerve.
6. Find the plantaris, a small muscle posteromedial to the lateral head of the gastrocnemius. Follow its tendon between the gastrocnemius and the soleus to the medial side of the tendo calcaneus.
7. Remove the fatty connective tissue in front of the tendo calcaneus. This uncovers the lower part of the intermuscular fascial septum which is close to the deep fascia here.
8. Cut across the lateral head of the gastrocnemius at the level of the knee joint. Turn its proximal part upwards. A small sesamoid bone may be felt within the head.
9. Turn both bellies of the gastrocnemius downwards, and note their insertion into the tendo calcaneus.
10. Expose the posterior surface of the soleus, and define its attachments.
11. Note the tibial nerve and popliteal vessels passing deep to a tendinous arch, from which the intermediate fibres of the soleus arise.

Dissection 22.8 describes the dissection of the back of the leg.

Superficial muscles of the calf

Gastrocnemius

Origin: the gastrocnemius takes origin by two heads from the femur—the **lateral head** from the lateral surface of the lateral condyle, and the **medial head** from the popliteal surface above the medial condyle [see Fig. 16.9]. The medial head is separated from the tendon of the semimembranosus and the articular capsule of the knee joint by bursae. The lateral head frequently contains a small sesamoid bone (**fabella**). The two fleshy bellies remain separate. **Insertion**: the bellies end near the middle of the leg on the posterior surface of a thin common tendon which fuses with the superficial surface of the tendon of the soleus to form the **tendo calcaneus**. The tendo calcaneus is inserted into the posterior surface of the calcaneus. **Nerve supply**: tibial nerve. **Actions**: it is a powerful plantar flexor of the ankle joint, together with the

soleus and plantaris. The gastrocnemius also flexes the knee joint, but it becomes ineffective as a plantar flexor of the ankle when the knee is bent.

Soleus

Origin: this powerful, flat muscle arises from the: (1) posterior surface of the head and upper third of the shaft of the fibula; (2) soleal line and middle third of the medial border of the tibia; and (3) tendinous arch posterior to the popliteal vessels and tibial nerve. **Insertion**: the thick tendon fuses with that of the gastrocnemius to form the tendo calcaneus [see Figs 22.15, 16.13] and is inserted into the posterior surface of the calcaneus. **Nerve supply**: tibial nerve. **Actions**: the gastrocnemius and soleus are powerful plantar flexors of the ankle. They act around the fulcrum of the heads of the metatarsals, mainly the first, to raise the weight of the body onto the toes—a position which the soleus maintains in running. They are responsible for the powerful push-off in running, jumping, and walking. They also act with the dorsiflexors of the ankle joint to stabilize the ankle. ➲ The gastrocne-

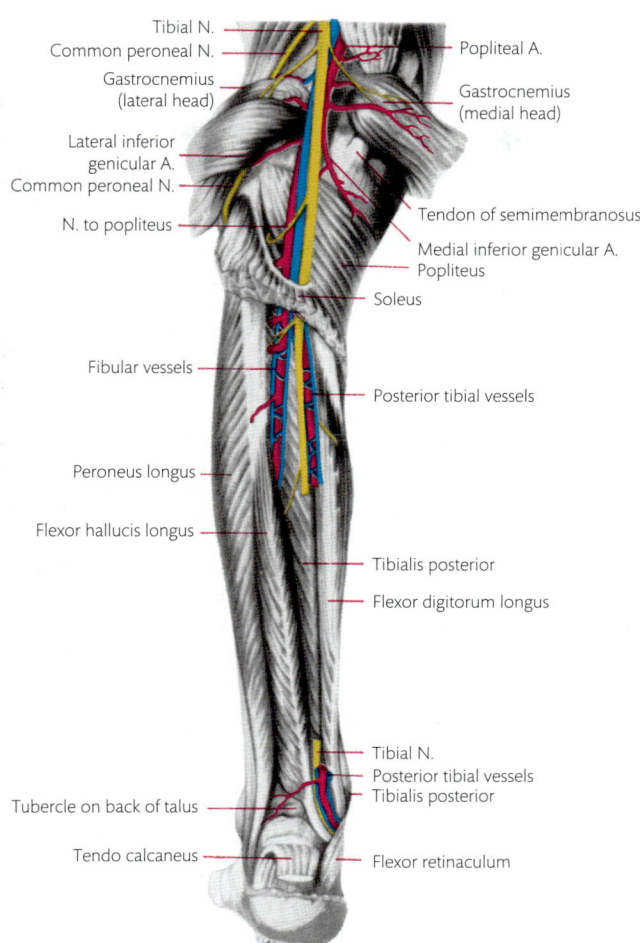

Tibial N.
Common peroneal N.
Gastrocnemius (lateral head)
Popliteal A.
Gastrocnemius (medial head)
Lateral inferior genicular A.
Common peroneal N.
N. to popliteus
Tendon of semimembranosus
Medial inferior genicular A.
Popliteus
Soleus
Fibular vessels
Posterior tibial vessels
Peroneus longus
Flexor hallucis longus
Tibialis posterior
Flexor digitorum longus
Tibial N.
Posterior tibial vessels
Tibialis posterior
Tubercle on back of talus
Tendo calcaneus
Flexor retinaculum

Fig. 22.15 Deep dissection of the back of the leg.

mius and soleus play an important role as 'muscle pumps' that return venous blood from the lower limbs. Both muscles have a considerable blood supply, and the soleus contains a large venous plexus. This plexus is emptied when the muscle contracts, an action which helps in the venous return from the lower limbs. For this reason, the soleus has been termed the 'peripheral heart'.

Plantaris

Origin: this small muscle (8–10 cm long) arises from the popliteal surface of the femur [see Fig. 16.9] and is partly hidden by the lateral head of the gastrocnemius. **Insertion**: the long, slender tendon passes between the gastrocnemius and soleus and along the medial side of the tendo calcaneus to the calcaneus. The plantaris is occasionally absent (compare with the palmaris longus). **Nerve supply**: like the gastrocnemius and soleus, the plantaris is supplied by the

tibial nerve. **Actions**: flexion of the knee joint and plantar flexion of the ankle. The plantaris adds very little to the strength of the gastrocnemius and soleus.

➲ If the femur is fractured a short distance proximal to the attachments of the gastrocnemius (supracondylar fracture), the gastrocnemius pulls the distal fragment backwards. For this reason, these fractures are treated with the knee flexed, to relax the gastrocnemius and prevent this displacement.

Tendo calcaneus

This powerful tendon is inserted into the smooth intermediate part of the posterior surface of the calcaneus. The upper part of this surface is separated from the tendon by a small bursa. ➲ If the tendon is ruptured, the disability in walking is severe and running is impossible.

Dissection 22.9 continues the dissection of the back of the leg.

DISSECTION 22.9 Back of the leg-2

Objectives

I. To dissect and study the flexor hallucis longus, flexor digitorum longus, and tibialis posterior.   **II.** To identify and trace the tibial nerve and posterior tibial vessels and fibular artery.

Instructions

1. Separate the soleus from the tibia, and turn it laterally. Divide the blood vessels to the muscle, noting the large veins which emerge from it. Retain the nerves.
2. Look for any communications between the long saphenous vein and the deep veins. Open these, and check for valves within them.
3. The first intermuscular septum (deep to the soleus) is now exposed. Divide the septum longitudinally in the middle, and reflect it to expose the second layer of muscles and the neurovascular bundles.
4. Trace the tibial nerve as far as the ankle. Find the muscular branches arising mainly in the upper part of the leg. The small nerve to the popliteus descends over the popliteus to enter the deep surface of the muscle [Fig. 22.15].
5. Define the lower border of the popliteus, and follow it to its tendon. Do not attempt to follow it into the knee joint at this stage.
6. Remove the fascia from the lowest part of the popliteal vessels. Find the anterior and posterior tibial branches of the artery and the corresponding veins. Check the continuity of the anterior tibial vessels with the vessels in the anterior compartment. Follow the posterior tibial artery as far as the ankle. Find and trace the fibular artery which arises from it. This artery descends posterior to the fibula, under cover of the flexor hallucis longus.
7. Define and separate the long flexor muscles of the toes. The flexor hallucis longus is lateral to, and larger than, the flexor digitorum longus. Follow their tendons deep to the flexor retinaculum. Push the flexor hallucis longus laterally, and separate its deep surface from the second intermuscular septum and the interosseous membrane.
8. Divide the intermuscular septum covering the tibialis posterior. Trace that muscle and its tendon as far as the flexor retinaculum. Note its close association with the medial malleolus. Palpate the tendon in your own foot between the medial malleolus and the navicular, first tightening it by plantar flexing and inverting the foot.

Tibial nerve (L. 4, 5; S. 1, 2, 3)

In the upper part of the leg, the tibial nerve lies deep to the first intermuscular septum with the posterior tibial vessels. In the lower third of the leg, it lies between the tendo calcaneus and the medial border of the tibia. Deep to the flexor retinaculum, it lies between the tendons of the flexor digitorum longus and flexor hallucis longus [Fig. 22.15]. The tibial nerve divides into medial and lateral plantar nerves, deep to the flexor retinaculum.

Branches in the leg

Muscular branches supply the tibialis posterior, flexor hallucis longus, flexor digitorum longus, and soleus. They arise in the upper part of the leg. The medial calcanean nerve (S. 1) arises at the ankle, pierces the flexor retinaculum, and supplies the skin on the posterior and lower surfaces of the heel. Small articular branches pass to the posterior aspect of the ankle joint.

Popliteal vessels

The popliteal artery starts at the adductor hiatus and ends at the distal border of the popliteus by dividing into the anterior and posterior tibial arteries. The anterior and posterior tibial veins unite to form the popliteal vein at the same point.

Anterior tibial artery

This vessel gives off a small posterior tibial recurrent artery to the back of the knee joint. It then passes forwards above the interosseous membrane to the anterior compartment of the leg.

Posterior tibial artery

The posterior tibial artery supplies the muscles of the back of the leg and is the main artery of the foot. It begins at the lower border of the popliteus and descends with the tibial nerve. It ends by dividing into the medial and lateral plantar arteries, deep to the flexor retinaculum.

Branches of the posterior tibial artery in the leg

1. The **fibular artery**: this large branch arises close to the origin of the parent artery and descends along the back of the fibula. It supplies: (1) the muscular branches; (2) the **nutrient artery to the fibula**; and (3) the **perforating branch** which pierces the interosseous membrane just above the inferior tibiofibular joint and anastomoses with the lateral tarsal branch of the dorsalis pedis artery. The fibular artery gives the lateral malleolar and calcanean branches and ends by anastomosing with the posterior tibial artery on the back of the ankle joint.

2. The circumflex fibular artery runs round the neck of the fibula to supply the muscles and skin.

3. A large **nutrient artery to the tibia** arises from the upper end of the posterior tibial artery. It enters the tibia a short distance below the soleal line. Like other nutrient arteries, it is the main supply to the bone.

4. **Muscular branches** to the muscles of the back of the leg.

5. **Cutaneous branches** to the medial side of the leg.

6. A **communicating branch** to the fibular artery behind the ankle joint.

7. **Medial calcanean branches** that run with the corresponding nerves.

Posterior tibial vein

The lateral and medial plantar veins unite to form the posterior tibial vein deep to the flexor retinaculum. The vein ends by emptying into the popliteal vein at the lower border of the popliteus. ⮕ The posterior tibial vein is a common site for deep vein thrombosis, especially at times of prolonged inactivity.

Deep muscles of the back of the leg

Popliteus

Origin: the popliteus is attached to the lateral condyle of the femur at the anterior end of the popliteal groove [see Fig. 16.8], and to the back of the lateral meniscus inside the capsule of the knee joint. It emerges through the posterior part of the capsule of the knee joint, below the **arcuate popliteal ligament**. **Insertion**: the popliteus tendon then expands into a triangular fleshy belly which is attached to the posterior surface of the tibia above the soleal line [see Fig. 16.13]. **Nerve supply**: tibial nerve. **Actions**: when the leg is free, it medially rotates the tibia on the femur at the beginning of

knee flexion. When the foot is on the ground, it laterally rotates the femur on the tibia. Both movements 'unlock' the extended knee joint and allow flexion to occur. The attachment to the meniscus ensures that the meniscus moves with the femoral condyle and is not caught between the femur and the tibia during movement.

The remaining deep muscles—flexor hallucis longus, flexor digitorum longus, and tibialis posterior—have considerable attachments to the intermuscular septa and interosseous membrane, in addition to the tibia and fibula. Note that, for muscles of the lower limb, the more distal attachments are often fixed because the foot is on the ground. The actions given for the following muscles are stated as though the distal attachment was free to move, but the reverse action occurs with equal frequency. This is especially true of muscles that stabilize the trunk on the hip joint and the femur on the tibia. The terms 'origin' and 'insertion' are used to differentiate the more proximal from the more distal attachment, rather than the more fixed from the more movable one.

Flexor hallucis longus

The flexor hallucis longus is much larger than the flexor digitorum longus—a feature determined by the relatively larger forces applied to the hallux, when compared to the other toes. **Origin**: it arises from the posterior surface of the fibula, below the origin of the soleus [see Fig. 16.13]. Its tendon descends obliquely over the back of the ankle joint and enters the sole of the foot. In its course, the tendon lies in almost continuous bony grooves on the posterior surfaces of the tibia [see Fig. 16.11], the talus, and the inferior surface of the **sustentaculum tali**. **Insertion**: it is inserted into the distal phalanx of the great toe. **Nerve supply**: tibial nerve. **Actions**: it flexes the metatarsophalangeal and interphalangeal joints of the great toe, and assists with plantar flexion of the ankle. These are important movements in the last phase of the 'push-off' in walking and running.

Flexor digitorum longus

Origin: the flexor digitorum longus arises from the medial part of the posterior surface of the tibia, distal to the soleal line [see Fig. 16.13]. It descends behind the tendon of the tibialis posterior, and its tendon grooves the back of the tibia, just medial to the medial malleolus. It passes deep to the

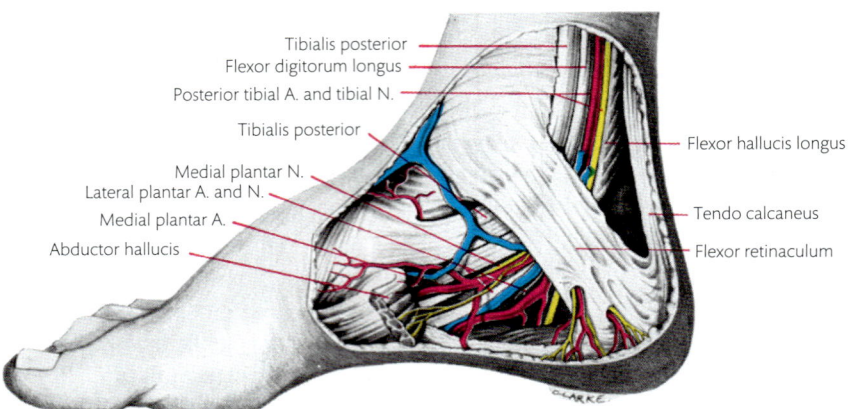

Fig. 22.16 Dissection of the medial side of the ankle, showing the structures deep to the flexor retinaculum.

flexor retinaculum, enters the sole of the foot, and divides into four tendons, one to the terminal phalanx of each of the lateral four toes [Figs 22.15, 22.16]. **Nerve supply**: tibial nerve. **Actions**: it flexes the metatarsophalangeal and interphalangeal joints of the lateral four toes, and assists with plantar flexion of the ankle joint. It may play a part in inversion of the foot.

Tibialis posterior

Origin: the tibialis posterior arises from the posterior surface of the interosseous membrane and the adjoining parts of the tibia and fibula [see Figs 22.15, 16.13]. Distally, its tendon grooves the posterior surface of the medial malleolus and passes deep to the flexor retinaculum. **Insertion**: in the foot, it crosses the inferior surface of the head of the talus and is inserted mainly into the tuberosity of the navicular bone. It also sends strong slips to all the other tarsal bones (except the talus) and to the middle three metatarsals. **Nerve supply**: tibial nerve. **Actions**: it plantar flexes the foot and also inverts it because of its extensions to the lateral tarsal bones.

Flexor retinaculum

This thick band of fascia passes from the medial malleolus to the medial process of the tubercle of the calcaneus. It is continuous proximally with the deep fascia of the leg and with the septum which covers the deep muscles [Fig. 22.16]. Distally, it is continuous with the deep fascia of the sole and gives attachment to the abductor hallucis muscle.

Beneath the retinaculum lie, in order, the tendons of the tibialis posterior, flexor digitorum longus, posterior tibial vessels, tibial nerve, and flexor

hallucis longus. They divide into the medial and lateral plantar vessels and nerves, deep to the retinaculum [Fig. 22.16]. The position of the muscles of the leg within the fascial compartments is shown in Fig. 22.17.

Synovial sheaths

The tendons deep to the flexor retinaculum are each surrounded by a synovial sheath. The synovial sheaths begin approximately 2 cm above the tip of the medial malleolus.

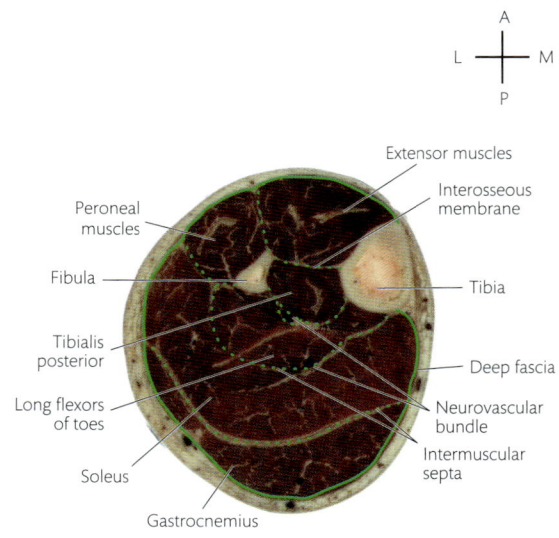

Fig. 22.17 Transverse section through the middle of the leg. A = anterior; P = posterior; M = medial; L = lateral.

Image courtesy of the Visible Human Project of the US National Library of Medicine.

Nerves and vessels of joints

Note these general features in all synovial joints: (1) the nerves which innervate the muscles that move the joint also supply the joint; and (2) all arteries in the region of the joint send branches to it. The arteries form a circular anastomosis around each bone taking part in the joint [Fig. 22.8] and supply the articular capsule and intracapsular bone.

Dissection 22.10 instructs on dissection of the nerves and vessels of the knee joint.

Nerves of the knee joint

The **femoral nerve** supplies the anterosuperior part of the joint through the nerves to the three vasti muscles. The nerve from the vastus medialis accompanies the descending genicular artery. The **common peroneal nerve** supplies the lateral part of the joint through the superior and inferior lateral genicular nerves and the recurrent genicular nerve. The **tibial nerve** supplies the articular capsule and the central structures within the capsule through the superior medial, middle, and inferior medial genicular nerves. And the **obturator nerve** supplies the posterior surface of the knee joint.

Anastomosis around the knee joint

Numerous arteries [see Figs 19.4, 22.8] contribute to the anastomosis around the knee joint: two lateral and two medial genicular arteries from the popliteal artery; the descending genicular artery from the femoral artery; the genicular artery from the lateral circumflex femoral artery; and the anterior and posterior tibial recurrent arteries. They anastomose with each other mainly in front of the joint. (The middle genicular artery plays little part in the anastomosis, as it mainly supplies structures within the joint cavity.) The anastomosis around the knee is insufficient to maintain blood flow to the distal part of the limb when the popliteal artery is blocked.

Anastomosis around the ankle joint

On the lateral side, the lateral malleolar branch of the anterior tibial artery and the lateral tarsal branch of the dorsalis pedis artery anastomose with the perforating and terminal branches of the fibular artery. On the medial side, the medial malleolar branch of the anterior tibial artery anastomoses with the medial calcanean branches of the posterior tibial artery. The posterior tibial artery also anastomoses with the fibular artery, posterior to the ankle joint.

Sole of the foot

Begin by revising the surface anatomy of the region on your own foot. Make certain that you can identify the palpable parts of the bones and the major tendons entering the foot from the leg. Note also the parts of the foot which are in contact with the ground in standing, in the various phases of walking, and in running.

The foot has many features in common with the hand. The differences are related to the function of

---

DISSECTION 22.10 Nerves and vessels of the knee joint

Objective

I. To clean and trace the genicular arteries and nerves.

Instructions

1. As most of the structures around the knee joint have been dissected, it is possible to follow the various nerves and arteries to the knee.
2. Find again the genicular branches of the popliteal artery and the tibial and common peroneal nerves. Follow the inferior lateral genicular artery to the fibular collateral ligament of the knee joint.
3. Cut through the tendon of the biceps above the knee, and turn it downwards to expose the fibular collateral ligament. Follow the artery and nerve between the ligament and the articular capsule.
4. Trace the inferior medial genicular artery and nerve along the upper border of the popliteus, till they disappear deep to the superficial part of the tibial collateral ligament of the knee joint.
5. Turn the tendons of the sartorius, gracilis, and semitendinosus forwards, and find the artery and nerve emerging from beneath the ligament.
6. Cut the popliteus near its tendon, and turn the muscle medially. This exposes the nerve to the popliteus.

the foot. The foot is a supporting structure carrying considerable loads in standing, and even greater loads in kicking, in pushing off in running, and on landing on the feet when jumping from a height. The foot is designed to be strong and resilient, unlike the hand which is designed for holding and grasping. Strength in the foot is obtained by having large tarsal and big toe bones held together by powerful ligaments. The binding of the metatarsal bone of the big toe to those of the other toes adds to the stability of the foot. Unlike in the hand, there is no opposition of the big or little toe and no muscles equivalent to the opponens muscles of the hand. Resilience is obtained by the presence of multiple joints, each of which has limited movement, and by arrangement of the bones in an arch where tension between the components is altered in different positions of the foot. When standing, the weight of the body is supported on the heel and on the heads of the metatarsals (mainly the first metatarsal), and to a lesser extent on the lateral border of the sole of the foot. When moving forwards, the force is carried principally on the head of the first metatarsal and the big toe. (The remaining metatarsals and toes are relatively weak and can be looked upon as a stabilizing flap.) The arched shape of the foot has another advantage. It gives protection to

the structures in the sole which would otherwise be subjected to the weight of the body. One special development to resist the pressure on the head of the first metatarsal is the presence of two **sesamoid bones** on its plantar surface. These transmit the pull of the small muscles of the big toe, without subjecting them to pressure, and also make a tunnel through which the tendon of the flexor hallucis longus can reach the toe. Plantar flex your ankle and appreciate the arch of the foot.

See Dissection 22.11 (and Fig. 22.18) for instructions on the dissection of the sole of the foot.

Arches of the foot

The tarsal and metatarsal bones of the foot together are shaped like half a dome. When the feet are placed together, the two half-domes form one single dome. The rim of each half-dome consists of the heel, the lateral border of the foot, and the heads of the metatarsal bones. It is these parts of the foot which imprint the ground and form the **footprint** of a bare foot. Each foot has **one longitudinal arch** and **two transverse arches**.

Longitudinal arch

This bony arch is higher on the medial side than on the lateral side of the foot. The arch has a posterior

---

DISSECTION 22.11 Sole of the foot-1

Objectives

I. To reflect the skin.   II. To clean the superficial fascia of the sole of the foot.

Instructions

1. Cut longitudinally through the skin and superficial fascia of the sole, from the heel to the root of the middle toe. Avoid cutting into the deep fascia (plantar aponeurosis).
2. Strip the skin and superficial fascia from the deep fascia with a knife. The superficial fascia, like that of the palm, is dense and firmly bound to the deep fascia. It has fat packed tightly in its interstices, so that it forms a firm pad, especially over weight-bearing areas. (Stripping the skin and superficial fascia in one piece removes the cutaneous vessels and nerves [Fig. 22.18], but they are difficult to follow through the dense fascia.)
3. On the lateral and medial sides, remove the fascia with care, so as to retain the digital nerves to

the medial side of the big toe and the lateral side of the little toe. These become superficial further proximally, when compared with the other plantar digital nerves [Fig. 22.18].

4. Make a longitudinal incision through the skin on the plantar surface of each toe. Reflect the skin, and find the plantar digital vessels and nerves. Follow them to the ends of the toes.
5. Expose the deep fascia of the toes. This is thickened to form the fibrous flexor sheath—a dense tunnel enclosing the flexor tendons in the toes.
6. Define the plantar aponeurosis, and note the furrows at its edges. The branches of the medial and lateral plantar vessels and nerves pass through these furrows to the skin of the sole. As you approach the toes, take care not to damage the plantar metatarsal arteries and common plantar digital nerves which become superficial between the slips of the plantar aponeurosis to the toes.

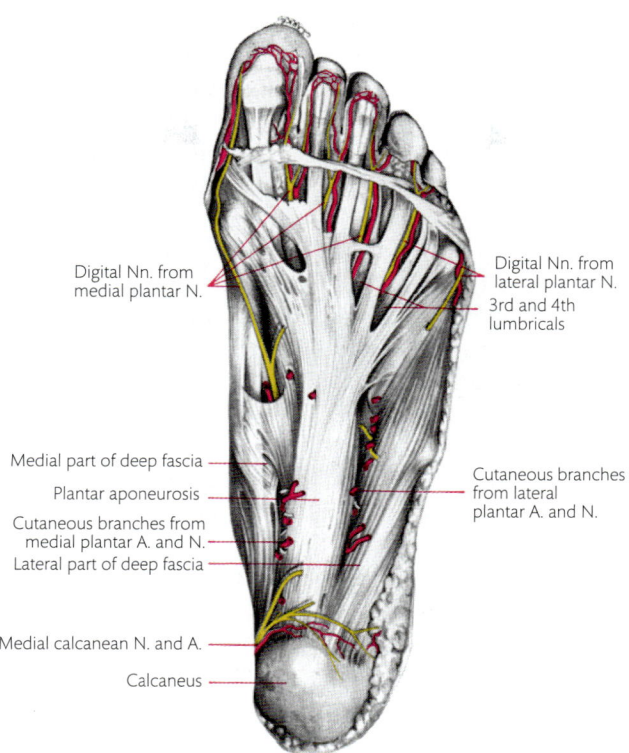

Digital Nn. from medial plantar N.

Digital Nn. from lateral plantar N.

3rd and 4th lumbricals

Medial part of deep fascia

Plantar aponeurosis

Cutaneous branches from medial plantar A. and N.

Lateral part of deep fascia

Cutaneous branches from lateral plantar A. and N.

Medial calcanean N. and A.

Calcaneus

Fig. 22.18 Superficial dissection of the sole of the foot to show plantar aponeurosis. The skin and superficial fascia have been removed, and the fibrous flexor sheaths partially opened.

pillar, a submit, and an anterior pillar. The **submit** of the arch is formed by the talus [Fig. 22.19]. The short, thick **posterior pillar** is formed by the calcaneus. The **anterior pillar** is formed by the remaining tarsals and metatarsals. For descriptive purposes, the longitudinal arch can be divided into a *medial column* where the anterior arch is formed by the navicular, cuneiforms, and medial three metatarsals; and a *lateral column* where the anterior arch is formed by the cuboid and lateral two metatarsals.

The talus transmits the weight of the body downwards: (1) posteriorly to the heel through its articulation with the calcaneus; (2) anteriorly and medially to the medial three metatarsals through its articulation with the navicular bone; and (3) anteriorly and laterally to the lateral two metatarsals through its articulations with the calcaneus and, indirectly, the cuboid.

Transverse arches

There are two transverse arches. The more prominent arch is in the region of the tarsometatarsal joints and is formed by the cuneiforms and the base

of the metatarsals. These bones are wedge-shaped, with the plantar surface being narrower than the dorsal surface. This arrangement results in the plantar surface having a much smaller radius of curvature than the dorsal surface, and so forming a well-defined transverse arch. A second and lower arch exists between the heads of the metatarsals.

Superficial fascia

This fascia is dense, especially over the heel and the ball of the foot. It contains loculi of fat in dense fascial pockets. The fat makes the sole firm and resilient.

The skin and superficial fascia of the sole are supplied by three sets of nerves and vessels: the **medial calcanean** nerves and vessels in the region of the heel; the branches of the **medial** and **lateral plantar** nerves and vessels in the greater part of the sole; and the **plantar digital** nerves and vessels in the toes [Fig. 22.18].

Deep fascia

In the sole, the deep fascia is extremely thick in the intermediate region and forms the **plantar**

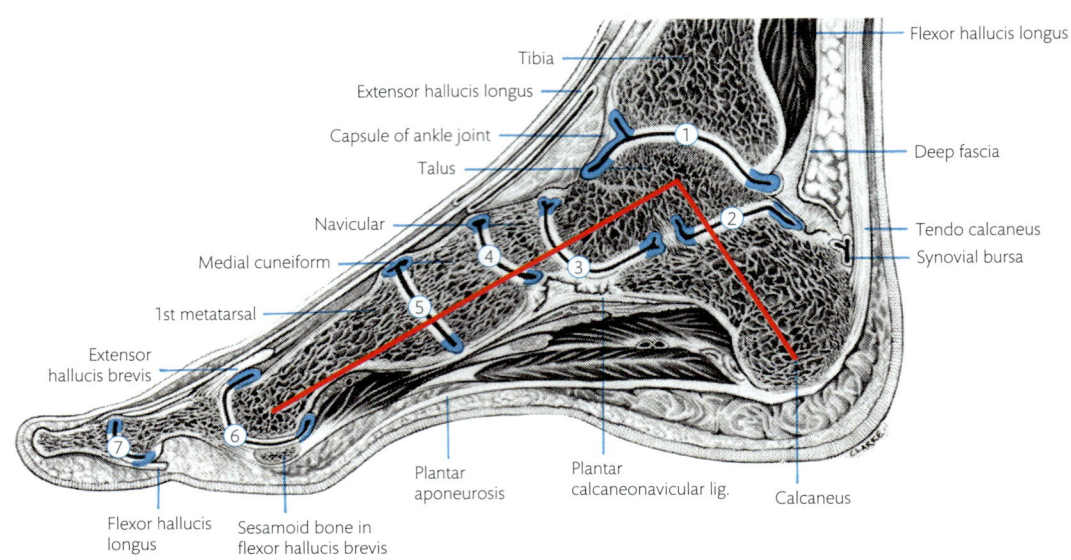

Fig. 22.19 Oblique sagittal section through the middle of the heel and the middle of the big toe. Synovial membrane, blue. The two red lines depict the anterior medial column and posterior pillar of the longitudinal arch. 1, ankle joint; 2, subtalar joint; 3, talocalcaneonavicular joint; 4, joint between navicular and medial cuneiform; 5, first tarsometatarsal joint; 6, first metatarsophalangeal joint; 7, interphalangeal joint of big toe.

aponeurosis. The deep fascia is thin medially and laterally where it covers the abductors of the big and little toes. The plantar aponeurosis is separated from the lateral and medial parts by shallow furrows, through which the cutaneous branches of the plantar vessels and nerves enter the skin of the sole [Fig. 22.18].

Plantar aponeurosis

The plantar aponeurosis is a thick layer of deep fascia attached posteriorly to the medial process on the plantar surface of the calcaneus. Anteriorly, it splits into five slips which pass to each toe. The margins of each slip curve dorsally over the sides of the flexor tendons and are attached to the plantar ligament of the metatarsophalangeal joint. The intermediate part is attached to the proximal end of the fibrous flexor sheath—the same arrangement as in the hand. Thus, each slip of the plantar aponeurosis is firmly bound to the proximal phalanx of a toe by its attachment to the plantar ligament and the fibrous flexor sheath. In being attached to the calcaneus behind and to the phalanges in front, the plantar aponeurosis acts as a flexible connector between the ends of the longitudinal arch of the foot. It is pulled distally when the toes are forcibly extended (e.g. in pushing off with the foot). This action tightens the aponeurosis

and pulls the ends of the arch together, so that it forms a rigid structure against which the push-off can be effective. Check this in your own foot by passively extending your toes with your hand. The plantar aponeurosis can then be seen and felt as a tight band which is relaxed when the toes are flexed [see Fig. 23.16].

➔ Inflammation of the plantar fascia results in a condition called **plantar fasciitis**, a common cause of heel pain. The pain can be of a stabbing or burning type and is often felt most acutely when one takes the first few steps in the morning.

Fibrous flexor sheaths

The fibrous flexor sheath in each toe is a thick, fibrous tunnel. It is attached to the margins of the proximal and middle phalanges (only proximal in the big toe), to the base of the distal phalanx, and to the plantar ligaments of the interphalangeal joints. It is relatively thin at the interphalangeal joints, so as not to restrict flexion. The sheath contains the long and short flexor tendons enclosed in a synovial sheath in the lateral four toes, and the long flexor only in the great toe. Proximally, each sheath is continuous with the plantar aponeurosis.

Dissection 22.12 continues the dissection of the sole of the foot.

Objective

I. To reflect the plantar aponeurosis and deep fascia.

Instructions

1. Cut across the plantar aponeurosis, 2–3 cm in front of the heel. Split the distal part longitudinally, and lift its parts away from the underlying flexor digitorum brevis. At the margins of this muscle, divide the intermuscular septa and reflect the aponeurosis distally. Avoid injury to the plantar digital vessels and nerves lying immediately deep to the distal part of the aponeurosis.
2. Remove the deep fascia from the abductor muscles of the hallux and little toe. Retain the plantar digital nerves already found to these toes.

Arrangement of structures in the sole of the foot

Structures in the sole of the foot are placed in a number of layers. These layers have no significance beyond that of description and are not clearly separated from one another. They represent the order in which the structures are uncovered and the general depth at which they lie.

From superficial to deep, the layers consist of the:

1. Abductor hallucis, flexor digitorum brevis, abductor digiti minimi, and plantar digital vessels and nerves distally. This is the first layer of muscles
2. Proximal parts of the medial and lateral plantar nerves and vessels
3. Long flexor tendons—tendons of the flexor hallucis longus and flexor digitorum longus, flexor accessories, and lumbrical muscles attached to the tendon of the flexor digitorum longus. This is the second layer of muscles
4. Flexor hallucis brevis, adductor hallucis, and flexor digiti minimi brevis. This is the third layer of muscles
5. Deep parts of the lateral plantar artery and nerve and their branches
6. Bones and ligaments of the foot, the tendon of the peroneus longus, insertions of the tibialis posterior to the tarsal and metatarsal bones, the interosseous muscles, and perforating branches of the plantar metatarsal arteries. This layer is the fourth layer of muscles.

In general, the more superficial a muscle is in the sole of the foot, the more likely it is to be long and to be attached to the ends of the arch (i.e. from heel to toe). The action of the superficial muscles helps to maintain the plantar arch. The muscles are important in holding the bones of the arch together when the arch is subjected to powerful forces during activity. They also protect the ligaments from more severe stresses.

Muscles in the sole of the foot–first layer

Flexor digitorum brevis

Origin: the flexor digitorum brevis takes origin from the medial process of the tubercle of the calcaneus and the plantar aponeurosis [see Fig. 16.17]. It gives rise to four tendons to the lateral four toes. **Insertion**: each tendon enters the fibrous flexor sheath of the toe and divides into two parts which curve over the long flexor tendon to the plantar surface of the middle phalanx [Fig. 22.20]. **Nerve supply**: medial plantar nerve. **Actions**: it flexes the metatarsophalangeal and proximal interphalangeal joints of the lateral four toes, and helps to reinforce the longitudinal arch of the foot.

Abductor hallucis

Origin: the abductor hallucis arises from the flexor retinaculum and the medial process of the tubercle of the calcaneus [see Fig. 16.17]. Part of the medial belly of the flexor hallucis brevis and the medial part of the extensor expansion formed by the extensor hallucis longus fuse with the tendon of the abductor hallucis. **Insertion**: it is inserted on the plantar aspect and adjacent medial surface of the proximal phalanx of the big toe, along with the medial belly of the flexor hallucis brevis and part of the extensor expansion. This plantar insertion of the abductor hallucis makes it less efficient as an abductor of the big toe. **Nerve supply**: medial plantar nerve. **Actions**: it moves the big toe away from the second toe (i.e. it abducts it at the metatarsophalangeal joint). It is often mainly a flexor of this joint.

Abductor digiti minimi

Origin: the abductor digiti minimi arises from both processes of the tubercle of the calcaneus [see Fig. 16.17]. **Insertion**: it is inserted into the lateral side of the base of the proximal phalanx of the lit-

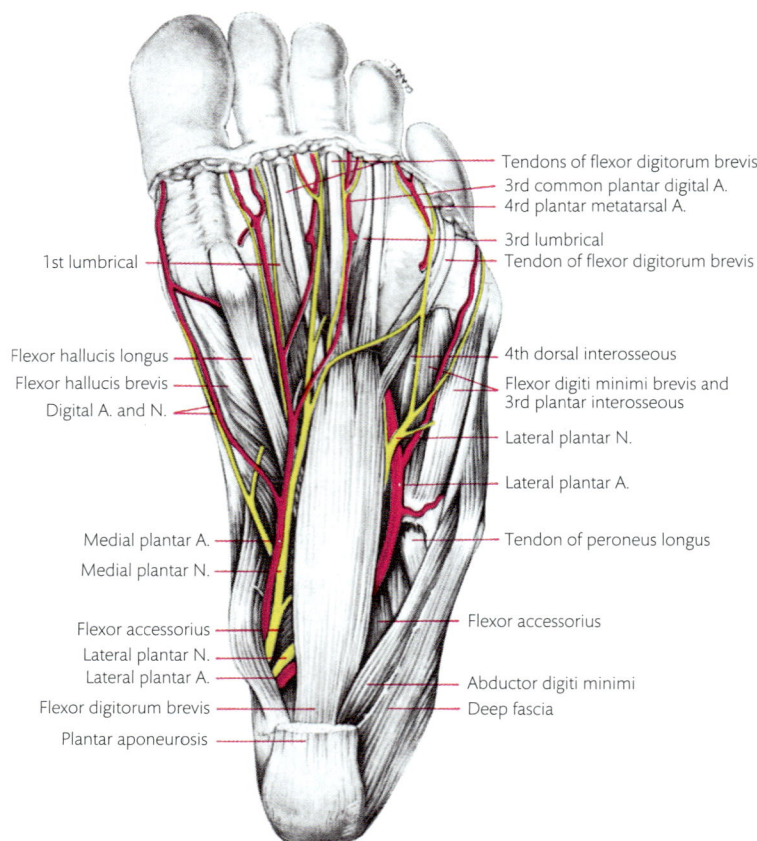

1st lumbrical

Flexor hallucis longus
Flexor hallucis brevis
Digital A. and N.

Medial plantar A.
Medial plantar N.

Flexor accessorius
Lateral plantar N.
Lateral plantar A.
Flexor digitorum brevis
Plantar aponeurosis

Tendons of flexor digitorum brevis
3rd common plantar digital A.
4rd plantar metatarsal A.

3rd lumbrical
Tendon of flexor digitorum brevis

4th dorsal interosseous
Flexor digiti minimi brevis and
3rd plantar interosseous
Lateral plantar N.

Lateral plantar A.

Tendon of peroneus longus

Flexor accessorius

Abductor digiti minimi
Deep fascia

Fig. 22.20 Superficial dissection of the sole of the foot. The plantar aponeurosis has been removed. The abductor digiti minimi and abductor hallucis have been pulled aside.

tle toe [Fig. 22.20]. **Nerve supply**: lateral plantar nerve. **Actions**: it abducts the little toe.

Dissection 22.13 (and Figs 22.21 and 22.22) continues the dissection of the sole of the foot.

Plantar nerves

The medial and lateral plantar nerves are terminal branches of the **tibial nerve** and arise deep to the flexor retinaculum. They enter the sole of the foot, with the corresponding branches of the posterior tibial artery deep to the abductor hallucis.

The **medial plantar nerve** gives branches to the abductor hallucis and flexor digitorum brevis, and runs forwards between them. Here it gives rise to: (1) the **proper plantar digital nerve** to the medial side of the great toe, which also supplies the flexor hallucis brevis; and (2) cutaneous branches to the medial part of the sole of the foot. Further distally, it divides into three **common plantar digital nerves** to the medial three interdigital clefts. These common digital nerves give rise to proper plantar nerves to the adjacent sides of the big, sec-

ond, third, and fourth toes [Fig. 22.21]. The distribution in the toes is similar to that of the median nerve in the fingers. The medial common plantar digital nerve also supplies the first lumbrical.

Note that four plantar muscles are supplied by this nerve: *from the trunk*—the nerve to the (1) abductor hallucis and (2) flexor digitorum brevis; *from the proper plantar digital nerve*—the nerve to the (3) flexor hallucis brevis; and *from the medial common plantar digital nerve*—the nerve to the (4) first lumbrical.

The **lateral plantar nerve** passes between the flexor digitorum brevis and flexor accessorius, giving branches to the flexor accessorius and abductor digiti minimi. It then passes forwards and gives cutaneous branches to the lateral part of the sole. It divides into superficial and deep branches.

The **superficial branch** divides into: (1) the **proper plantar digital nerve** to the lateral side of the little toe; and (2) the **common plantar digital nerve** to the fourth interdigital cleft. The nerve to the lateral side of the little toe gives mus-

DISSECTION 22.13 Sole of the foot-3

Objectives

I. To identify and study the flexor digitorum brevis, abductor hallucis, abductor digiti minimi, flexor accessorius, lumbricals, and long flexor tendons.   II. To identify and trace the plantar arteries and nerves, and their branches.

Instructions

1. Find the proper plantar digital nerves in the toes [Fig. 22.21], and follow them proximally. At the metatarsals, these nerves unite to form the common plantar digital nerves. Nerves from the medial three interdigital clefts are branches of the medial plantar nerve. The nerve from the fourth space is from the lateral plantar nerve.
2. Lift the flexor digitorum brevis, and cut across the muscle near its middle. Reflect its parts forwards and backwards, avoiding injury to the common plantar digital nerves which pass superficial to its distal part [Fig. 22.20]. Follow at least one of the tendons to its insertion. Cut the fibrous flexor sheath longitudinally to expose the tendon in the toe.
3. Turn the abductor hallucis medially, and expose the medial and lateral plantar arteries and nerves. Follow the nerves and arteries distally in the foot. Trace their branches into continuity with the: (1) digital branches already exposed; and (2) branches to the medial side of the big toe and the lateral side of the little toe.
4. Identify the long flexor tendons and flexor accessorius, deep to the vessels and nerves [Fig. 22.21].
5. Remove the abductor hallucis from the flexor retinaculum.
6. Cut the retinaculum, and follow the plantar nerves and arteries to their origins from the tibial nerve and posterior tibial artery deep to the retinaculum.
7. Identify and follow the tendon of the tibialis posterior to its insertion into the navicular bone.
8. Follow the tendons of the flexor digitorum longus and flexor hallucis longus into the sole of the foot. As the tendons are separated in the foot, note the small extension from the flexor hallucis longus to the flexor digitorum longus [Fig. 22.22].
9. Note the insertion of the flexor accessorius into the tendon of the flexor digitorum longus and its origin from the calcaneus. Note the branch from the lateral plantar nerve entering the flexor accessorius.
10. Lift the superficial branch of the lateral plantar nerve, and trace its branch to the flexor digiti minimi brevis. Do the same with the medial two digital branches of the medial plantar nerve. The most medial sends a branch to the flexor hallucis brevis; the second sends a branch to the first lumbrical [Fig. 22.20].

cular branches to the flexor digiti minimi brevis, and the third plantar and fourth dorsal interossei. The common plantar digital nerve gives proper digital nerves to adjacent sides of the fourth and fifth toes. These are distributed in the same manner as the corresponding branches of the medial plantar nerve. The common plantar digital nerve of the third space (branch of the medial plantar nerve) and that of the fourth space (branch of the lateral plantar nerve) communicate. Because of these communications, there is considerable overlap of the areas supplied by each nerve.

The **deep branch** of the lateral plantar nerve runs medially across the proximal parts of the metatarsals and supplies the remaining small muscles of the foot—adductor hallucis, flexor digiti minimi brevis, lateral three lumbrical muscles, medial two plantar interossei, and medial three dorsal interossei. The distribution of the lateral plantar nerve in the foot is very similar to that of the ulnar nerve in the hand.

Note the plantar muscles supplied by this nerve are: *from the trunk*, the nerve to the (1) flexor accessorius and (2) abductor digiti minimi; *from the proper plantar digital nerve*—the nerve to the (3) flexor digiti minimi brevis, and (4) third plantar and (5) fourth dorsal interossei; *from the deep branch*—the nerve to the (6) adductor hallucis, (7) flexor digiti minimi brevis, (8–10) lateral three lumbricals, (11, 12) medial two plantar interossei, and (13–15) medial three dorsal interossei.

Plantar arteries

The **medial** and **lateral plantar** arteries begin as terminal branches of the posterior tibial artery, and are given off deep to the flexor retinaculum. The **medial plantar artery** runs with the medial plantar nerve supplying the surrounding structures, and gives branches corresponding to those of the nerve. The medial plantar artery ends by anastomosing with the branch of the first plantar metatarsal artery to the medial side of the big toe [Fig. 22.9].

The **lateral plantar artery** runs with the lateral plantar nerve. It gives branches to the surrounding skin, muscles, and bones, and forms the **plantar arch**

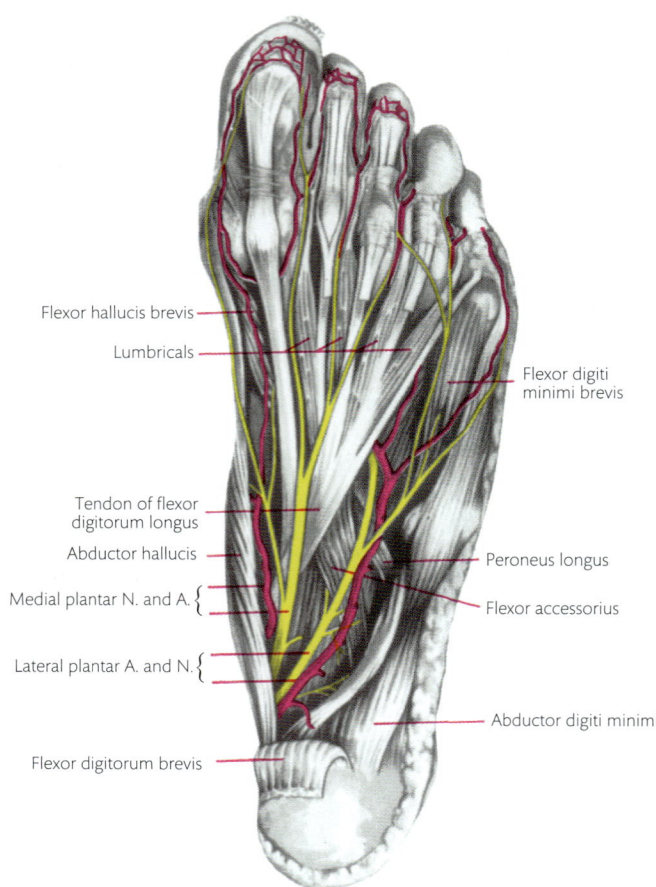

Fig. 22.21 Dissection of the sole of the foot. Most of the flexor digitorum brevis has been removed.

Labels on Fig. 22.21:
Flexor hallucis brevis
Lumbricals
Flexor digiti minimi brevis
Tendon of flexor digitorum longus
Abductor hallucis
Peroneus longus
Medial plantar N. and A. {
Flexor accessorius
Lateral plantar A. and N. {
Abductor digiti minimi
Flexor digitorum brevis

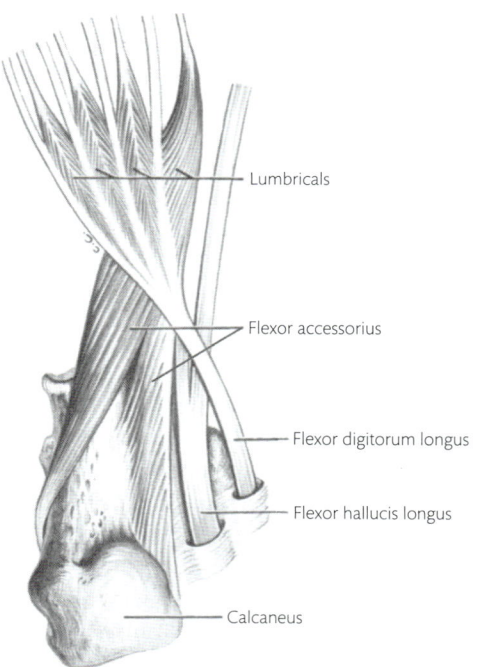

Fig. 22.22 Long flexor tendons in the sole of the foot.

Labels on Fig. 22.22:
Lumbricals
Flexor accessorius
Flexor digitorum longus
Flexor hallucis longus
Calcaneus

beside the deep branch of the nerve. The arch gives the **proper plantar digital artery** to the lateral side of the little toe, and the **plantar metatarsal arteries** on each intermetatarsal space. Each plantar metatarsal artery communicates with the corresponding **dorsal metatarsal artery** by a **perforating branch** in the proximal part of the space. The plantar metatarsal arteries form the **common plantar digital artery** which divides into the **proper plantar digital arteries** to the adjacent sides of two toes.

Dissection of the sole of the foot continues in Dissection 22.14.

Muscles in the sole of the foot—second layer

This layer consists of the long flexor tendons and associated muscles [Fig. 22.22]. Note particularly the pulley-like arrangement of the flexor hallucis longus on the plantar surface of the sustentaculum tali and the slip which it gives to the flexor digitorum longus. Also note that the long flexor tendons cross each other, inferior to the head of the talus.

Flexor accessorius

Origin: the flexor accessorius takes origin from both margins of the plantar surface of the calcaneus [see Fig. 16.16]. **Insertion**: it is inserted into the tendon of the flexor digitorum longus. **Nerve supply**: lateral plantar nerve. **Actions**: it assists in flexing the toes by aligning the tendon of the flexor digitorum longus to the toes.

Lumbrical muscles

Origin: these four small muscles arise from the tendons of the flexor digitorum longus [Fig. 22.22]. **Insertion**: one tendon enters the medial side of each of the lateral four toes and is attached to the base of the proximal phalanx and to the extensor expansion. The tendons are inferior to the deep transverse metatarsal ligament and the transverse head of the adductor hallucis [Fig. 22.23]. **Nerve supply**: the first lumbrical is innervated by the medial plantar nerve, and the other three by the lateral plantar

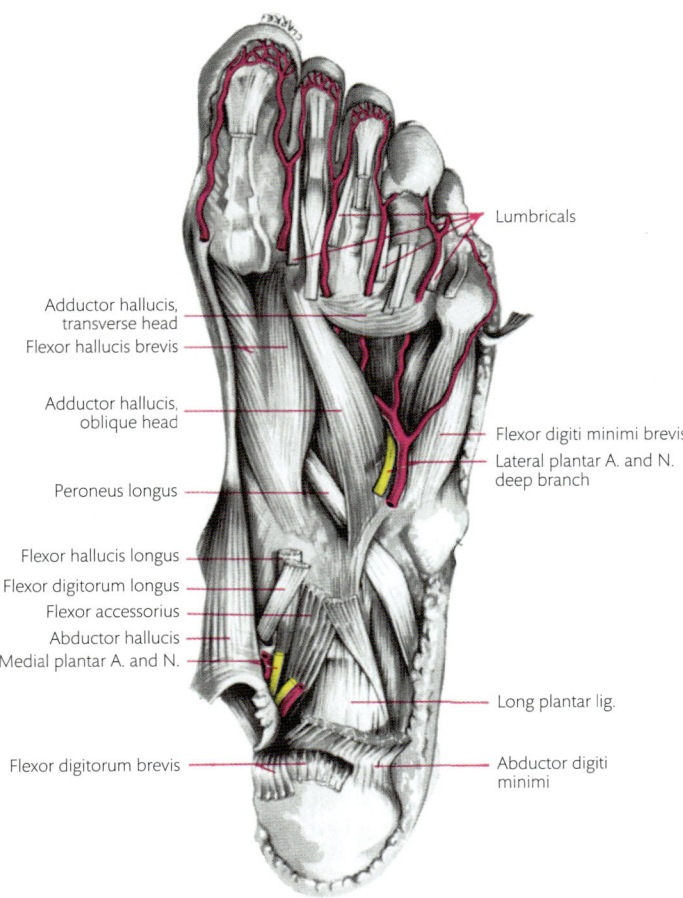

Lumbricals

Adductor hallucis, transverse head
Flexor hallucis brevis

Adductor hallucis, oblique head

Peroneus longus

Flexor hallucis longus
Flexor digitorum longus
Flexor accessorius
Abductor hallucis
Medial plantar A. and N.

Flexor digitorum brevis

Flexor digiti minimi brevis
Lateral plantar A. and N. deep branch

Long plantar lig.

Abductor digiti minimi

Fig. 22.23 Deep dissection of the sole of the foot.

nerve. **Actions**: they are weak muscles, but they play a part in flexion of the metatarsophalangeal joints of the lateral four toes. They can also extend the interphalangeal joints. ➲ As in the hand, paralysis of the lumbricals prevents extension of the interphalangeal joints when the metatarsophalangeal joints are fully extended. This leads to a condition known as '**hammer toe**', in which the metatarsophalangeal joints of the lateral four toes are fully extended, the proximal interphalangeal joints are flexed (by the flexor digitorum brevis), and the distal interphalangeal joints are extended due to pressure on the ground.

Flexor tendons in the toes

The arrangement of the flexor tendons and fibrous flexor sheaths of the toes is the same as in the fingers. The **synovial sheaths** begin distal to the attachment of the lumbrical muscles, and extend to the base of the distal phalanx where these tendons are inserted. The sheath of the little toe may be continuous with the synovial sheath around the main tendon of the flexor digitorum longus. The sheath of the flexor hallucis longus extends from the lower part of the leg to the insertion of the tendon into the distal phalanx of the great toe. It is incomplete where it connects with the flexor digitorum longus. Tendons in the toes have the same arrangement of **vinculae** as tendons in the fingers.

Dissection of the sole of the foot continues in Dissection 22.15.

DISSECTION 22.15 Sole of the foot-5

Objective

I. To transect the muscles and tendons of the second layer—flexor accessorius, flexor digitorum longus, and flexor hallucis longus.

Instructions

1. Divide the flexor accessorius and the tendons of the flexor digitorum longus and flexor hallucis longus close to where they unite. Reflect the distal parts of the tendons to uncover the three muscles which form the fourth layer [Fig. 22.23].
2. If needed, cut across the medial plantar nerve, and reflect it distally to permit sufficient reflection of the tendons.
3. As the flexor digitorum is turned back, look for the branches of the deep branch of the lateral plantar nerve to the lateral three lumbrical muscles.

Muscles in the sole of the foot—third layer

Flexor hallucis brevis

Origin: this powerful muscle arises from the plantar surface of the cuboid bone [see Fig. 16.16] and the adjoining fascia. **Insertion**: it divides into two bellies which are inserted into the medial and lateral margins of the plantar surface of the proximal phalanx of the big toe. The medial belly is inserted with the tendon of the abductor hallucis, and the lateral with that of the adductor hallucis. Each tendon contains a sesamoid bone [Fig. 22.19]. **Nerve supply**: medial plantar nerve. **Actions**: it flexes the metatarsophalangeal joint of the big toe and produces slight adduction because of its obliquity.

Adductor hallucis

Origin: the adductor hallucis takes origin from an oblique and a transverse head. The oblique head is from the base of the middle three metatarsals and from the tendon of the peroneus longus. The transverse head is from the deep transverse metatarsal ligament and the plantar ligaments of the lateral four metatarsophalangeal joints [Fig. 22.23]. The two heads fuse together and are inserted into the lateral margin of the proximal phalanx of the big toe, along with the lateral head of the flexor hallucis brevis. **Nerve supply**: deep branch of the lateral plantar nerve. **Actions**: the oblique head adducts and flexes the metatarsophalangeal joint of the big toe. The transverse head draws the plantar surface of the roots of the toes together and helps to increases the transverse metatarsal arch.

Flexor digiti minimi brevis

Origin: the flexor digiti minimi brevis arises from the base of the fifth metatarsal [see Fig. 16.16]. **Insertion**: it is inserted into the lateral side of the base of the proximal phalanx of the little toe. **Nerve supply**: superficial branch of the lateral plantar nerve. **Actions**: flexion of the metatarsophalangeal joint of the little toe.

See Dissection 22.16 for further instructions on dissection of the sole of the foot.

Sesamoid bones of the foot

There are two small sesamoid bones in each of the tendons of the flexor hallucis brevis. They extend beyond the tendons, through the plantar ligament of the metatarsophalangeal joint of the big toe, and articulate with the plantar surface of the head of

DISSECTION 22.16 Sole of the foot-6

Objectives

I. To expose and study the flexor hallucis brevis, adductor hallucis, and flexor digiti minimi brevis. II. To identify and trace the deep branch of the lateral plantar nerve, the plantar arch, and their branches.

Instructions

1. Define the attachments of the flexor hallucis brevis, adductor hallucis, and flexor digiti minimi brevis. Avoid injury to the deep branch of the lateral plantar nerve, the plantar arch, and their branches.

2. Detach the flexor hallucis brevis and the oblique head of the adductor from their origins. Turn them forwards to expose the nerve and artery.

3. Cut across the abductor hallucis. Turn its distal part forwards with the flexor hallucis brevis. Define their common insertion, and identify the sesamoid bone in their tendon. Cut this tendon and the bone away from the articular capsule of the first metatarsophalangeal joint. Note that the sesamoid bone rests directly on the head of the first metatarsal.

4. Find the nerves to the oblique and transverse heads of the adductor. Then trace the deep branch of the lateral plantar nerve, the plantar arch, and their branches.

the metatarsal. This arrangement adds to the size of the metatarsal head of the big toe. The bones also protect against compression and devascularization of the tendons of the flexor hallucis brevis and longus. (Sesamoid bones or cartilages are also found in the tendons of the peroneus longus and tibialis posterior as they enter the sole.)

Deep branch of the lateral plantar nerve

The deep branch arises from the lateral plantar nerve at the base of the fifth metatarsal and runs medially on the proximal parts of the metatarsal bones and the interosseous muscles [Fig. 22.9]. It is deep to the oblique head of the adductor hallucis and ends in it. It supplies the adductor hallucis, the lateral three lumbrical muscles, the medial two plantar interossei, and the medial three dorsal interossei. It sends branches to the distal intertarsal, tarsometatarsal, and intermetatarsal joints.

Plantar arch

The plantar arch is a continuation of the lateral plantar artery. It runs with the deep branch of the lateral plantar nerve to the proximal end of the first intermetatarsal space. The arch gives rise to a **plantar metatarsal artery** to each intermetatarsal spaces. Each of these arteries communicates with the corresponding dorsal metatarsal artery by a **perforating branch** through the space. The plantar metatarsal arteries form the **common plantar digital arteries**. In the first space, the

perforating branch comes from the arcuate artery (**deep plantar branch**) and is mainly responsible for the formation of the plantar metatarsal artery of that space. Each common plantar digital artery divides into two **proper plantar digital arteries** to the adjacent sides of two toes. The proper plantar digital artery to the medial side of the big toe arises from the common plantar digital artery in the first space [Fig. 22.9]. The arch and its branches supply all the surrounding structures, including the bones and joints. The proper digital arteries anastomose with each other and with the dorsal digital arteries in the distal parts of the toes. Thus, there is a very free anastomosis between the branches of the dorsalis pedis artery and the plantar arteries. This is well illustrated in the angiogram of the foot [Fig. 22.10].

See Dissection 22.17 for further investigation of the sole of the foot.

Muscles and ligaments in the sole of the foot—fourth layer

Deep transverse metatarsal ligament

These strong bands of dense connective tissue unite the plantar ligaments of the metatarsophalangeal joints and are attached to the proximal phalanges through them. The ligaments help to prevent the bases of the toes from spreading out, and so maintain the transverse metatarsal arch. The interossei enter the toes, dorsal to these ligaments; the lumbricals and plantar digital nerves and vessels are on their plantar surface.

Interossei

The general arrangement of the interossei is the same as in the hand, except that the second (and not the middle) toe has two dorsal interossei. There is a plantar interosseous to each of the lateral three toes, and four dorsal interossei to the middle three toes. **Origin, dorsal interossei**: each dorsal interosseus muscle takes origin from two adjacent metatarsals—the first and second metatarsals give origin to the first dorsal interosseus, the second and third to the second, the third and fourth to the third, and the fourth and fifth to the fourth dorsal interosseus. **Origin, palmar interossei**: the first, second, and third palmar interossei take origin from the medial aspect of the third, fourth, and fifth metatarsals. **Insertion**: all the interossei are inserted into the bases of the proximal phalanges, with little attachment to the extensor expansion [Fig. 22.24]. **Nerve supply**: lateral plantar nerve. **Actions**: the plantar interossei adduct the lateral three toes to the line of the second toe—the axis of the foot. The dorsal interossei abduct the middle three toes from the axis of the foot. They also flex the metatarsophalangeal joints but usually do not extend the interphalangeal joints.

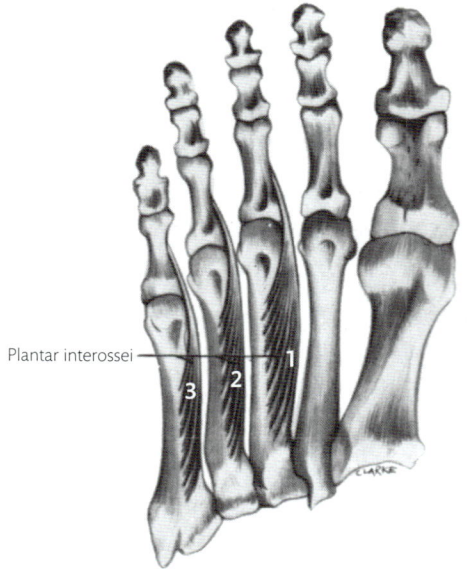
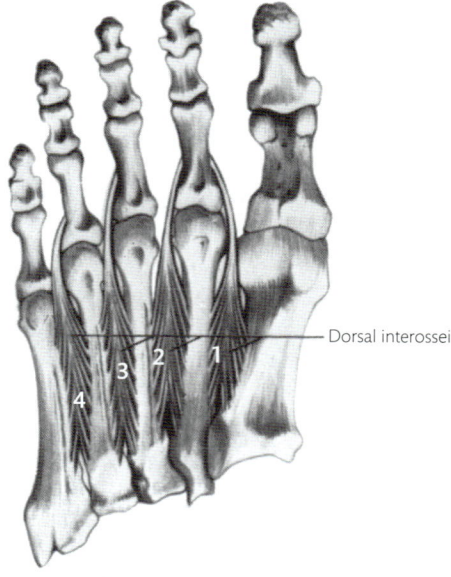

Plantar interossei

Dorsal interossei

Fig. 22.24 Interosseous muscles of the right foot.

Tendon of tibialis posterior

This tendon runs over the posterior surface of the medial malleolus (which acts as a pulley for it) and beneath the medial part of the plantar calcaneonavicular ligament to the tuberosity of the navicular bone [Fig. 22.25]. It spreads out to all the other tarsal bones (except the talus) and to the bases of the middle three metatarsals.

The powerful **plantar calcaneonavicular ligament** runs from the sustentaculum tali to the navicular. It supports the head of the talus which carries the weight of the body. The downward force exerted by the talus tends to separate the calcaneus and navicular, and stretch the ligament. The tibialis posterior pulls the navicular posteriorly and helps to support some of the load on the ligament. Under the ligament, the tibialis posterior tendon contains a piece of fibrocartilage, which is sometimes ossified to form a sesamoid bone. The extensions of the tendon to the other bones pull the plantar surfaces of the bones together, strengthen the arches, and protect the plantar ligaments.

Tendon of peroneus longus

The tendon of the peroneus longus contains a sesamoid bone or cartilage. It turns around the lateral border of the cuboid bone and runs anteromedially in the groove on its plantar surface. The **long plantar ligament** [Fig. 22.25] converts the groove into a tunnel, in which the tendon slides in its synovial sheath. The tendon is inserted into the base of the first metatarsal and the adjacent part of the medial cuneiform bone, close to the insertion of the tibialis anterior. When the peroneus longus contracts, it everts the foot. Tension developed in its tendon prevents stretching of the ligaments which maintain the transverse arch of the tarsal bones.

➲ The presentation of pain and swelling at the head of the metatarsal is known as **metatarsalgia**. It may be accompanied by pain and tingling in the toes.

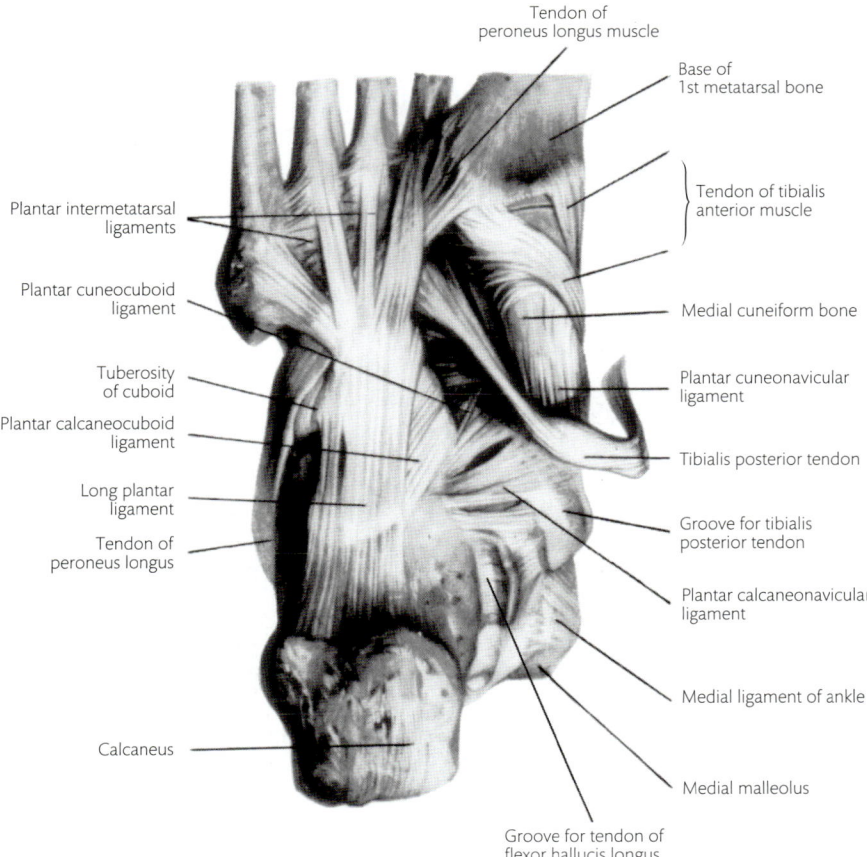

Tendon of
peroneus longus muscle

Base of
1st metatarsal bone

Plantar intermetatarsal
ligaments

Plantar cuneocuboid
ligament

Tuberosity
of cuboid

Plantar calcaneocuboid
ligament

Long plantar
ligament

Tendon of
peroneus longus

Calcaneus

Groove for tendon of
flexor hallucis longus

Tendon of tibialis
anterior muscle

Medial cuneiform bone

Plantar cuneonavicular
ligament

Tibialis posterior tendon

Groove for tibialis
posterior tendon

Plantar calcaneonavicular
ligament

Medial ligament of ankle

Medial malleolus

Fig. 22.25 Plantar aspect of the tarsal and tarsometatarsal joints.

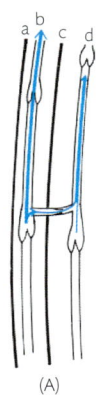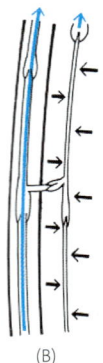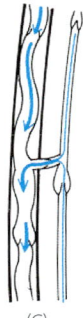

(A)　　　　　　　　(B)　　　　　　　　(C)

Fig. 22.26 Diagrams to show the flow of blood in the superficial and deep veins. (A) Valves in superficial and deep veins preventing downward flow of blood. (B) Deep veins are emptied by contraction of surrounding muscles. (C) Dilatation of the superficial and communicating veins makes their valves incompetent, so that blood flows distally in the superficial veins and from the deep veins to the superficial veins when the muscles contract. a. Skin. b. Superficial vein in the superficial fascia. c. Deep fascia. d. Deep vein.

Supports of the arches of the foot

The arches of the foot are maintained by the shape of the bones, the ligaments holding the bones in position, and muscles. The wedge shape of the cuneiform and base of the metatarsals help in maintaining the arch, like a keystone, and prevent downward displacement.

The main **ligaments** that maintain the arches of the foot are: (1) the plantar calcaneonavicular ligament, extending from the sustentaculum tali to the navicular; (2) the long plantar ligament, extending from the calcaneus to the cuboid and the base of the middle three metatarsals; (3) the plantar calcaneocuboid ligament; and (4) the plantar aponeurosis, which stretches between the lower parts of the anterior and posterior pillars (calcaneus and heads of the metatarsals) and holds them together.

The muscles supporting the arches are the tibialis anterior, tibialis posterior, peroneus longus, and long and short flexors of the toes. The tibialis anterior raises the medial longitudinal arch through its insertion on the upper surface of the navicular and base of the first metatarsal. The tibialis posterior holds the navicular against the head of the talus, and the other tarsal and metatarsal bones together by its insertion. The peroneus longus runs across the sole of the foot from lateral to medial, and supports the transverse arch in this position. The long and short flexors draw the ends of the longitudinal arches together when they contract.

➲ The essential deformity in '**flat foot**' is eversion of the heel relative to the anterior part of the foot. In the reverse deformity of '**club foot**', the heel is strongly inverted relative to the forepart of the foot and the arch is greatly exaggerated.

See Clinical Applications 22.1 (and Fig. 22.26), 22.2, 22.3, and 22.4, and 22.5 for some practical implications of the anatomy discussed in this chapter.

CLINICAL APPLICATION 22.1 Varicose veins of the leg

In the long and short saphenous veins, valves ensure that the blood flows upwards towards the heart and is prevented from flowing downwards under the influence of gravity [Fig. 22.26A]. The valves in the perforators ensure that blood flows from superficial to deep, taking advantage of the fact that the deep veins (but not the superficial ones) are emptied by contraction of the surrounding muscles [Fig. 22.26B]. Commonly, one or both saphenous veins become tortuous and dilated, due to incompetency of their valves [Fig. 22.26C]. Venous blood stagnates and pools in the superficial veins, causing discomfort and pain. As the valves in the perforators are incompetent, blood from the deep veins enters the superficial ones during muscle contraction. As the valves in the superficial veins are incompetent, blood flows distally under the influence of gravity. Varicose veins prevent proper venous return in the lower limb and dam back the entry of arterial blood to the region. This leads to poor nutrition of tissue and non-healing of injuries. Small accidental cuts into the vein wall can lead to rapid exsanguination.

CLINICAL APPLICATION 22.2 Intragluteal injection with damage to the common peroneal nerve

A 39-year-old man developed sensory loss and foot drop after a poorly administered intragluteal injection. On examination, the patient was found to have sensory loss on the outer side of the right calf and the dorsum of the right foot. He was unable to dorsiflex the ankle and evert the foot. When he was asked to flex his knee and lift his leg, there was noticeable plantar flexion—a condition known as foot drop. He also had difficulty extending his toes.

Study question 1: of the two parts of the sciatic nerve—the tibial and common peroneal—which part seems solely involved in this case? (Answer: common peroneal nerve.)

Study question 2a: how do you account for the loss of sensation to the lateral side of the calf? (Answer: the lateral cutaneous nerve of the calf and the peroneal communicating branch of the sural nerve arise from the common peroneal nerve and are damaged.)

Study question 2b: how do you account for the loss of sensation over the dorsum of the foot? (Answer: the dorsum of the foot is supplied by the cutaneous branches of the deep peroneal nerve—the first interdigital cleft and adjacent sides of the big and second toes—and by the superficial peroneal nerve—the medial side of the big toe, adjacent sides of the second and third toes, adjacent sides of the third and fourth toes, and medial side of the little toe. Deep and superficial peroneal nerves are

branches of the common peroneal nerve which is damaged in the gluteal region.)

Study question 3: why is the patient unable to evert his foot? Name the muscles affected and their nerve supply. (Answer: eversion is lost, because the muscles causing eversion—peroneus longus and brevis—supplied by the superficial peroneal nerve are denervated.)

Study question 4: why is the patient unable to dorsiflex his foot? Name the muscles affected and their nerve supply. (Answer: dorsiflexion is lost, because the muscles causing dorsiflexion—tibialis anterior, extensor hallucis longus, extensor digitorum longus, peroneus tertius—supplied by the deep peroneal nerve are denervated.)

Study question 5: if this were a case of injury to the deep peroneal nerve alone, with what neurological signs would the patient present? (Answer: the patient would present with foot drop and an inability to dorsiflex his foot. Sensory loss would be confined to the first interdigital cleft. Eversion and sensation on the rest of the dorsum of the foot would be normal.)

Study question 6: explain why the patient had difficulty extending his toes. (Answer: the patient has lost the use of the extensor hallucis longus and extensor digitorum longus, so his extension is weakened. Extension is still possible, because the interossei extend the toes through their insertion into the extensor expansion.)

CLINICAL APPLICATION 22.3 Anterior compartment syndrome

During football practice, a 20-year-old student experienced severe pain over the anterolateral aspect of his right leg, which radiated down to his ankle. On examination, there was redness and swelling over the anterolateral aspect of his right leg. On palpation, the area was extremely tender and hard, and corresponded to the area of the tibialis anterior. Dorsiflexion of the foot and toes was limited. The dorsalis pedis pulse was well felt. A diagnosis of 'anterior tibial syndrome' was made. The condition is caused by impairment of blood flow to the muscles during strenuous exercise and possibly tearing of muscle fibres and microhaemorrhage. As there was no improvement with conservative treatment, surgical measures were resorted

to. The fascia over the anterior aspect of the leg was incised under general anaesthesia, with good results.

Study question 1: name the bony surfaces and fascia which enclose the tibialis anterior. (Answer: the lateral surface of the shaft of the tibia, the anterior surface of the shaft of the fibula, the deep fascia of the leg, and the interosseous membrane [Fig. 22.17].)

Study question 2: name the vessels and nerves in this compartment. (Answer: deep fibular nerve and anterior tibial artery.)

Study question 3: from the history given, how do we know that the artery is not compressed? (Answer: the dorsalis pedis pulse is well felt.)

CLINICAL APPLICATION 22.4 Stress fractures

Weight-bearing bones, such as the tibia and the metatarsals, can develop 'stress fractures'. A stress fracture is a fatigue-induced fracture of the bone, caused by repeated stress over time, and may result from pro-

longed exercise or from repeated submaximal loading in running or jumping. Because of this mechanism, stress fractures are common overuse injuries in athletes.

CLINICAL APPLICATION 22.5 Flat foot

A 35-year-old housewife presents with pain in her feet when standing or walking, which becomes worse as the day progresses. The pain radiates up her leg and to her back, and by the evening, her feet are swollen. She feels her gait has changed, as she tries to minimize the pain by putting the whole foot on the ground, rather than moving her weight from the heel to the toes at every step. Prints of her feet while weight-bearing show that the entire sole of the foot leaves an imprint on the ground. The patient is diagnosed to have 'flat foot', possibly due to chronic strain.

Study question 1: which parts of the foot normally touch the ground when a person stands? Why does the entire foot not touch the ground normally? (Answer: the heel, the lateral border of the foot, and the heads of the metatarsals imprint the ground. The medial side of the foot is raised off the ground by the longitudinal arch of the foot.)

Central to the development of flat foot due to chronic strain is the sagging of the head of the talus. Study question 2: what structure/structures normally support the head of the talus? (Answer: the plantar calcaneonavicular ligament and the tendon of the tibialis posterior.) When the head of the talus sags under the weight of the body, the plantar calcaneonavicular ligament is stretched and the unsupported head is displaced medially and downwards. The foot is chronically everted, and pressure on the nerves and blood vessels of the sole causes pain and discomfort.

CHAPTER 23
The joints of the lower limb

Hip joint

The hip joint is described in Chapter 21.

Knee joint

Type and articular surfaces

The massive knee joint is a **synovial condylar joint** between the condyles of the femur, the condyles of the tibia, and the patella. In all positions of the joint, the femur articulates with the tibia and patella.

The articular surface on the femoral condyles is long and extends on the anterior, inferior, and posterior surfaces [see Figs 16.6, 16.7]. The articular area of the tibial condyles is much smaller, and

at any stage of movement, only a relatively small area of the convex femoral condyle articulates with the corresponding slightly concave tibial condyle. On the tibia, the periphery of the condyles has C-shaped rims of fibrocartilage—the **medial** and **lateral menisci** [Fig. 23.1]. The menisci deepen the articular surfaces on the tibia. The strength of the joint depends on ligaments and muscles, rather than on the close fitting of bones.

The cavity of the knee joint may be described as four separate communicating cavities. There is one cavity between the medial femoral and medial tibial condyles, and one between the lateral femoral and lateral tibial condyles. The **patellofemoral cavity** lies between the anterior surface of the femur and the patella. The fourth cavity is the **suprapatellar bursa** between the quadriceps tendon and the femur [Fig. 23.2A].

See Dissection 23.1.

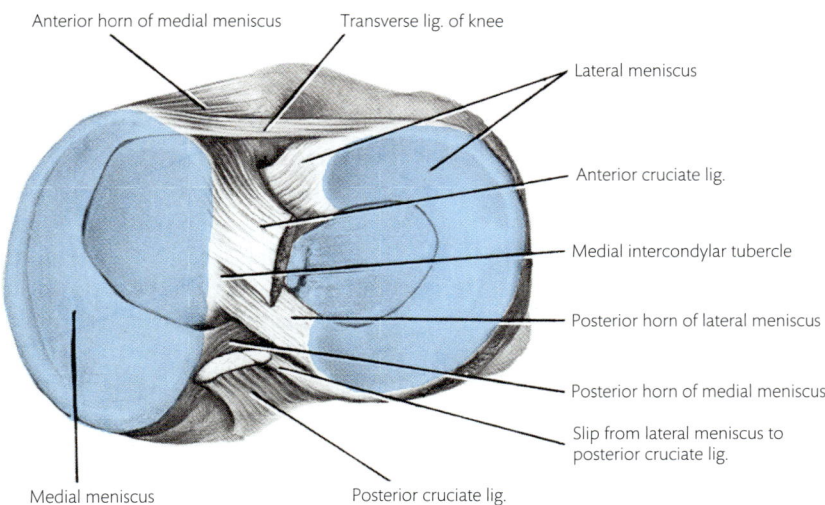

Fig. 23.1 Upper end of the tibia with menisci and portions of the cruciate ligaments.

Anterior horn of medial meniscus

Transverse lig. of knee

Lateral meniscus

Anterior cruciate lig.

Medial intercondylar tubercle

Posterior horn of lateral meniscus

Posterior horn of medial meniscus

Slip from lateral meniscus to posterior cruciate lig.

Medial meniscus

Posterior cruciate lig.

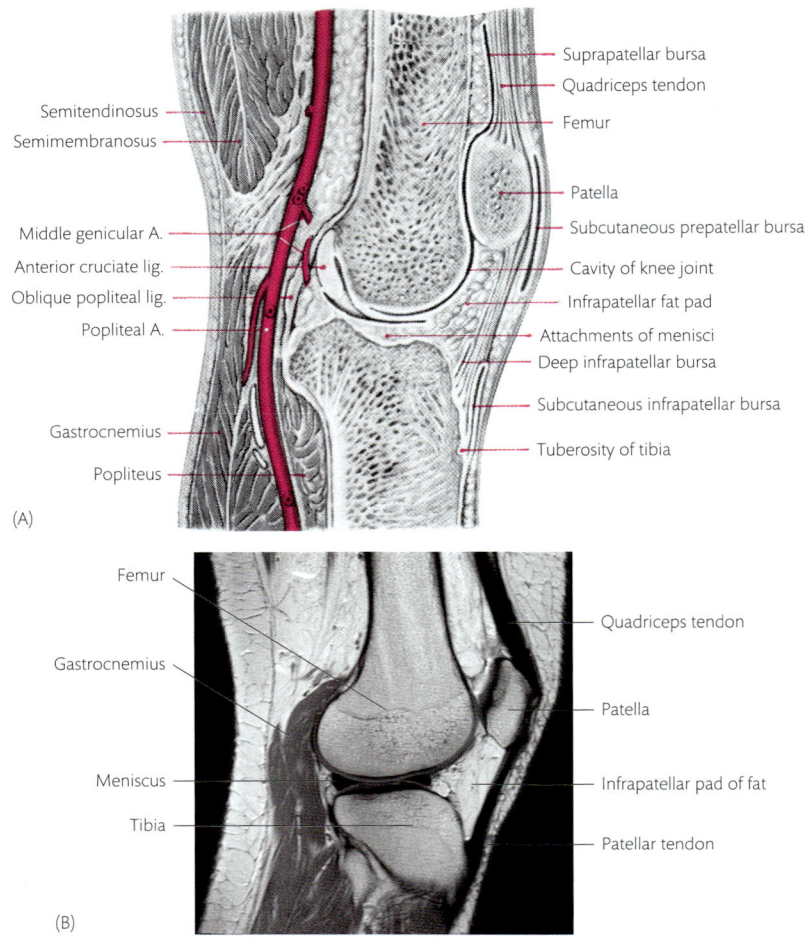

Semitendinosus
Semimembranosus

Middle genicular A.
Anterior cruciate lig.
Oblique popliteal lig.
Popliteal A.

Gastrocnemius
Popliteus

(A)

Suprapatellar bursa
Quadriceps tendon
Femur

Patella
Subcutaneous prepatellar bursa
Cavity of knee joint
Infrapatellar fat pad
Attachments of menisci
Deep infrapatellar bursa
Subcutaneous infrapatellar bursa
Tuberosity of tibia

Femur

Gastrocnemius

Meniscus
Tibia

(B)

Quadriceps tendon

Patella

Infrapatellar pad of fat

Patellar tendon

Fig. 23.2 (A) Sagittal section of the right knee joint. The popliteal vein is not shown. (B) Magnetic resonance imaging of the knee, sagittal section.

Articular capsule

The fibrous capsule is thin and extensive at the back, but thicker and shorter at the sides. In front, it is replaced from above downwards by the patellar tendon, patella, and tendon of the quadriceps.

This arrangement allows full range of flexion of the knee, yet maintains necessary tension in the capsule in all positions by contraction of the quadriceps.

Attachments of the articular capsule

Posteriorly and at the sides, the fibrous capsule is attached close to the articular margins of the femoral condyles, the intercondylar line of the femur, and the tibial condyles. **Anteriorly**, it follows the oblique lines on the tibia downwards to the sides of the tibial tuberosity and blends with the patellar tendon, the sides of the patella, and the tendon of the quadriceps.

The fibrous capsule is perforated in two places: (1) where the tendon of the popliteus emerges from the capsule posterior to the lateral tibial condyle; and (2) where the bursa under the medial head of the gastrocnemius continues with the synovial

---

DISSECTION 23.1 Knee joint-1

Objective

I. To clean the capsule of the knee joint.

Instructions

1. Remove the structures surrounding the knee joint. Retain the fibrous articular capsule, collateral ligaments, and tendons that have connections with the ligaments.

membrane of the joint at the back of the medial femoral condyle.

Extracapsular ligaments of the knee joint

Patellar tendon/patellar ligament

This powerful ligament is the continuation of the quadriceps tendon, inferior to the patella. **Attachments**: it extends from the apex and lower parts of the patella to the smooth upper part of the tibial tuberosity. Two important structures lie deep to the ligament. The **infrapatellar pad of fat** separates the upper part of the ligament from the synovial membrane of the knee joint. The **deep infrapatellar bursa** separates the lower part of the ligament from the upper part of the tibia [Fig. 23.2].

Fibular collateral ligament

The fibular collateral ligament is cord-like. **Attachments**: it extends from the lateral epicondyle of the femur to the head of the fibula. It pierces the tendon of the biceps femoris. It is separated from the fibrous capsule of the joint by fatty tissue, in which the inferior lateral genicular vessels and nerve run. Deep to the ligament, the fibrous capsule is separated from the meniscus by the tendon of the popliteus [Fig. 23.3].

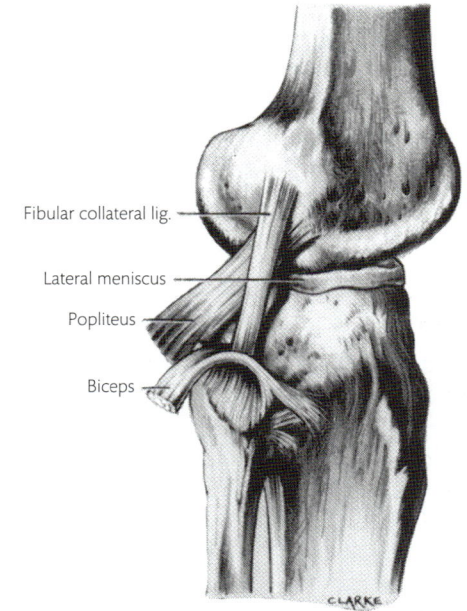

Fibular collateral lig.

Lateral meniscus

Popliteus

Biceps

CLARKE

Fig. 23.3 Fibular collateral ligament of the right knee joint.

Tibial collateral ligament

The tibial collateral ligament is broad and flat. **Attachments**: it arises from the medial epicondyle of the femur, in continuation with fibres from the tendon of the adductor magnus. Inferiorly, it splits into two layers. The **deep layer** fuses with the fibrous capsule and the underlying medial meniscus, and passes to the articular margin of the medial condyle of the tibia. The **superficial layer** is inserted into the medial surface of the tibia, anterior to the insertions of the sartorius, gracilis, and semitendinosus. It is separated from the medial condyle of the tibia by the inferior medial genicular vessels and nerve and by the insertion of the semimembranosus [see Fig. 16.12].

Oblique popliteal ligament

This is an extension of the semimembranosus tendon. It arises from the tendon close to its insertion, and runs upwards and laterally to fuse with the fibrous capsule [Fig. 23.4].

Patellar retinacula

These are fibrous expansions from the vastus medialis and lateralis into the fibrous capsule on the medial and lateral sides of the patellar tendon.

Iliotibial tract

The iliotibial tract fuses with the fibrous capsule between the fibular collateral ligament and the patellar tendon.

Arcuate ligament

The fibres of the posterior part of the fibrous capsule which arch over the aperture for the tendon of the popliteus form the **arcuate ligament**.

See Dissection 23.2 (and Figs 23.5 and 23.6) which continues with the dissection of the knee joint.

Intracapsular structures of the knee joint

Infrapatellar fold of the synovial membrane

The **infrapatellar fold** of the synovial membrane extends from the posterior surface of the patellar tendon to the anterior margin of the intercondylar fossa of the femur. The infrapatellar fold has a free crescentic margin directed posteriorly [Fig. 23.6].

Synovial membrane

The synovial membrane lines the capsule of the knee joint, except in the posterior part where it turns forwards to surround the cruciate ligaments

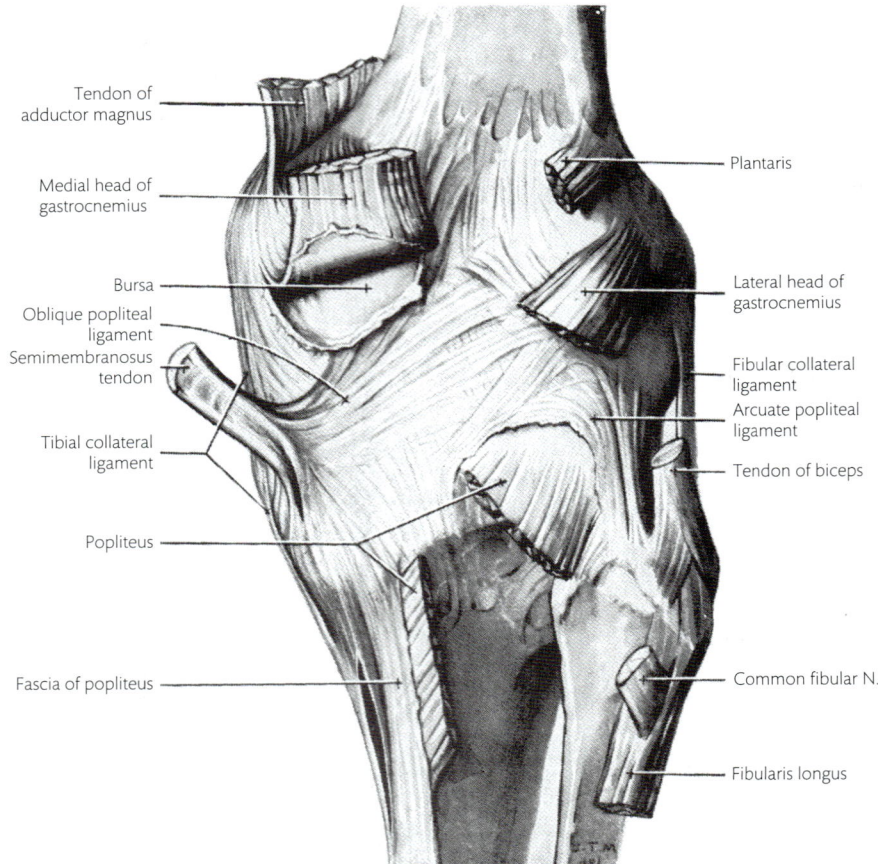

Fig. 23.4 Posterior aspect of right knee joint.

[Fig. 23.6]. It also covers the non-articular structures within the fibrous capsule (i.e. the infrapatellar pad of fat and the tendon of the popliteus). It separates the popliteus tendon from the lateral condyle of the femur. The synovial membrane does not cover the articular surfaces of the bones and the menisci, but lines the **suprapatellar bursa**.

See Dissection 23.3 (and Fig. 23.7) for instructions on continuing dissection of the knee joint.

Intracapsular ligaments of the knee joint

Cruciate ligaments

These ligaments cross each other like the letter 'X', which accounts for their name.

DISSECTION 23.2 Knee joint-2

Objective

I. To open the knee joint by incising the quadriceps tendon.

Instructions

1. Cut across the quadriceps tendon, immediately proximal to the patella. Carry the ends of this incision downwards to the tibial condyles, passing 2–3 cm on either side of the patellar tendon. Turn

the patella down, and expose the cavity of the knee joint [Fig. 23.5].

2. Lift the tendon of the quadriceps, and note that the cavity of the joint extends upwards, deep to it, to form the suprapatellar bursa. Split the lower part of the quadriceps longitudinally, and examine the extent of the bursa.

3. Flex and extend the joint. Note the type of movement which occurs between the tibia and the femur.

4. Examine the infrapatellar fold [Fig. 23.6].

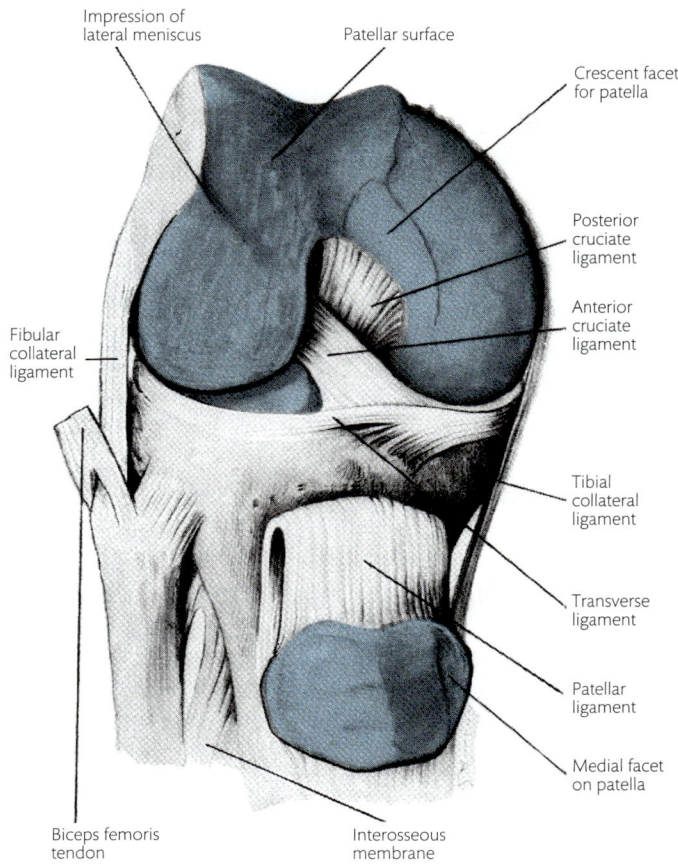

Impression of
lateral meniscus

Patellar surface

Crescent facet
for patella

Posterior
cruciate
ligament

Anterior
cruciate
ligament

Fibular
collateral
ligament

Tibial
collateral
ligament

Transverse
ligament

Patellar
ligament

Medial facet
on patella

Biceps femoris
tendon

Interosseous
membrane

Fig. 23.5 Dissection of the right knee from the front—the patella and patellar ligament turned down.

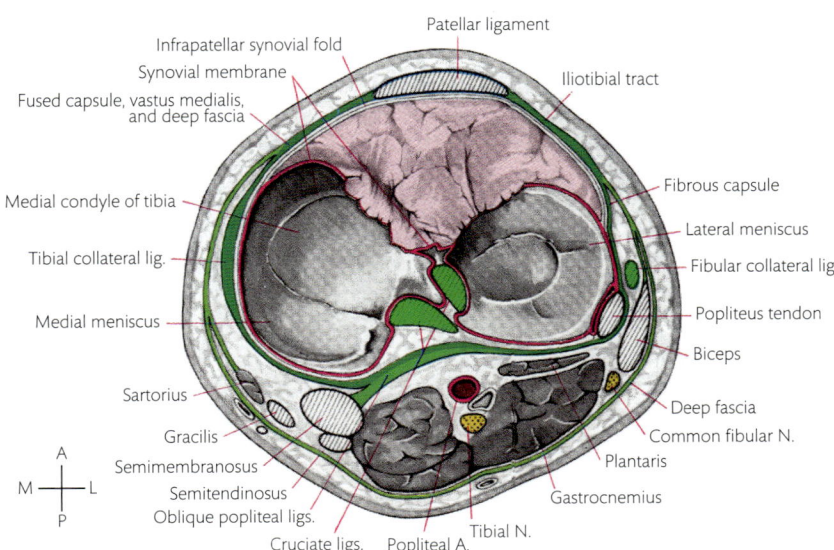

Patellar ligament

Infrapatellar synovial fold

Synovial membrane

Iliotibial tract

Fused capsule, vastus medialis,
and deep fascia

Medial condyle of tibia

Fibrous capsule

Lateral meniscus

Tibial collateral lig.

Fibular collateral lig.

Medial meniscus

Popliteus tendon

Biceps

Sartorius

Deep fascia

Gracilis

Common fibular N.

Semimembranosus

Plantaris

Semitendinosus

Gastrocnemius

Oblique popliteal ligs.

Tibial N.

Cruciate ligs.

Popliteal A.

A
M ⊢ L
P

Fig. 23.6 Transverse section through the right knee joint and its surroundings, showing the synovial membrane, red; fascia and ligaments, green. A = anterior; P = posterior; M = medial; L = lateral.

Attachment: the **anterior cruciate ligament** passes from the anterior part of the intercondylar area of the tibia [Fig. 23.1] to the posterior part of the medial surface of the lateral condyle of the femur [Fig. 23.7].

Attachment: the **posterior cruciate ligament** passes from the posterior part of the tibial intercondylar area [Figs 23.1, 23.7] to the anterior part of the lateral surface of the medial condyle of the femur. It receives one or more slips from the posterior part of the lateral meniscus (**meniscofemoral ligaments**).

The cruciate ligaments hold the femur to the tibia and prevent it from sliding forwards (posterior cruciate) or backwards (anterior cruciate) on the flat upper surface of the tibia. Both ligaments remain relatively tight throughout flexion and extension of the knee joint, but do not prevent these movements because their attachments to the femur are close to the axis of the movements.

Towards the end of extension, the anterior cruciate ligament becomes tight and prevents the lateral femoral condyle from sliding back on the tibia. At

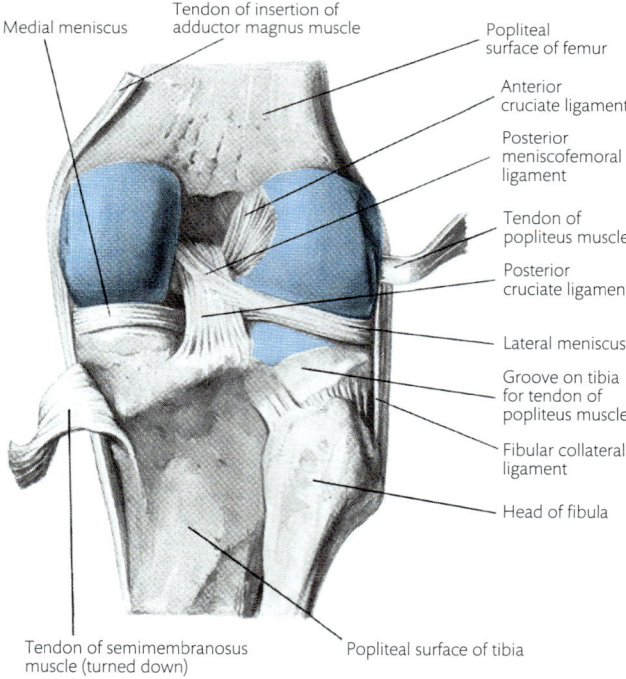

Medial meniscus
Tendon of insertion of adductor magnus muscle
Popliteal surface of femur
Anterior cruciate ligament
Posterior meniscofemoral ligament
Tendon of popliteus muscle
Posterior cruciate ligament
Lateral meniscus
Groove on tibia for tendon of popliteus muscle
Fibular collateral ligament
Head of fibula
Tendon of semimembranosus muscle (turned down)
Popliteal surface of tibia

Fig. 23.7 Right knee joint opened from behind by removing the posterior part of the articular capsule.

this stage of extension, the medial condyle continues to slide back on the tibia, producing medial rotation of the femur on the tibia. This process screws home the joint and 'locks' it in extended position.

Transverse ligament of the knee

The transverse ligament of the knee extends between the anterior margins of the medial and lateral menisci.

Menisci

The medial and lateral menisci are C-shaped plates of fibrocartilage which extend inwards from the articular capsule between the articular surfaces of the femur and tibia. They are thick at the periphery, but thin at the free concave edge internally [Fig. 23.1].

The ends of the 'C' of both menisci are attached to the median non-articular intercondylar area of the tibia. The synovial cavity extends over the thin internal edges of the menisci, and between them and the articular surfaces of the bones. The menisci are free to slide on these surfaces as far as their attachments to the articular capsule and tibia permit. The menisci deepen the articular surfaces on the tibia and help to spread the synovial fluid between the weight-bearing surfaces of the femur and tibia.

Attachments: the menisci are attached to the tibia at the following points: (1) to the intercondylar area by their fibrous extremities called **horns**; and (2) to the margins of the tibial condyles close to where the peripheral part is fused with the articular capsule of the knee joint. Anteriorly, they are connected to each other by the **transverse ligament of the knee** [Fig. 23.1]. Posterolaterally, the **tendon of the popliteus** lies between the lateral meniscus and the capsule.

The **lateral meniscus** is nearly circular (in keeping with the more spherical lateral condyle of the femur). The anterior and posterior horns are fibrous extensions from the ends of the meniscus which attach it to the superior surface of the tibia [see Fig. 23.1]. They lie close to each other in the intercondylar area of the tibia. The **medial meniscus** is elongated anteroposteriorly (in keeping with the shape and movements of the medial femoral condyle). Because of the shape of the menisci, the anterior and posterior horns of the medial meniscus are attached further apart on the anterior and posterior parts of the tibial intercondylar area [Fig. 23.1].

A fibrous slip from the posterior part of the lateral meniscus joins the posterior cruciate ligament and is attached to the medial condyle of the femur. This is the **meniscofemoral ligament** [Fig. 23.7]. It enables the lateral meniscus to move with the femur and, in so doing, makes it less vulnerable to mechanical injury than the medial meniscus. The medial meniscus is less free to move on the tibia than the lateral meniscus. This limited mobility and the greater anteroposterior movement of the medial condyle of the femur make the medial meniscus more likely to be trapped and injured between the moving surfaces of the tibia and femur. ➲ Medial meniscus injury occurs especially in sudden turning movements with the foot fixed on the ground (e.g. in sudden changes of direction when running). In such movements, the medial condyle of the femur pivots around the spherical lateral condyle, slides violently on the tibia while under pressure, and may cause the medial meniscus to tear or be wrenched from its attachment. If the torn piece of cartilage becomes wedged between the tibia and the femur, the joint becomes 'locked' (unable to move) because the ligaments cannot stretch sufficiently to allow the bones to be forced apart. (See Clinical Application 23.1.)

Relations of the knee joint

Anterior: the tendon of the quadriceps femoris, patella, and ligamentum patellae replace the fibrous capsule of the knee joint. In front of these lie the peripatellar plexus of nerves, the subcutaneous prepatellar bursa, and the superficial infrapatellar bursa.

Anteromedial and anterolateral to the knee joint lie the retinacula of the medial and lateral patellar retinacula, which are extensions of the fibres of the vastus medialis and lateralis, respectively. The iliotibial tract also lies anterolateral to the knee.

Medial: the inferior medial genicular artery and the medial collateral ligament form the immediate medial relation of the joint. The medial collateral ligament is overlapped by the saphenous nerve and long saphenous vein of the medial side. The tendons of the sartorius, gracilis, and semitendinosus and the bursa anserina lie in the inferior medial part of the knee.

Posterior: the posterior capsule of the knee joint is closely related to the popliteal vessels, tibial nerve, and the two heads of the gastrocnemius. Posteromedial to the joint are the semitendinosus and semimembranosus. Posterolateral is the tendon of the biceps femoris and the common peroneal nerve.

Lateral: the tendon of the popliteus lies immediately lateral to the knee joint. The fibular collateral ligament lies lateral to it with the inferior lateral genicular artery.

Movements of the knee joint

The main movements possible at the knee are flexion, which is limited only by contact between the leg and the thigh, and extension. Extension stops when the thigh and leg are in a straight line. In this position, the joint is firmly 'locked'. The anterior cruciate, tibial and fibular collateral ligaments, oblique popliteal ligament, and the posterior part of the capsule are taut, and the leg and thigh are converted into a rigid column. Medial rotation of the femur occurs during the end of extension, and lateral rotation during the start of flexion [Fig. 23.8].

Figs 23.8A, B, and C show movement of the femoral condyles on the articulating surface of the tibia. The shadow of the medial femoral condyle (continuous line) and the lateral femoral condyle (interrupted line) are marked. To appreciate the movements of the knee, note that, in full flexion, the posterior surfaces of the femoral condyles articulate with the tibia. In this position, the patella articulates with the crescent-shaped facet on the medial condyle of the femur. The femur is rotated laterally, as evidenced by the relative position of the medial condyle (continuous line) to the lateral condyle (interrupted line) [Figs 23.5, 23.8A]. As the knee is extended, the femoral condyles roll forwards and slide backwards on the tibial condyles. (The sliding back is essential, so that the femur can continue to articulate with the limited area for articulation on the tibia.) The points of contact of the femur with the tibia move steadily forwards on the femoral condyles [Fig. 23.8B]. When the lateral condyle reaches its maximum extension (an event which occurs earlier in the medial condyle), the groove between its patellar and tibial surfaces [Fig. 23.5] comes into contact with the lateral meniscus and the condylar movement is stopped by the taut, fully stretched anterior cruciate ligament. Because the anteroposterior curvature of the medial femoral condyle is flatter anteriorly than the more spherical lateral condyle, the medial condyle continues to slide backwards, after the lateral condyle has stopped. Extension is completed by medial rotation of the femur [Fig. 23.8C]. This screws the femur home on the tibia, tightens the ligaments, and 'locks' the knee joint.

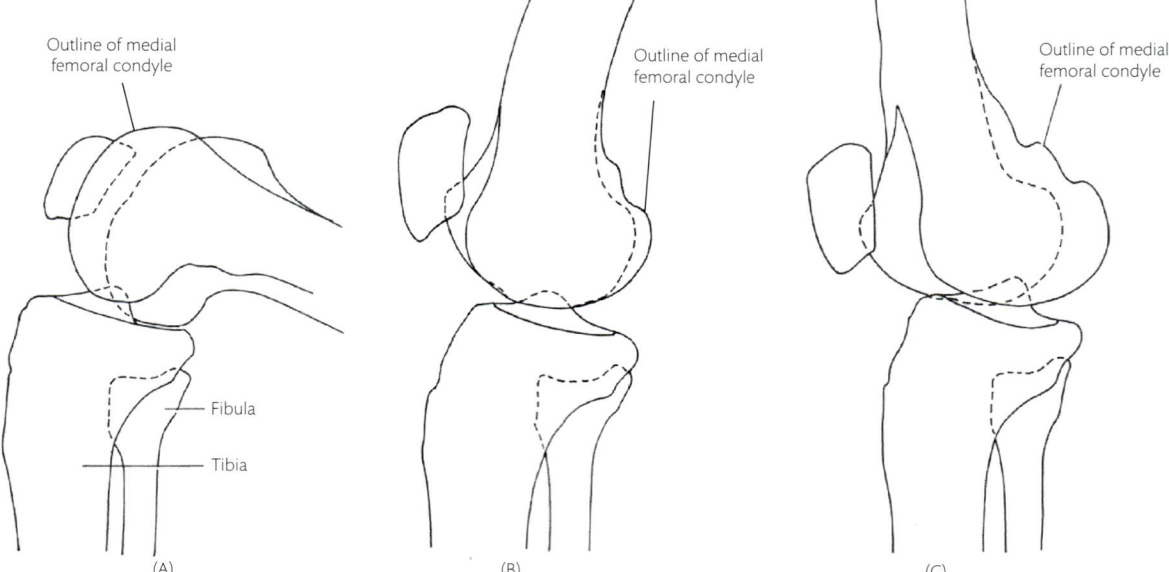

Fig. 23.8 Tracings of three positions of the right knee joint taken from radiographs, showing three phases of flexion of the knee with the foot firmly fixed on the ground. In this way, there is no movement of the tibia and fibula, and the full effects of rotation are visible in the femur. The tracings are viewed from the medial side; the parts that are hidden by the more medial structures are shown as broken lines. (A) Considerable flexion of the knee with lateral rotation of the femur on the tibia. (B) Slight extension. Note that there has been considerable medial rotation of the femur, so that the outlines of the condyles are nearly superimposed. This movement occurs at the very outset of flexion. (C) Position of full extension with the femur fully rotated medially and the knee joint locked.

Flexion is produced by the same movements in the reverse order. It begins with lateral rotation of the femur because of differences in shape of the two condyles. This 'unlocks' the knee joint, an action done by the **popliteus**. Once the knee is unlocked, flexion proceeds by the action of the knee flexors.

Rotation of the flexed joint may be produced independently of flexion and extension by the muscles at the sides of the joint. The biceps femoris is the principal lateral rotator of the tibia; the sartorius, gracilis, semitendinosus, and semimembranosus are the main medial rotators.

In addition to the movements between the femur and the tibia, during extension, the patella rises to progressively higher levels on the patellar surface of the femur [Fig. 23.8]. Because the pull of the quadriceps on the patella is parallel to the obliquity of the femur, the patella tends to deviate laterally as it ascends. This lateral displacement of the patella is prevented by: (1) the greater anterior projection of the lateral femoral condyle; and (2) the lowest fibres of the vastus medialis which are inserted horizontally into the medial surface of the patella. These factors help to keep the pull of the quadriceps at right angles to the axis of flexion and extension of the knee.

In learning these details, you should not lose sight of the fact that the knee joint is essentially a hinge joint. The axis for the movement passes through an almost straight line through the femoral attachments of the collateral and cruciate ligaments. Study the anteroposterior and lateral X-rays of the knee joint by using Fig. 23.9.

Tables 23.1 and 23.2 summarize the muscles and movements of the knee joint. The origin, insertion, nerve supply, and action of muscles acting on the knee are listed in Table 23.1. (Only details of muscle attachments that are relevant to the understanding of the muscle actions are listed.)

Blood supply of the knee joint

The knee joint is supplied by: (1) five genicular branches from the popliteal artery; (2) the descending genicular branch of the femoral artery; (3) the genicular branch of the lateral circumflex artery; and (4) recurrent branches from the anterior and posterior tibial arteries. The middle genicular artery pierces the posterior capsule to supply the intracapsular structures. The other arteries communicate with each other to form the anastomosis around the knee joint [see Fig. 22.8].

Nerve supply of the knee joint

1. The **femoral nerve** supplies the anterosuperior part of the joint through the nerves to the three vasti muscles. The nerve from the vastus medialis accompanies the descending genicular artery.

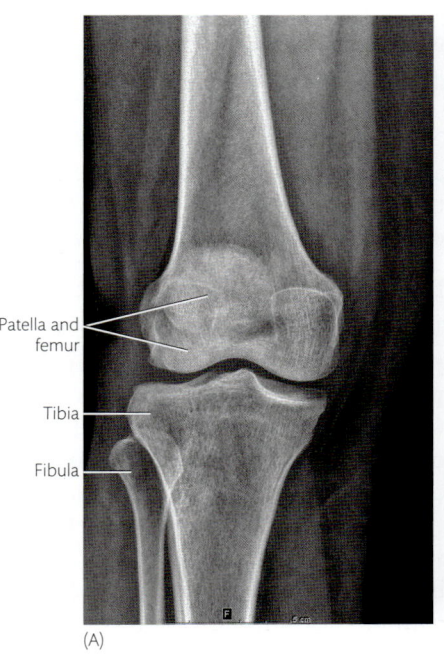
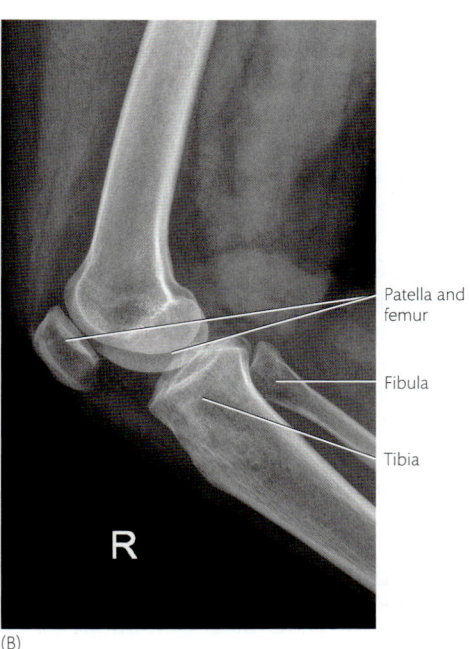

(A)   (B)

Fig. 23.9 Normal X-ray of the knee. (A) Anteroposterior view. (B) Lateral view.

Table 23.1 Muscles acting on the knee joint

Muscle	Origin	Insertion	Action	Nerve supply
Biceps femoris, long head	Ischial tuberosity	Fibula, head	Flexion Lateral rotation of leg	Sciatic, tibial part
Semimembranosus	Ischial tuberosity	Tibia, medial condyle	Flexion Medial rotation of leg	Sciatic, tibial part
Semitendinosus	Ischial tuberosity	Tibia, medial surface	Flexion Medial rotation of leg	Sciatic, tibial part
Gracilis	Pubis, body, and inferior ramus	Tibia, medial surface	Flexion Medial rotation of leg	Obturator
Sartorius	Ilium, anterior superior spine	Tibia, medial surface	Flexion Medial rotation of leg	Femoral
Tensor fasciae latae	Ilium, crest	Tibia, lateral condyle	Extension Stabilizes pelvis on thigh	Superior gluteal
Gluteus maximus, superficial three-quarters	Ilium, posterior to posterior gluteal line Sacrum Sacrotuberous ligament	Tibia, lateral condyle	Extension Stabilizes pelvis on thigh	Inferior gluteal
Rectus femoris	Ilium, anterior inferior spine, area above acetabulum	Patella	Extension	Femoral
Vastus medialis	Femur, intertrochanteric line, linea aspera, medial supracondylar line, tendon of adductor magnus	Patella, capsule of knee joint	Extension Medial displacement of patella	Femoral
Vastus intermedius	Femur, body anterior and lateral surfaces	Patella, suprapatellar bursa	Extension Elevates suprapatellar bursa	Femoral
Vastus lateralis	Femur, greater trochanter, linea aspera	Patella, capsule of knee joint	Extension	Femoral
Biceps femoris, short head	Linea aspera	Head of fibula	Flexion Lateral rotation of leg	Sciatic, common fibular part
Popliteus	Femur, lateral condyle	Tibia, posterior surface	Medial rotation of leg Flexion	Tibial
Gastrocnemius	Femur, lateral condyle, medial condyle	Calcaneus	Flexion	Tibial
Plantaris	Femur, lateral condyle	Calcaneus	Acts with gastrocnemius	Tibial

2. The **common peroneal nerve** supplies the lateral part of the joint through the superior and inferior lateral genicular nerves and the recurrent genicular nerve.
3. The **tibial nerve** supplies the medial and posterior parts of the articular capsule and the central structures within the capsule through the superior medial, middle, and inferior medial genicular nerves. They run with the corresponding arteries [see Fig. 22.8].
4. The **obturator nerve** sends a branch to the posterior surface of the knee joint.

Bursae around the knee joint

The many tendons crossing the knee joint are protected from wear and tear by bursae between them and the surfaces over which they move. In addition, there are bursae between the skin and the deeper tissues in front of the knee joint.

Bursae on the anterior aspect

Two subcutaneous bursae—the **subcutaneous prepatellar bursa** and the **superficial infrapatellar bursa**—are located deep to the skin in front of the knee. The two deep bursae in front of the knee are the **deep infrapatellar bursa**, between the ligamentum patellae and the tibial tuberosity, and the **suprapatellar bursa**. The suprapatellar bursa is a large consistent bursa between the femur and the quadriceps tendon. It is in direct communication with the joint cavity [Fig. 23.2].

Table 23.2 Movements at the knee joint

Movement	Muscles	Nerve supply
Flexion	Semimembranosus	Sciatic, tibial part
	Semitendinosus	Sciatic, tibial part
	Biceps femoris, long head	Sciatic, tibial part
	Biceps femoris, short head	Sciatic, common fibular part
	Gracilis	Obturator
	Sartorius	Femoral
	Popliteus	Tibial
	Gastrocnemius	Tibial
Extension	Vastus medialis	Femoral
	Vastus lateralis	Femoral
	Vastus intermedius	Femoral
	Rectus femoris	Femoral
	Gluteus maximus	Inferior gluteal
	Tensor fasciae latae	Superior gluteal
Lateral rotation of leg	Biceps femoris	Sciatic, tibial and common fibular parts
	Gluteus maximus, anterior fibres	Inferior gluteal
	Tensor fasciae latae	Superior gluteal
Medial rotation of leg when knee extended or lateral rotation of thigh when knee flexed	Popliteus	Tibial
	Semimembranosus	Sciatic, tibial part
	Semitendinosus	Sciatic, tibial part
	Gracilis	Obturator
	Sartorius	Femoral

Bursae on the posterior aspect

Two bursae are found deep to the medial and lateral heads of the **gastrocnemius**. Another bursa lies deep to the insertion of the **semimembranosus** and may communicate with the knee joint. ⮕ Painful swelling and inflammation of this bursa is known as a popliteal cyst.

Bursae on the medial aspect

On the medial side of the knee, there is a bursa deep to the tendons of the sartorius, gracilis, and semitendinosus—the **bursa anserina**—and a bursa between the medial collateral ligament and the **semimembranosus**.

Bursae on the lateral aspect

There are bursae deep to the **fibular collateral ligament**, between this ligament and the **tendon of the popliteus**, and between the ligament and the **tendon of the biceps femoris**. One more bursa lies between the tendon of the popliteus and the lateral condyle of the femur.

Osteoarthritis

⮕ Osteoarthritis is a degenerative joint disease. In osteoarthritis of the knee, the cartilage covering the articular surfaces of the femur and tibia gradually wears away. As it wears away, it becomes frayed and rough. The joint space is reduced, and the underlying bones are exposed. The exposed bones rub against each other during movements of the joint and cause pain. Osteophytes (or bone spurs) grow on the exposed bone.

See Clinical Application 23.1 for injury of the knee joint.

Ankle joint

Type and articular surfaces

The ankle joint is a strong **synovial hinge joint** between the inferior aspect and the medial malleo-

Capsule of the ankle joint

As the ankle joint is a hinge joint, the main ligaments are lateral and medial. The anterior part of the fibrous capsule is thin and consists mainly of transverse fibres. It extends from the anterior margin of the distal end of the tibia to the superior surface of the neck of the talus. The posterior part of the capsule extends from the posterior margin of the distal end of the tibia and the posterior tibiofibular ligament to the posterior surface of the body of the talus.

Ligaments

Medial (deltoid) ligament

This very strong ligament radiates from the distal border of the medial malleolus to the medial side of the talus, the sustentaculum tali, the medial edge of the plantar calcaneonavicular (spring) ligament, the navicular bone, and the neck of the talus [Fig. 23.10]. As such, the medial ligament not only strengthens the ankle joint, but also holds the calcaneus and the navicular against the talus.

Lateral ligament

The lateral ligament consists of three bands—the **anterior talofibular ligament**, the **posterior**

lus of the tibia, the lateral malleolus of the fibula, and the trochlea of the talus.

See Dissection 23.4 and Figs 23.10 and 23.11.

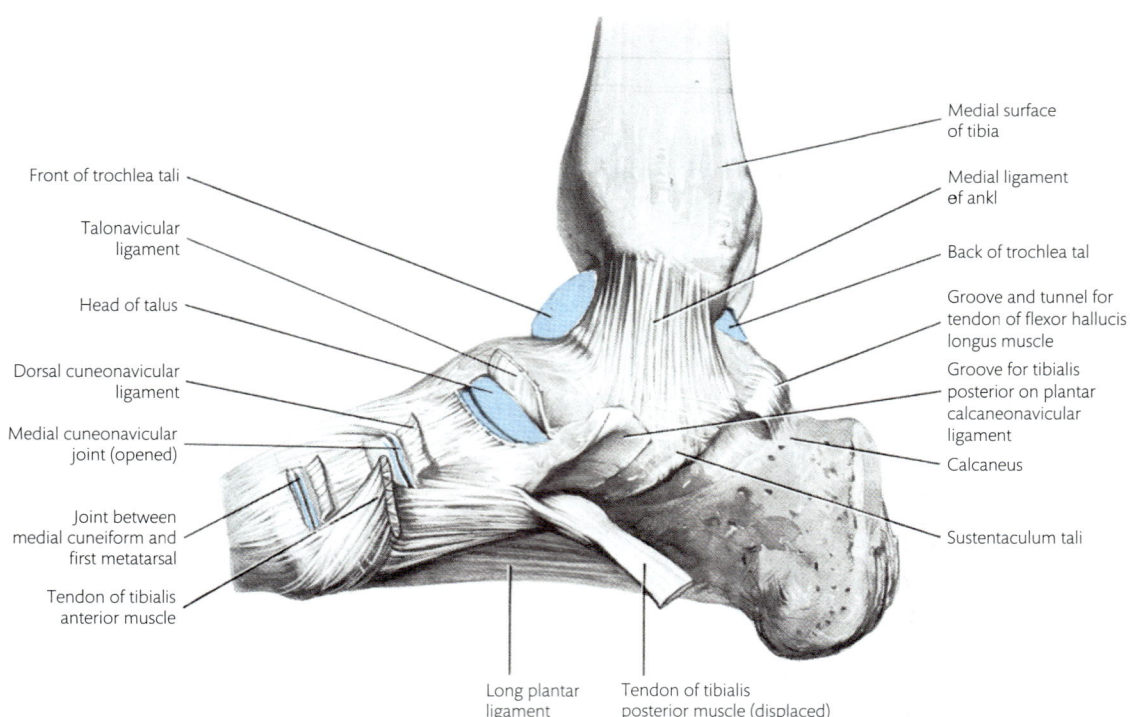

Medial surface of tibia

Medial ligament ef ankl

Back of trochlea tal

Groove and tunnel for tendon of flexor hallucis longus muscle

Groove for tibialis posterior on plantar calcaneonavicular ligament

Calcaneus

Sustentaculum tali

Front of trochlea tali

Talonavicular ligament

Head of talus

Dorsal cuneonavicular ligament

Medial cuneonavicular joint (opened)

Joint between medial cuneiform and first metatarsal

Tendon of tibialis anterior muscle

Long plantar ligament

Tendon of tibialis posterior muscle (displaced)

Fig. 23.10 Ankle joint and tarsal joints from the medial side.

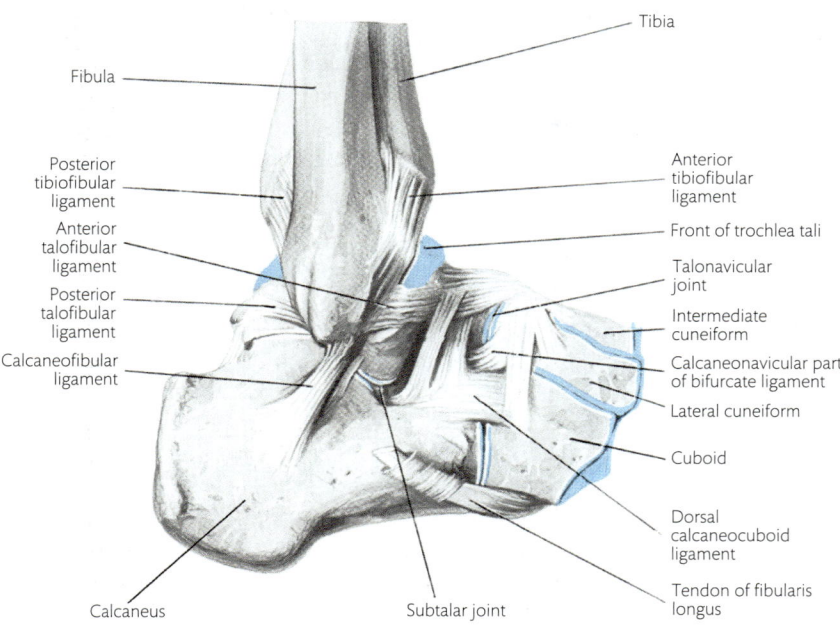

Labels on figure:
- Fibula
- Tibia
- Posterior tibiofibular ligament
- Anterior tibiofibular ligament
- Anterior talofibular ligament
- Front of trochlea tali
- Posterior talofibular ligament
- Talonavicular joint
- Calcaneofibular ligament
- Intermediate cuneiform
- Calcaneonavicular part of bifurcate ligament
- Lateral cuneiform
- Cuboid
- Dorsal calcaneocuboid ligament
- Tendon of fibularis longus
- Calcaneus
- Subtalar joint

Fig. 23.11 Ligaments of the lateral side of the ankle joint and the dorsum of the tarsus.

talofibular ligament, and the **calcaneofibular ligament**. The anterior and posterior ligaments are thickenings of the fibrous capsule. The **anterior talofibular ligament** passes anteriorly from the anterior border of the lateral malleolus to the neck of the talus. The posterior talofibular ligament is much stronger and runs back from the fossa of the lateral malleolus to the **posterior tubercle of the talus** [Figs 23.11, 23.12]. The calcaneofibular ligament is a round cord which passes inferiorly from the distal end of the lateral malleolus to the lateral surface of the calcaneus [Figs 23.11, 23.12]. It is separate from the articular capsule of the ankle joint and functions also as a ligament of the talocalcanean or subtalar joint which it crosses.

Synovial membrane

The synovial membrane lines the fibrous capsule but is separated from it by fat pads which lie deep to the anterior and posterior parts of the capsule. There is a short extension of the synovial membrane between the tibia and the fibula, inferior to the thickened lower end of the interosseous membrane.

Relations

All structures passing from the leg into the foot (except the tendo calcaneus) lie close to the ankle joint. Tendons, vessels, and nerves from the anterior compartment lie on the anterior surface; those from the posterior compartment lie on the posteromedial surface, and those from the lateral compartment (peroneal tendons) lie on the posterolateral surface [Fig. 23.13].

Movements and muscles involved

Plantar flexion and **dorsiflexion** are the only significant movements at the ankle joint. The trochlea of the talus is wide anteriorly, and the socket formed by the malleoli is also broader at the front. Thus, the talus fits tightly into the socket when the foot is dorsiflexed. When the foot is plantar flexed, the talus is slightly loose and some lateral movement is possible.

Tables 23.3 and 23.4 summarize the muscles and movements of the ankle joint.

The ankle joint is stable because of: (1) the powerful ligaments and tendons around it; and (2) the relatively great depth of the socket formed by the medial and lateral malleoli within which the trochlea of the talus is fitted. The socket offers flexibility because of the body of the fibula. If the talus is forced laterally, the lateral malleolus moves outwards (a movement accompanied by medial movement of the shaft of the fibula acting with the tibiofibular ligaments as a fulcrum). In extreme cases, this may lead to fracture of the fibula in the leg. The socket of the ankle joint is deepened posteriorly by the inferior part of the posterior tibiofibular ligament (transverse tibiofibular ligament [Fig. 23.12]).

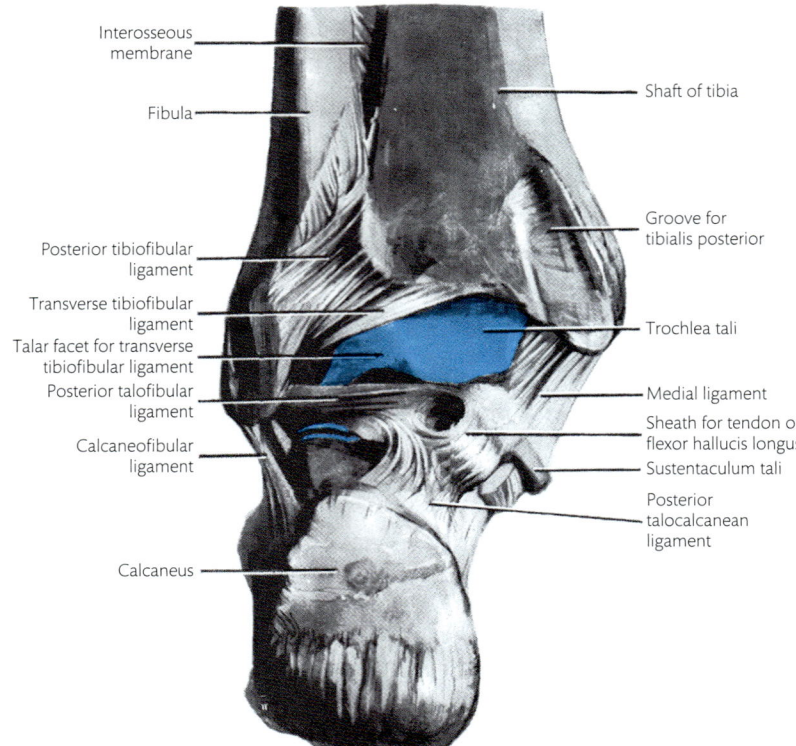

Fig. 23.12 Ankle joint dissected from behind.

Interosseous membrane

Fibula

Shaft of tibia

Posterior tibiofibular ligament

Groove for tibialis posterior

Transverse tibiofibular ligament

Talar facet for transverse tibiofibular ligament

Posterior talofibular ligament

Calcaneofibular ligament

Trochlea tali

Medial ligament

Sheath for tendon of flexor hallucis longus

Sustentaculum tali

Posterior talocalcanean ligament

Calcaneus

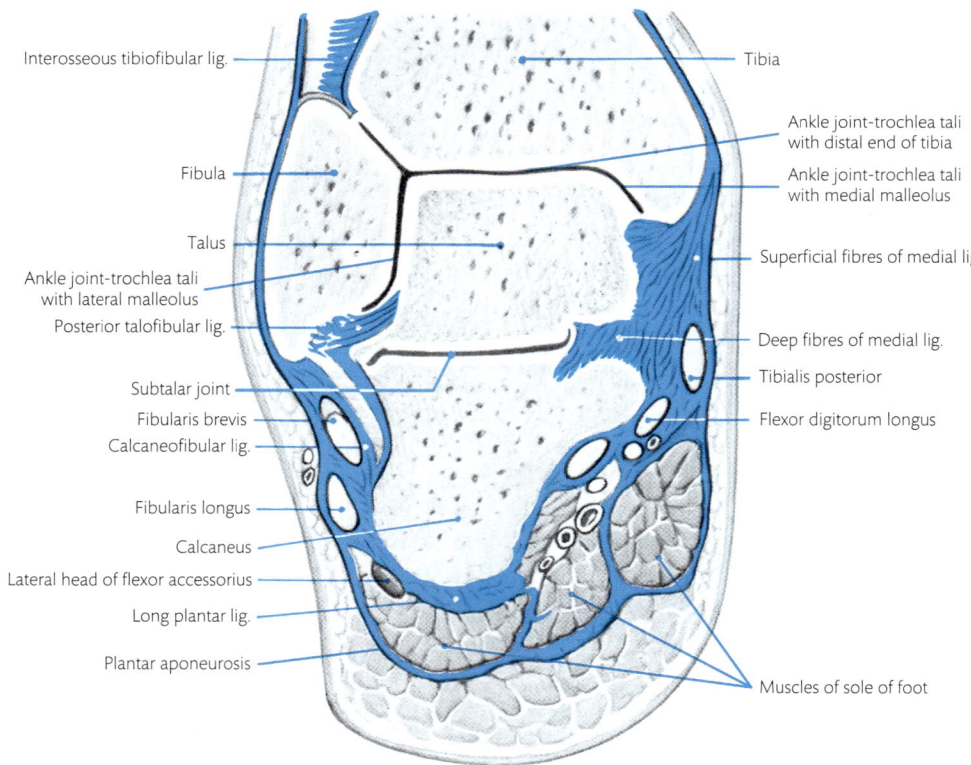

Fig. 23.13 Oblique coronal section through the ankle and subtalar joints.

Interosseous tibiofibular lig.

Tibia

Ankle joint-trochlea tali with distal end of tibia

Ankle joint-trochlea tali with medial malleolus

Fibula

Talus

Ankle joint-trochlea tali with lateral malleolus

Posterior talofibular lig.

Superficial fibres of medial lig.

Deep fibres of medial lig.

Tibialis posterior

Subtalar joint

Fibularis brevis

Calcaneofibular lig.

Flexor digitorum longus

Fibularis longus

Calcaneus

Lateral head of flexor accessorius

Long plantar lig.

Plantar aponeurosis

Muscles of sole of foot

Table 23.3 Muscles acting on the ankle joint

Muscle	Origin	Insertion	Action	Nerve supply
Gastrocnemius	Femur, lateral condyle, medial condyle	Calcaneus	Plantar flexion	Tibial
Plantaris	Femur, lateral condyle	Calcaneus	Plantar flexion	Tibial
Soleus	Tibia, soleal line and medial border Fibula, posterior surface	Calcaneus	Plantar flexion	Tibial
Peroneus longus	Fibula, lateral surface	First metatarsal, base Medial cuneiform	Plantar flexion	Superficial fibular
Peroneus brevis	Fibula, lateral surface	Fifth metatarsal, base	Plantar flexion	Superficial fibular
Tibialis anterior	Tibia, lateral surface Interosseous membrane	First metatarsal, base Medial cuneiform	Dorsiflexion	Tibial
Tibialis posterior	Interosseous membrane Adjacent parts of tibia and fibula	Navicular, tuberosity Medial cuneiform All tarsals, except talus Metatarsal bases, 2–4	Plantar flexion	Tibial
Peroneus tertius	Fibula, anterior surface	Fifth metatarsal base	Dorsiflexion	Deep fibular
Extensor hallucis longus	Fibula, anterior surface	Hallux, base of distal phalanx	Ankle, dorsiflexion Tarsal, inversion Hallux, extension	Deep fibular
Extensor digitorum longus	Fibula, anterior surface	Extensor expansion, toes 2–5	Ankle, dorsiflexion Toes 2–5, extension all joints	Deep fibular
Flexor hallucis longus	Fibula, posterior surface	Hallux, distal phalanx, base	Ankle, plantar flexion Hallux, flexion all joints	Tibial
Flexor digitorum longus	Tibia, posterior surface	Terminal phalanges, toes 2–5	Ankle, plantar flexion Toes 2–5, flexion all joints	Tibial

Blood supply and nerve supply

The arteries around the ankle joint form as anastomosis around it. On the lateral side, the lateral malleolar branch of the anterior tibial artery and the lateral tarsal branch of the dorsalis pedis artery anastomose with the perforating and terminal branches of the fibular artery. On the medial side, the medial malleolar branch of the anterior tibial artery anastomoses with the medial calcanean branches of the posterior tibial artery. The posterior tibial artery also anastomoses with the fibular artery, posterior to the ankle joint.

The deep peroneal and tibial nerves send articular branches to the ankle joint.

Table 23.4 Movements at the ankle joint

Movement	Muscles	Nerve supply
Plantar flexion	Gastrocnemius	Tibial
	Soleus	Tibial
	Plantaris	Tibial
	Tibialis posterior	Tibial
	Flexor digitorum longus	Tibial
	Flexor hallucis longus	Tibial
	Peroneus longus	Superficial fibular
	Peroneus brevis	Superficial fibular
Dorsiflexion	Tibialis anterior	Deep fibular
	Extensor hallucis longus	Deep fibular
	Extensor digitorum longus	Deep fibular
	Peroneus tertius	Deep fibular

Tibiofibular joints

The tibia and fibula articulate with each other at the two ends and along the shaft.

Proximal tibiofibular joint

Type and articular surfaces

This joint is between the head of the fibula and the posteroinferior surface of the lateral condyle of the tibia.

Articular capsule

The articular capsule is attached at the articular margins and is strengthened anteriorly and posteriorly by fibres running down and laterally from the tibia to the head of the fibula.

Relations

The fibular collateral ligament of the knee joint and the tendon of the biceps femoris cross the lateral surface of the joint. The tendon of the popliteus with a synovial extension from the knee joint crosses the posteromedial surface. At this point, the synovial cavities of the two joints may communicate.

Nerve supply

Nerve supply is from the nerve to the popliteus and the recurrent genicular nerve.

Blood supply

Blood supply is from the inferior lateral genicular and recurrent genicular arteries.

See Dissection 23.5.

Interosseous membrane of the leg

The strong interosseous membrane stretches between the interosseous borders of the tibia and fibula [see Figs 16.10, 16.11]. It consists of fibres which run downwards and laterally from the tibia to the fibula. There is an oval opening in the upper part of the membrane for the **anterior tibial vessels**, and a small opening lower down for the **perforating branch of the fibular artery**. The tibialis posterior, the flexor hallucis longus, and all the muscles of the anterior compartment of the leg originate, in part, from the interosseous membrane.

Distal tibiofibular joint

Type

The inferior tibiofibular joint is a **fibrous joint**. The tibia and fibula are firmly bound together by a strong **interosseous tibiofibular ligament** which extends from the longitudinal groove on the lateral side of the tibia to the rough medial side of the distal part of the fibula.

Ligaments

The interosseous tibiofibular ligament is strengthened and hidden by the **anterior** and **posterior tibiofibular ligaments**. These pass upwards and medially from the corresponding surfaces of the lateral malleolus to the distal end of the tibia and are continuous with the fibrous capsule of the ankle joint. The **transverse tibiofibular ligament** passes from the malleolar fossa of the fibula to the inferior margin of the posterior surface of the tibia. It forms part of the socket for the ankle joint [Fig. 23.12] and articulates with the body of the talus.

Movements

Only a small amount of movement of the fibula on the tibia is possible. Slight medial movement of the shaft of the fibula occurs when the lateral malleolus is forced laterally by the trochlea of the talus. This movement takes place around the powerful tibiofibular ligaments and increases the stability of the ankle joint.

Joints of the foot

Small joints exist between the tarsal, metatarsal, and phalangeal bones. They are the: (1) intertarsal joints; (2) tarsometatarsal joints; (3) intermetatarsal joints; (4) metatarsophalangeal joints; and (5) interphalangeal joints.

Talocalcaneonavicular joint

Type and articulating surfaces

This is a complex **'ball-and-socket' type of synovial joint**. The ball is formed by the head and adjacent part of the body of the talus. The socket

---

DISSECTION 23.5 Tibiofibular joints

Objective

I. To study the superior tibiofibular joint and the interosseous membrane.

Instructions

1. Define the anterior and posterior tibiofibular ligaments of the inferior tibiofibular joint [Fig. 23.12].
2. Strip the muscles from the front of the interosseous membrane, and define its attachments. Follow it as far inferiorly as possible. The posterior surface may also be exposed.
3. Define the fibrous capsule of the proximal tibiofibular joint. Reflect the tendon of the popliteus from its posterior surface.
4. Note that the synovial extension from the knee joint deep to the popliteus may be in continuity with the cavity of the proximal tibiofibular joint.
5. Open the joint, and confirm its synovial nature.

is formed by the proximal surface of the navicular bone, the plantar calcaneonavicular ligament, the anteromedial part of the calcaneus, and the sustentaculum tali [Figs 23.14, 23.15, 23.16].

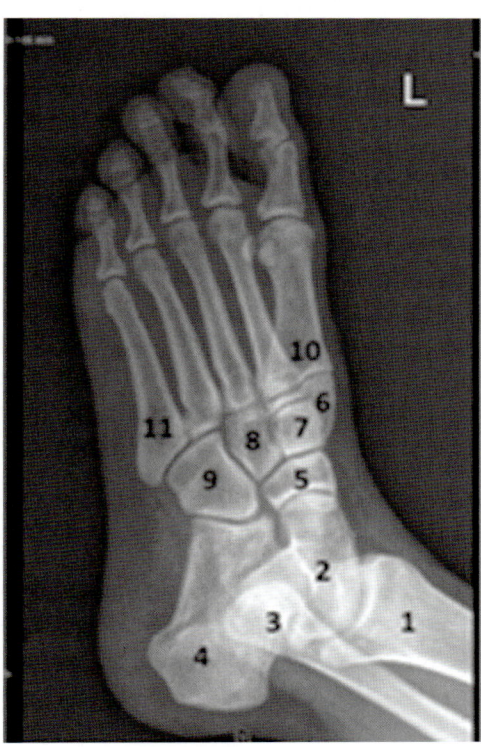

(A)

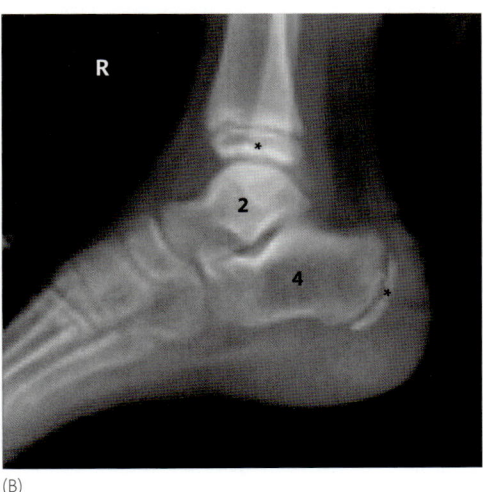

(B)

Fig. 23.14 (A) Oblique radiograph of the left foot. 1. Distal end of the tibia. 2. Talus. 3. Distal end of the fibula. 4. Calcaneus. 5. Navicular. 6. Medial cuneiform. 7. Intermediate cuneiform. 8. Lateral cuneiform. 9. Cuboid. 10. Base of the first metatarsal. 11. Base of the fifth metatarsal. (B) Ankle X-ray of a person younger than 14–16 years of age. The epiphyses for the posterior aspect of the calcaneus and the distal end of the tibia (*) have not yet united.

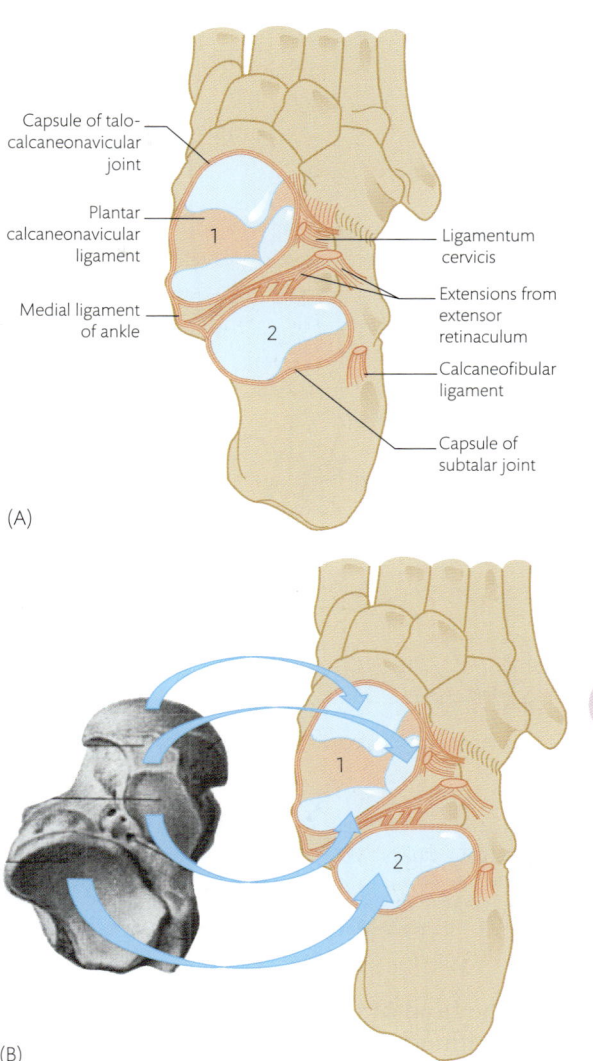

(A)

(B)

Fig. 23.15 (A) The ligaments and inferior articular surfaces of the subtalar and talocalcaneonavicular joints as seen from above after removal of the talus. (B) The arrows connect articulating surfaces of the talus to articular facets of the calcaneus and navicular. 1. Talocalcaneonavicular joint. 2. Subtalar joint.

Fibrous capsule

A single fibrous capsule encloses all components of the talocalcaneonavicular joint. The capsule is continuous with the plantar calcaneonavicular ligament and is strengthened by ligaments.

Ligaments

The **plantar calcaneonavicular** ('spring') **ligament** is a thick, triangular ligament which stretches from the sustentaculum tali to the plantar surface of the navicular bone [see Fig. 22.25]. It fills the triangular gap between these bones. Its tough fibrocartilaginous upper surface articulates

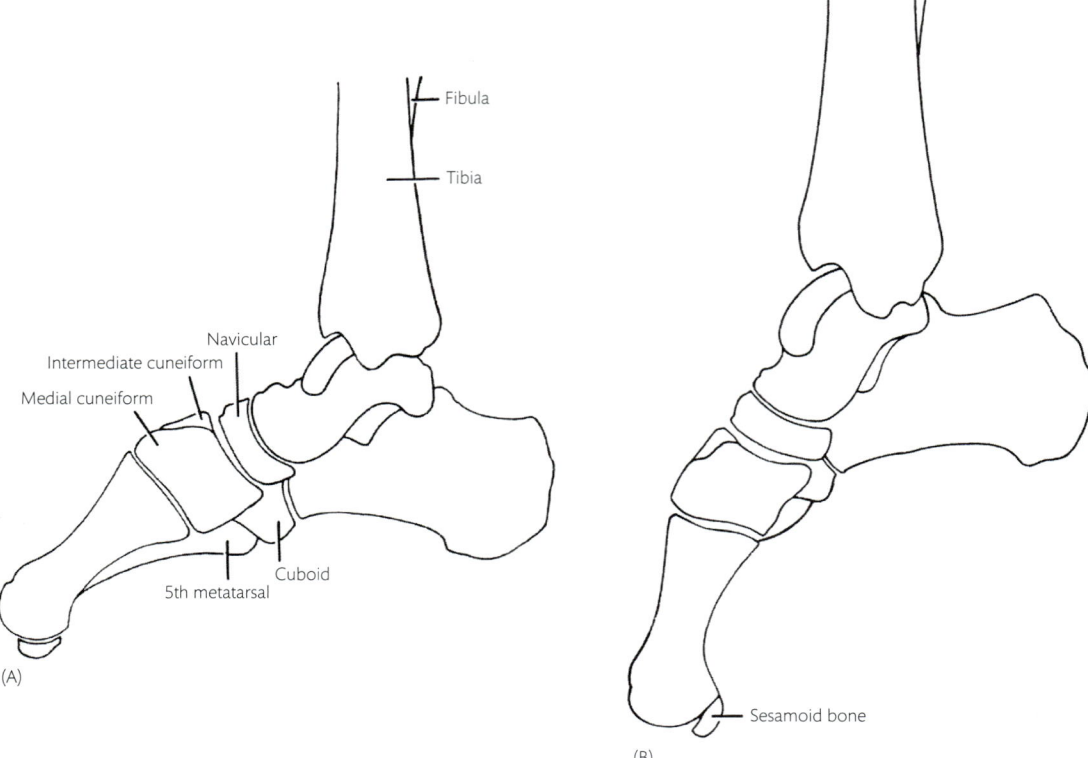

Fibula

Tibia

Navicular

Intermediate cuneiform

Medial cuneiform

Cuboid

5th metatarsal

(A)

Sesamoid bone

(B)

Fig. 23.16 Outline drawings of two radiographs of the same foot. In (B), the foot has been plantar flexed without any other movement. Note that virtually all of this movement takes place at the ankle joint, but that there is also slight flexion of the bones forming the medial longitudinal arch. The distance between the medial process of the tuber calcanei and the head of the first metatarsal is 7 mm less in (B) than in (A). This movement, produced by the plantar flexors, increases the height of the longitudinal arch.

with the head of the talus [Fig. 23.15]. Medially, it is continuous with the fibrous capsule of the talocalcaneonavicular joint and the **medial ligament of the ankle joint**. Inferiorly, the tendon of the tibialis posterior supports it. This ligament plays an important part in maintaining the medial longitudinal arch of the foot and prevents the development of flat foot.

The anterior part of the superior surface of the calcaneus, lateral to its articulation with the talus, is bound to the lateral surface of the navicular and to the dorsal surface of the cuboid by the **bifurcate ligament**. The two parts of the bifurcate ligament are the **calcaneocuboid** and **calcaneonavicular** parts [Fig. 23.11].

Movements possible

This joint moves with the subtalar joint in inversion and eversion. The axis for these movements passes upwards, forwards, and medially through

the calcaneus and the head of the talus. The talus remains stationary, whereas the other tarsal bones move around it.

The dissection of the joints of the foot is explained in Dissection 23.6.

DISSECTION 23.6 Joints of the foot-1

Objective

I. To demonstrate the ligaments and joints of the foot.

Instructions

1. Remove all muscles and tendons from the tarsal and metatarsal bones. Define the ligaments between the various bones.
2. Note that the ligaments on the plantar surfaces are much thicker than those on the dorsal surfaces.

Subtalar joint

Type and articulating surfaces

This is a **synovial plane joint**. It is a cylindrical joint between the lower surface of the body of the talus and the upper surface of the middle of the calcaneus [Figs 23.13, 23.15]. The joint surface of the talus is concave, and that of the calcaneus is convex. These curvatures run transversely, allowing the calcaneus to turn around on the inferior surface of the talus in movements of inversion and eversion. The fibrous capsule is attached near the margins of the articular surfaces [Fig. 23.15]. The deltoid and calcaneofibular ligaments of the ankle joint strengthen the capsule. The **ligament of the neck of the talus** (**ligamentum cervicis**) passes from the neck of the talus to the calcaneus [Fig. 23.15] and strengthens the joint.

Movements of the subtalar and talocalcaneonavicular joints

Acting together, the joints increase the resilience of the foot, but the main movements of inversion and eversion occur at the **subtalar** and **talocalcaneonavicular joints**. **Inversion** is the movement which raises the medial border of the foot and turns the sole medially. **Eversion** is the opposite movement which raises the lateral border of the foot and turns the sole inferolaterally. Eversion is more restricted than inversion. Inversion and eversion are important movements, while walking on uneven ground. The main evertors are the peroneus longus, brevis, and tertius. The main invertors are the tibialis anterior and posterior.

The dissection of the joints of the foot is continued in Dissection 23.7.

Tables 23.5 and 23.6 summarize the muscles and movements of the tarsal bones, mainly those occurring in the talocalcaneonavicular joint.

Calcaneocuboid joint

This is an oblique, saddle-shaped synovial joint which permits rotatory sliding of the cuboid on the distal surface of the calcaneus. This joint lies almost in the same transverse plane as the **talonavicular joint**. The two together constitute the **transverse tarsal joint**. Movement of the transverse tarsal joint adds little to the movements of inversion and eversion.

The calcaneocuboid joint is strengthened by the dorsal calcaneocuboid ligament, and the **long plantar** and **plantar calcaneocuboid ligaments** [see Figs 16.17, 22.25, 23.11]. The plantar calcaneocuboid ligament is attached close to the joint. The long plantar ligament lies inferior to the plantar calcaneocuboid ligament. It extends from the calcaneus to both lips of the groove for the peroneus longus tendon on the cuboid and to the bases of the middle three metatarsals. Both these ligaments are important in maintaining the lateral longitudinal arch of the foot.

Dissection of the foot is continued in Dissection 23.8.

Smaller intertarsal joints

The three cuneiform bones articulate with one another and with the distal surface of the navicular bone by a single joint. The lateral cuneiform and navicular articulate with the cuboid. The articular surfaces of all these joints are flat, and all have dorsal and plantar ligaments. The intercuneiform, cuneocuboid, and cuboideonavicular joints also have interosseous ligaments. These joints give resilience to the tarsal bones, but the amount of movement at each joint is small.

Table 23.5 Muscles acting on tarsal joints, particularly the talocalcaneonavicular joint

Muscle	Origin	Insertion	Action	Nerve supply
Peroneus longus	Fibula, lateral surface	First metatarsal, base Medial cuneiform	Eversion	Superficial fibular
Peroneus brevis	Fibula, lateral surface	Fifth metatarsal, base	Eversion	Superficial fibular
Tibialis anterior	Tibia, lateral surface Interosseous membrane	First metatarsal, base Medial cuneiform	Inversion	Deep fibular
Tibialis posterior	Interosseous membrane Adjacent parts of tibia and fibula	Navicular, tuberosity Medial cuneiform All tarsals, except talus Metatarsal, bases 2–4	Inversion	Tibial
Peroneus tertius	Fibula, anterior surface	Fifth metatarsal, base	Eversion	Deep fibular
Extensor hallucis longus	Fibula, anterior surface	Hallux, base distal phalanx	Inversion	Deep fibular
Extensor digitorum longus	Fibula, anterior surface	Extensor expansion, 2–5	Inversion	Deep fibular

Tarsometatarsal joints

The medial three metatarsals articulate with the three cuneiform bones. The lateral two metatarsals articulate with the cuboid bone [see Figs 16.14, 16.15]. All these joints have strong plantar and dorsal ligaments.

The joint surfaces at the base of the metatarsal bones vary. In the middle, three metatarsals, the surfaces that articulate with the tarsal bones, are flat. The metatarsal bases are wedge-shaped and firmly fitted together. They have minimal mobility on the tarsus and on each other. The **second metatarsal** is the least mobile. In addition to articulating with the intermediate cuneiform, it articulates with the sides of the medial and lateral cuneiforms [see Fig. 16.14]. The base of the second metatarsal is so firmly fixed that its thin body is liable to fracture when sudden stresses are applied to the distal part of the foot. The middle three metatarsals form a relatively rigid beam in the centre of the foot. The articular surfaces of the base of the first and fifth

metatarsals are slightly curved. As such, they have a greater degree of mobility than the middle three.

Intermetatarsal joints

The bases of the lateral four metatarsal bones articulate with one another. They are firmly bound together by plantar, dorsal, and interosseous ligaments.

Dissection 23.9 looks at the joints of the foot.

Metatarsophalangeal joints

In these joints, the base of the proximal phalanx articulates with the head of the metatarsal. The fibrous capsule is attached close to the articular surfaces. It is thickened at the sides to form collateral ligaments, and on the plantar surface to form the plantar ligament. The dorsal part of the capsule is formed by the extensor expansion.

Table 23.6 Movements at tarsal joints, particularly the talocalcaneonavicular joint

Movement	Muscles	Nerve supply
Inversion	Tibialis anterior	Deep fibular
	Extensor hallucis longus	Deep fibular
	Tibialis posterior	Tibial
Eversion	Peroneus longus	Superficial fibular
	Peroneus brevis	Superficial fibular
	Extensor digitorum longus (lateral)	Deep fibular
	Peroneus tertius	Deep fibular

DISSECTION 23.8 Joints of the foot-3

Objectives

I. To define the long plantar ligament. II. To cut the long plantar ligament and expose the plantar calcaneocuboid ligament.

Instructions

1. Define the margins of the long plantar ligament [see Fig. 22.25]. Lift it from the calcaneus by passing a knife between it and the plantar calcaneocuboid ligament which projects beyond its medial edge. Detach the long plantar ligament from the calcaneus to expose the plantar calcaneocuboid ligament.

DISSECTION 23.9 Joints of the foot-4

Objective

I. To dissect the first and second metatarsophalangeal joints.

Instructions

1. Remove the short muscles of the big toe by detaching them from the sesamoid bones. Note the connection between the tendon of the abductor hallucis and the extensor expansion, and the attachment of the abductor to the medial sesamoid bone.
2. Identify and cut through the deep transverse metatarsal ligaments on each side of the second toe. Trace the tendons of the interosseous muscles of the first two spaces to their insertions, and review the course of the lumbrical tendons.
3. Lift the extensor expansions from the dorsal surfaces of the metatarsophalangeal joints of the big and second toes. It forms the fibrous capsule on the dorsal aspect and is continuous on either side with the thickened lateral part of the capsule—the collateral ligaments.
4. Define these collateral ligaments, and then identify the plantar ligament between the long flexor tendon and the joint.

The thick fibrous **plantar ligament** is attached firmly to the plantar margin of the base of the proximal phalanx and loosely to the plantar surface of the neck of the metatarsal bone. The plantar ligament separates the long flexor tendon from the joint. The margins of the plantar ligaments are attached to: (1) the fibrous flexor sheath; (2) the slips of the plantar aponeurosis; (3) the deep transverse metatarsal ligament; (4) the collateral ligaments; and (5) the margins of the extensor expansions. The plantar ligament of the metatarsophalangeal joint of the big toe contains sesamoid bones which articulate superiorly with the grooves on the plantar surface of the head of the first metatarsal.

Articular surfaces

The distal surface of the head of each metatarsal has two continuous articular surfaces, one for articulation with the proximal phalanx and the other for articulation with the plantar ligament of the metatarsophalangeal joint. The articular surface for the proximal phalanx is circular and convex, and fits into the concavity of the base of the proximal phalanx to make a shallow ball-and-socket joint [Fig. 23.17]. The articular surface for the plantar ligament is on the plantar surface. In the first metatarsal, this surface is deeply grooved on each side by a sesamoid bone.

Movements

The metatarsophalangeal joints permit flexion, extension, abduction, and adduction. Rotation is prevented by collateral ligaments. Flexion and extension are produced by the long and short flexor and extensor muscles. The flexors are assisted by the interossei and the lumbricals. When the joints are flexed, the **plantar ligament** slides proximally towards the neck of the metatarsal, and, when they are extended, the ligament moves onto the distal surface of the head of the metatarsal. This tightens the plantar aponeurosis and the extensor expansion. When the metatarsophalangeal joints are extended, the extensor expansion is not able to act on the interphalangeal joints. The lumbricals then become the sole extensors of the interphalangeal joints if they reach the extensor expansion.

Abduction and adduction take place from the line of the second toe. These movements are produced by the interossei, the abductor of the little toe, and the abductor and adductor of the big toe.

Interphalangeal joints

These are hinge-type synovial joints at which only flexion and extension take place. They are similar to the interphalangeal joints in the hand but have a smaller range of movements. The articular surfaces are sinusoidal, not circular. The single interphalangeal joint of the big toe is much larger than any of the others. It also has a flatter curvature and a smaller range of movements. This makes the joint more rigid—a necessary feature to bear the forces which are applied to it.

Movements

At rest, the interphalangeal joints are in a position of partial flexion by comparison with the extended position of the metatarsophalangeal joints. In the big toe, movements are produced by the long and short flexors and the long and short extensor muscles. In the other toes, there is the additional action of the short flexor on the proximal interphalangeal

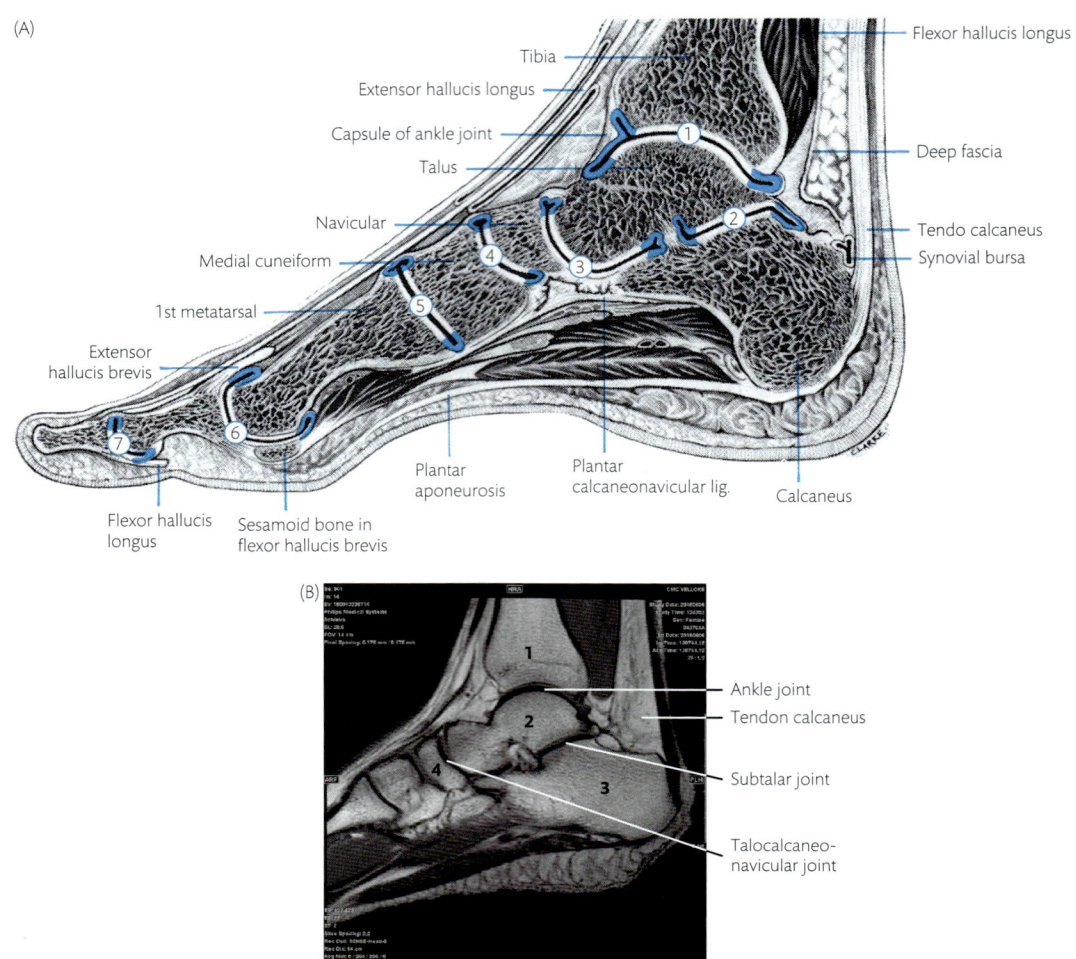

Fig. 23.17 (A) Oblique sagittal section through the middle of the heel and the middle of the big toe. Synovial membrane, blue. 1, Ankle joint; 2, subtalar joint; 3, talocalcaneonavicular joint; 4, joint between navicular and medial cuneiform; 5, first tarsometatarsal joint; 6, first metatarsophalangeal joint; 7, interphalangeal joint of big toe. (B) Magnetic resonance imaging showing ankle, subtalar, and talocalcaneonavicular joints. 1. Lower end of the tibia. 2. Talus. 3. Calcaneus. 4. Navicular.

joint. The lumbricals extend the interphalangeal joints of the lateral four toes and are the only muscles responsible for this action when the metatarsophalangeal joints are fully extended. ◑ Thus, when the lumbricals are weakened or paralysed and the metatarsophalangeal joints are fully extended, the proximal interphalangeal joints are flexed by the long and short flexor muscles, and the distal interphalangeal joints are extended because of pressure on the ground. This position of the toes is known as 'hammer toe'.

Tables 23.7 and 23.8 summarize the muscles and movements of the joints of the toes (MP = metatarsophalangeal; IP = interphalangeal).

Hallux valgus

◑ This deformity is more common in women. It consists of a fixed adduction of the big toe at the metatarsophalangeal joint. Adduction of the big toe occurs because of the obliquity of the tendons inserted into it and because the attachment of the abductor hallucis is more to the plantar than to the medial surface of the proximal phalanx—a position which makes it less efficient as an abductor.

Clinical Application 23.1 looks at some of the practical implications of the anatomy discussed in this chapter.

Table 23.7 Muscles acting on the joints of the toes

Muscle	Origin	Insertion	Action	Nerve supply
Extensor digitorum brevis	Calcaneus, superior surface	Hallux, proximal phalanx Extensor expansion, toes 2–4	Hallux, MP extension Toes 2–4, extension all joints	Deep fibular
Flexor digitorum brevis	Calcaneus, tuber medial process	Middle phalanx, toes 2–5	Tarsus, support Toes 2–5, MP and proximal IP flexion	Medial plantar
Flexor accessorius	Calcaneus, plantar aspect	Tendon of flexor digitorum longus	Straightens pull of flexor digitorum longus	Lateral plantar
Abductor hallucis	Calcaneus, tuber medial process Flexor retinaculum	Hallux, proximal phalanx, medial side Medial sesamoid bone	Hallux, MP abduction or flexion	Medial plantar
Abductor digiti minimi	Calcaneus, tuber both processes	Fifth toe, proximal phalanx, lateral side Fifth metatarsal, base	Fifth toe, MP abduction	Lateral plantar
Dorsal interossei	Metatarsals, adjacent sides of two	Proximal phalanx, lateral side, toes 2–4; medial side toe 2	Toes 2–4, abduction at MP	Lateral plantar
Plantar interossei	Metatarsals 3–5, medial side	Proximal phalanx, medial side, toes 3–5	Toes 3–5, adduction at MP	Lateral plantar
Lumbricals	Tendons, flexor digitorum longus	Extensor expansion, toes 2–5	Toes 2–5, MP flexion, IP extension	Medial and lateral plantar
Flexor hallucis brevis	Cuboid, plantar surface, medial side	Hallux, proximal phalanx, medial and lateral sides	Hallux, MP flexion	Medial plantar
Adductor hallucis	Metatarsals 2–4, bases Plantar ligaments, MP 2–4	Hallux, proximal phalanx, lateral side	Hallux, MP adduction	Lateral plantar
Flexor digiti minimi brevis	Metatarsal 5, base	Fifth toe, proximal phalanx, lateral side	Fifth toe, MP flexion	Lateral plantar

Table 23.8 Movements of toes

Movement	Joints	Muscles	Nerve supply
Flexion	All joints	Flexor hallucis longus Flexor digitorum longus Flexor accessorius	Tibial Tibial Lateral plantar
	MP and proximal IP (2–5)	Flexor digitorum brevis	Medial plantar
	MP only	Flexor hallucis brevis Flexor digiti minimi brevis	Medial plantar Lateral plantar
	Toes 2–5	Lumbricals	Medial plantar, toe 2 Lateral plantar, toes 3–5
	Toes 2–5	Interossei	Lateral plantar
Extension	All joints	Extensor digitorum longus*	Deep fibular
	Toes 2–4	Extensor digitorum brevis*	Deep fibular
	Hallux	Extensor hallucis longus	Deep fibular
	MP only	Extensor hallucis brevis	Deep fibular
	IP, only toes 2–5	Lumbricals*	Lateral plantar, toes 3–5 Medial plantar, toe 2
Abduction at MP	Toe 1	Abductor hallucis**	Medial plantar
	Toes 2–4	Dorsal interossei	Lateral plantar
	Toe 5	Abductor digiti minimi	Lateral plantar
Adduction at MP	Toe 1	Adductor hallucis	Lateral plantar
	Toes 3–5	Plantar interossei	Lateral plantar
	Toe 2	Dorsal interossei when toe 2 already abducted by other dorsal interosseous	Lateral plantar

* Full extension of MP prevents these muscles from extending the IP joints, which are then only extended by the lumbricals. If the lumbricals are weak or paralysed, or fail to reach the extensor expansion, 'hammer toes' result.
** This muscle is often inserted with the flexor hallucis brevis through the medial sesamoid to the plantar aspect of the proximal phalanx, so that it acts as a flexor, rather than as an abductor. Thus, it may fail to prevent the tendency of the adductor hallucis and the oblique tendons of the extensor hallucis brevis and longus to cause the proximal phalanx of the great toe to deviate laterally. This condition, known as hallux valgus, exposes the medial side of the head of the metatarsal of the great toe to rubbing on the shoe. A bursa tends to form between it and the skin, and this becomes inflamed and swollen, forming a bunion.

CLINICAL APPLICATION 23.1 Knee joint injury

An 18-year-old student twisted his right knee, while playing football. He experienced excruciating pain and felt as though something had torn inside his knee. On examination, he had tenderness on the medial side of the knee joint. When the knee was flexed at right angles and an attempt was made to pull the tibia forwards, there was increased mobility of the tibia. Surgical exploration under anaesthesia showed that the superficial and deep layers of the medial collateral ligament were torn. The medial meniscus was torn and detached, and was lying on the intercondylar area of the tibia. The anterior cruciate ligament was torn from its posterior attachment on the femur.

Study question 1: name the ligaments of the knee joint which support the knee in all movements and yet prevent excessive mobility. Which ones are intracapsular? (Answer: medial and lateral collateral ligaments, anterior and posterior cruciate ligaments, patellar ligament (tendon), and oblique popliteal ligament. The cruciate ligaments are intracapsular; the others are extracapsular.)

The medial collateral ligament is frequently torn when the knee is forcefully abducted, and the tibia is laterally rotated on the femur while running. Review the attachments of the medial collateral ligament.

Study question 2: what are the normal attachments of the anterior cruciate ligament? (Answer: on the tibia, the anterior cruciate ligament is attached to the anterior intercondylar area. It runs upwards, backwards, and laterally to the medial surface of the lateral condyle of the femur.) The increased mobility observed when the tibia is pulled forwards on a flexed knee is indicative of a torn anterior cruciate ligament. It is called the 'anterior drawer sign'.

Study question 3: explain why the medial meniscus is more prone to injury than the lateral meniscus. (Answer: the medial meniscus is less mobile, when compared to the lateral meniscus, as it is fused to the medial collateral ligament. (The lateral meniscus is not attached to the capsule laterally.) Also, through the attachment of the meniscofemoral ligament, the lateral meniscus is able to move with movements of the femur. It is less likely to be trapped between the moving condyles in forceful movements of the knee. In addition, due to the larger anteroposterior diameter of the medial condyle, the medial meniscus is subject to greater movement of the medial femoral condyle on it.) Injury to the medial collateral ligament, medial meniscus, and anterior cruciate ligament is seen to occur together frequently.

CHAPTER 24

The nerves and nerve injuries of the lower limb

Effects of nerve injury

A **lower limb neurological examination** is part of the general neurological examination done to assess the integrity of motor and sensory nerves of the upper limb. Most peripheral nerves of the lower limb are mixed nerves, and damage to one would result in both sensory and motor loss. In addition, loss of sympathetic innervation would result in changes in skin texture, absence or decreased sweating, and inability to regulate blood flow in response to changes in temperature.

Sensory distribution

Sensory loss is experienced in the skin supplied by the damaged nerve. For example, damage to the deep peroneal nerve results in sensory loss over the first interdigital cleft. As cutaneous nerves supplying adjacent areas of skin overlap to a considerable degree, total destruction of one nerve produces an area of complete sensory loss much smaller than the sum of the areas supplied by its individual branches. Fig. 24.1 shows the cutaneous distribution of the main nerves of the lower limb, and Fig. 24.2 shows the dermatomal pattern.

Motor loss after nerve injury

Injury to a motor nerve will result in paralysis of the muscles supplied by it and inability to move the joint on which they act. Where a nerve innervates muscles in more than one segment of the limb (thigh, leg, foot), the effects of injury to the nerve depend on the level of injury. Thus, when the sciatic nerve is destroyed in the popliteal fossa, the hamstring muscles supplied by it in the thigh are not paralysed, though those in the leg and foot are. In general, the more proximal the nerve lesions, the greater will be the motor loss.

The results of lower limb nerve injuries are shown in Tables 24.1 to 24.10. (Abbreviations: the toes are numbered 1 to 5, from the medial side—1 is the big toe, and 5 the little toe. MP=metatarsophalangeal joint; IP=interphalangeal joints; PIP=proximal interphalangeal joint(s); DIP=distal interphalangeal joint(s).)

Femoral nerve injury

Table 24.1 shows the effects of injury on the femoral nerve.

Obturator nerve injury

Table 24.2 shows the effects of injury on the obturator nerve.

Superior gluteal nerve injury

Table 24.3 shows the effects of injury on the superior gluteal nerve.

Inferior gluteal nerve injury

Table 24.4 shows the effects of injury on the inferior gluteal nerve.

Sciatic nerve injury

Table 24.5 shows the effects of injury on the sciatic nerve.

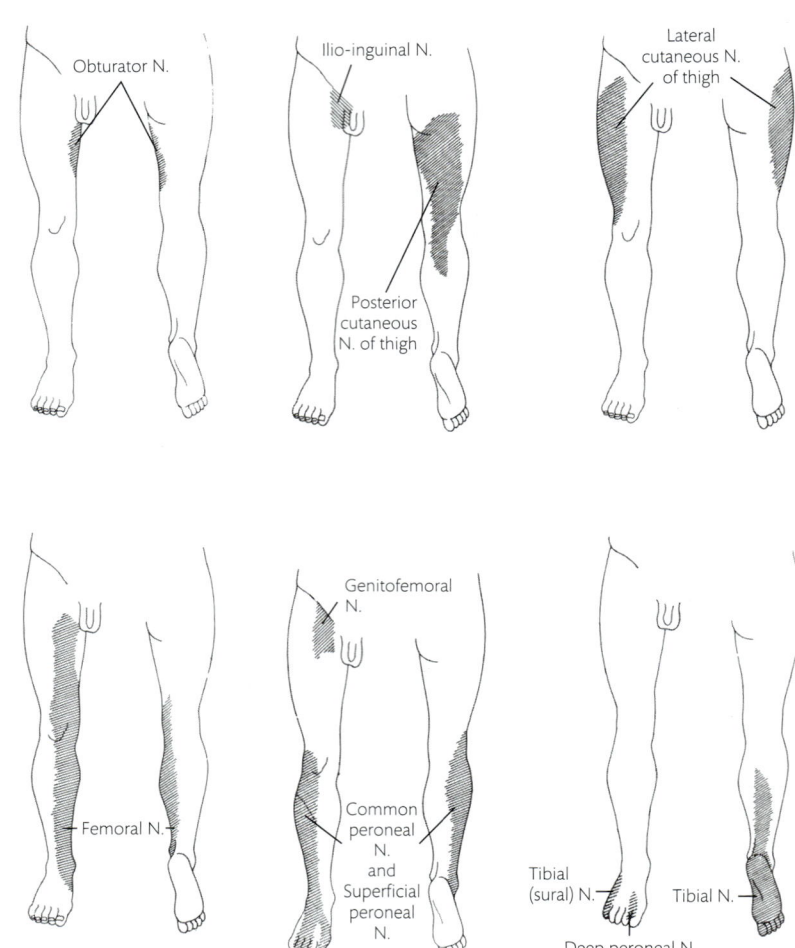

Fig. 24.1 Diagrams of the cutaneous distribution of nerves in the lower limb.

Deep peroneal nerve injury

Table 24.6 shows the effects of injury on the deep fibular nerve.

Superficial peroneal nerve injury

Table 24.7 shows the effects of injury on the superficial fibular nerve.

Tibial nerve injury

Table 24.8 shows the effects of injury on the tibial nerve.

Medial plantar nerve injury

Table 24.9 shows the effects of injury on the medial plantar nerve.

Lateral plantar nerve injury

Table 24.10 shows the effects of injury on the lateral plantar nerve.

Clinical Applications 24.1 and 24.2 (and Table 24.11) look at some of the practical implications of the anatomy described in this chapter.

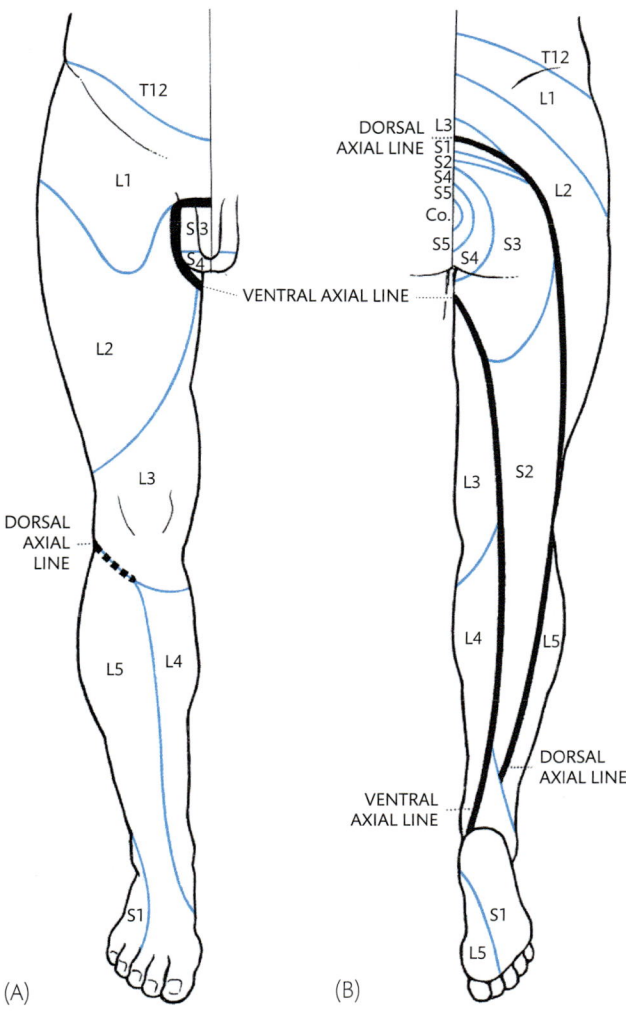

Fig. 24.2 Dermatomes of the lower limb, showing the segmental cutaneous distribution of spinal nerves (T.12, L.I to 5, S.1 to 5, Co.) on (A) the front and (B) the back of the limb and lower part of the trunk.

After Head, 1893, and Foerster, 1933.

Table 24.1 Effects of injury to the femoral nerve

Joint involved	Movement affected	Explanation for loss/weakness of movement
Effect on hip joint	Flexion weakened	Sartorius, rectus femoris, and pectineus are paralysed. Weak flexion brought about by iliopsoas
Effect on knee joint	Extension lost or severely weakened	Quadriceps femoris is paralysed. Weak extension maintained by tensor fascia latae and gluteus maximus

Table 24.2 Effects of injury to the obturator nerve

Joint involved	Movement affected	Explanation for loss/weakness of movement
Effect on hip joint	Adduction lost	Obturator externus, adductor longus, adductor brevis, adductor magnus (adductor part), and gracilis are paralysed. Weak adduction is brought about by pectineus
	Instability in standing and walking	Due to weakness of adduction

Table 24.3 Effects of injury to the superior gluteal nerve

Joint involved	Movement affected/compensatory movement	Explanation for loss/weakness of movement
Effect on hip joint	Abduction lost. Compensatory flexion of the body to the paralysed side, to enable the opposite limb to be raised from the ground	Gluteus medius, gluteus minimus, and tensor fascia latae are paralysed

Table 24.4 Effects of injury to the inferior gluteal nerve

Joint involved	Movement affected	Explanation for loss/weakness of movement
Effect on hip joint	Extension weakened	Gluteus maximus is paralysed. Weak extension is brought about by hamstrings

Table 24.5 Effects of injury to the sciatic nerve

Joint involved	Movement affected	Explanation for loss/weakness of movement
Effect on hip joint	Extension weakened	Adductor magnus (hamstring part), biceps femoris long head, semitendinosus, and semimembranosus are paralysed. Weak extension is brought about by gluteus maximus
Effect on knee joint	Flexion severely weakened	Biceps femoris, semitendinosus, and semimembranosus are paralysed. Sartorius and gracilis are responsible for weak flexion
	Lateral rotation of leg on femur weakened	Biceps femoris is paralysed. Iliotibial tract is responsible for weak lateral rotation

Table 24.6 Effects of injury to the deep peroneal nerve

Joint involved	Movement affected	Explanation for loss/weakness of movement
Effect on ankle joint	Dorsiflexion lost	Tibialis anterior, extensor hallucis longus, and extensor digitorum longus are paralysed
Effect on talocalcaneonavicular and calcaneocuboid joints	Inversion is weakened	Tibialis anterior and extensor hallucis longus are paralysed Weak inversion by tibialis posterior
	Eversion slightly weakened	Peroneus tertius is paralysed. Peroneus longus and brevis are responsible for eversion
Effect on first MP joint	Extension lost	Extensor hallucis longus and extensor hallucis brevis are paralysed
Effect on lateral four MP joints	Extension lost	Extensor digitorum longus and extensor digitorum brevis are paralysed
Effect on IP joint of big toe	Extension lost	Extensor hallucis longus is paralysed
Effect on IP joint of lateral four toes	Extension weakened in middle three toes Extension lost in little toe	Extensor digitorum longus and extensor digitorum brevis are paralysed. Lumbricals in the middle three toes are responsible for the weak extension

Table 24.7 Effects of injury to the superficial peroneal nerve

Joint involved	Movement affected	Explanation for loss/weakness of movement
Effect on talocalcaneonavicular and calcaneocuboid joints	Eversion weakened	Peroneus longus and peroneus brevis are paralysed. Peroneus tertius and lateral part of extensor digitorum are responsible for weak eversion

Table 24.8 Effects of injury to the tibial nerve

Joint involved	Movement affected	Explanation for loss/weakness of movement
Effect on knee joint	Flexion slightly weakened	Popliteus and gastrocnemius are paralysed. Hamstrings, sartorius, and gracilis are responsible for flexion
Effect on ankle joint	Plantar flexion markedly weakened	Gastrocnemius, soleus, plantaris, flexor hallucis longus, flexor digitorum longus, and tibialis posterior are paralysed. Peroneus longus and brevis are responsible for weak flexion
Effect on first MP joint	Flexion weakened	Flexor hallucis longus and flexor hallucis brevis are paralysed. Lumbrical is responsible for weak flexion
Effect on lateral four MP joints	Flexion weakened	Flexor digitorum longus is paralysed. Lumbricals are responsible for some flexion
Effect on first IP joint	Flexion lost	Flexor digitorum longus is paralysed
Effect on lateral four PIP joints	Flexion weakened	Flexor digitorum longus is paralysed. Lumbricals are responsible for weak flexion
Effect on lateral four DIP joints	Flexion lost	Flexor digitorum longus is paralysed

Table 24.9 Effects of injury to the medial plantar nerve

Joint involved	Movement affected	Explanation for loss/weakness of movement
Effect on first MP joint	Abduction lost	Abductor hallucis is paralysed
	Flexion weakened	Flexor hallucis brevis is paralysed. Flexor hallucis longus is responsible for some flexion
Effect on MP joints of lateral four toes	Flexion weakened	Flexor hallucis brevis is paralysed. Lumbricals and interossei are responsible for some flexion

Table 24.10 Effects of injury to the lateral plantar nerve

Joint involved	Movement affected	Explanation for loss/weakness of movement
Effect on first MP joint	Adduction lost	Adductor hallucis is paralysed
Effect on MP joints, toes 3–5	Adduction lost	Plantar interossei are paralysed
Effect on MP joints, toes 2–4	Abduction lost	Dorsal interossei are paralysed
Effect on MP joint of fifth toe	Abduction lost	Abductor digiti minimi is paralysed
Effect on IP joints, toes 2–5	Extension lost if MP fully extended	Lumbricals and interossei are paralysed
Effect on IP joint of fifth toe	Flexion lost	Flexor digiti minimi brevis is paralysed

Table 24.11 Motor assessment of lower limb musculature

Spinal segment (myotome)	Primary movement	Prime muscle causing movement
L. 2	Hip flexion	Iliopsoas
L. 3	Knee extension	Quadriceps
L. 4	Ankle dorsiflexion	Tibialis anterior
L. 5	Big toe extension	Extensor hallucis longus
S. 1	Ankle plantar flexion	Gastrocnemius—soleus

Reference: *Standard neurological classification of spinal cord injury* by American Spinal Injury Association (ASIA).

CLINICAL APPLICATION 24.1 Peripheral neuropathy (polyneuropathy)

Diseases such as diabetes can affect nerves of the periphery, causing a condition called peripheral neuropathy where nerve functions are lost. The initial symptoms are most often sensory, and the longest nerves are affected first. Because this is a systemic disease, the condition produces symmetrical symptoms.

The sensory loss on the feet has a typical 'stocking' distribution.

Study question: list all the nerves damaged in a patient who has peripheral neuropathy extending up to mid calf. (Answer: saphenous, sural, superficial fibular, deep fibular, tibial, lateral, and medial plantar nerves [Fig. 24.1].)

CLINICAL APPLICATION 24.2 Motor assessment of lower limb musculature

Motor examination of the lower limb in a patient with spinal cord injury provides a reliable and quick way to localize the level of the lesion. Five muscles of the lower limb, one primarily supplied by each of the five segmental nerves L.2 to S.1, are tested.

The integrity of each spinal segment is evaluated by the ability of a muscle supplied by it to bring about a particular movement of a joint [Table 24.11]. (The strength of the muscle is scored on a 5-point scale not included here.)

CHAPTER 25
Surface marking of the lower limb

Introduction

Having studied the important structures of the lower limb, it is essential to have the knowledge and skills to mark their position on the surface of the body. Before you begin the study of surface marking, review the surface anatomy and palpate the bony landmarks of the gluteal region, thigh, knee, leg, and foot, as discussed in the preceding chapters.

Mid-inguinal point

The mid-inguinal point is marked on the midpoint of the line joining the anterior superior iliac spine to the pubic symphysis. It should be differentiated from the *midpoint of the inguinal ligament* which is the midpoint between the anterior superior iliac spine and the pubic tubercle [Fig. 25.1].

Saphenous opening

The centre of the saphenous opening is marked by a point lying 3–4 cm inferolateral to the pubic tubercle. Draw an oval 3 cm long and 1.5 cm wide around this point [Fig. 25.2].

Long saphenous vein

The long saphenous vein begins at the medial side of the dorsum of the foot and ends at the saphenous opening. It can be marked on the surface of the body by joining the points shown in Fig. 25.3:

- 2.5 cm anterior to the medial malleolus (A)
- At the knee, a hand's breath posterior to the medial margin of the patella (B)
- Saphenous opening (C).

The vein crosses obliquely on the medial surface of the lower third of the tibia.

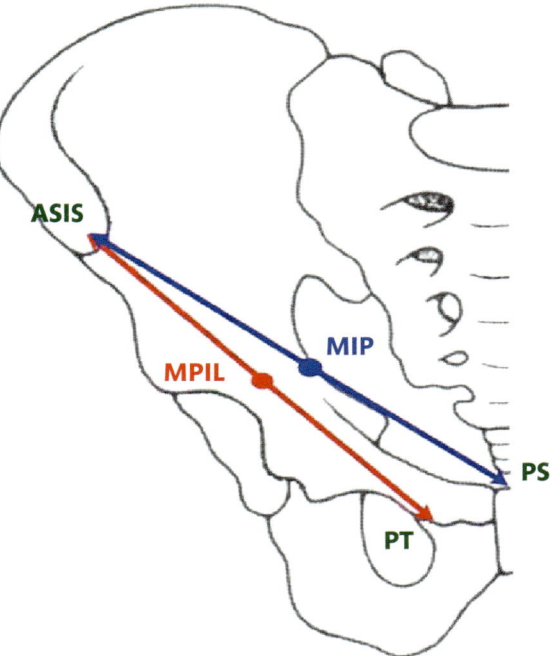

Fig. 25.1 Diagram showing the mid-inguinal point (MIP) and the mid point of the inguinal ligament (MPIL). ASIS = Anterior superior iliac spine. PS = Pubic symphysis. PT = Pubic tubercle.

Reproduced with permission from P. Sanjay et al., 'Defining the position of deep inguinal ring in patients with indirect inguinal hernias', *Surgical and Radiologic Anatomy*, 28, pp. 121–124, copyright Springer-Verlag 2006. DOI 10.1007/s00276-006-0105-0 https://link.springer.com/journal/276

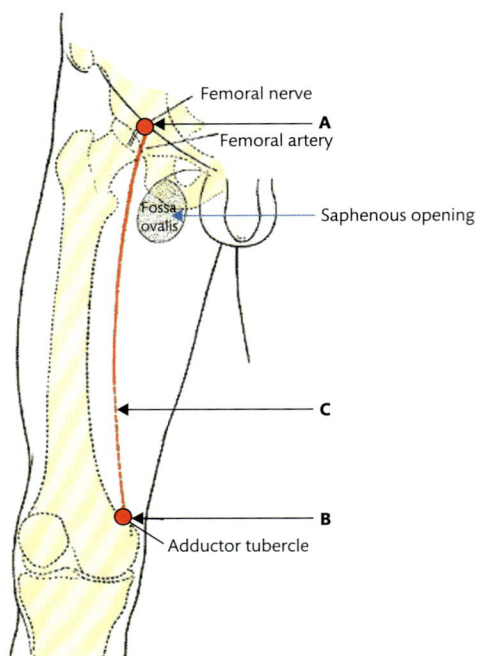

Fig. 25.2 Surface marking of the saphenous opening and femoral artery. A = mid-inguinal point; B = adductor tubercle; C = lower extent of the femoral artery.

Reproduced from Henry Gray (1918) *Anatomy of the Human Body*, Philadelphia: Lea & Febiger. Plate 1245, public domain.

Short saphenous vein

The short saphenous vein begins on the lateral side of the dorsum of the foot and ends in the popliteal fossa. It can be marked on the surface of the body by joining the following points:

- Behind the lateral malleolus
- Centre of the popliteal fossa.

Femoral artery

The femoral artery begins at the mid-inguinal point and ends at the adductor hiatus. It can be marked on the surface of the body by joining the points shown in Fig. 25.2, on a limb that is slightly abducted and laterally rotated:

- Mid-inguinal point (A)
- Adductor tubercle (B).

The upper two-thirds of this line marks the femoral artery (A–C).

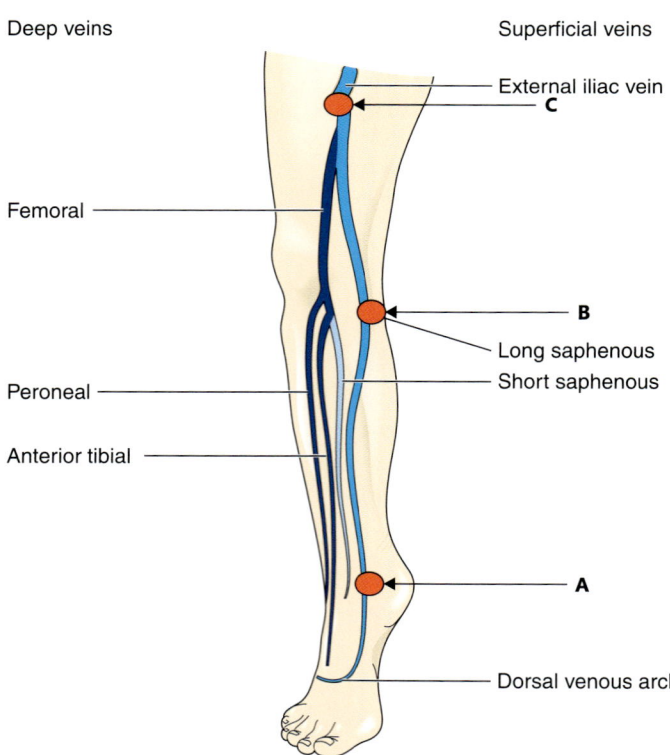

Fig. 25.3 Surface marking of the long saphenous vein. A, B, and C indicate relevant bony and soft tissue landmarks.

Blamb/Shutterstock.com.

Popliteal artery

The popliteal artery begins at the adductor hiatus and ends at the lower border of the popliteus. It can be marked on the surface of the body by joining the points shown in Fig. 25.4:

- Junction of the upper two-thirds and lower one-third of the thigh, just medial to the midline (A)
- Middle of the popliteal fossa (middle of the back of the knee) (B)
- At the level of the tibial tuberosity on the midline of the leg posteriorly (C).

Posterior tibial artery

The posterior tibial artery extends from the lower border of the popliteus to the midpoint of the flexor retinaculum of the ankle. It can be marked on the surface of the body by joining the points shown in Fig. 25.4:

- At the level of the tibial tuberosity on the midline of the leg posteriorly (B)
- Midpoint between the medial malleolus and the tendo calcaneus (D).

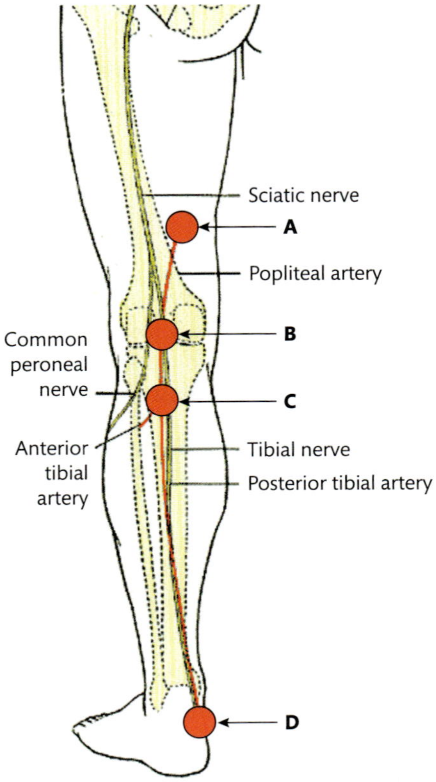

Anterior tibial artery

The anterior tibial artery extends from the lower border of the popliteus to the midpoint of the superior extensor retinaculum of the ankle. It can be marked on the surface of the body by joining the points shown in Fig. 25.5:

- At the level of the tibial tuberosity on the midline of the leg anteriorly (A)
- Midpoint between the medial and lateral malleoli at the front of the ankle (B).

Dorsalis pedis

The dorsalis pedis artery extends from the midpoint of the superior extensor retinaculum of the ankle to

Fig. 25.4 Surface marking of the popliteal artery and posterior tibial artery. A, B, C, and D indicate relevant bony and soft tissue landmarks.

Reproduced from Henry Gray (1918) *Anatomy of the Human Body*, Philadelphia: Lea & Febiger. Plate 1247, public domain.

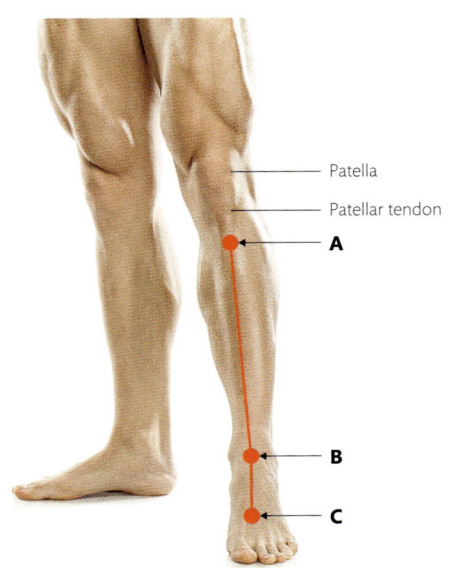

Fig. 25.5 Bony and soft tissue landmarks for marking the anterior tibial artery and dorsalis pedis.

I T A L O/ Shutterstock.com..

the proximal end of the first intermetatarsal space. It can be marked on the surface of the body by joining the points shown in Fig. 25.5:

- Midpoint between the medial and lateral malleoli at the front of the ankle (B)
- Proximal end of the first intermetatarsal space (C).

Femoral nerve

The femoral nerve has a short course in the thigh (Fig. 25.6). It can be marked on the surface of the body by joining the following points:

- 1 cm lateral to the mid-inguinal point
- 2 cm below and slightly lateral to the first point.

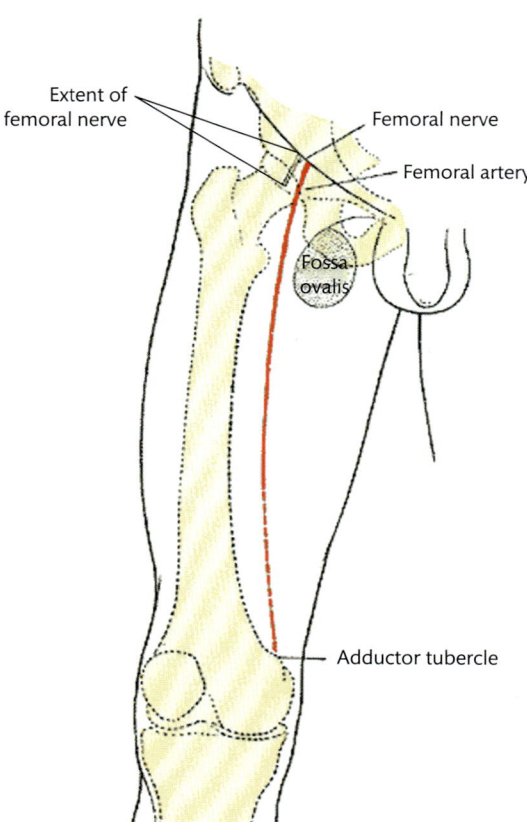

Fig. 25.6 Diagram showing the femoral nerve in relation to the inguinal ligament and femoral artery.

Reproduced from Henry Gray (1918) *Anatomy of the Human Body*, Philadelphia: Lea & Febiger. Plate 1245, public domain.

Sciatic nerve

The sciatic nerve traverses the gluteal region and back of the thigh. It can be marked on the surface of the body by joining the points shown in Figs 25.7A, B.

- Midpoint on a line joining the posterior superior iliac spine and the ischial tuberosity (A)
- Midpoint on a line joining the tip of the greater trochanter and the ischial tuberosity (B)
- At the upper angle of the popliteal fossa (junction of the upper two-thirds and lower one-third of the thigh posteriorly) (C).

Tibial nerve

The tibial nerve traverses the popliteal fossa and back of the leg to end deep to the flexor retinaculum. It can be marked on the surface of the body by joining the points shown in Fig. 25.7B:

- At the upper angle of the popliteal fossa (C)
- Midline on the back of the leg at the level of the tibial tuberosity (D)
- Midpoint between the medial malleolus and the tendo calcaneus (E).

Common peroneal nerve

The common peroneal nerve extends from the upper border of the popliteal fossa to the neck of the fibula. It can be marked on the surface of the body by joining the points shown in Fig. 25.8:

- At the upper angle of the popliteal fossa (A)
- Back of the neck of the fibula (B).

Deep peroneal nerve

The deep peroneal nerve extends from neck of the fibula into the anterior compartment of the leg and on the dorsum of the foot. It can be marked on the surface of the body by joining the points shown in Fig. 25.9:

- Lateral aspect of the neck of the fibula (A)
- Midpoint of the line joining the medial and lateral malleoli (B).

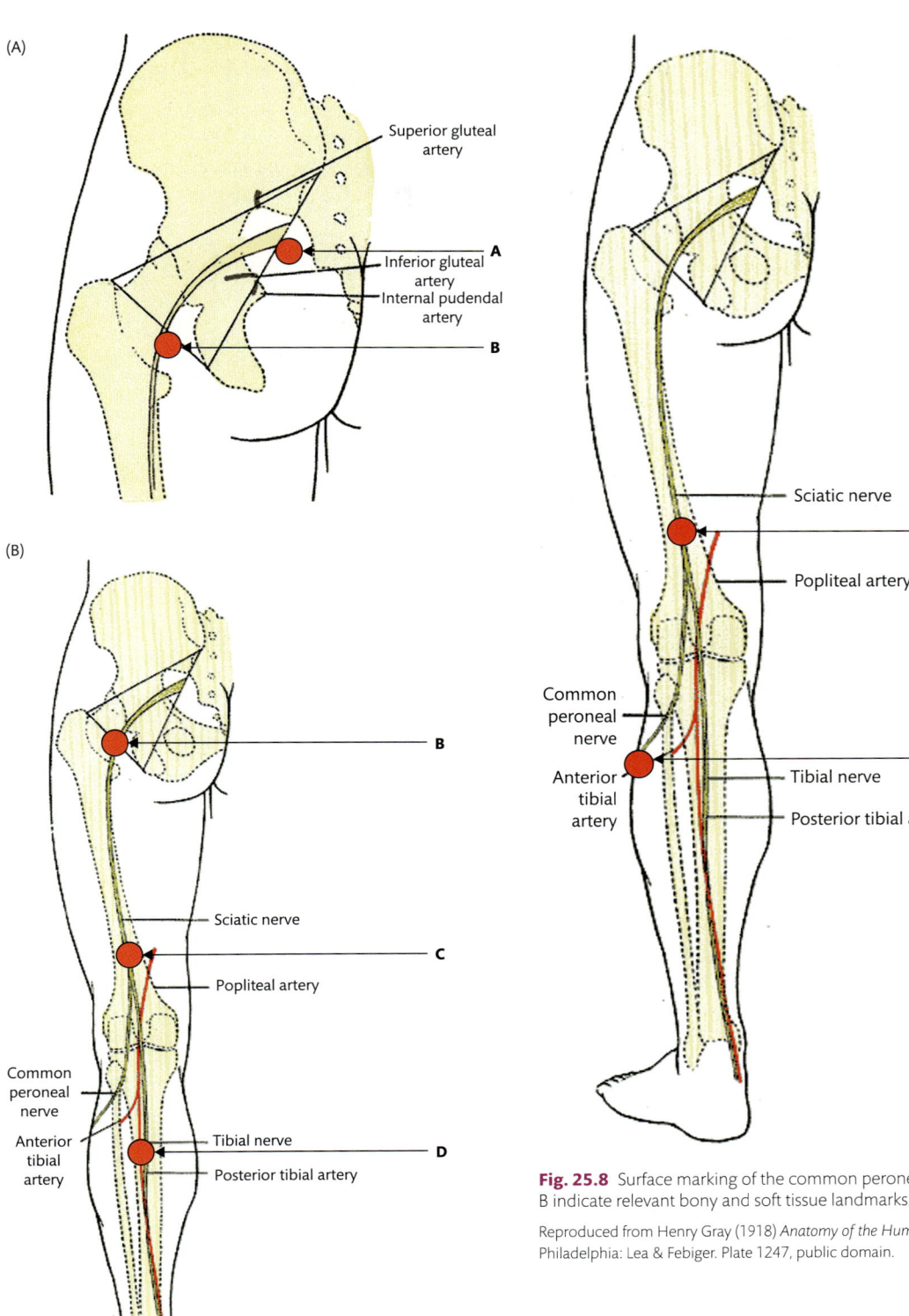

(A)

Superior gluteal artery

Inferior gluteal artery

Internal pudendal artery

A

B

(B)

Sciatic nerve

Popliteal artery

Common peroneal nerve

Anterior tibial artery

Tibial nerve

Posterior tibial artery

B

C

D

E

Sciatic nerve

Popliteal artery

Common peroneal nerve

Anterior tibial artery

Tibial nerve

Posterior tibial artery

A

B

Fig. 25.8 Surface marking of the common peroneal nerve. A and B indicate relevant bony and soft tissue landmarks.

Reproduced from Henry Gray (1918) *Anatomy of the Human Body*, Philadelphia: Lea & Febiger. Plate 1247, public domain.

Fig. 25.7 Panels (A) and (B) show surface marking of the sciatic nerve. A, B, and C indicate relevant bony and soft tissue landmarks.

Reproduced from Henry Gray (1918) *Anatomy of the Human Body*, Philadelphia: Lea & Febiger. Plate 1244 and 1247, public domain.

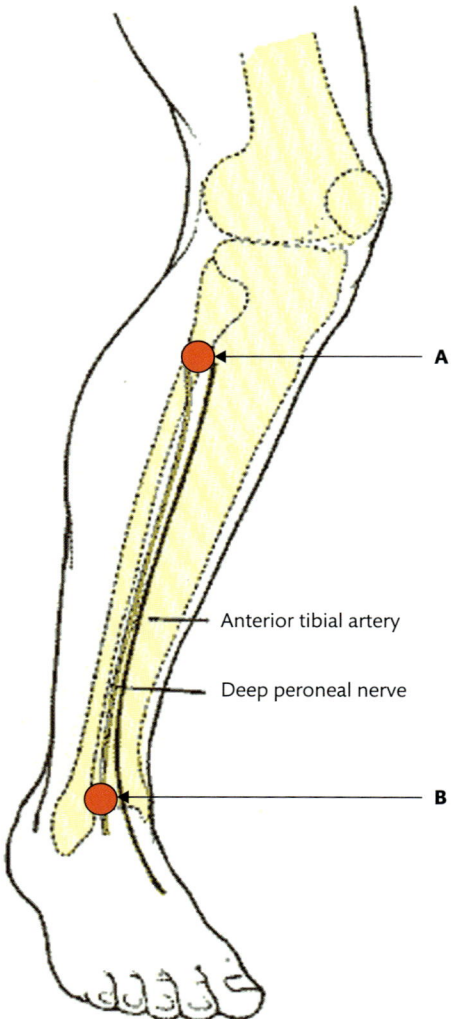

Anterior tibial artery

Deep peroneal nerve

Fig. 25.9 Surface marking of the deep peroneal nerve. A and B indicate relevant bony and soft tissue landmarks.

Reproduced from Henry Gray (1918) *Anatomy of the Human Body*, Philadelphia: Lea & Febiger. Plate 1246, public domain.

CHAPTER 26
MCQs on the lower limb

The following questions have four options. You are required to choose the most correct answer.

1. **The saphenous opening**

   A. is a deficiency in the superficial fascia of the thigh
   B. is limited on the medial side by the falciform margin
   C. transmits the long saphenous vein and superficial inguinal arteries
   D. is present 4 cm inferomedial to the pubic tubercle

2. **The femoral canal contains the**

   A. femoral artery
   B. femoral branch of the genitofemoral nerve
   C. genital branch of the genitofemoral nerve
   D. lymph node

3. **All of the following statements are true regarding the sartorius, EXCEPT**

   A. it forms the lateral boundary of the femoral triangle
   B. it covers the roof of the adductor canal
   C. it is supplied by the femoral nerve
   D. it flexes the hip joint and extends the knee joint

4. **The descending genicular artery is a branch of the**

   A. femoral artery
   B. profunda femoral artery
   C. obturator artery
   D. deep external pudendal artery

5. **The nerve supply of the tensor fasciae latae is the**

   A. femoral nerve
   B. sciatic nerve
   C. superior gluteal nerve
   D. inferior gluteal nerve

6. **The articularis genu muscle is part of the**

    A. vastus medialis
    B. vastus intermedius
    C. vastus lateralis
    D. rectus femoris

7. **The pectineus is supplied by the**

    A. femoral branch of the genitofemoral nerve
    B. posterior division of the obturator nerve
    C. femoral nerve
    D. ventral rami of L. 2 and L. 3

8. **The posterior superior iliac spine lies at the level of the**

    A. fourth lumbar spine
    B. fifth lumbar spine
    C. first sacral spine
    D. second sacral spine

9. **The companion artery of the sciatic nerve arises from the**

    A. medial circumflex femoral artery
    B. lateral circumflex femoral artery
    C. inferior gluteal artery
    D. superior gluteal artery

10. **The sural nerve is the cutaneous branch arising from the**

    A. tibial nerve
    B. common fibular nerve
    C. superficial fibular nerve
    D. deep fibular nerve

11. **The structures that are attached to the anterior inferior iliac spine are the**

    A. sartorius and inguinal ligament
    B. sartorius and iliofemoral ligament
    C. rectus femoris and inguinal ligament
    D. rectus femoris and iliofemoral ligament

12. **The tendon of the peroneus longus grooves the inferior surface of the**

    A. calcaneus
    B. talus
    C. cuboid
    D. navicular

13. **The skin of the first interdigital cleft of the leg is innervated by the**

    A. deep fibular nerve
    B. superficial fibular nerve
    C. saphenous nerve
    D. sural nerve

14. **The actions of the tibialis posterior are**

    A. dorsiflexion and inversion of the foot

    B. plantar flexion and inversion of the foot

    C. dorsiflexion and eversion of the foot

    D. plantar flexion and eversion of the foot

15. **The oblique popliteal ligament is an extension of the**

    A. semimembranosus tendon

    B. semitendinosus tendon

    C. biceps femoris tendon

    D. gastrocnemius tendon

Please see Chapter 27 for the answers.

Find additional MCQs online by searching for *Cunningham's Manual of Practical Anatomy Volume 1 General Anatomy, Upper and Lower Limbs*, 17th edition at https://academic.oup.com/, and go to the online appendix at the end of the book. Use your scratch-off code on the inside cover to access the material. The code will work for 12 months.

CHAPTER 27
Answers to MCQs

Answers to Chapter 2 MCQs on general anatomy

1. C	5. C	9. D
2. B	6. D	10. A
3. C	7. D	
4. A	8. A	

Answers to Chapter 14 MCQs on the upper limb

1. C	6. D	11. A
2. D	7. C	12. B
3. A	8. B	13. B
4. B	9. D	14. A
5. C	10. A	15. A

Answers to Chapter 26 MCQs on the lower limb

1. C	6. B	11. D
2. D	7. C	12. C
3. D	8. D	13. A
4. A	9. C	14. B
5. C	10. A	15. A

Index

For the benefit of digital users, indexed terms that span two pages (e.g., 52-53) may, on occasion, appear on only one of those pages.

Tables, figures, and boxes are indicated by an italic *t*, *f*, and *b* following the page number.

Index

Index